HANDBOOK OF IMAGING MULTIPLE SCLEROSIS

HANDBOOK OF IMAGING IN MULTIPLE SCLEROSIS

Edited by

DEJAN JAKIMOVSKI
Buffalo Neuroimaging Analysis Center (BNAC), Department of Neurology, Jacobs School of Medicine and Biomedical Sciences, University at Buffalo, State University of New York, Buffalo, NY, United States

ROBERT ZIVADINOV
Buffalo Neuroimaging Analysis Center (BNAC), Department of Neurology, Jacobs School of Medicine and Biomedical Sciences, University at Buffalo, State University of New York, Buffalo, NY, United States

Academic Press is an imprint of Elsevier
125 London Wall, London EC2Y 5AS, United Kingdom
525 B Street, Suite 1650, San Diego, CA 92101, United States
50 Hampshire Street, 5th Floor, Cambridge, MA 02139, United States

Copyright © 2025 Elsevier Inc. All rights are reserved, including those for text and data mining, AI training, and similar technologies.

Publisher's note: Elsevier takes a neutral position with respect to territorial disputes or jurisdictional claims in its published content, including in maps and institutional affiliations.

No part of this publication may be reproduced or transmitted in any form or by any means, electronic or mechanical, including photocopying, recording, or any information storage and retrieval system, without permission in writing from the publisher. Details on how to seek permission, further information about the Publisher's permissions policies and our arrangements with organizations such as the Copyright Clearance Center and the Copyright Licensing Agency, can be found at our website: www.elsevier.com/permissions.

This book and the individual contributions contained in it are protected under copyright by the Publisher (other than as may be noted herein).

Notices

Knowledge and best practice in this field are constantly changing. As new research and experience broaden our understanding, changes in research methods, professional practices, or medical treatment may become necessary.

Practitioners and researchers must always rely on their own experience and knowledge in evaluating and using any information, methods, compounds, or experiments described herein. In using such information or methods they should be mindful of their own safety and the safety of others, including parties for whom they have a professional responsibility.

To the fullest extent of the law, neither the Publisher nor the authors, contributors, or editors, assume any liability for any injury and/or damage to persons or property as a matter of products liability, negligence or otherwise, or from any use or operation of any methods, products, instructions, or ideas contained in the material herein.

ISBN: 978-0-323-95739-7

For Information on all Academic Press publications
visit our website at https://www.elsevier.com/books-and-journals

Publisher: Stacy Masucci
Acquisitions Editor: Joslyn Chaiprasert-Paguio
Editorial Project Manager: Debarati Roy
Production Project Manager: Erragounta Saibabu Rao
Cover Designer: Christian Bilbow

Typeset by MPS Limited, Chennai, India

Dedication

To all those living with multiple sclerosis: You inspire us all. This book is dedicated to your journey, your hope, and your unwavering spirit.

—**Dr. Dejan Jakimovski and Dr. Robert Zivadinov**

Contents

List of contributors xi

1. Introduction
Dejan Jakimovski and Robert Zivadinov

Introduction 1

2. The immunology and pathophysiology of multiple sclerosis
Nil Saez Calveras and Olaf Stuve

Introduction 3
The immunology of multiple sclerosis 4
The pathophysiology of multiple sclerosis 18
Conclusion 25
References 25

3. Diagnosis and clinical features of multiple sclerosis
Svetlana Eckert, Channa Kolb and Bianca Weinstock-Guttman

What is multiple sclerosis 38
Clinical symptoms of multiple sclerosis 38
Development and mechanisms of multiple sclerosis 38
Ethnic and racial disparities in multiple sclerosis 39
Clinical course of multiple sclerosis and multiple sclerosis types 40
Linking multiple sclerosis pathophysiology and in vivo imaging 41
Multiple sclerosis diagnostic criteria 43
Precursors to multiple sclerosis: clinically isolated syndrome and radiologically isolated syndrome 45
Recommendations on the timing of imaging for monitoring of multiple sclerosis disease activity and progression 45
Recommendations on the use of newer magnetic resonance imaging techniques for evaluation of multiple sclerosis 46
Summary 47
References 47

4. Cognition in multiple sclerosis
Zachary L. Weinstock and Ralph H.B. Benedict

Introduction 51
A historical review of cognition in multiple sclerosis 52
Acute cognitive decline as a marker of active disease in multiple sclerosis 58
Treatment of cognitive impairment in multiple sclerosis 61
Open questions and future directions 61
Conclusion 64
References 65

5. Multiple sclerosis treatment
Hernan Inojosa and Tjalf Ziemssen

Introduction 71
Escalation versus induction approach? 72
Interferon beta 1-a and interferon beta 1-b 74
Glatiramer acetate 77
Dimethyl fumarate and diroximel fumarate (monomethyl fumarate) 78
Teriflunomide 79
Sphingosine 1-phosphate–receptor modulators (ozanimod, ponesimod, fingolimod, siponimod) 80
B-cell depletion 82
Natalizumab 84
Alemtuzumab 85
Conflict of interest 88
References 88

6. Use of magnetic resonance imaging and quantitative imaging reports in clinical care of multiple sclerosis

Tomas Uher and Manuela Vaneckova

Introduction 94
Magnetic resonance imaging in multiple sclerosis diagnosis 94
Magnetic resonance imaging in differential diagnosis of multiple sclerosis 98
Magnetic resonance imaging in the prediction of disease activity 100
Magnetic resonance imaging for monitoring disease activity 100
Magnetic resonance imaging in drug safety surveillance 106
Quantitative magnetic resonance imaging reports in clinical care 109
Disclosures 113
References 114

7. Magnetic resonance imaging markers in multiple sclerosis clinical trials and emerging imaging biomarkers

Eleonora Tavazzi and Niels Bergsland

Introduction 123
Conventional magnetic resonance imaging markers in multiple sclerosis 124
Emerging imaging biomarkers 128
Conclusions 130
References 130

I
Imaging of multiple sclerosis

1
Conventional MRI use in multiple sclerosis

8. Magnetic resonance imaging physics and image acquisition

Junghun Cho and Alexey Dimov

Introduction 139

Magnetic resonance imaging image formation 140
Relationship between magnetic resonance imaging signal and tissue properties 144
Summary 155
References 155

9. Magnetic resonance imaging of the multiple sclerosis lesions

Massimiliano Calabrese and Agnese Tamanti

Introduction 159
White matter lesions 160
Black holes 163
Contrast-enhancing lesions 164
Chronic active lesions 165
Cortical lesions 166
Conclusions 169
References 169

10. Spinal cord imaging in multiple sclerosis

Lorena Lorefice and Giuseppe Fenu

Introduction 175
Characteristics of spinal cord lesions in multiple sclerosis: multiple sclerosis diagnosis and differential diagnosis 176
Practical considerations in spinal cord clinical imaging 180
Spinal cord magnetic resonance imaging assessment for patient’s prognosis and disease monitoring 182
Future directions 183
Conclusions 185
References 185

2
Non-conventional MRI use in multiple sclerosis

11. Magnetization transfer imaging in multiple sclerosis

Matteo Mancini and Mara Cercignani

Introduction 192
Clinical use of magnetization transfer contrast 193
Quantifying the magnetization transfer effect 193

Validation of magnetization transfer–derived
  parameters as myelin markers 198
Magnetization transfer ratio in multiple
  sclerosis 198
Quantitative magnetization transfer in multiple
  sclerosis 200
Magnetization transfer saturation in multiple
  sclerosis 200
Inhomogeneous magnetization transfer in multiple
  sclerosis 200
Magnetization transfer in spinal cord and optic
  nerve 201
Conclusions 202
References 202

12. Susceptibility weighted imaging in multiple sclerosis

Sagar Buch and E. Mark Haacke

Introduction 207
Imaging biomarkers in multiple sclerosis 208
SWI-FLAIR or FLAIR* 209
Quantitative susceptibility mapping 210
Introduction to STrategically Acquired Gradient
  Echo Imaging 211
Water content as a new biomarker for multiple
  sclerosis lesions 211
Microvascular in vivo contrast revealed origins 214
Conclusion 216
References 216

13. Quantitative susceptibility mapping in multiple sclerosis

Ferdinand Schweser and Alexander Rauscher

Introduction 222
Fundamentals of quantitative susceptibility
  mapping 223
Nonheme iron concentrations in the deep gray
  matter 227
Regional content of nonheme iron in the deep gray
  matter 229
Subvoxel distribution of iron in the deep gray
  matter 230
Gadolinium retention in the deep gray matter 230
Confounding effects of myelin in the deep gray
  matter 231
Normal appearing white matter 231

Focal white matter damage 232
Other applications of quantitative susceptibility
  mapping in multiple sclerosis 235
Summary and outlook 236
Acknowledgments 237
References 237

14. Functional magnetic resonance imaging in multiple sclerosis

Eva A. Krijnen and Menno M. Schoonheim

Introduction 249
Functional reorganization: task-based functional
  magnetic resonance imaging 250
Functional connectivity: resting-state functional
  magnetic resonance imaging 253
Functional brain changes over time 255
Dynamic functional connectivity 257
The network collapse 257
Advanced network analyses: network
  efficiency 257
Conclusion 258
References 259

15. Perfusion-weighted imaging in multiple sclerosis

Maria Marcella Laganà and Laura Pelizzari

Magnetic resonance imaging techniques for
  estimating cerebral perfusion 268
Magnetic resonance imaging perfusion studies in
  multiple sclerosis 274
Conclusions 281
References 281

16. Magnetic resonance spectroscopy and myelin water fraction in multiple sclerosis

Cornelia Laule and Irene M. Vavasour

Magnetic resonance spectroscopy 288
Metabolites of interest for multiple sclerosis
  studies 289
Magnetic resonance spectroscopy data
  acquisition 292
Magnetic resonance spectroscopy data analysis 293
Factors affecting reproducibility of magnetic
  resonance spectroscopy and consensus
  protcols 294

Magnetic resonance spectroscopy findings in multiple sclerosis lesions 295
Magnetic resonance spectroscopy findings in multiple sclerosis normal-appearing white matter and gray matter 298
Magnetic resonance spectroscopy and multiple sclerosis cognitive impairment 299
Magnetic resonance spectroscopy in multiple sclerosis clinical trials and disease modifying therapy evaluation 300
Limitations of magnetic resonance spectroscopy studies in multiple sclerosis 301
Future directions in magnetic resonance spectroscopy for multiple sclerosis research and clinical care 301
Myelin water fraction 302
What is myelin water? 302
Measurement of myelin water fraction 303
Myelin water fraction validation 305
Myelin water fraction in different multiple sclerosis tissues 305
Myelin water fraction in multiple sclerosis subtypes 307
Myelin water fraction correlations with clinical measures 307
Myelin water fraction and disease modifying treatments 307
Limitations 308
Future directions 308
Summary 311
References 311

17. High-field imaging in multiple sclerosis 321

Francesca Bagnato, Kelsey Barter, Chloe Cho, Carynn Koch, Zachery Rohm and Colin McKnight

Introduction 322
Multiple sclerosis–induced disease under the microscope of high-field imaging 323
Clinical application of high-field imaging 333
Technical challenges associated with the use of high-field imaging 334
Summary and conclusions 335
Acknowledgments 335
Conflict of interest 335
References 335

3

Use of other imaging acquisition and analysis modalities in multiple sclerosis

18. Positron emission tomography imaging in multiple sclerosis

Steven Cicero, Caleb Hansel, Eero Rissanen and Tarun Singhal

Introduction 344
Positron emission tomography imaging 344
Microglial activation and translocator protein–positron emission tomography 345
Translocator protein–positron emission tomography ligands 346
Translocator protein–positron emission tomography and gray matter pathology in multiple sclerosis 347
Translocator protein–positron emission tomography and white matter 347
Translocator protein–positron emission tomography and symptom pathogenesis 350
Translocator protein–positron emission tomography and prognostication in multiple sclerosis 351
Translocator protein–positron emission tomography and treatment effects 353
Beyond translocator protein–positron emission tomography: other glial imaging targets and PET ligands 356
Beyond translocator protein–positron emission tomography: PET imaging of nonimmune mechanisms in MS 356
Future directions and conclusion 356
References 357

19. Optical coherence tomography in multiple sclerosis

Nik Krajnc and Gabriel Bsteh

Introduction 361
Technical principles of optical coherence tomography 362
Retina as a window to the brain 363
Optical coherence tomography in optic neuritis 365

Optical coherence tomography is a marker of multiple sclerosis–associated neuroaxonal damage 367
Optical coherence tomography in treatment monitoring 369
Practical issues and limitations affecting clinical application of optical coherence tomography in multiple sclerosis 370
Conclusion 371
References 372

20. Imaging of multiple sclerosis in resource-poor settings

Avinash Chandra

Optimization strategies 379
Conclusion 380
References 380

21. Use of artificial intelligence in multiple sclerosis imaging

Ceren Tozlu, Amy Kuceyeski and Michael G. Dwyer

Introduction 384
Basics of AI in medical imaging 385
AI in MS neuroimaging: survey of current applications 393
AI pitfalls and ethical concerns 408
Future directions 410
Conclusions 413
References 413

Index 421

List of contributors

Francesca Bagnato Neuroimaging Unit, Neuroimmunology Division, Department of Neurology, Vanderbilt University Medical Center, Nashville, TN, United States; Department of Neurology, VA Hospital, TN Valley Healthcare Center, Nashville, TN, United States

Kelsey Barter Neuroimaging Unit, Neuroimmunology Division, Department of Neurology, Vanderbilt University Medical Center, Nashville, TN, United States; Division of Pediatric and Developmental Neurology, Washington University School of Medicine, St. Louis, MO, United States; Vanderbilt University School of Medicine, Nashville, TN, United States

Ralph H.B. Benedict Department of Neurology, Jacobs MS Center for Treatment and Research, Jacobs School of Medicine and Biomedical Sciences, University at Buffalo, State University of New York, Buffalo, NY, United States

Niels Bergsland Buffalo Neuroimaging Analysis Center (BNAC), Department of Neurology, Jacobs School of Medicine and Biomedical Sciences, University at Buffalo, State University of New York, Buffalo, NY, United States

Gabriel Bsteh Department of Neurology, Medical University of Vienna, Vienna, Austria; Comprehensive Center for Clinical Neurosciences and Mental Health, Medical University of Vienna, Vienna, Austria

Sagar Buch Department of Neurology, Wayne State University, Detroit, MI, United States

Massimiliano Calabrese University of Verona, Verona, Italy

Nil Saez Calveras Department of Neurology, University of Texas Southwestern Medical Center, Dallas, TX, United States; Peter O'Donnell Jr. Brain Institute, University of Texas Southwestern Medical Center, Dallas, TX, United States

Mara Cercignani Cardiff University Brain Research Imaging Centre, Cardiff University, Cardiff, Wales, United Kingdom

Avinash Chandra Consultant Neurologist Neurology NAMS, Bir Hospital, Kathmandu, Bagmati, Nepal; Consultant Neurologist Neurology Annapurna Neurological Institute and Allied Sciences, Kathmandu, Bagmati, Nepal

Chloe Cho Vanderbilt University School of Medicine, Nashville, TN, United States

Junghun Cho Department of Biomedical Engineering, State University of New York at Buffalo, Buffalo, NY, United States

Steven Cicero Department of Neurology, Ann Romney Center for Neurologic Diseases, Brigham and Women's Hospital, Harvard Medical School, Boston, MA, United States

Alexey Dimov Department of Radiology, Weill Cornell Medicine, New York, NY, United States

Michael G. Dwyer Buffalo Neuroimaging Analysis Center (BNAC), Department of Neurology, Jacobs School of Medicine and Biomedical Sciences, University at Buffalo, State University of New York, Buffalo, NY, United States; Center for Biomedical Imaging at Clinical and Translational Science Institute, University of Buffalo, State University of New York, New York, NY, United States

Svetlana Eckert Department of Neurology, Jacobs Comprehensive Ms Treatment and Research Center, Jacobs School of Medicine and Biomedical Sciences, University at Buffalo, State University of New York, Buffalo, NY, United States

Giuseppe Fenu Department of Neurosciences, ARNAS Brotzu, Cagliari, Italy

E. Mark Haacke Department of Neurology, Wayne State University, Detroit, MI, United States

Caleb Hansel Department of Neurology, Ann Romney Center for Neurologic Diseases, Brigham and Women's Hospital, Harvard Medical School, Boston, MA, United States

Hernan Inojosa Department of Neurology, Center of Clinical Neuroscience, University Hospital Carl Gustav Carus, Technical University of Dresden, Dresden, Saxony, Germany

Dejan Jakimovski Buffalo Neuroimaging Analysis Center (BNAC), Department of Neurology, Jacobs School of Medicine and Biomedical Sciences, University at Buffalo, State University of New York, Buffalo, NY, United States

Carynn Koch Neuroimaging Unit, Neuroimmunology Division, Department of Neurology, Vanderbilt University Medical Center, Nashville, TN, United States

Channa Kolb Department of Neurology, Jacobs Comprehensive Ms Treatment and Research Center, Jacobs School of Medicine and Biomedical Sciences, University at Buffalo, State University of New York, Buffalo, NY, United States

Nik Krajnc Department of Neurology, Medical University of Vienna, Vienna, Austria; Comprehensive Center for Clinical Neurosciences and Mental Health, Medical University of Vienna, Vienna, Austria

Eva A. Krijnen MS Center Amsterdam, Anatomy and Neurosciences, Amsterdam Neuroscience, Amsterdam UMC Location VUmc, Amsterdam, The Netherlands; Department of Neurology, Massachusetts General Hospital, Harvard Medical School, Boston, MA, United States

Amy Kuceyeski Department of Radiology, Weill Cornell Medicine, New York, NY, United States

Maria Marcella Laganà IRCCS Fondazione Don Carlo Gnocchi ONLUS, Milan, Italy

Cornelia Laule Department of Radiology, University of British Columbia, Vancouver, Canada; Department of Pathology & Laboratory Medicine, University of British Columbia, Vancouver, Canada; Department of Physics & Astronomy, University of British Columbia, Vancouver, Canada; International Collaboration on Repair Discoveries (ICORD), University of British Columbia, Vancouver, Canada

Lorena Lorefice Department of Medical Sciences and Public Health, Multiple Sclerosis Center, Binaghi Hospital, ASL Cagliari, University of Cagliari, Cagliari, Italy

Matteo Mancini Cardiff University Brain Research Imaging Centre, Cardiff University, Cardiff, Wales, United Kingdom

Colin McKnight Department of Radiology and Radiological Sciences, Vanderbilt University Medical Center, Nashville, TN, United States

Laura Pelizzari IRCCS Fondazione Don Carlo Gnocchi ONLUS, Milan, Italy

Alexander Rauscher UBC MRI Research Centre, The University of British Columbia, Vancouver, BC, Canada; Department of Physics and Astronomy, The University of British Columbia, Vancouver, BC, Canada; Department of Pediatrics, The University of British Columbia, Vancouver, BC, Canada

Eero Rissanen Department of Neurology, Ann Romney Center for Neurologic Diseases, Brigham and Women's Hospital, Harvard Medical School, Boston, MA, United States

Zachery Rohm Neuroimaging Unit, Neuroimmunology Division, Department of Neurology, Vanderbilt University Medical Center, Nashville, TN, United States

Menno M. Schoonheim MS Center Amsterdam, Anatomy and Neurosciences, Amsterdam Neuroscience, Amsterdam UMC Location VUmc, Amsterdam, The Netherlands

Ferdinand Schweser Department of Neurology, Buffalo Neuroimaging Analysis Center, Jacobs School of Medicine and Biomedical Sciences at the University at Buffalo, Buffalo, NY, United States; Center for Biomedical Imaging, Clinical and Translational Science Institute at the University at Buffalo, Buffalo, NY, United States

Tarun Singhal Department of Neurology, Ann Romney Center for Neurologic Diseases, Brigham and Women's Hospital, Harvard Medical School, Boston, MA, United States

Olaf Stuve Department of Neurology, University of Texas Southwestern Medical Center, Dallas, TX, United States; Peter O'Donnell Jr. Brain Institute, University of Texas Southwestern Medical Center, Dallas, TX, United States; Neurology Section, VA North Texas Health Care System, Dallas, TX, United States

Agnese Tamanti University of Verona, Verona, Italy

Eleonora Tavazzi Multiple Sclerosis Centre, IRCCS Mondino Foundation, Pavia, Italy

Ceren Tozlu Department of Radiology, Weill Cornell Medicine, New York, NY, United States

Tomas Uher Department of Neurology and Center of Clinical Neuroscience, First Faculty of Medicine and General University Hospital in Prague, Charles University, Prague, Czech Republic

Manuela Vaneckova Department of Radiology, First Faculty of Medicine and General University Hospital in Prague, Charles University, Prague, Czech Republic

Irene M. Vavasour Department of Radiology, University of British Columbia, Vancouver, Canada; International Collaboration on Repair Discoveries (ICORD), University of British Columbia, Vancouver, Canada

Zachary L. Weinstock Buffalo Neuroimaging Analysis Center (BNAC), Department of Neurology, Jacobs School of Medicine and Biomedical Sciences, University at Buffalo, State University of New York, Buffalo, NY, United States

Bianca Weinstock-Guttman Department of Neurology, Jacobs Comprehensive Ms Treatment and Research Center, Jacobs School of Medicine and Biomedical Sciences, University at Buffalo, State University of New York, Buffalo, NY, United States

Tjalf Ziemssen Department of Neurology, Center of Clinical Neuroscience, University Hospital Carl Gustav Carus, Technical University of Dresden, Dresden, Saxony, Germany

Robert Zivadinov Buffalo Neuroimaging Analysis Center (BNAC), Department of Neurology, Jacobs School of Medicine and Biomedical Sciences, University at Buffalo, State University of New York, Buffalo, NY, United States

CHAPTER 1

Introduction

Dejan Jakimovski[1] and Robert Zivadinov[1,2]

[1]Buffalo Neuroimaging Analysis Center (BNAC), Department of Neurology, Jacobs School of Medicine and Biomedical Sciences, University at Buffalo, State University of New York, Buffalo, NY, United States [2]Center for Biomedical Imaging at the Clinical Translational Science Institute, University at Buffalo, State University of New York, Buffalo, NY, United States

OUTLINE

Introduction 1

Introduction

Multiple sclerosis (MS) is recognized as one of the most common reasons for neurological disability in the young and working population. Latest epidemiological studies have estimated that up to 2.7 million people live with MS worldwide, with highest prevalence in the northern parts of the world such as the United States, Canada, and Western Europe. The disability accrued as a result of the disease substantially impacts the quality of life of people with MS with both physical and cognitive impairments contributing to professional and social limitations. Being life-long disease, it also results with a significant societal economic cost. For example, the annual burden of MS in the United States is estimated over US$85 billion. Similar economic and social burden is described around the world.

Due to the unpredictability and complexity of the symptoms that may emerge in people with MS, this disease has relatively short medical history. Few descriptions of people potentially suffering from MS can be recognized in the history books, ranging from Saint Lidwina van Schiedam (1380–1433) and Augustus d'Esté (1794–1848), the grandson of King George III. The emergence of MS in the medical field as an established neurological disease was brought by Jean-Martin Charcot in 1868 during a series of lectures describing the symptoms of this progressive disease. The term *sclérose en plaque disseminé* was coined by his colleague and coauthor Edme Felix Alfred Vulpian 2 years earlier in 1866.

Throughout the 20th century, multiple theories of its etiology were raised, with viral, immunological, genetic, or vascular links being suggested.

Over the last 30 decades, there has been exponential increase in our ability to detect, monitor, and treat MS. One aspect that played major part in the renaissance of MS knowledge was the widespread adoption and utilization of magnetic resonance imaging (MRI). It allows noninvasive and most importantly, an in vivo visualization of the MS pathophysiological processes. Today, MRI remains a cornerstone of the MS field, being essential for implementing the MS diagnostic criteria, its routine use in monitoring of the MS progression and understanding the effectiveness of MS-targeted interventions.

This handbook is written targeting a vast profile of readers including Neurologists, Radiologists, Residents in these respective fields, Researchers in the field of Neurology and Radiology, Physicists, Immunologists, Neuroscience researchers, Biomedical engineering PhD students, and Physics and Neuroscience students. It aims at filling the current gap of a handbook that could introduce and delve deeper the latest and most pertinent information needed for MS care providers, radiologists, and researchers that would like to integrate imaging as a new modality into their research repertoire. It is written by exceptional international list of authors that are highly regarded in the very specific field of their respective chapter and represent leaders in the field of imaging in MS.

This handbook contains chapters that focus on the conventional and nonconventional MRI techniques that have been essential part of the clinical practice and to the expanding frontier of MS knowledge. Greater number of technologies are being integrated into the routine clinical care of MS patients, with providing unprecedented amount of information that can be utilized for better disease diagnosis, prognosis and monitoring of therapy responsiveness. The first part of this book contains chapters that provide the latest overview of the pathophysiological MS processes, the current clinical MS knowledge and how can we modify the disease by using ever increasing portfolio of disease modifying therapies. These chapters should serve as the background for the subsequent imaging chapters. As mentioned earlier, the MRI represents the cornerstone of MS imaging, with chapters being split by whether this technology is currently utilized in the clinical routine (conventional MRI use) or is actively being developed and researched (nonconventional MRI use). The world of MS imaging contains greater arsenal than just MRI, with chapters dedicated to role of positron emission tomography, optical coherence tomography, and use of newly developed artificial intelligence—based technologies.

We are very grateful to have the unique opportunity to compile and edit this book with a concise list of chapters that can serve as a building block for every new MS provider and researcher in the field of imaging in MS. It can serve as a source of introductory texts and references that would lead to a deeper exploration of each MRI technique. None of the work presented in this book would have been possible without the expertise of each author that contributed to its completion.

CHAPTER 2

The immunology and pathophysiology of multiple sclerosis

Nil Saez Calveras[1,2] and Olaf Stuve[1,2,3]

[1]Department of Neurology, University of Texas Southwestern Medical Center, Dallas, TX, United States [2]Peter O'Donnell Jr. Brain Institute, University of Texas Southwestern Medical Center, Dallas, TX, United States [3]Neurology Section, VA North Texas Health Care System, Dallas, TX, United States

OUTLINE

Introduction	3	From autoimmunity to pathology	18
The immunology of multiple sclerosis	4	Inflammatory lesion formation and evolution	21
Autoimmunity in multiple sclerosis: genetic and environmental risk factors	4	Pathogenesis in the context of progressive multiple sclerosis	22
Meet the players: adaptive and innate immune cell responses in multiple sclerosis	5	Remyelination	24
		Conclusion	25
The pathophysiology of multiple sclerosis	18	References	25

Introduction

Multiple sclerosis (MS) is a complex autoimmune disease of the central nervous system (CNS) that affects more than 2.5 million people worldwide and represents the most common neurological disease in young adults [1]. This very heterogeneous disease exhibits a variety of clinical manifestations that differ widely between patients. In the first few years after MS is diagnosed, the patients typically experience a relapsing-remitting multiple sclerosis (RRMS) characterized by fluctuating episodes of acute neurological symptoms, followed by a partial or complete resolution in the remission phase. As the disease course advances, patients can develop a progressive disease, known as secondary progressive MS (SPMS). In this stage, patients accumulate neurological deficits in a steady manner while the acute exacerbations become less frequent. In a small subset of patients, MS has a relentless progressive course from the onset, in what is known as primary progressive MS [2]. All currently approved therapeutics consist of immune modulatory agents that are effective at preventing the relapses associated with RRMS or SPMS with relapses. Immunotherapy agents, including steroids or intravenous immunoglobulin, are also useful in the setting of an acute exacerbation. Despite this, the options to manage progressive disease are limited, with the current available agents having limited efficacy in curbing its course.

This chapter delves into the intricate immunological and pathophysiological mechanisms underlying MS, shedding light on the risk factors, immune cells and pathogenic events that contribute to its development and progression.

The immunology of multiple sclerosis

Autoimmunity in multiple sclerosis: genetic and environmental risk factors

MS is characterized by immune dysregulation, where the immune system mistakenly targets CNS autoantigens leading to damage to the myelin sheath of axons causing demyelination. The underlying etiology of this autoimmune response is not well defined but is thought to be triggered by both genetic and environmental factors.

MS has a genetic component, and certain genes are associated with an increased susceptibility to the disease. The HLA-Dr2 serotype within the major histocompatibility complex (MHC) on chromosome 6 has been consistently linked to MS susceptibility [3]. In particular, harboring the HLA-DRB1*15:01 allele increased the risk of MS threefold [4]. A genome-wide association study of over 40,000 MS patients identified multiple susceptibility loci in the extended MHC region as well as variants outside of the MHC [5]. Weighted genetic risk scores compiling these susceptibility alleles have been shown to modestly predict MS risk [6,7].

As genetic factors alone do not solely account for the risk of MS, it is intuitive that environmental factors play an important pathogenic role. These factors include vitamin D deficiency, viral infections, diet and early life obesity, and cigarette smoking. Viral infections, particularly Epstein–Barr virus (EBV), have been associated with the clinical onset of MS [8]. The potential causal relationship between viral infections and MS remains an active area of research, including molecular mimicry between EBV determinants and CNS autoantigens [9]. Low vitamin D levels have been associated not only with an increased risk of developing MS but also with a higher risk of long-term disease activity and progression [10]. When compared to nonsmokers, MS patients who smoke also have more motor symptoms [11], increased magnetic resonance imaging (MRI) disease activity [12], brain atrophy and cognitive impairment [13], earlier onset of secondary progressive disease [14], and even earlier death [15]. Interestingly, smoking cessation at any time can slow the rate of motor deterioration [16].

The relationship between genetics and the environment in the pathogenesis of MS is a complex one, and further research is needed. However, certain lifestyle interventions including smoking cessation [16] and adequate vitamin D intake [17] might have a beneficial effect both for primary prevention and prevention of disease progression [16].

Meet the players: adaptive and innate immune cell responses in multiple sclerosis

Genetic and pathological evidence appears to suggest a concurrent role of the innate and adaptive immune system in the pathogenesis of MS. In this section we will provide an overview of the cell types that have been involved in the onset and progression of MS, starting with an overview of the innate immune system, which primarily includes infiltrating bone-marrow–derived myeloid cells and CNS-resident microglia, and followed by the adaptive immune system, mainly consisting of autoreactive T and B cells.

The role of innate immune cells
Bone marrow-derived myeloid cells

Monocytes and macrophages Monocytes constitute an essential component of the innate immune response and are part of the first line of defense against infections and other noxious stimuli. These immune cells are produced in the bone marrow and released into the blood stream where they can then phagocytose exogenous agents or cellular debris. Two main populations of monocytes have been described in mammalian species. Mouse $Ly6C^-$ monocytes (or $CD14^{lo}CD16^{hi}$ in humans) patrol blood vessels and remove damaged endothelial cells, while $Ly6C^+$ monocytes (or $CD14^{hi}CD16^{lo}$ in humans) can egress from the circulation into tissues where they give rise to a variety of monocyte-derived cells. Once they have infiltrated, these cells can in turn differentiate into macrophages or monocyte-derived dendritic cells (DCs) to exert phagocytic functions, act as antigen presenting cells (APCs) to present antigens to autoreactive T cells via their MHC-II molecules, produce reactive oxygen species (ROS), release cytokines, and recruit other immune cells to the affected tissue.

In patients with MS, the blood monocyte count during early stages of the disease was robustly associated with its clinical severity, and served as a prognostic factor [18]. In addition, the percentage of Interleukin-12+ (IL-12+) monocytes in the blood also correlated with the presence of active MRI lesions [19]. IL-12 is a cytokine secreted by monocytes that can induce the differentiation of T cells into T helper 1 (Th1) cells. As it will be described in more detail below (see "The role of adaptive immune cells" section), autoreactive Th1 cells are important players in the inflammatory adaptive immune response in MS.

In the CNS of MS patients, along with tissue-resident microglia, bone-marrow–derived macrophages constitute the most abundant cell type in inflammatory lesions. Their presence and activation correlates with tissue damage [20], and many of them can be found closely related to damaged axons or phagocytosing myelin [21]. There is also abundant evidence that monocyte-derived macrophages as well as DCs, reside and act as APCs in the cerebral perivascular spaces (CPVS). These myeloid cells in this compartment are thought to play a role in the initiation and perpetuation of CNS autoimmune disease through the reactivation of autoreactive T cells [22].

The importance of macrophages in the pathogenesis of MS is also highlighted in studies employing the experimental autoimmune encephalomyelitis (EAE) animal model. In this model, a CNS inflammatory demyelinating disease is triggered by immunization of these animals, typically rodents, with brain extracts, CNS proteins or peptides emulsified in an adjuvant. One group of investigators found that macrophage depletion prevented CNS lesion development in the EAE Lewis rat model [23]. In addition, interfering with the influx of monocytes into the CNS also curtailed disease progression in EAE [24]. A study also identified that the entry of bone-marrow derived $CD11c^+CD88^+CD317^+$ myeloid cells into the CNS was required for the establishment of clinical EAE after active immunization [25]. The single-cell RNA and chromatin accessibility profile of these cells revealed that they have a unique signature and distinct functions when compared with CNS-resident microglia [26]. This study also challenged that the entry of these immune cells into the CNS may not be solely dependent on α4-integrin expression but may rather be facilitated by complementary molecules such as CD317, which can be induced by interferon-γ (IFN-γ) in the setting of inflammation. As it will be described in more detail below, IFN-γ is a proinflammatory cytokine secreted by activated Th1 cells, one of the principal T cell subsets involved in the pathogenesis of MS [25]. Beyond IFN-γ, the growth factor granulocyte–macrophage colony stimulating factor (GM-CSF) also appears to play a role in the recruitment of myeloid cells into the CNS in EAE. Immediately before EAE relapses, $Ly6C^+$ monocytes are mobilized into the blood stream by this growth factor, and then traffic across the blood–brain barrier (BBB). This blood enrichment of $Ly6C^+$ monocytes was associated with an earlier onset and increased severity of EAE [27].

The use of single-cell RNA-sequencing has significantly aided in characterizing several molecularly distinct bone-marrow–derived myeloid cell subsets in MS and EAE models [28,29]. Along this line, a study identified a particular monocyte subset characterized by Cxcl10 positivity that was pathogenic in EAE, and their specific depletion reduced clinical disease symptoms in mice. These cells arose from monocyte precursor cells and were independent of the classical circulating $Ly6C^+$ monocytes [30].

Beyond their role in inflammatory lesions, monocytes and macrophages can also have an important role in lesion recovery, supporting axonal regeneration and myelin repair, and driving oligodendrocyte differentiation during CNS remyelination [31]. These seemingly opposing effects can be attributed to the fact that the local CNS environment and lesion stage may influence the polarization of these cells toward a pro- or an antiinflammatory phenotype [28]. As mentioned, IFN-γ can promote the recruitment of monocytes into the CNS in the context of EAE. However, this cytokine is not exclusively pathogenic in this disease. The expansion and recruitment of programmed death-ligand 1 (PD-L1) expressing myeloid cells with an antiinflammatory phenotype also appears to be dependent on IFN-γ [32].

Dendritic cells DCs can be broadly divided into conventional DCs, the main function of which is to recognize and present antigens as APCs through their MHC-II molecules to adaptive immune cells, and plasmacytoid DCs which characteristically act to sense intracellular antigens (i.e., DNA, RNA) through their Toll-like receptors TLR7 and TLR9. In addition, and as mentioned earlier, another subpopulation of DC known as monocyte-derived DCs can be derived from monocytes as they become tissue-resident cells.

In the context of MS, circulating conventional DC exhibit decreased expression of thymic stromal lymphopoietin receptor, which determines the differentiation of regulatory T cells (Treg) in the thymus. As it will be detailed later, Treg cells serve an important immune modulatory function. This suggests that conventional DCs may impair Treg development in the context of MS contributing to immune dysregulation [33]. In this disease, conventional DCs also had increased IL-12 production upon TLR binding and increased expression of migratory molecules (CCR5 and CR7) that allowed them to migrate into the CNS of these patients [34]. As mentioned, IL-12 contributes to T cell differentiation into Th1 cells.

DCs may play an important role in progressive MS. One study identified that subjects with SPMS had an increased percentage of IL-12 and TNFα-producing DCs when compared to RRMS and controls. These cells also had a decreased expression of the immune response coinhibitory factor PD-L1 and predominantly induced a Th1 cell-polarized response [35]. Along this line, monocyte-derived DCs, but not conventional DCs, were critical in the induction of differentiation of Th17 lymphocytes in EAE, an effect that was promoted by GM-CSF [36]. As detailed below, IL-17 secreting Th17 cells, along with Th1 cells, also constitute an important part of the autoreactive adaptive immune response in MS.

Plasmacytoid DCs are also instrumental in maintaining peripheral T cell tolerance and innate immunity, and their imbalance is thought to contribute to the development of autoimmunity. One study identified that these cells consist of two populations, namely pDC1 and pDC2, which could differently induce IL-10 and IL-17 producing T cells. In the setting of MS, plasmacytoid DCs had an overall propensity to prime IL-17 secreting cells over IL-10 secreting CD4 + T cells, an effect that was rescued by IFNβ administration [37]. When stimulated, these cells exhibited an inefficient maturation, surface marker expression, IFNα secretion, and failed to upregulate the proliferation of peripheral blood mononuclear cells. This function was partly restored by glatiramer acetate administration, an immune modulatory therapy agent used in MS [38].

Microglia

Microglia constitute the tissue-resident macrophages of the CNS. In normal brains, these cells appear to play an important role in the regulation of CNS homeostasis and inflammation. Their dysregulation in MS appears to favor an activated, tissue-destructive function [39]. As with bone-marrow derived myeloid cells, single cell analysis has revealed time and region-dependent subtypes of microglia in the context of MS. In the brains of MS patients, distinct clusters of microglia with unique molecular hallmarks have been identified [40]. As an example, one cluster was characterized by the increased expression of chemokine and cytokine genes, another showed increased expression of MHC class II molecules, and they all exhibited different degrees of microglial core gene expression [40]. Interestingly, in the cuprizone-induced demyelination mouse model, MHC-II deficiency in microglia, which led to absent microglial antigen presentation, did not appear to affect the establishment, progression and severity of EAE [41].

In another study, single cell profiling of $CD88^+CD317^+$ myeloid cells in the CNS of EAE mice showed that this signature is shared between microglia and infiltrating bone-marrow—derived myeloid cells alike, which likely points toward these being markers of an "activated" state. Both bone-marrow derived myeloid cells and microglia exhibited patterns of gene expression along a spectrum of immune-regulatory to proinflammatory functions. At the same time, they harbored unique transcriptional and epigenetic signatures that allowed to distinguish between the two [26].

In progressive MS, low-grade inflammation and microglial activation are a characteristic feature. It is thus thought that this microglial activation contributes to active demyelination and axonal injury in the white matter, as well as synaptic loss and neuronal degeneration in the gray matter of these patients [42]. As it is the case in bone-marrow−derived macrophages, microglia can also mediate the phagocytosis of myelin debris, and other targets in MS [43]. Another characteristic feature of progressive MS is the presence of diffuse microglial activation in normal-appearing white and gray matter [44]. In these patients, the diffuse injury associated with microglial activation in the normal-appearing white matter contributes to cortical volume loss more than focal demyelinating lesions. This microglia express Nicotinamide Adenine Dinucleotide Phosphate (NADPH) oxidase and myeloperoxidase which are involved in the production of ROS that lead to oxidative damage [45]. Histological evidence shows that these cells are found in close contact with degenerating axons, at the edge of active and slowly expanding lesions and with degenerating cortical neurons [46]. Despite the importance of this diffuse oxidative injury generated by microglia in human MS, the animal models of inflammatory demyelination do not appear to fully replicate these findings [47].

Reactive astrocytes are strongly induced by CNS injury and disease. One study identified that activated microglia can induce the generation of a subtype of astrocytes known as A1, which secrete proinflammatory cytokines, namely IL-1α, TNF, and complement factor C1q. These A1 astrocytes promote neuronal and oligodendrocyte death, as well as synaptic loss [48]. Microglia may in turn also mediate synaptic loss in the gray matter of MS patients through the opsonization of complement-bound synaptic terminals [49]. Thus this portrays another mechanism by which microglia can induce tissue injury in MS.

Overall, the contribution of microglia, as well as bone-marrow derived myeloid cells, to the disease phenotype in MS remains a topic of ongoing debate and active research. The full understanding of these cells throughout the disease course and their differential contribution at the different stages of RRMS and progressive MS is hampered by the availability of animal models that can faithfully replicate this disease. EAE, despite being a good model of autoimmune demyelination, constitutes a largely monophasic disease. Thus it cannot fully elucidate the spectrum of phenotypes that these cells have in MS.

Natural killer cells and other innate lymphoid cells

Natural killer (NK) cells are a heterogeneous cellular population that contribute to both innate immune effector functions via their cytotoxic activity as well as regulatory functions through the secretion of cytokines and growth factors [50]. They belong to the family of innate lymphoid cells (ILCs) and their population can be distinguished into two main groups, immature ($CD56^{bright}$) and mature ($CD56^{dim}$) NK cells. Immature NK cells have been proposed to have an immune-regulatory function. In one study, having a high proportion of immature NK cells in the blood was associated with stable Mr imaging in MS patients, suggesting that an expansion of these cells may serve as an attempt to counteract the immune activation seen in MS [51]. Along these lines, another study found that the proportion of these immature NK regulatory cells was higher in the CSF of patients with MS when compared with other neurological diseases [52]. However, these cells may have a defective immune-regulatory function in this disease, exhibiting a reduced cytolytic activity against auto-reactive CD4 T cells [53]. Interestingly, the numbers of these immature NK cells can increase with immune modulatory treatments [54–56], again supporting the regulatory role of these cells in the setting of autoimmunity. In contrast to immature NK cells, the role of mature NK cells in the context of MS is less well defined.

The family of ILCs has been expanded with the discovery of group 1, 2, and 3 ILCs, namely ILC1, ILC2, and ILC3. ILC1 and ILC3 resemble Th1 and Th17 cells, respectively, which constitute important players in MS pathogenesis. Within the ILC family, a subset of T-bet-dependent NKp46 + ILC3 was found to be associated with the initiation of CD4 + Th17 cell-mediated neuroinflammation. In an EAE mouse model, loss of these cells in the meninges impaired the ability of myelin-reactive Th17 cells to enter the CNS and develop autoimmunity [57]. Treatment with fingolimod, an agonist of sphingosine-phosphate receptor 1, which is expressed by lymphocytes and regulates their egression from secondary lymphoid organs, led to a decrease in peripheral blood ILCs in MS patients. Ex vivo studies also showed that exposure of both ILC1 and ILC3 cells to fingolimod reduced the production of cytokines by these cells [58]. This portrays a possible mechanism of action that contributes to the effectiveness of this immune modulatory drug in MS. In contrast to these classes, ILC2 cells appeared to promote and sustain a nonpathogenic Th2 myelin-specific response in a model of EAE serving as attenuators of the pathogenic Th response seen in CNS inflammation [59]. However, in the HSV-IL2 mouse model of CNS demyelination, these cells were necessary to promote CNS demyelination [60]. Overall, all this evidence suggests a role of ILC in CNS autoimmunity and MS. However, the specific function of each cellular subtype appears to be conditioned by the environment and disease state.

Neutrophils

Neutrophils are the most abundant cell population in the circulating leukocyte fraction, and constitute a first line of defense against bacterial and fungal infections [61]. Once recruited to the tissue site, these polymorphonuclear cells can eliminate pathogens through engulfment into phagosomes [62], ROS production, secretion of granules containing myeloperoxidases, proteases and other enzymes through degranulation, and formation of neutrophil extracellular traps that bind organisms and prevent them from spreading [63]. These cells can also recruit and activate monocytes, DCs, and lymphocytes to the tissue site. To perform their function appropriately, neutrophils undergo a closely regulated activation process that involves a preactivation process known as "priming." These primed neutrophils exhibit decreased apoptosis, enhanced adhesion molecule expression, and effector mechanisms. In the context of MS, neutrophils have been found to exhibit inappropriate priming [64]. These elevated numbers of preactivated or primed neutrophils could contribute to MS pathogenesis by increasing inflammation and tissue injury. In addition, these cells exhibit a heightened activation during infection, which could explain the association between infection and relapses in MS [65].

Animal studies have revealed multiple mechanisms whereby which neutrophils could be contributing to disease pathogenesis in EAE and MS. First, neutrophils appear to be key regulators of BBB permeability and are able to induce BBB breakdown through a decrease in tight junction adhesion molecules [66]. Second, CNS-infiltrating neutrophils in

EAE secreted proinflammatory cytokines that contributed to the maturation of APCs [67]. In turn, neutrophils can act as APCs themselves to regulate T cell activity [68]. CXCR2 is the main receptor mediating the chemotaxis of neutrophils, as well as its effector function. CXCR2 expression is stimulated by IL-8 (or CXCL1), which is increased in MS patients [69]. Depletion of CXCR2 led to resistance to demyelination and CNS neurodegeneration in EAE [70,71]. In these animals, injection of CXCR2 + neutrophils restored their susceptibility to demyelination [71]. Interestingly, in an adoptive transfer EAE model, the frequency of CNS neutrophils as well as neutrophil-mobilizing factors was higher in the CNS of middle aged to elderly mice when compared to younger mice. These findings could be explained by an increased production of neutrophil-attracting chemokines CXCL2 and CXCL10 by aged microglia [72]. Serum neutrophil-related factor levels (i.e., neutrophil elastase) have also been found to be elevated in patients with progressive MS when compared with RRMS [73]. This evidence points toward a potentially more preponderant role of these cells in the progressive forms of this disease.

However, as it was the case for the other innate immune cells, certain neutrophil subsets may actually have protective functions after neuronal injury. One study identified that a granulocyte subset with characteristics of an immature neutrophil had neuroprotective properties and was able to drive CNS axon regeneration after transection in vivo in the optic nerve and spinal cord [74] (Fig. 2.1).

The role of adaptive immune cells
CD4 + T cells

Autoreactive CD4 + T cells play a central role in MS pathogenesis. Before the onset of MS, these cells are activated in secondary lymphoid organs by APCs which present peptide antigens through their MHC-II molecules, to which T cells can bind via their T-cell receptor (TCR). From there, these activated T cells then migrate into the CNS. Once in the CNS, they are locally reactivated and secrete cytokines and chemokines that trigger and modulate the classical inflammatory lesions seen in MS [75,76]. As mentioned earlier, favoring the role of these cells in this disease, the HLA-DRB*15:01 allele, which constitutes a risk factor for MS, encodes a MHC class II molecule involved in the presentation of self-antigens to CD4 + T cells [77,78]. Activated CD4 + T cells proliferate and differentiate into multiple subsets. The predominant cell subsets that have been associated with MS are Th1 and Th17 cells [79,80]. Once in the CNS, these cells can promote the activation of resident glial cells and other infiltrating cells. The main proposed autoimmune targets of autoreactive CD4 + T cells in MS are components of the myelin sheath (i.e., myelin basic protein [MBP], myelin oligodendrocyte glycoprotein, or proteolipid protein), which can lead to the characteristic demyelinating lesions seen in patients with MS. However, Th cells that are self-reactive against gray matter antigens have also been identified, which may contribute to the cortical neurodegeneration seen in progressive MS patients [81]. Several mechanisms that do not require the recognition of an auto-antigen by the TCR may contribute CNS tissue damage in patients with MS. Bystander T cell activation refers to the activation of T cells that are not antigen-specific through cytokines and other soluble inflammatory mediators, and that constitute a second wave of adaptive immune response following the initial antigen recognition [82–84]. These cells can than amplify the

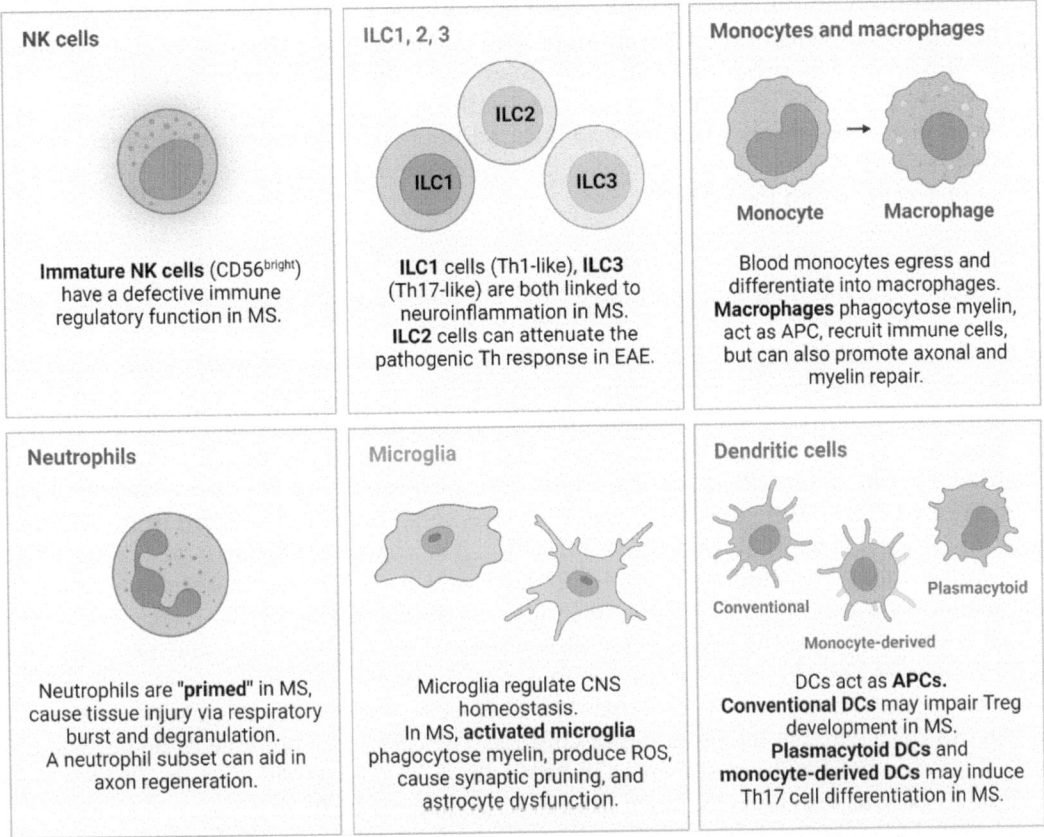

FIGURE 2.1 **Schematic review of the innate immune system in MS.** Note: Monocyte-derived dendritic cells (DCs) are classified under DCs although this population derives from circulating blood monocytes.

inflammatory response by also expressing cytokines. Antigen-specific T cells can also bystander-activate innate immune responses [85]. These activated innate immune cells in turn perpetuate systemic and localized inflammation.

Th1 cells Autoreactive CD4+ Th1 cells have been classically heralded as a main driving factor in the pathogenesis of MS. IL-12 signaling drives the differentiation of naïve CD4+ T cells into Th1 cells. These cells are characterized by the secretion of IFN-γ, a cytokine that promotes inflammation and the recruitment of immune cells into the CNS [78]. They can be identified by the surface expression of the IL-12 receptor, and the CXC chemokine receptor 3, as well as the intracytoplasmic master transcription factor T-bet [86]. The evidence for the role of Th1 cells in MS is also supported by the studies on the EAE animal model of MS, where Th1 cells constitute the most frequent cell type in CNS lesions [87]. The adoptive transfer of these auto-reactive cells to another animal also leads to the development of a disease phenotype [88]. Elevated levels of IFN-γ have also been identified in the brains of MS patients, and administration of this cytokine in human MS patients led to an exacerbation of the disease phenotype [89].

The role of Th1 cells in the CNS immune pathogenesis of MS remains to be fully elucidated. However, it is presumed that the cytokines secreted by these cells can lead to the differentiation of CNS resident microglia and infiltrating bone-marrow–derived myeloid cells to a proinflammatory phenotype [85]. In addition, Th1 cells can also upregulate the expression of MHC class II on myeloid cells, which can in turn further potentiate the activation of T cells within the CNS [90].

Th17 cells Th1 cells were once thought to be the main T-cell driver of MS. However, the observation that mice deficient in the IL-12 p35 subunit were still susceptible to EAE [89,91] questioned this paradigm. The apparent paradox that IL-12 p40 deficient mice were resistant to the disease, but that those deficient in IL-12 p35 were not, was solved when IL-23 was identified [92]. This cytokine is structurally related to IL-12 and shares the p40 subunit with it. However, IL-23 drives the expansion of IL-17 secreting Th17 cells, and mice deficient in the unique IL-23 p19 chain or treated with anti-IL-23 are resistant to EAE [93,94].

Th17 cells can be identified based on the surface expression of the chemokine receptors CCR6 and CCR4, CD161, the IL-23 and IL-1 receptors, and by the secretion of the cytokines IL-17, as well as IL-21, and IL-22. While IL-23 is important for the expansion of Th17 cells, other cytokines including IL-1β, IL-6, TGFβ, and IL-21 are also important for the differentiation of these cells [95,96]. Interestingly, mice deficient in the IL-1 receptor type I, which is expressed on Th17 cells, were also resistant to EAE induction and had defective Th17 responses [97]. In addition, proteolipid protein peptide ($PLP_{139-151}$)-specific T cells cultured in the presence of IL-23 generated Th17 cells which induced EAE after passive transfer in mice [98]. On the other hand, cultured Th1 cells were not able to do so. The use of IL-17 antibodies has been shown to improve the EAE course [99], and the induction of EAE in IL-17 knockout mice led to an improved phenotype [100].

As with Th1 cells, the role of Th17 cells in MS pathogenesis requires further analysis. In MS patients, high levels of Th17 cells have been found in the blood and CSF during active disease [101], and their preferential expansion correlated with the number of plaques on MRI [102]. Th17 cells appear to interact with astrocytes in the CNS. These cells constitute an essential component of the BBB and regulate the transport of molecules in and out of the CNS. They express IL-17 receptor A, which is upregulated in MS [103], and disruption of IL-17 Act1-mediated signaling in astrocytes in EAE improved the disease phenotype [104]. Th17 cells in conjunction with Th1, appear to upregulate the production of inflammatory factors by astrocytes, while reducing their neurotrophic activity [105]. Thus it seems that Th17 cells can promote BBB disruption in MS through their effect on astrocyte function [106]. In addition, in RRMS patients, CSF IL-17A levels correlated with measures of BBB dysfunction. The combination of IL-17 and IL-6 also reduced the expression of tight junction-associated genes and disrupted the integrity of the BBB cell line [107]. Beyond its effect on astrocytes and the BBB, Th17 cells also appear to influence oligodendrocytes, inhibiting their maturation and promoting their oxidative-stress mediated apoptosis [108].

All this evidence points toward a concurrent and potentially additive action of Th1 and Th17 cells in MS. One study suggested that Th1 cells may be the initial Th cells recruited into the noninflamed CNS, after which these cells may facilitate the recruitment of Th17 cells from the periphery. This is supported by the fact that adoptively transferred Th17 cells in the absence of IFN-γ producing cells are not able to induce EAE [109]. Given these findings, one can envision a similar relationship of these cell subtypes in the context of human MS. However, further research is necessary to characterize this interaction. More recently, a subset of Th1 cells, known as Th1-like Th17 cells, have been described. These cells coexpress markers of both Th cell subsets and are thought to arise from Th17 cells that were exposed to cytokines involved in Th1 differentiation [110]. These cells have been identified in the brain parenchyma [111], blood and CSF of MS patients [112], and appear to have a neuroinflammatory role that overlaps that of Th1 and Th17 cells.

Th22 and Th9 cells   Beyond Th1 and Th17 cells, other Th cells have been posited to be involved in the course of MS. Th22 cells have been identified in the brain tissue sections of MS patients [113]. Along with Th17 cells, Th22 cells primarily secrete IL-22, and this cytokine levels were found to be higher in the serum of patients during relapses [114]. Through its effects on astrocytes and on endothelial cells, IL-22 may trigger BBB disruption in MS [115]. IL-22 also induced Fas expression and promoted oligodendrocyte apoptosis through the activation of the NF-kb pathway in these cells [116].

In contrast to Th22 cells, a different Th cell subtype known as IL-9 producing Th9 cells, potentially exert an immune modulatory role and contribute to the maintenance of the remission phase in patients with MS [117]. This effect could be mediated through an interference with IL-17 secretion by Th17 cells [117]. Despite this finding, EAE animal studies have shown conflicting evidence regarding the beneficial effect of IL-9 and Th9 cells in this disease [118,119].

T follicular helper cells   T follicular helper (Tfh) cells were first identified in the human tonsil. These cells are primarily located in secondary lymphoid organs where they are involved in the formation of germinal centers and support B cells in lymphoid follicles. A study found that a subset of Tfh cells expands in the CSF of MS patients and worsens EAE [120]. In EAE mice, these Tfh cells drive the local infiltration and expansion of B cells [120]. In some patients with progressive MS, a tertiary lymphoid tissue containing follicle-like structures has been described in the meninges [121,122]. In these patients, this tissue is thought to play a role in this disease and contribute to the development of subpial cortical lesions. These meningeal follicular structures are thought to be maintained with the aid of both Tfh as well as Th17 cells [120,123]. A T cell population with mixed features between Tfh and Th17 cells has also been observed in patients with progressive MS and is thought to contribute to the pathogenesis of this disease [124].

Treg cells   In normal conditions, the maintenance of peripheral immune tolerance is dependent on a population of $CD4^+CD25^+FOXP3^+$ regulatory T (Treg) cells. The adoptive transfer of these Treg cells in a MOG_{35-55} EAE mouse model led to protection from clinical disease, an increase in MOG_{35-55}-specific Th2 cells and decreased T cell CNS infiltration [125]. Interestingly, in MS patients, these cells exhibit reduced FOXP3 expression, display Th1-like features, and have reduced suppressive function [126,127]. Treg cells can be separated into different subsets. One subset in particular, known as Tr1 cells, secretes IL-10 and was found to reduce EAE severity upon adoptive transfer to mice 10 days after disease induction [128].

The continuum phenotype spectrum of CD4+ T cells It appears that the polarization of CD4+ T cells into different phenotypes in MS is dependent on multiple complex cues from the environment, and these cell populations exhibit a high complexity of subtypes and plasticity that needs to be better elucidated. CD4+ T cell phenotypes in MS may indeed represent a continuum that extends beyond the classical CD4+ T lineage classification [129]. In the context of autoimmunity, both the nature of the stimuli and the location of priming may determine the T cell differentiation, functionality, and effects [130].

Therapeutic targeting Given the importance of CD4+ T cells in MS, a wide array of immune modulatory treatments that deplete, block, or delay the entry of these cells into the CNS have been shown to provide therapeutic benefit in these patients. As an example, one study identified an enriched Th cell population characterized by the expression of IFN-γ, GM-CSF, IL-2, and CXCR4 in the blood, CSF, and brain tissue of RRMS patients. Treatment with dimethyl fumarate, an immune modulatory agent used in MS, led to a specific depletion of these cells [131]. Interestingly, treatment with natalizumab, an integrin VLA4 inhibitor increased the circulating levels of a T cell population with mixed features of Th17 and Tfh cells [124]. This same cell population was then identified in the brains of SPMS patients. This suggests that the inhibition of entry of these cells into the CNS in MS is not only important for RRMS but can also aid in preventing their homing into the CNS and development of secondary progressive disease.

Similarly, promoting a switch in the phenotype of CD4+ T cells may also serve as a potential intervention for patients with MS. Active vitamin D exposure has been found to switch Th17 cells into regulatory IL-10 expressing cells [132]. IL-10 expression can also be induced by melatonin and augmented by interferon I administration [133,134].

As it has been described earlier, proinflammatory cytokines secreted by Th cells such as IFN-γ and IL-17 play an important part in the pathogenesis of MS. Given this, targeted approaches aimed at tackling these cytokines have been trialed in MS. One of these approaches involved the use of secukinumab, an anti-IL17A therapy. A proof-of-concept study showed that this treatment could reduce Mr lesion activity in MS [135]. However, the subsequent phase II trial (MABINGO study) was halted by the company before its completion [136,137]. Targeting cytokines IL-12 and IL-23, which promote Th1 and Th17 cell differentiation, respectively, also appears as a feasible option in this disease. However, the use of ustekinumab, an antibody targeting the p40 subunit of IL-12 and IL-23, showed no clinical or radiological improvement in RRMS patients. One limitation of this study was the inclusion of patients with advanced disease, where Th1 and Th17 cell differentiation has already likely occurred [138]. Overall, it seems that targeted at approaches aimed at a singular cytokines may have limited efficacy in the context of MS. This may be due to the complex network of interactions that cytokines have in this disease, where their role may also vary depending on the clinical context. Thus developing a better understanding of cytokine functions along the clinical spectrum of MS may allow for better directed therapies.

CD8+ T cells

CD8+ T cells have emerged as key players in the autoimmune cascade driving MS. With their ability to recognize and attack self-antigens, CD8+ T cells engage in an interplay with other immune cells that leads to demyelination in MS. These cells selectively respond to peptide antigens presented on MHC-I receptors. In contrast to MHC-II, these receptors are ubiquitously expressed, thus allowing for CD8+ T cells to interact and potentially mount a cytotoxic response against any cell that expresses their target antigen [139,140]. These cells can include neurons and oligodendrocytes [141]. Clonal CD8+ T cell populations have been identified in the parenchymal, perivascular, and meningeal immune cell infiltrates of MS patients [142]. Phenotypic analysis of these CD8+ T cells suggested that part of them proliferate and have an activated cytotoxic phenotype [143], while the remaining part appear to have features of tissue-resident memory cells that can be reactivated in relapses or contribute to progressive disease. In progressive MS cases, the clustering of CD8+ T cells in the perivascular space was associated with an increased inflammatory activity and demyelinating lesion load [144]. In an EAE mouse model, adoptive transfer of MBP-specific CD8+ cytotoxic T cells recapitulated the pathological and clinical features of myelin-specific CD4+ T cell-mediated EAE [145]. Interestingly, as it was the case for CD4+ T cells, the antigenic target of CD8+ T cells is a subject of active investigation. In one study, MBP was found to be presented to naive CD8+ T cells by a subset of DCs known as Tip-DCs in CD4+ T cell—mediated EAE [146]. These myelin peptide-specific cytotoxic CD8 T cells were then able to lyse human oligodendrocytes that expressed these peptides through their MHC class I molecules [147]. Isolated CD8+ T cells from MS patients can also be activated by B cells infected with EBV [148]. Cross-reactivity between EBV antigens and autologous CNS antigens offers another potential explanation for the cytotoxic effect of these cells in MS. Early in the disease course in EAE, these cells also exhibit a classic antiviral profile before switching to an exhausted phenotype characterized by increased expression of inhibitory receptors such as PD-1 and a loss of effector function [149].

CD8+ T cells can also exhibit a regulatory phenotype (CD8+ Treg). CD8+ T cells that are specific for CNS autoantigens can suppress CD4+ T cells upon stimulation by their antigen. In an EAE model, the induction of MOG-specific autoreactive CD4+ T cells triggered an opposing mobilization of regulatory CD8+ T cells that suppressed the proliferation of CD4+ T cells [150]. Interestingly, during an MS exacerbation, the function of these cells is deficient and a rescue of their function correlates with relapse recovery [151]. These regulatory CD8+ T cells can be upregulated with the use of glatiramer acetate [152], an approved immune modulatory drug for MS.

Unconventional T cells

While most classic T cells are only reactive to complexes of peptide and MHC proteins, there are a few of these that do not fit this model. These cells include Mr1-restricted mucosal associated invariant T cells (MAIT cells), $\gamma\delta$ T cells, CD1-restricted T cells, and MHC class Ib-reactive T cells. These unconventional T cells are non-MHC restricted, they do not recognize classical peptide antigens, and are not donor restricted. These cells have a more limited TCR diversity and tend to localize in nonlymphoid tissues but are primed to rapidly respond to antigenic triggers in the early response to an infectious or other inflammatory stimuli [153]. Unconventional T cells have recently received attention in the context of MS. The frequency of one of their subtypes, $\gamma\delta$ T cells, in the periphery correlates with disease activity on Mr [154]. These $\gamma\delta$ T cells are present in chronic active lesions of MS patients [155], and their pathogenic effects could be mediated through IL-17 production

[156] as well as a direct cytotoxic effect on oligodendrocytes [157]. Similar to γδ T cells, MAIT cells from MS patients also exhibit increased IL-17 production suggesting a proinflammatory role in MS complementing that of Th17 cells [158].

B cells

B cells are increasingly recognized as key players in MS [159]. These cells constitute the humoral component of adaptive immunity and are involved in the production and secretion of antibodies. In the setting of MS, these antibodies can potentially bind CNS antigen targets, which can then be recognized and phagocytosed by myeloid cells. At the same time, B cells can also act as APCs for T cells, exacerbating the immune response. B-cell-targeted therapies, such as anti-CD20 monoclonal antibodies, have shown promise in MS treatment, and further support the importance of these cells in CNS autoimmunity.

Anatomopathological analysis of MS lesions revealed the presence of CD20 + B cells in all disease courses and lesion stages, including in those patients with acute disease of very short duration [143]. As noted earlier, lymphoid follicle-like structures containing B-cells and plasma cells have been described in the meninges of SPMS patients [121] and may contribute to cortical pathology [160]. The accumulation of CSF mature B cells and plasmablasts is associated with acute brain inflammation measured by MRI as well as with inflammatory CSF parameters [161].

These B cells appear to mature in the draining cervical lymph nodes before migrating into the CNS of MS patients [162]. However, other studies appear to suggest that these clonal B cells have a strong bidirectional exchange across the BBB with expansion, maturation and diversification occurring on both the CNS and peripheral compartments [163,164]. Tfh seem to drive the infiltration and expansion of B cells [120]. At the same time, and as described above, both Tfh and Th17 have a critical role in the formation of ectopic B-cell follicles in the meningeal tertiary lymphoid tissue of MS patients [122,123]. Interestingly, one study pointed toward the existence of a meningeal B cell population directly derived from the calvaria bone marrow that may contribute to maintaining the immune privilege within the CNS. This population appears to dwindle in ageing mice and is then replaced by peripheral B cells [165]. Although the function of these calvaria meningeal B cells needs to be better elucidated, a disruption of their function in the setting of CNS autoimmunity seems plausible.

The antigen target(s) of B cells and their antibodies in MS is a subject of intensive research. Studies examining oligoclonal bands from the CSF of MS patients have shown that these antibodies can bind a broad array of self-protein, some intracellular [166]. One study identified that a B cell antibody against EBV transcription factor EBV nuclear antigen 1 (EBNA1) cross reacted with the CNS protein glial cell adhesion molecule (GlialCAM) given the molecular mimicry between both antigens. Immunization with EBNA1 also led to worsened EAE in a mouse model. These findings provided a potential mechanistic explanation for the role of EBV infection in MS and suggest an involvement of B cells as mediators of this cross-reaction [167]. Of note, in patients harboring the HLA-DRB1*15:01 allele, B cells can directly activate CD4 + T cells through presentation of autologous peptides without the need for any exogenous antigen [168]. Independently of its immunoglobulin-dependent effects, in vitro studies have identified that secretory products derived from B cells from RRMS patients can also be toxic to rat oligodendrocytes [169] and rat and human neurons [170].

As we have described in the setting of T cells, certain B cells can also exhibit a regulatory or antiinflammatory phenotype. In the brains of MS patients, plasma cells were found to be a prominent source of the antiinflammatory cytokine IL-10 [144]. About 20% of B cells in meningeal tertiary lymphoid tissue produce either IL-10 or IL-35 suggesting that

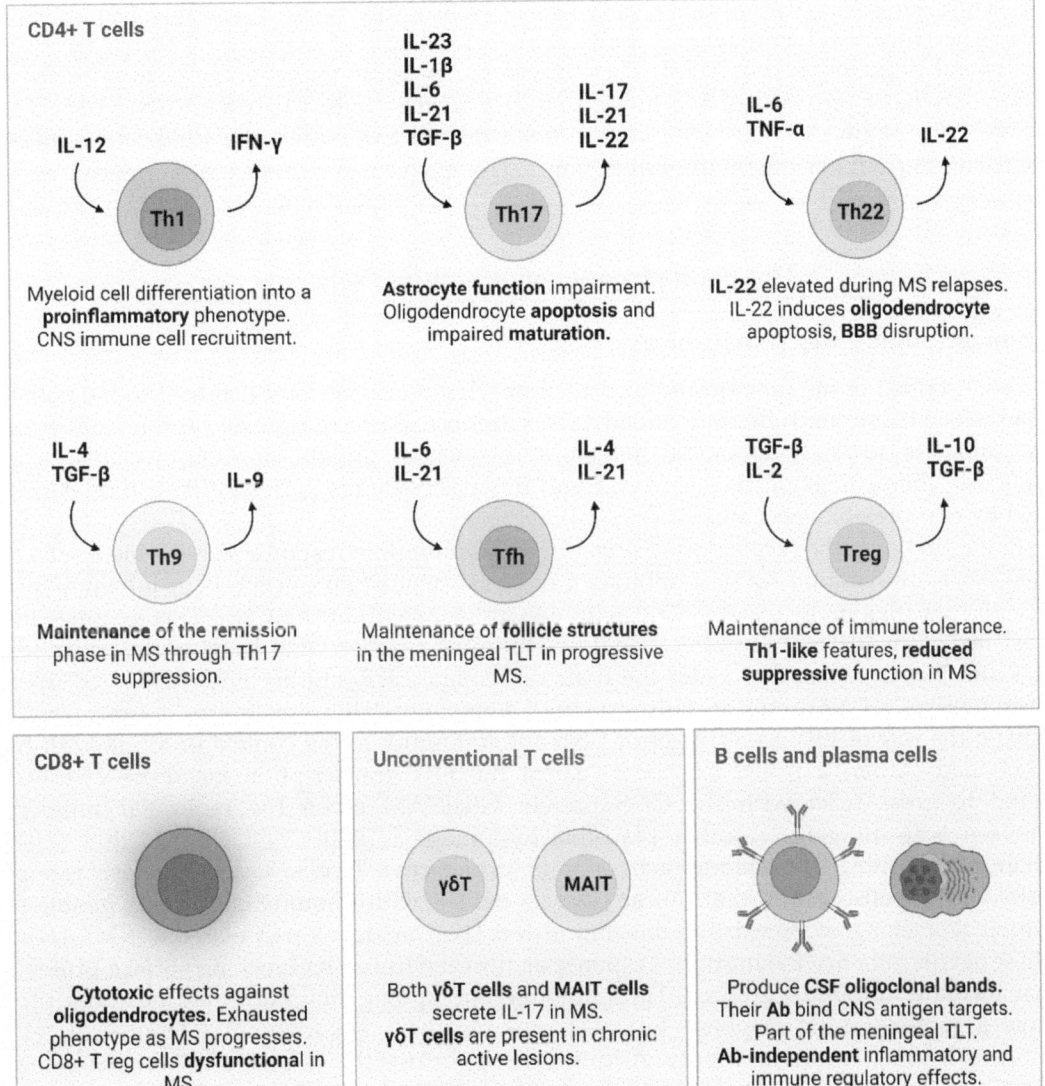

FIGURE 2.2 Schematic review of the adaptive immune system in MS. *MHC II*, major histocompatibility complex class II; *Ab*, antibody; *TLT*, tertiary lymphoid tissue.

meningeal B cell aggregates may also exert an immune-regulatory role. Depletion of these meningeal B lymphocyte aggregates led to worsened disease in opticospinal encephalomyelitis mice [171]. As hypothesized in the extensive review by Attfield et al. the selective depletion of CD20+ cells by immune therapies (i.e., ocrelizumab, rituximab) can spare plasma cells, and this may boost B lineage cells with a regulatory phenotype, thus contributing to the beneficial effect of these treatments [1].

One limitation of anti-CD20 therapies, as well as any other antibody-based therapy, in the setting of MS is their limited CNS bioavailability due to their inability to cross the BBB. Therefore the therapeutic benefit of these antibodies is largely mediated by their peripheral effect on B cells [172]. This provides a possible explanation as to why these treatments have proven to have limited efficacy in the context of progressive MS [173], where the immune response becomes compartmentalized within the CNS with limited contribution from peripheral immunity (Fig. 2.2).

The pathophysiology of multiple sclerosis

From autoimmunity to pathology

As described in the previous section, the underlying mechanisms that lead to the pathogenesis of MS are not fully understood. APCs are thought to recognize, capture, and present antigens to T lymphocytes which subsequently initiate an adaptive response. However, the pathogenic autoreactive T and B lymphocyte responses in MS could be initiated by two possible mechanisms (Fig. 2.3):

In the first one (the "Outside-In" dogma), an autoimmune response is generated outside the CNS as the immune system mounts a response against an infectious or inflammatory stimuli that triggers a cross-reaction with autoantigens within the CNS. This assumes the existence of an aberrant peripheral adaptive immune response that leads to the activation of T and B cells which then enter the brain and spinal cord and become reactivated upon encountering a CNS antigen with structural homology. This theory could for example explain the role of EBV as a trigger of CNS autoimmunity in the context of MS [8,174]. As mentioned earlier, a B cell antibody against the EBV transcription factor EBNA1 was found to cross react with the CNS protein GlialCAM given the molecular mimicry between both antigens. Another plausible hypothesis is that CNS inflammation could occur in the setting of bystander activation of autoreactive T cells. This hypothesis is supported by the observation that cancer patients treated with immune checkpoint inhibitors exhibit worsening of preexisting subclinical MS [175] or developed new-onset MS [176]. These agents enhance the immune response against the tumor by blocking surface proteins that mediate immune tolerance. Thus immune checkpoint inhibitors may not only promote a tumor-specific response but also promote the breakdown of tolerance to self-antigens leading to autoimmunity. Additional support for the "Outside-In" dogma comes from both the active and adoptive transfer EAE animal models of MS. In the case of actively induced EAE, an autoreactive CD4+ T cell immune response can be mounted in the periphery by the administration of a protein or peptide antigen along with an adjuvant, typically Freund's Complete Adjuvant composed of inactivated and dried

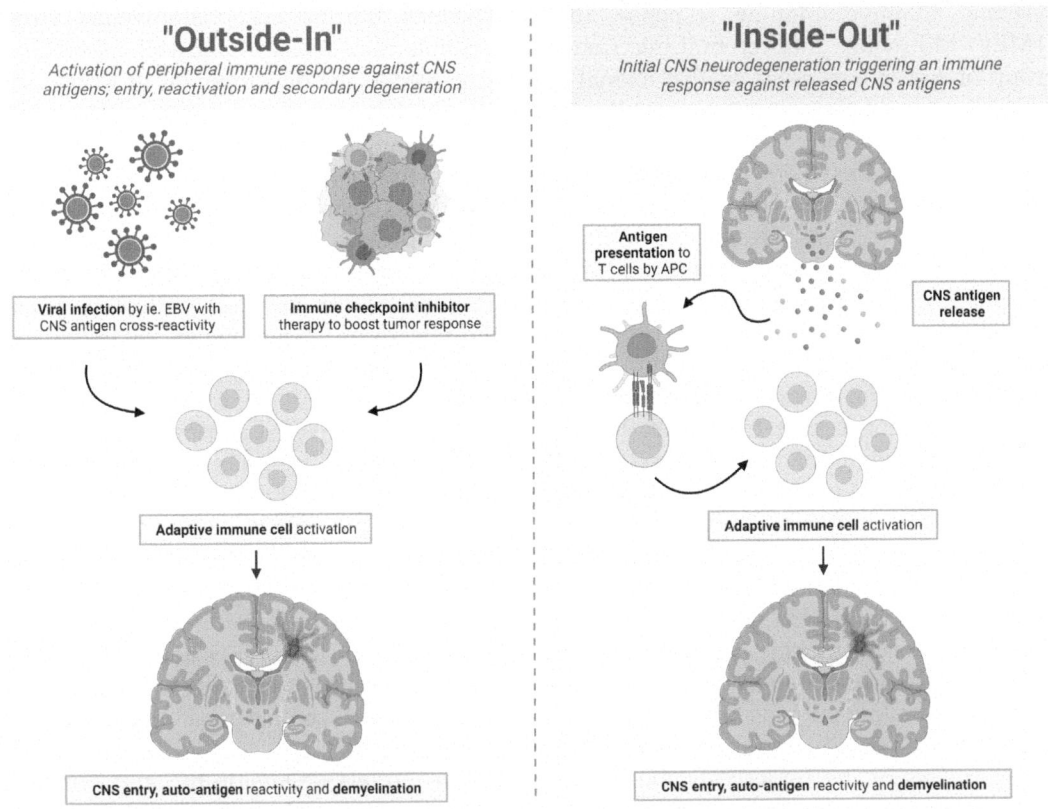

FIGURE 2.3 Graphic depiction for the proposed "Outside-In" and "Inside-Out" dogmas that could contribute to the initial pathogenesis of MS.

mycobacteria emulsified in mineral oil. After induction, there is an egression of these autoreactive cells to the CNS which can trigger demyelination in these animals. In adoptive transfer EAE, the transfer of these autoreactive T cells to a healthy animal, can lead to demyelination. Thus this supports the hypothesis that a peripheral antigen autoimmune response can by itself trigger demyelination resembling that seen in MS.

In the second one (the "Inside-Out" dogma), it is proposed that a CNS-intrinsic disorder results in the release of CNS antigens into the periphery. These antigens are then presented by APCs in secondary lymphoid organs where they activate T cells that are not educated to recognize these antigens as self. Using a mice model in which oligodendrocytes are killed in adult mice by the genetic activation of diphtheria toxin fragment A expression (DTA mouse) [177], a group evidenced the development of a secondary disease characterized by extensive demyelination, axonal loss, increased numbers of T cells in the CNS and myelin-specific T cells in lymphoid organs [178]. Adoptive transfer of T cells derived from these DTA mice to naïve recipients was also associated with clinical deficits and white matter inflammation. Although the pathology in these animals is diffuse and may not accurately reflect MS pathology, it provides evidence that primary CNS oligodendrocyte death can trigger a systemic adaptive immune response. In the Theiler's murine encephalomyelitis virus (TMEV) MS model, intracerebral injection of susceptible mice leads to demyelination. One group of researchers revealed that axonal damage preceded the development of secondary demyelination in these mice, potentially supporting the hypothesis that axonal destruction triggered by viral infection can then contribute to CNS antigenic release that elicits autoimmune demyelination [179]. One pitfall is that, up until this point, no neurotropic pathogen has yet been described which acts through a similar mechanism to the TMEV model in the setting of MS. However, despite constituting a different disease, support for this theory also comes from patients who develop NMDA receptor encephalitis, a type of autoimmune encephalitis, after having suffered monophasic HSV encephalitis [180]. One theory for this association is that CNS HSV infection leads to the release of CNS antigens including the NMDA receptor, with a subsequent development of antibodies against this target. This proposed mechanism, however, remains to be confirmed.

Human endogenous retroviruses (HERVs) have also been posited as potentially pathogenic in MS. The human genome contains DNA sequences that are mobile, some of which (retrotransposons) use an RNA sequence to self-amplify back into DNA [181]. These retrotransposons contain endogenous retroviruses. While usually dormant, certain factors such as infections [182] or inflammatory signals can lead to the expression of HERVs, and some of these HERVs encode for proteins that could contribute to the pathogenesis of MS [183]. In one instance, the HERV-W-encoded envelope (Env) protein was found to stimulate proinflammatory cytokine production by monocytes in vitro [184,185], stimulate Th1-like CD4+ T cell differentiation by DC [184], and induce an abnormal response in T cells similar to that of superantigens [186]. This protein was able to induce increased autoimmune T cell reactivity when used as an adjuvant in an EAE model [187].

Regardless of the underlying mechanism, in both models ("Outside-In" and "Inside-Out"), a response against an antigen mounted in a secondary lymphoid tissue guides adaptive immune cells into the CNS [188]. The precise identity of the APC that sample, process, and present these antigens to T cells at the onset of MS is not clearly elucidated. One study pointed toward a subset of conventional DCs as critical APCs in CNS autoimmunity. These cells were necessary to sample and present myelin antigens to T cells, which allowed their entry into the CNS to initiate neuroinflammation in an EAE model [189]. In addition, as mentioned EARLIER, myeloid cells occupying the CPVS are also though to play a role in the perpetuation of CNS autoimmune disease through antigen presentation and reactivation of T cells [22]. Beyond their role as APCs, infiltrating bone marrow derived myeloid cells can also act as effectors from the early stages of the disease through both their phagocytic activity and inflammatory signaling. CNS-resident microglial cells also contribute to this proinflammatory signaling and axonal damage. Interestingly, data generated in one animal model showed that there is no requirement for T cells to exhibit antigen specificity to cause CNS inflammation [190]. In these mice, induction of GM-CSF expression by peripheral Th cells led by itself to the egression of peripheral inflammatory myeloid cells into the CNS causing severe neurological deficits [190].

With time, and as the disease evolves from the early stages of RRMS to a progressive course, the peripheral immune cell contribution in MS decreases, and the immune response becomes more compartmentalized within the CNS, where the damage is more diffuse, with presence of diffuse chronic microglial activation [44,191]. In addition, the formation of a tertiary lymphoid tissue characterized by meningeal immune cell infiltrates contributes to the development of cortical lesions [122,192]. In this progressive disease stage, chronic axonal demyelination and failure of remyelination lead to axonal loss [193]. All these factors combined contribute to the development of brain atrophy.

Inflammatory lesion formation and evolution

Patients with MS suffer from a variety of neurological symptoms related to the location of the demyelinating lesions within the CNS. In most cases, early in the course the disease exhibits an RRMS phenotype, where the recurrence periods are followed by remission phases in which there is either complete or partial recovery from the symptoms [194]. This disease typically affects young-aged females more than males. With time, patients evolve to a secondary progressive disease, where they experience a continuous neurological impairment in between relapses which in turn become less frequent with increased disease duration. A small subset of patients exhibits a primary progressive disease course from the onset. The onset of this disease form typically occurs in patients of an older age than in RRMS, and affects both genders equally [195].

MS is characterized by the formation of demyelinating lesions or plaques, with inflammation, axonal injury, and loss (Fig. 2.4). These plaques can typically form in the

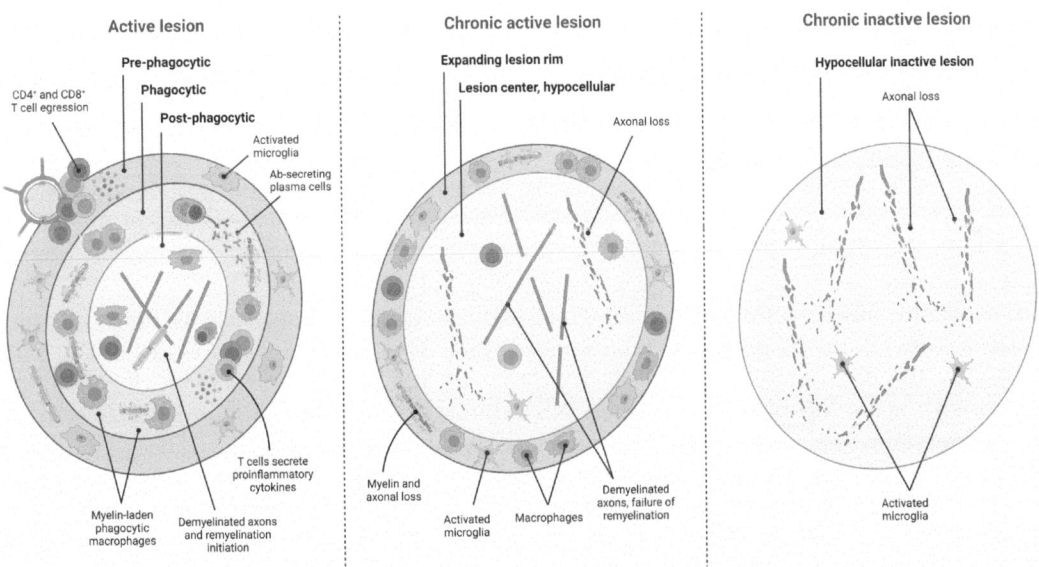

FIGURE 2.4 **Types of demyelinating plaques seen in MS**. *Active lesions* are common in RRMS. They can contain a prephagocytic area consisting mostly of activated microglia and egressing lymphoid cells; the phagocytic rim mostly consists of myelin-laden phagocytic macrophages. There, antibody (Ab)-secreting plasma cells can produce antibodies against CNS antigens that can be opsonized by these phagocytic cells. In the lesion core (post-phagocytic), demyelinated axons with initial signs of remyelination are observed. *Chronic active lesions* are more commonly seen in progressive MS and consist of an expanding lesional rim containing active microglia and macrophages, and a hypocellular lesion center with demyelinated axons and failure of remyelination. *Chronic inactive lesions* are hypocellular and exhibit significant axonal loss.

periventricular white matter, the optic nerve and tract, the subpial region of the spinal cord and brainstem, as well as in juxtacortical and cortical areas. The location of the inflammatory lesions determines the neurological presentation. In addition, these lesions vary over time with marked heterogeneity in its structure and immunopathology between relapsing-remitting and progressive disease [21].

Active focal inflammatory plaques are more frequent in the early stages of the disease in patients with RRMS. These plaques contain demyelinated axons, dense periventricular and parenchymal lymphocyte infiltrates, and myelin-laden phagocytic cells [195,196]. Transected axons also represent an abundant feature of MS lesions, even in the early stages of the disease [197,198]. Transected axonal damage may even potentially precede demyelination [193]. Active lesions typically progress from a prephagocytic state with few immune cell infiltrates, activated microglia and little demyelination, to a phagocytic lesion where myelin-laden phagocytes are prominent, and finally to a postphagocytic state where the demyelination has completed and remyelination efforts are starting [1]. BBB compromise is present in these lesions. Given this, these active lesions exhibit an enhancing pattern on contrast Mr imaging, which serves as a useful tool to detect disease activity.

Pathogenesis in the context of progressive multiple sclerosis

New focal inflammatory lesions become less frequent as patients get older and have a longer disease duration. In progressive disease, inflammation is more common in the form of chronic active and slowly expanding (or smoldering) lesions. When compared to acute active inflammatory plaques, these exhibit a thinner margin of activated microglia and few myelin-containing macrophages [199]. Yet, these lesions are still active and produce ongoing tissue damage, and its presence is associated with a more aggressive disease course [200]. Chronic inactive lesions are those characterized by the presence of demyelination with the absence of macrophages or microglia. A longer disease duration is associated with increased frequency of these lesions (Fig. 2.4).

One of the most prominent pathological features of progressive MS is brain atrophy that can affect both the gray and white matter [42]. This brain atrophy is at least in part caused by the degeneration of chronically demyelinated axons. This axonal degeneration begins with acute MS plaques, but the CNS exhibits a remarkable ability to recover from acute demyelination. However, in chronic demyelinating lesions like those commonly occurring in progressive MS, demyelinated axons ultimately degenerate and lead to irreversible neurological disability [42]. As it will be described later, failure of remyelination has also been suggested as a cause of disease progression in these patients [201]. This axonal loss can be seen as hypointense lesions on T1-weighted spin-echo Mr [202].

Another typical feature of progressive MS is cortical demyelination, including juxtacortical, intracortical, and subpial lesions (Fig. 2.5). Cortical lesions have been associated with the cognitive impairment seen in MS [203]. Juxtacortical lesions are most commonly due to subcortical white matter lesions that extend outwardly to involve the cortex, and can be seen even in early MS [204]. Intracortical lesions are typically small and centered around a blood vessel. Finally, subpial lesions are the most common subtype and are most prominent in progressive MS. The tertiary lymphoid tissue consisting of leptomeningeal

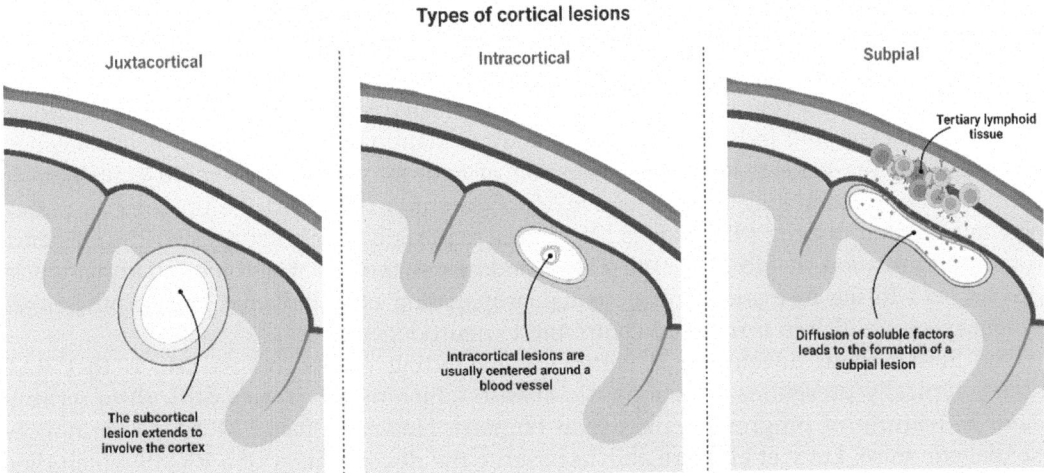

FIGURE 2.5 **Types of cortical lesions in MS.** Patients can exhibit juxtacortical, intracortical, and subpial lesions. *Juxtacortical lesions* arise from white matter lesions that extend into the cortex. *Intracortical lesions* are typically centered around a blood vessel. *Subpial lesions* are thought to arise due to the soluble inflammatory factors secreted by the meningeal inflammatory infiltrates. They are more commonly seen in progressive MS.

inflammatory infiltrates seen in these patients has been implicated in their pathogenesis [122,205]. It is thought that the diffusion of inflammatory soluble factors secreted by these cells from the leptomeninges to the cortex is responsible for the development of these subpial lesions.

In addition to demyelination and oligodendrocyte loss, cortical lesions are typically characterized by the presence of axonal and dendritic transections, synaptic terminal loss, and neural degeneration with apoptotic neurons [206]. Abnormal complement deposition and complement-mediated synaptic pruning by microglia have been proposed as potential contributors to the synaptic loss that occurs in the gray matter of these progressive MS patients [207–209]. Perivascular fibrin deposition, potentially related to BBB dysfunction, has also been observed in the cortex of these patients. In areas where fibrin(ogen) deposition occurs, neuronal density is significantly reduced, suggesting that fibrin itself could also act as a neurotoxic agent [210]. In addition, fibrin deposition is also thought to promote microglial activation.

Finally, another characteristic feature of progressive MS is the presence of diffuse pathology involving the so-called normal-appearing white matter (NAWM) and normal-appearing gray matter where no demyelinating plaques are evident. This type of injury seems to be associated with diffuse microglial activation and is deemed to contribute to the progression of cortical atrophy more than focal white matter lesions [211]. This diffuse pathology can also occur in the spinal cord and is related to the degree of meningeal T cell infiltration [212]. It seems that soluble factors including proinflammatory cytokines, complement factors, and others produced by immune cells in the meningeal follicles and that diffuse into the brain parenchyma may act to activate microglia [213]. In the NAWM, where regions are deemed pathology free, changes at the epigenomic level can still be detected in genes that affect oligodendrocyte susceptibility to damage [214].

Brain volume atrophy can be evidenced on Mr and is more severe in those with progressive MS, but can even be seen in clinically isolated syndrome and RRMS [215]. Another proposed mechanism for the development of neurodegeneration and atrophy in MS is that neuronal injury may eventually progress in a pattern that resembles that of other neurodegenerative diseases and independently from active inflammation. As in other neurodegenerative diseases, the generation and propagation of toxic protein species could contribute to neuronal loss in the setting of progressive MS. One study identified the presence of pathologic tau seeding outside MS plaques [216]. Prior reports have also suggested the existence of tau accumulation in MS [217,218]. These findings suggest that chronic inflammation in MS can lead to the formation and potential propagation of pathogenic Tau or other toxic protein species which in turn could contribute to neurodegeneration.

One limitation in developing a better understanding of progressive MS is that EAE models typically present as a monophasic disease which makes it very difficult to recapitulate the features of progressive disease in humans. However, using MOG_{35-55} in nonobese diabetic mice, Levy et al. were able to observe the development of a RRMS which then progressed into a chronic progressive stage [219]. In these animals, Mr fractional anisotropy showed demyelinating and axonal damage in gray and white matter areas resembling those seen in humans, and Mr contrast-imaging showed BBB permeability in the affected areas. The damage seen in these mice was deemed secondary to astrocyte toxicity. Despite its imperfections, this model was proposed as a feasible candidate to study the development of progressive MS.

Remyelination

As mentioned, the capacity to repair damaged myelin through remyelination, a process involving oligodendrocyte precursor cells, is crucial for the recovery from neurological deficits after an MS lesion. Complete remyelinated lesions, the so-called shadow plaques, are commonly seen in RRMS and constitute sharply demarcated areas with reduced myelin density and thin myelin sheaths representing late stage remyelination [220]. Although patients with RRMS can have significant remyelination in the resolution phase of an exacerbation, in patients with progressive MS, failure or incomplete remyelination is more common. In these patients, chronic inactive plaques typically show demyelinated axons with absent or minimal remyelination. However, a subset of progressive MS patients of older age and with longer disease duration, where inflammation and tissue injury have resolved, were found to have significantly more remyelinated lesions [221]. This remyelination is more commonly seen in cortical than in white matter lesions [222].

The presence of ongoing demyelination, especially in the setting of slowly expanding chronic active plaques appears to significantly curtail the remyelination effort [223], and remyelinated areas also appear more vulnerable than pathology-free areas to recurrence of demyelination. Therefore therapeutic strategies that combine the use of immune modulatory agents that taper active demyelination along with agents that could boost the remyelinating capacity of axons are an appealing MS treatment. Several candidate remyelinating agents that mobilize oligodendrocyte precursor cells or boost the function of mature oligodendrocytes are currently on preclinical and clinical trials for use in MS [201,224].

Conclusion

MS is a complex autoimmune disease with a multifaceted immunological and pathophysiological basis. While much progress has been made in understanding the mechanisms driving MS, many questions remain unanswered. Advances in research continue to shed light on potential therapeutic targets and strategies to improve the lives of those living with MS. The main unmet need remains the identification of interventions that can stop disease progression and repair tissue damage that has already been inflicted. This will require a better understanding of the immunological and pathophysiological aspects of MS.

References

[1] Attfield KE, Jensen LT, Kaufmann M, Friese MA, Fugger L. The immunology of multiple sclerosis. Nat Rev Immunol 2022;22(12):734–50. Available from: https://doi.org/10.1038/s41577-022-00718-z.

[2] Scalfari A, Romualdi C, Nicholas RS, et al. The cortical damage, early relapses, and onset of the progressive phase in multiple sclerosis. Neurology 2018;90(24):e2107–18. Available from: https://doi.org/10.1212/WNL.0000000000005685.

[3] Haines JL, Terwedow HA, Burgess K, et al. Linkage of the MHC to familial multiple sclerosis suggests genetic heterogeneity. The Multiple Sclerosis Genetics Group. Hum Mol Genet 1998;7(8):1229–34. Available from: https://doi.org/10.1093/hmg/7.8.1229.

[4] Patsopoulos NA, Barcellos LF, Hintzen RQ, et al. Fine-mapping the genetic association of the major histocompatibility complex in multiple sclerosis: HLA and non-HLA effects. PLoS Genet 2013;9(11):e1003926. Available from: https://doi.org/10.1371/journal.pgen.1003926.

[5] International Multiple Sclerosis Genetics Consortium. Multiple sclerosis genomic map implicates peripheral immune cells and microglia in susceptibility. Science 2019;365(6460):eaav7188. Available from: https://doi.org/10.1126/science.aav7188.

[6] De Jager PL, Chibnik LB, Cui J, et al. Integration of genetic risk factors into a clinical algorithm for multiple sclerosis susceptibility: a weighted genetic risk score. Lancet Neurol 2009;8(12):1111–19. Available from: https://doi.org/10.1016/S1474-4422(09)70275-3.

[7] Gourraud PA, McElroy JP, Caillier SJ, et al. Aggregation of MS genetic risk variants in multiple and single case families. Ann Neurol 2011;69(1):65–74. Available from: https://doi.org/10.1002/ana.22323.

[8] Epstein–Barr virus and multiple sclerosis | Nature Reviews Microbiology. https://www-nature-com.foyer.swmed.edu/articles/s41579-022-00770-5; 2023 [Accessed September 5, 2023].

[9] Clonally expanded B cells in multiple sclerosis bind EBV EBNA1 and GlialCAM | Nature. https://www.nature.com/articles/s41586-022-04432-7; 2023 [Accessed September 13, 2023].

[10] Ascherio A, Munger KL, White R, et al. Vitamin D as an early predictor of multiple sclerosis activity and progression. JAMA Neurol 2014;71(3):306–14. Available from: https://doi.org/10.1001/jamaneurol.2013.5993.

[11] Emre M, de Decker C. Effects of cigarette smoking on motor functions in patients with multiple sclerosis. Arch Neurol 1992;49(12):1243–7. Available from: https://doi.org/10.1001/archneur.1992.00530360041015.

[12] Smoking is associated with increased lesion volumes and brain atrophy in multiple sclerosis | PMC. https://www-ncbi-nlm-nih-gov.foyer.swmed.edu/pmc/articles/PMC2833095/; 2023 [Accessed September 5, 2023].

[13] Association between smoking and cognitive impairment in multiple sclerosis | PubMed. https://pubmed-ncbi-nlm-nih-gov.foyer.swmed.edu/25246792/; 2023 [Accessed September 5, 2023].

[14] Hernán MA, Jick SS, Logroscino G, Olek MJ, Ascherio A, Jick H. Cigarette smoking and the progression of multiple sclerosis. Brain J Neurol 2005;128(Pt 6):1461–5. Available from: https://doi.org/10.1093/brain/awh471.

[15] Research paper: Tobacco smoking and excess mortality in multiple sclerosis: a cohort study | PMC. https://www-ncbi-nlm-nih-gov.foyer.swmed.edu/pmc/articles/PMC4173752/; 2023 [Accessed September 5, 2023].

[16] Rodgers J, Friede T, Vonberg FW, et al. The impact of smoking cessation on multiple sclerosis disease progression. Brain J Neurol 2022;145(4):1368–78. Available from: https://doi.org/10.1093/brain/awab385.

[17] Gombash SE, Lee PW, Sawdai E, Lovett-Racke AE. Vitamin D as a risk factor for multiple sclerosis: immunoregulatory or neuroprotective Front Neurol 2022;13, Accessed September 13, 2023. Available from: https://www.frontiersin.org/articles/10.3389/fneur.2022.796933.

[18] Akaishi T, Takahashi T, Nakashima I. Peripheral blood monocyte count at onset may affect the prognosis in multiple sclerosis. J Neuroimmunol 2018;319:37−40. Available from: https://doi.org/10.1016/j.jneuroim.2018.03.016.

[19] Makhlouf K, Weiner HL, Khoury SJ. Increased percentage of IL-12 + monocytes in the blood correlates with the presence of active MRI lesions in MS. J Neuroimmunol 2001;119(1):145−9. Available from: https://doi.org/10.1016/s0165-5728(01)00371-x.

[20] Bitsch A, Schuchardt J, Bunkowski S, Kuhlmann T, Brück W. Acute axonal injury in multiple sclerosis. Correlation with demyelination and inflammation. Brain J Neurol 2000;123(Pt 6):1174−83. Available from: https://doi.org/10.1093/brain/123.6.1174.

[21] Lucchinetti C, Brück W, Parisi J, Scheithauer B, Rodriguez M, Lassmann H. Heterogeneity of multiple sclerosis lesions: implications for the pathogenesis of demyelination. Ann Neurol 2000;47(6):707−17. Available from: https://doi.org/10.1002/1531-8249(200006)47:6 < 707::aid-ana3 > 3.0.co;2-q.

[22] Hussain RZ, Hayardeny L, Cravens PC, et al. Immune surveillance of the central nervous system in multiple sclerosis − relevance for therapy and experimental models. J Neuroimmunol 2014;276(0):9−17. Available from: https://doi.org/10.1016/j.jneuroim.2014.08.622.

[23] Huitinga I, van Rooijen N, de Groot CJ, Uitdehaag BM, Dijkstra CD. Suppression of experimental allergic encephalomyelitis in Lewis rats after elimination of macrophages. J Exp Med 1990;172(4):1025−33. Available from: https://doi.org/10.1084/jem.172.4.1025.

[24] Ajami B, Bennett JL, Krieger C, McNagny KM, Rossi FMV. Infiltrating monocytes trigger EAE progression, but do not contribute to the resident microglia pool. Nat Neurosci 2011;14(9):1142−9. Available from: https://doi.org/10.1038/nn.2887.

[25] Manouchehri N, Hussain RZ, Cravens PD, et al. CD11c + CD88 + CD317 + myeloid cells are critical mediators of persistent CNS autoimmunity. Proc Natl Acad Sci U S A 2021;118(14). Available from: https://doi.org/10.1073/pnas.2014492118, e2014492118.

[26] Manouchehri N, Salinas VH, Hussain RZ, Stüve O. Distinctive transcriptomic and epigenomic signatures of bone marrow-derived myeloid cells and microglia in CNS autoimmunity. Proc Natl Acad Sci U S A 2023;120 (6). Available from: https://doi.org/10.1073/pnas.2212696120, e2212696120.

[27] King IL, Dickendesher TL, Segal BM. Circulating Ly-6C + myeloid precursors migrate to the CNS and play a pathogenic role during autoimmune demyelinating disease. Blood 2009;113(14):3190−7. Available from: https://doi.org/10.1182/blood-2008-07-168575.

[28] Locatelli G, Theodorou D, Kendirli A, et al. Mononuclear phagocytes locally specify and adapt their phenotype in a multiple sclerosis model. Nat Neurosci 2018;21(9):1196−208. Available from: https://doi.org/10.1038/s41593-018-0212-3.

[29] Caravagna C, Jaouën A, Desplat-Jégo S, et al. Diversity of innate immune cell subsets across spatial and temporal scales in an EAE mouse model. Sci Rep 2018;8(1):5146. Available from: https://doi.org/10.1038/s41598-018-22872-y.

[30] Giladi A, Wagner LK, Li H, et al. Cxcl10 + monocytes define a pathogenic subset in the central nervous system during autoimmune neuroinflammation. Nat Immunol 2020;21(5):525−34. Available from: https://doi.org/10.1038/s41590-020-0661-1.

[31] M2 microglia and macrophages drive oligodendrocyte differentiation during CNS remyelination | Nature Neuroscience. https://www-nature-com.foyer.swmed.edu/articles/nn.3469; 2023 [Accessed September 4, 2023].

[32] White MPJ, Webster G, Leonard F, La Flamme AC. Innate IFN-γ ameliorates experimental autoimmune encephalomyelitis and promotes myeloid expansion and PDL-1 expression. Sci Rep 2018;8(1):259. Available from: https://doi.org/10.1038/s41598-017-18543-z.

[33] Haas J, Schwarz A, Korporal-Kuhnke M, Jarius S, Wildemann B. Myeloid dendritic cells exhibit defects in activation and function in patients with multiple sclerosis. J Neuroimmunol 2016;301:53−60. Available from: https://doi.org/10.1016/j.jneuroim.2016.10.007.

[34] Thewissen K, Nuyts AH, Deckx N, et al. Circulating dendritic cells of multiple sclerosis patients are proinflammatory and their frequency is correlated with MS-associated genetic risk factors. Mult Scler Houndmills Basingstoke Engl 2014;20(5):548−57. Available from: https://doi.org/10.1177/1352458513505352.

[35] Karni A, Abraham M, Monsonego A, et al. Innate immunity in multiple sclerosis: myeloid dendritic cells in secondary progressive multiple sclerosis are activated and drive a proinflammatory immune response. J Immunol Baltim Md 1950 2006;177(6):4196–202. Available from: https://doi.org/10.4049/jimmunol.177.6.4196.

[36] Ko HJ, Brady JL, Ryg-Cornejo V, et al. GM-CSF-responsive monocyte-derived dendritic cells are pivotal in Th17 pathogenesis. J Immunol Baltim Md 1950 2014;192(5):2202–9. Available from: https://doi.org/10.4049/jimmunol.1302040.

[37] Schwab N, Zozulya AL, Kieseier BC, Toyka KV, Wiendl H. An imbalance of two functionally and phenotypically different subsets of plasmacytoid dendritic cells characterizes the dysfunctional immune regulation in multiple sclerosis. J Immunol Baltim Md 1950 2010;184(9):5368–74. Available from: https://doi.org/10.4049/jimmunol.0903662.

[38] Stasiolek M, Bayas A, Kruse N, et al. Impaired maturation and altered regulatory function of plasmacytoid dendritic cells in multiple sclerosis. Brain J Neurol 2006;129(Pt 5):1293–305. Available from: https://doi.org/10.1093/brain/awl043.

[39] Prinz M, Jung S, Priller J. Microglia biology: one century of evolving concepts. Cell 2019;179(2):292–311. Available from: https://doi.org/10.1016/j.cell.2019.08.053.

[40] Masuda T, Sankowski R, Staszewski O, et al. Spatial and temporal heterogeneity of mouse and human microglia at single-cell resolution. Nature 2019;566(7744):388–92. Available from: https://doi.org/10.1038/s41586-019-0924-x.

[41] Wolf Y, Shemer A, Levy-Efrati L, et al. Microglial MHC class II is dispensable for experimental autoimmune encephalomyelitis and cuprizone-induced demyelination. Eur J Immunol 2018;48(8):1308–18. Available from: https://doi.org/10.1002/eji.201847540.

[42] Mahad DH, Trapp BD, Lassmann H. Pathological mechanisms in progressive multiple sclerosis. Lancet Neurol 2015;14(2):183–93. Available from: https://doi.org/10.1016/S1474-4422(14)70256-X.

[43] Pinto MV, Fernandes A. Microglial phagocytosis—rational but challenging therapeutic target in multiple sclerosis. Int J Mol Sci 2020;21(17):5960. Available from: https://doi.org/10.3390/ijms21175960.

[44] Kutzelnigg A, Lucchinetti CF, Stadelmann C, et al. Cortical demyelination and diffuse white matter injury in multiple sclerosis. Brain J Neurol 2005;128(Pt 11):2705–12. Available from: https://doi.org/10.1093/brain/awh641.

[45] Fischer MT, Sharma R, Lim JL, et al. NADPH oxidase expression in active multiple sclerosis lesions in relation to oxidative tissue damage and mitochondrial injury. Brain J Neurol 2012;135(Pt 3):886–99. Available from: https://doi.org/10.1093/brain/aws012.

[46] Prineas JW, Kwon EE, Cho ES, et al. Immunopathology of secondary-progressive multiple sclerosis. Ann Neurol 2001;50(5):646–57. Available from: https://doi.org/10.1002/ana.1255.

[47] Schuh C, Wimmer I, Hametner S, et al. Oxidative tissue injury in multiple sclerosis is only partly reflected in experimental disease models. Acta Neuropathol (Berl) 2014;128(2):247–66. Available from: https://doi.org/10.1007/s00401-014-1263-5.

[48] Liddelow SA, Guttenplan KA, Clarke LE, et al. Neurotoxic reactive astrocytes are induced by activated microglia. Nature 2017;541(7638):481–7. Available from: https://doi.org/10.1038/nature21029.

[49] Geloso MC, D'Ambrosi N. Microglial pruning: relevance for synaptic dysfunction in multiple sclerosis and related experimental models. Cells 2021;10(3):686. Available from: https://doi.org/10.3390/cells10030686.

[50] Mayo L, Quintana FJ, Weiner HL. The innate immune system in demyelinating disease. Immunol Rev 2012;248(1):170–87. Available from: https://doi.org/10.1111/j.1600-065X.2012.01135.x.

[51] Caruana P, Lemmert K, Ribbons K, Lea R, Lechner-Scott J. Natural killer cell subpopulations are associated with MRI activity in a relapsing-remitting multiple sclerosis patient cohort from Australia. Mult Scler Houndmills Basingstoke Engl 2017;23(11):1479–87. Available from: https://doi.org/10.1177/1352458516679267.

[52] Rodríguez-Martín E, Picón C, Costa-Frossard L, et al. Natural killer cell subsets in cerebrospinal fluid of patients with multiple sclerosis. Clin Exp Immunol 2015;180(2):243–9. Available from: https://doi.org/10.1111/cei.12580.

[53] Gross CC, Schulte-Mecklenbeck A, Rünzi A, et al. Impaired NK-mediated regulation of T-cell activity in multiple sclerosis is reconstituted by IL-2 receptor modulation. Proc Natl Acad Sci U S A 2016;113(21):E2973–82. Available from: https://doi.org/10.1073/pnas.1524924113.

[54] Saraste M, Irjala H, Airas L. Expansion of CD56Bright natural killer cells in the peripheral blood of multiple sclerosis patients treated with interferon-beta. Neurol Sci J Ital Neurol Soc Ital Soc Clin Neurophysiol 2007;28(3):121–6. Available from: https://doi.org/10.1007/s10072-007-0803-3.

[55] Darlington PJ, Stopnicki B, Touil T, et al. Natural killer cells regulate Th17 cells after autologous hematopoietic stem cell transplantation for relapsing remitting multiple sclerosis. Front Immunol 2018;9:834. Available from: https://doi.org/10.3389/fimmu.2018.00834.

[56] Montes Diaz G, Fraussen J, Van Wijmeersch B, Hupperts R, Somers V. Dimethyl fumarate induces a persistent change in the composition of the innate and adaptive immune system in multiple sclerosis patients. Sci Rep 2018;8(1):8194. Available from: https://doi.org/10.1038/s41598-018-26519-w.

[57] Kwong B, Rua R, Gao Y, et al. T-bet-dependent NKp46+ innate lymphoid cells regulate the onset of TH17-induced neuroinflammation. Nat Immunol 2017;18(10):1117−27. Available from: https://doi.org/10.1038/ni.3816.

[58] Eken A, Yetkin MF, Vural A, et al. Fingolimod alters tissue distribution and cytokine production of human and murine innate lymphoid cells Front Immunol 2019;10, Accessed September 3, 2023. Available from: https://www.frontiersin.org/articles/10.3389/fimmu.2019.00217.

[59] Male-specific IL-33 expression regulates sex-dimorphic EAE susceptibility | PNAS. https://www-pnas-org.foyer.swmed.edu/doi/abs/10.1073/pnas.1710401115; 2023 [Accessed September 3, 2023].

[60] Hirose S, Jahani PS, Wang S, et al. Type 2 innate lymphoid cells induce CNS demyelination in an HSV-IL-2 mouse model of multiple sclerosis. iScience 2020;23(10):101549. Available from: https://doi.org/10.1016/j.isci.2020.101549.

[61] Nathan C. Neutrophils and immunity: challenges and opportunities. Nat Rev Immunol 2006;6(3):173−82. Available from: https://doi.org/10.1038/nri1785.

[62] Hampton MB, Kettle AJ, Winterbourn CC. Inside the neutrophil phagosome: oxidants, myeloperoxidase, and bacterial killing. Blood 1998;92(9):3007−17. Available from: https://doi.org/10.1182/blood.V92.9.3007.

[63] Brinkmann V, Reichard U, Goosmann C, et al. Neutrophil extracellular traps kill bacteria. Science 2004;303 (5663):1532−5. Available from: https://doi.org/10.1126/science.1092385.

[64] Naegele M, Tillack K, Reinhardt S, Schippling S, Martin R, Sospedra M. Neutrophils in multiple sclerosis are characterized by a primed phenotype. J Neuroimmunol 2012;242(1):60−71. Available from: https://doi.org/10.1016/j.jneuroim.2011.11.009.

[65] Granieri E, Casetta I, Tola MR, Ferrante P. Multiple sclerosis: infectious hypothesis. Neurol Sci J Ital Neurol Soc Ital Soc Clin Neurophysiol 2001;22(2):179−85. Available from: https://doi.org/10.1007/s100720170021.

[66] Bolton SJ, Anthony DC, Perry VH. Loss of the tight junction proteins occludin and zonula occludens-1 from cerebral vascular endothelium during neutrophil-induced blood-brain barrier breakdown in vivo. Neuroscience 1998;86(4):1245−57. Available from: https://doi.org/10.1016/s0306-4522(98)00058-x.

[67] Steinbach K, Piedavent M, Bauer S, Neumann JT, Friese MA. Neutrophils amplify autoimmune central nervous system infiltrates by maturing local APCs. J Immunol Baltim Md 2013;191(9):4531−9. Available from: https://doi.org/10.4049/jimmunol.1202613, *1950*.

[68] Vono M, Lin A, Norrby-Teglund A, Koup RA, Liang F, Loré K. Neutrophils acquire the capacity for antigen presentation to memory CD4+ T cells in vitro and ex vivo. Blood 2017;129(14):1991−2001. Available from: https://doi.org/10.1182/blood-2016-10-744441.

[69] Campbell SJ, Meier U, Mardiguian S, et al. Sickness behaviour is induced by a peripheral CXC-chemokine also expressed in multiple sclerosis and EAE. Brain Behav Immun 2010;24(5):738−46. Available from: https://doi.org/10.1016/j.bbi.2010.01.011.

[70] Liu L, Belkadi A, Darnall L, et al. CXCR2-positive neutrophils are essential for cuprizone-induced demyelination: relevance to multiple sclerosis. Nat Neurosci 2010;13(3):319−26. Available from: https://doi.org/10.1038/nn.2491.

[71] Carlson T, Kroenke M, Rao P, Lane TE, Segal B. The Th17-ELR+ CXC chemokine pathway is essential for the development of central nervous system autoimmune disease. J Exp Med 2008;205(4):811−23. Available from: https://doi.org/10.1084/jem.20072404.

[72] Atkinson JR, Jerome AD, Sas AR, et al. Biological aging of CNS-resident cells alters the clinical course and immunopathology of autoimmune demyelinating disease. JCI Insight. 7(12):e158153. Available from: https://doi.org/10.1172/jci.insight.158153.

[73] Huber AK, Wang L, Han P, et al. Dysregulation of the IL-23/IL-17 axis and myeloid factors in secondary progressive MS. Neurology 2014;83(17):1500−7. Available from: https://doi.org/10.1212/WNL.0000000000000908.

[74] Sas AR, Carbajal KS, Jerome AD, et al. A new neutrophil subset promotes CNS neuron survival and axon regeneration. Nat Immunol 2020;21(12):1496−505. Available from: https://doi.org/10.1038/s41590-020-00813-0.

[75] Ota K, Matsui M, Milford EL, Mackin GA, Weiner HL, Hafler DA. T-cell recognition of an immunodominant myelin basic protein epitope in multiple sclerosis. Nature 1990;346(6280):183–7. Available from: https://doi.org/10.1038/346183a0.
[76] Zamvil S, Nelson P, Trotter J, et al. T-cell clones specific for myelin basic protein induce chronic relapsing paralysis and demyelination. Nature 1985;317(6035):355–8. Available from: https://doi.org/10.1038/317355a0.
[77] Genetic risk and a primary role for cell-mediated immune mechanisms in multiple sclerosis | Nature. https://www-nature-com.foyer.swmed.edu/articles/nature10251; 2023 [Accessed September 2, 2023].
[78] Kunkl M, Frascolla S, Amormino C, Volpe E, Tuosto L. T helper cells: the modulators of inflammation in multiple sclerosis. Cells 2020;9(2):482. Available from: https://doi.org/10.3390/cells9020482.
[79] Kaskow BJ, Baecher-Allan C. Effector T cells in multiple sclerosis. Cold Spring Harb Perspect Med 2018;8(4): a029025. Available from: https://doi.org/10.1101/cshperspect.a029025.
[80] Domingues HS, Mues M, Lassmann H, Wekerle H, Krishnamoorthy G. Functional and pathogenic differences of Th1 and Th17 cells in experimental autoimmune encephalomyelitis. PLoS One 2010;5(11):e15531. Available from: https://doi.org/10.1371/journal.pone.0015531.
[81] Lodygin D, Hermann M, Schweingruber N, et al. β-Synuclein-reactive T cells induce autoimmune CNS grey matter degeneration. Nature 2019;566(7745):503–8. Available from: https://doi.org/10.1038/s41586-019-0964-2.
[82] Lee HG, Cho MJ, Choi JM. Bystander CD4 + T cells: crossroads between innate and adaptive immunity. Exp Mol Med 2020;52(8):1255–63. Available from: https://doi.org/10.1038/s12276-020-00486-7.
[83] Lees JR, Sim J, Russell JH. Encephalitogenic T-cells increase numbers of CNS T-cells regardless of antigen specificity by both increasing T-cell entry and preventing egress. J Neuroimmunol 2010;220(1-2):10–16. Available from: https://doi.org/10.1016/j.jneuroim.2009.11.017.
[84] Van Kaer L, Postoak JL, Wang C, Yang G, Wu L. Innate, innate-like and adaptive lymphocytes in the pathogenesis of MS and EAE. Cell Mol Immunol 2019;16(6):531–9. Available from: https://doi.org/10.1038/s41423-019-0221-5.
[85] Prajeeth CK, Löhr K, Floess S, et al. Effector molecules released by Th1 but not Th17 cells drive an M1 response in microglia. Brain Behav Immun 2014;37:248–59. Available from: https://doi.org/10.1016/j.bbi.2014.01.001.
[86] Human Th1 dichotomy: origin, phenotype and biologic activities | PubMed. https://pubmed-ncbi-nlm-nih-gov.foyer.swmed.edu/25284714/; 2023 [Accessed September 2, 2023].
[87] Ando DG, Clayton J, Kono D, Urban JL, Sercarz EE. Encephalitogenic T cells in the B10.PL model of experimental allergic encephalomyelitis (EAE) are of the Th-1 lymphokine subtype. Cell Immunol 1989;124 (1):132–43. Available from: https://doi.org/10.1016/0008-8749(89)90117-2.
[88] Baron JL, Madri JA, Ruddle NH, Hashim G, Janeway CA. Surface expression of alpha 4 integrin by CD4 T cells is required for their entry into brain parenchyma. J Exp Med 1993;177(1):57–68. Available from: https://doi.org/10.1084/jem.177.1.57.
[89] Panitch HS, Hirsch RL, Haley AS, Johnson KP. Exacerbations of multiple sclerosis in patients treated with gamma interferon. Lancet Lond Engl 1987;1(8538):893–5. Available from: https://doi.org/10.1016/s0140-6736(87)92863-7.
[90] Effector molecules released by Th1 but not Th17 cells drive an M1 response in microglia | PubMed. https://pubmed-ncbi-nlm-nih-gov.foyer.swmed.edu/24412213/; 2023 [Accessed September 2, 2023].
[91] Loss of T-bet, but not STAT1, prevents the development of experimental autoimmune encephalomyelitis | PubMed. https://pubmed-ncbi-nlm-nih-gov.foyer.swmed.edu/15238607/; 2023 [Accessed September 2, 2023].
[92] Interleukin-23 rather than interleukin-12 is the critical cytokine for autoimmune inflammation of the brain | PubMed. https://pubmed-ncbi-nlm-nih-gov.foyer.swmed.edu/12610626/; 2023 [Accessed September 2, 2023].
[93] IL-12– and IL-23–modulated T cells induce distinct types of EAE based on histology, CNS chemokine profile, and response to cytokine inhibition | PMC. https://www-ncbi-nlm-nih-gov.foyer.swmed.edu/pmc/articles/PMC2442630/; 2023 [Accessed September 6, 2023].
[94] Chen Y, Langrish CL, Mckenzie B, et al. Anti–IL-23 therapy inhibits multiple inflammatory pathways and ameliorates autoimmune encephalomyelitis. J Clin Invest 2006;116(5):1317–26. Available from: https://doi.org/10.1172/JCI25308.

[95] IL-21 and TGF-beta are required for differentiation of human T(H)17 cells | PubMed. https://pubmed-ncbi-nlm-nih-gov.foyer.swmed.edu/18469800/; 2023 [Accessed September 2, 2023].

[96] Wilson NJ, Boniface K, Chan JR, et al. Development, cytokine profile and function of human interleukin 17-producing helper T cells. Nat Immunol 2007;8(9):950–7. Available from: https://doi.org/10.1038/ni1497.

[97] Sutton C, Brereton C, Keogh B, Mills KHG, Lavelle EC. A crucial role for interleukin (IL)-1 in the induction of IL-17−producing T cells that mediate autoimmune encephalomyelitis. J Exp Med 2006;203(7):1685–91. Available from: https://doi.org/10.1084/jem.20060285.

[98] Langrish CL, Chen Y, Blumenschein WM, et al. IL-23 drives a pathogenic T cell population that induces autoimmune inflammation. J Exp Med 2005;201(2):233–40. Available from: https://doi.org/10.1084/jem.20041257.

[99] IL-23 drives a pathogenic T cell population that induces autoimmune inflammation | PubMed. https://pubmed-ncbi-nlm-nih-gov.foyer.swmed.edu/15657292/; 2023 [Accessed September 2, 2023].

[100] IL-17 plays an important role in the development of experimental autoimmune encephalomyelitis | PubMed. https://pubmed-ncbi-nlm-nih-gov.foyer.swmed.edu/16785554/; 2023 [Accessed September 2, 2023].

[101] Matusevicius D, Kivisäkk P, He B, et al. Interleukin-17 mRNA expression in blood and CSF mononuclear cells is augmented in multiple sclerosis. Mult Scler Houndmills Basingstoke Engl 1999;5(2):101–4. Available from: https://doi.org/10.1177/135245859900500206.

[102] T helper cell type 1 (Th1), Th2 and Th17 responses to myelin basic protein and disease activity in multiple sclerosis | PubMed. https://pubmed.ncbi.nlm.nih.gov/18397264/; 2023 [Accessed September 2, 2023].

[103] Functional interleukin-17 receptor A is expressed in central nervous system glia and upregulated in experimental autoimmune encephalomyelitis | Journal of Neuroinflammation | Full Text. https://jneuroinflammation-biomedcentral-com.foyer.swmed.edu/articles/10.1186/1742-2094-6-14; 2023 [Accessed September 2, 2023].

[104] Kang Z, Altuntas CZ, Gulen MF, et al. Astrocyte-restricted ablation of interleukin-17-induced Act1-mediated signaling ameliorates autoimmune encephalomyelitis. Immunity 2010;32(3). Available from: https://doi.org/10.1016/j.immuni.2010.03.004.

[105] Prajeeth CK, Kronisch J, Khorooshi R, et al. Effectors of Th1 and Th17 cells act on astrocytes and augment their neuroinflammatory properties. J Neuroinflammat 2017;14(1):1–14. Available from: https://doi.org/10.1186/s12974-017-0978-3.

[106] The role of Th17 cells in auto-inflammatory neurological disorders | PubMed. https://pubmed.ncbi.nlm.nih.gov/28760387/; 2023 [Accessed September 2, 2023].

[107] Setiadi AF, Abbas AR, Jeet S, et al. IL-17A is associated with the breakdown of the blood-brain barrier in relapsing-remitting multiple sclerosis. J Neuroimmunol 2019;332:147–54. Available from: https://doi.org/10.1016/j.jneuroim.2019.04.011.

[108] Paintlia MK, Paintlia AS, Singh AK, Singh I. Synergistic activity of interleukin-17 and tumor necrosis factor-α enhances oxidative stress-mediated oligodendrocyte apoptosis. J Neurochem 2011;116(4):508–21. Available from: https://doi.org/10.1111/j.1471-4159.2010.07136.x.

[109] O'Connor RA, Prendergast CT, Sabatos CA, et al. Cutting edge: Th1 cells facilitate the entry of Th17 cells to the central nervous system during experimental autoimmune encephalomyelitis. J Immunol Baltim Md 1950 2008;181(6):3750–4. Available from: https://doi.org/10.4049/jimmunol.181.6.3750.

[110] Duhen T, Campbell DJ. IL-1β promotes the differentiation of polyfunctional human CCR6 + CXCR3 + Th1/17 cells that are specific for pathogenic and commensal microbes. J Immunol Baltim Md 1950 2014;193(1):120–9. Available from: https://doi.org/10.4049/jimmunol.1302734.

[111] Kebir H, Ifergan I, Alvarez JI, et al. Preferential recruitment of interferon-gamma-expressing TH17 cells in multiple sclerosis. Ann Neurol 2009;66(3):390–402. Available from: https://doi.org/10.1002/ana.21748.

[112] Functional inflammatory profiles distinguish myelin-reactive T cells from patients with multiple sclerosis | PubMed. https://pubmed-ncbi-nlm-nih-gov.foyer.swmed.edu/25972006/; 2023 [Accessed September 2, 2023].

[113] Xu W, Li R, Dai Y, et al. IL-22 secreting CD4 + T cells in the patients with neuromyelitis optica and multiple sclerosis. J Neuroimmunol 2013;261(1-2):87–91. Available from: https://doi.org/10.1016/j.jneuroim.2013.04.021.

[114] Th22 cells are expanded in multiple sclerosis and are resistant to IFN-β | PubMed. https://pubmed-ncbi-nlm-nih-gov.foyer.swmed.edu/25097195/; 2023 [Accessed September 2, 2023].

[115] Perriard G, Mathias A, Enz L, et al. Interleukin-22 is increased in multiple sclerosis patients and targets astrocytes. J Neuroinflammation 2015;12(1):1–18. Available from: https://doi.org/10.1186/s12974-015-0335-3.

[116] Zhen J, Yuan J, Fu Y, et al. IL-22 promotes Fas expression in oligodendrocytes and inhibits FOXP3 expression in T cells by activating the NF-κB pathway in multiple sclerosis. Mol Immunol 2017;82:84–93. Available from: https://doi.org/10.1016/j.molimm.2016.12.020.

[117] Ruocco G, Rossi S, Motta C, et al. T helper 9 cells induced by plasmacytoid dendritic cells regulate interleukin-17 in multiple sclerosis. Clin Sci Lond Engl 1979 2015;129(4):291–303. Available from: https://doi.org/10.1042/CS20140608.

[118] Neutralization of IL-9 ameliorates experimental autoimmune encephalomyelitis by decreasing the effector T cell population | PMC. https://www-ncbi-nlm-nih-gov.foyer.swmed.edu/pmc/articles/PMC2978501/; 2023 [Accessed September 10, 2023].

[119] Elyaman W, Bradshaw EM, Uyttenhove C, et al. IL-9 induces differentiation of TH17 cells and enhances function of FoxP3 + natural regulatory T cells. Proc Natl Acad Sci U S A 2009;106(31):12885–90. Available from: https://doi.org/10.1073/pnas.0812530106.

[120] Schafflick D, Xu CA, Hartlehnert M, et al. Integrated single cell analysis of blood and cerebrospinal fluid leukocytes in multiple sclerosis. Nat Commun 2020;11:247. Available from: https://doi.org/10.1038/s41467-019-14118-w.

[121] Serafini B, Rosicarelli B, Magliozzi R, Stigliano E, Aloisi F. Detection of ectopic B-cell follicles with germinal centers in the meninges of patients with secondary progressive multiple sclerosis. Brain Pathol Zur Switz 2004;14(2):164–74. Available from: https://doi.org/10.1111/j.1750-3639.2004.tb00049.x.

[122] Zhan J, Kipp M, Han W, Kaddatz H. Ectopic lymphoid follicles in progressive multiple sclerosis: From patients to animal models. Immunology 2021;164(3):450–66. Available from: https://doi.org/10.1111/imm.13395.

[123] Quinn JL, Kumar G, Agasing A, Ko RM, Axtell RC. Role of TFH cells in promoting T helper 17-induced neuroinflammation Front Immunol 2018;9, Accessed September 7, 2023. Available from: https://www.frontiersin.org/articles/10.3389/fimmu.2018.00382.

[124] Kaufmann M, Evans H, Schaupp AL, et al. Identifying CNS-colonizing T cells as potential therapeutic targets to prevent progression of multiple sclerosis. Med N Y N 2021;2(3):296–312. Available from: https://doi.org/10.1016/j.medj.2021.01.006, e8.

[125] Kohm AP, Carpentier PA, Anger HA, Miller SD. Cutting edge: CD4 + CD25 + regulatory T cells suppress antigen-specific autoreactive immune responses and central nervous system inflammation during active experimental autoimmune encephalomyelitis. J Immunol Baltim Md 1950 2002;169(9):4712–16. Available from: https://doi.org/10.4049/jimmunol.169.9.4712.

[126] Venken K, Hellings N, Thewissen M, et al. Compromised CD4 + CD25high regulatory T-cell function in patients with relapsing-remitting multiple sclerosis is correlated with a reduced frequency of FOXP3-positive cells and reduced FOXP3 expression at the single-cell level. Immunology 2008;123(1):79–89. Available from: https://doi.org/10.1111/j.1365-2567.2007.02690.x.

[127] Dominguez-Villar M, Baecher-Allan CM, Hafler DA. Identification of T helper type 1-like, Foxp3 + regulatory T cells in human autoimmune disease. Nat Med 2011;17(6):673–5. Available from: https://doi.org/10.1038/nm.2389.

[128] Mascanfroni ID, Takenaka MC, Yeste A, et al. Metabolic control of type 1 regulatory T cell differentiation by AHR and HIF1-α. Nat Med 2015;21(6):638–46. Available from: https://doi.org/10.1038/nm.3868.

[129] Cano-Gamez E, Soskic B, Roumeliotis TI, et al. Single-cell transcriptomics identifies an effectorness gradient shaping the response of CD4 + T cells to cytokines. Nat Commun 2020;11(1):1801. Available from: https://doi.org/10.1038/s41467-020-15543-y.

[130] Hiltensperger M, Beltrán E, Kant R, et al. Skin and gut imprinted helper T cell subsets exhibit distinct functional phenotypes in central nervous system autoimmunity. Nat Immunol 2021;22(7):880–92. Available from: https://doi.org/10.1038/s41590-021-00948-8.

[131] Galli E, Hartmann FJ, Schreiner B, et al. GM-CSF and CXCR4 define a T helper cell signature in multiple sclerosis. Nat Med 2019;25(8):1290–300. Available from: https://doi.org/10.1038/s41591-019-0521-4.

[132] Dankers W, Davelaar N, van Hamburg JP, van de Peppel J, Colin EM, Lubberts E. Human memory Th17 cell populations change into anti-inflammatory cells with regulatory capacity upon exposure to active vitamin D. Front Immunol 2019;10:1504. Available from: https://doi.org/10.3389/fimmu.2019.01504.

[133] Farez MF, Mascanfroni ID, Méndez-Huergo SP, et al. Melatonin contributes to the seasonality of multiple sclerosis relapses. Cell 2015;162(6):1338–52. Available from: https://doi.org/10.1016/j.cell.2015.08.025.

[134] Mitsdoerffer M, Kuchroo V. New pieces in the puzzle: how does interferon-beta really work in multiple sclerosis? Ann Neurol 2009;65(5):487–8. Available from: https://doi.org/10.1002/ana.21722.
[135] Havrdová E, Belova A, Goloborodko A, et al. Activity of secukinumab, an anti-IL-17A antibody, on brain lesions in RRMS: results from a randomized, proof-of-concept study. J Neurol 2016;263(7):1287–95. Available from: https://doi.org/10.1007/s00415-016-8128-x.
[136] Krämer J, Wiendl H. What have failed, interrupted, and withdrawn antibody therapies in multiple sclerosis taught us. Neurotherapeutics 2022;19(3):785–807. Available from: https://doi.org/10.1007/s13311-022-01246-3.
[137] IL-17 neutralization by subcutaneous CJM112, a fully human anti IL-17A monoclonal antibody for the treatment of relapsing-remitting multiple sclerosis: study design of a phase 2 trial | Cochrane Library. Available from: https://doi.org/10.1002/central/CN-01476018.
[138] Longbrake EE, Racke MK. Why did IL-12/IL-23 antibody therapy fail in multiple sclerosis. Expert Rev Neurother 2009;9(3):319–21. Available from: https://doi.org/10.1586/14737175.9.3.319.
[139] Requirement for CD8-major histocompatibility complex class I interaction in positive and negative selection of developing T cells. J Exp Med 1992;176(1):89–97.
[140] Bergmann C, Lowenstein P. MHC class I expression and CD8 T cell function: towards the cell biology of T-APC interactions in the infected brain. In: Lane TE, Carson M, Bergmann C, Wyss-Coray T, editors. Central nervous system diseases and inflammation. Springer US; 2008, p. 277–306. Available from: http://doi.org/10.1007/978-0-387-73894-9_14.
[141] Redwine JM, Buchmeier MJ, Evans CF. In vivo expression of major histocompatibility complex molecules on oligodendrocytes and neurons during viral infection. Am J Pathol 2001;159(4):1219–24.
[142] Babbe H, Roers A, Waisman A, et al. Clonal expansions of CD8(+) T cells dominate the T cell infiltrate in active multiple sclerosis lesions as shown by micromanipulation and single cell polymerase chain reaction. J Exp Med 2000;192(3):393–404. Available from: https://doi.org/10.1084/jem.192.3.393.
[143] Machado-Santos J, Saji E, Tröscher AR, et al. The compartmentalized inflammatory response in the multiple sclerosis brain is composed of tissue-resident CD8 + T lymphocytes and B cells. Brain J Neurol 2018;141(7):2066–82. Available from: https://doi.org/10.1093/brain/awy151.
[144] Fransen NL, Hsiao CC, van der Poel M, et al. Tissue-resident memory T cells invade the brain parenchyma in multiple sclerosis white matter lesions. Brain J Neurol 2020;143(6):1714–30. Available from: https://doi.org/10.1093/brain/awaa117.
[145] Huseby ES, Liggitt D, Brabb T, Schnabel B, Ohlén C, Goverman J. A pathogenic role for myelin-specific CD8(+) T cells in a model for multiple sclerosis. J Exp Med 2001;194(5):669–76. Available from: https://doi.org/10.1084/jem.194.5.669.
[146] Ji Q, Castelli L, Goverman JM. MHC class I-restricted myelin epitopes are cross-presented by Tip-DCs that promote determinant spreading to CD8[+] T cells. Nat Immunol 2013;14(3):254–61. Available from: https://doi.org/10.1038/ni.2513.
[147] Jurewicz A, Biddison WE, Antel JP. MHC class I-restricted lysis of human oligodendrocytes by myelin basic protein peptide-specific CD8 T lymphocytes. J Immunol Baltim Md 1950 1998;160(6):3056–9.
[148] van Nierop GP, van Luijn MM, Michels SS, et al. Phenotypic and functional characterization of T cells in white matter lesions of multiple sclerosis patients. Acta Neuropathol (Berl) 2017;134(3):383–401. Available from: https://doi.org/10.1007/s00401-017-1744-4.
[149] Page N, Lemeille S, Vincenti I, et al. Persistence of self-reactive CD8 + T cells in the CNS requires TOX-dependent chromatin remodeling. Nat Commun 2021;12(1):1009. Available from: https://doi.org/10.1038/s41467-021-21109-3.
[150] Saligrama N, Zhao F, Sikora MJ, et al. Opposing T cell responses in experimental autoimmune encephalomyelitis. Nature 2019;572(7770):481–7. Available from: https://doi.org/10.1038/s41586-019-1467-x.
[151] Baughman EJ, Mendoza JP, Ortega SB, et al. Neuroantigen-specific CD8 + regulatory T-cell function is deficient during acute exacerbation of multiple sclerosis. J Autoimmun 2011;36(2):115–24. Available from: https://doi.org/10.1016/j.jaut.2010.12.003.
[152] Tennakoon DK, Mehta RS, Ortega SB, Bhoj V, Racke MK, Karandikar NJ. Therapeutic induction of regulatory, cytotoxic CD8 + T cells in multiple sclerosis. J Immunol Baltim Md 2006;176(11):7119–29. Available from: https://doi.org/10.4049/jimmunol.176.11.7119, *1950*.
[153] Godfrey DI, Uldrich AP, McCluskey J, Rossjohn J, Moody DB. The burgeoning family of unconventional T cells. Nat Immunol 2015;16(11):1114–23. Available from: https://doi.org/10.1038/ni.3298.

[154] Longitudinal analysis of immune cell phenotypes in early stage multiple sclerosis: distinctive patterns characterize MRI-active patients | PubMed. https://pubmed-ncbi-nlm-nih-gov.foyer.swmed.edu/16870883/; 2023 [Accessed September 3, 2023].
[155] Hvas J, Oksenberg JR, Fernando R, Steinman L, Bernard CC. Gamma delta T cell receptor repertoire in brain lesions of patients with multiple sclerosis. J Neuroimmunol 1993;46(1-2):225–34. Available from: https://doi.org/10.1016/0165-5728(93)90253-u.
[156] Schirmer L, Rothhammer V, Hemmer B, Korn T. Enriched CD161high CCR6 + γδ T cells in the cerebrospinal fluid of patients with multiple sclerosis. JAMA Neurol 2013;70(3):345–51. Available from: https://doi.org/10.1001/2013.jamaneurol.409.
[157] Zeine R, Pon R, Ladiwala U, Antel JP, Filion LG, Freedman MS. Mechanism of gammadelta T cell-induced human oligodendrocyte cytotoxicity: relevance to multiple sclerosis. J Neuroimmunol 1998;87(1-2):49–61. Available from: https://doi.org/10.1016/s0165-5728(98)00047-2.
[158] Willing A, Jäger J, Reinhardt S, Kursawe N, Friese MA. Production of IL-17 by MAIT cells is increased in multiple sclerosis and is associated with IL-7 receptor expression. J Immunol Baltim Md 1950 2018;200(3):974–82. Available from: https://doi.org/10.4049/jimmunol.1701213.
[159] Häusser-Kinzel S, Weber MS. The role of B cells and antibodies in multiple sclerosis, neuromyelitis optica, and related disorders Front Immunol 2019;10, Accessed September 6, 2023. Available from: https://www.frontiersin.org/articles/10.3389/fimmu.2019.00201.
[160] Magliozzi R, Howell O, Vora A, et al. Meningeal B-cell follicles in secondary progressive multiple sclerosis associate with early onset of disease and severe cortical pathology. Brain J Neurol 2007;130(Pt 4):1089–104. Available from: https://doi.org/10.1093/brain/awm038.
[161] Kuenz B, Lutterotti A, Ehling R, et al. Cerebrospinal fluid B cells correlate with early brain inflammation in multiple sclerosis. PLoS ONE 2008;3(7):e2559. Available from: https://doi.org/10.1371/journal.pone.0002559.
[162] Stern JNH, Yaari G, Vander Heiden JA, et al. B cells populating the multiple sclerosis brain mature in the draining cervical lymph nodes. Sci Transl Med 2014;6(248):248ra107. Available from: https://doi.org/10.1126/scitranslmed.3008879.
[163] von Büdingen HC, Kuo TC, Sirota M, et al. B cell exchange across the blood-brain barrier in multiple sclerosis. J Clin Invest 2012;122(12):4533–43. Available from: https://doi.org/10.1172/JCI63842.
[164] Haas J, Bekeredjian-Ding I, Milkova M, et al. B cells undergo unique compartmentalized redistribution in multiple sclerosis. J Autoimmun 2011;37(4):289–99. Available from: https://doi.org/10.1016/j.jaut.2011.08.003.
[165] Brioschi S, Wang WL, Peng V, et al. Heterogeneity of meningeal B cells reveals a lymphopoietic niche at the CNS borders. Science 2021;373(6553):eabf9277. Available from: https://doi.org/10.1126/science.abf9277.
[166] Brändle SM, Obermeier B, Senel M, et al. Distinct oligoclonal band antibodies in multiple sclerosis recognize ubiquitous self-proteins. Proc Natl Acad Sci U S A 2016;113(28):7864–9. Available from: https://doi.org/10.1073/pnas.1522730113.
[167] Lanz TV, Brewer RC, Ho PP, et al. Clonally expanded B cells in multiple sclerosis bind EBV EBNA1 and GlialCAM. Nature 2022;603(7900):321–7. Available from: https://doi.org/10.1038/s41586-022-04432-7.
[168] Wang J, Jelcic I, Mühlenbruch L, et al. HLA-DR15 molecules jointly shape an autoreactive T cell repertoire in multiple sclerosis. Cell 2020;183(5):1264–81. Available from: https://doi.org/10.1016/j.cell.2020.09.054, e20.
[169] Lisak RP, Benjamins JA, Nedelkoska L, et al. Secretory products of multiple sclerosis B cells are cytotoxic to oligodendroglia in vitro. J Neuroimmunol 2012;246(1-2):85–95. Available from: https://doi.org/10.1016/j.jneuroim.2012.02.015.
[170] Lisak RP, Nedelkoska L, Benjamins JA, et al. B cells from patients with multiple sclerosis induce cell death via apoptosis in neurons in vitro. J Neuroimmunol 2017;309:88–99. Available from: https://doi.org/10.1016/j.jneuroim.2017.05.004.
[171] Mitsdoerffer M, Di Liberto G, Dötsch S, et al. Formation and immunomodulatory function of meningeal B cell aggregates in progressive CNS autoimmunity. Brain J Neurol 2021;144(6):1697–710. Available from: https://doi.org/10.1093/brain/awab093.
[172] Heming M, Wiendl H. Learning multiple sclerosis immunopathogenesis from anti-CD20 therapy. Proc Natl Acad Sci U S A 2023;120(6). Available from: https://doi.org/10.1073/pnas.2221544120, e2221544120.
[173] Sellebjerg F, Blinkenberg M, Sorensen PS. Anti-CD20 monoclonal antibodies for relapsing and progressive multiple sclerosis. CNS Drugs 2020;34(3):269–80. Available from: https://doi.org/10.1007/s40263-020-00704-w.

[174] Increased frequency and broadened specificity of latent EBV nuclear antigen-1-specific T cells in multiple sclerosis | PubMed. https://pubmed-ncbi-nlm-nih-gov.foyer.swmed.edu/16569670/; 2023 [Accessed September 5, 2023].
[175] Gettings EJ, Hackett CT, Scott TF. Severe relapse in a multiple sclerosis patient associated with ipilimumab treatment of melanoma. Mult Scler Houndmills Basingstoke Engl 2015;21(5):670. Available from: https://doi.org/10.1177/1352458514549403.
[176] CNS demyelination and enhanced myelin-reactive responses after ipilimumab treatment | PubMed. https://pubmed-ncbi-nlm-nih-gov.foyer.swmed.edu/26984943/; 2023 [Accessed September 5, 2023].
[177] A genetic mouse model of adult-onset, pervasive central nervous system demyelination with robust remyelination | PubMed. https://pubmed-ncbi-nlm-nih-gov.foyer.swmed.edu/20851998/; 2023 [Accessed September 5, 2023].
[178] Traka M, Podojil JR, McCarthy DP, Miller SD, Popko B. Oligodendrocyte death results in immune-mediated CNS demyelination. Nat Neurosci 2016;19(1):65–74. Available from: https://doi.org/10.1038/nn.4193.
[179] Tsunoda I, Fujinami RS. Inside-Out versus Outside-In models for virus induced demyelination: axonal damage triggering demyelination. Springer Semin Immunopathol 2002;24(2):105–25. Available from: https://doi.org/10.1007/s00281-002-0105-z.
[180] Leypoldt F, Titulaer MJ, Aguilar E, et al. Herpes simplex virus–1 encephalitis can trigger anti-NMDA receptor encephalitis: case report. Neurology 2013;81(18):1637–9. Available from: https://doi.org/10.1212/WNL.0b013e3182a9f531.
[181] Hancks DC, Kazazian HH. Roles for retrotransposon insertions in human disease. Mob DNA 2016;7:9. Available from: https://doi.org/10.1186/s13100-016-0065-9.
[182] Uleri E, Mei A, Mameli G, Poddighe L, Serra C, Dolei A. HIV Tat acts on endogenous retroviruses of the W family and this occurs via Toll-like receptor 4: inference for neuroAIDS. AIDS Lond Engl 2014;28(18):2659–70. Available from: https://doi.org/10.1097/QAD.0000000000000477.
[183] Morandi E, Tanasescu R, Tarlinton RE, et al. The association between human endogenous retroviruses and multiple sclerosis: a systematic review and meta-analysis. PLoS One 2017;12(2):e0172415. Available from: https://doi.org/10.1371/journal.pone.0172415.
[184] The envelope protein of a human endogenous retrovirus-W family activates innate immunity through CD14/TLR4 and promotes Th1-like responses | PubMed. https://pubmed-ncbi-nlm-nih-gov.foyer.swmed.edu/16751411/; 2023 [Accessed September 5, 2023].
[185] Saresella M, Rolland A, Marventano I, et al. Multiple sclerosis-associated retroviral agent (MSRV)-stimulated cytokine production in patients with relapsing-remitting multiple sclerosis. Mult Scler Houndmills Basingstoke Engl 2009;15(4):443–7. Available from: https://doi.org/10.1177/1352458508100840.
[186] Perron H, Jouvin-Marche E, Michel M, et al. Multiple sclerosis retrovirus particles and recombinant envelope trigger an abnormal immune response in vitro, by inducing polyclonal Vbeta16 T-lymphocyte activation. Virology 2001;287(2):321–32. Available from: https://doi.org/10.1006/viro.2001.1045.
[187] Human endogenous retrovirus protein activates innate immunity and promotes experimental allergic encephalomyelitis in mice | PubMed. https://pubmed-ncbi-nlm-nih-gov.foyer.swmed.edu/24324591/; 2023 [Accessed September 5, 2023].
[188] Thompson AJ, Baranzini SE, Geurts J, Hemmer B, Ciccarelli O. Multiple sclerosis. Lancet 2018;391(10130):1622–36. Available from: https://doi.org/10.1016/S0140-6736(18)30481-1.
[189] Mundt S, Mrdjen D, Utz SG, Greter M, Schreiner B, Becher B. Conventional DCs sample and present myelin antigens in the healthy CNS and allow parenchymal T cell entry to initiate neuroinflammation. Sci Immunol 2019;4(31):eaau8380. Available from: https://doi.org/10.1126/sciimmunol.aau8380.
[190] Spath S, Komuczki J, Hermann M, et al. Dysregulation of the cytokine GM-CSF induces spontaneous phagocyte invasion and immunopathology in the central nervous system. Immunity 2017;46(2):245–60. Available from: https://doi.org/10.1016/j.immuni.2017.01.007.
[191] van Horssen J, Singh S, van der Pol S, et al. Clusters of activated microglia in normal-appearing white matter show signs of innate immune activation. J Neuroinflammat 2012;9:156. Available from: https://doi.org/10.1186/1742-2094-9-156.
[192] Howell OW, Reeves CA, Nicholas R, et al. Meningeal inflammation is widespread and linked to cortical pathology in multiple sclerosis. Brain J Neurol 2011;134(Pt 9):2755–71. Available from: https://doi.org/10.1093/brain/awr182.

[193] Kornek B, Storch MK, Weissert R, et al. Multiple sclerosis and chronic autoimmune encephalomyelitis: a comparative quantitative study of axonal injury in active, inactive, and remyelinated lesions. Am J Pathol 2000;157(1):267–76. Available from: https://doi.org/10.1016/S0002-9440(10)64537-3.

[194] Defining the clinical course of multiple sclerosis: results of an international survey. National Multiple Sclerosis Society (USA) Advisory Committee on Clinical Trials of New Agents in Multiple Sclerosis | PubMed. https://pubmed-ncbi-nlm-nih-gov.foyer.swmed.edu/8780061/; 2023 [Accessed September 4, 2023].

[195] Huang WJ, Chen WW, Zhang X. Multiple sclerosis: pathology, diagnosis and treatments. Exp Ther Med 2017;13(6):3163–6. Available from: https://doi.org/10.3892/etm.2017.4410.

[196] Staging of multiple sclerosis (MS) lesions: pathology of the time frame of MS | PubMed. https://pubmed-ncbi-nlm-nih-gov.foyer.swmed.edu/10736062/; 2023 [Accessed September 4, 2023].

[197] Axonal transection in the lesions of multiple sclerosis | PubMed. https://pubmed-ncbi-nlm-nih-gov.foyer.swmed.edu/9445407/; 2023 [Accessed September 5, 2023].

[198] Milo R, Korczyn AD, Manouchehri N, Stüve O. The temporal and causal relationship between inflammation and neurodegeneration in multiple sclerosis. Mult Scler J 2020;26(8):876–86. Available from: https://doi.org/10.1177/1352458519886943.

[199] Absinta M, Sati P, Masuzzo F, et al. Association of chronic active multiple sclerosis lesions with disability in vivo. JAMA Neurol 2019;76(12):1474–83. Available from: https://doi.org/10.1001/jamaneurol.2019.2399.

[200] Elliott C, Belachew S, Wolinsky JS, et al. Chronic white matter lesion activity predicts clinical progression in primary progressive multiple sclerosis. Brain J Neurol 2019;142(9):2787–99. Available from: https://doi.org/10.1093/brain/awz212.

[201] Remyelination in the CNS: from biology to therapy | PubMed. https://pubmed-ncbi-nlm-nih-gov.foyer.swmed.edu/18931697/; 2023 [Accessed September 4, 2023].

[202] van Walderveen MA, Kamphorst W, Scheltens P, et al. Histopathologic correlate of hypointense lesions on T1-weighted spin-echo MRI in multiple sclerosis. Neurology 1998;50(5):1282–8. Available from: https://doi.org/10.1212/wnl.50.5.1282.

[203] Cortical lesions and atrophy associated with cognitive impairment in relapsing-remitting multiple sclerosis | PubMed. https://pubmed-ncbi-nlm-nih-gov.foyer.swmed.edu/19752305/; 2023 [Accessed September 4, 2023].

[204] Lucchinetti CF, Popescu BFG, Bunyan RF, et al. Inflammatory cortical demyelination in early multiple sclerosis. N Engl J Med 2011;365(23):2188–97. Available from: https://doi.org/10.1056/NEJMoa1100648.

[205] Bø L, Vedeler CA, Nyland HI, Trapp BD, Mørk SJ. Subpial demyelination in the cerebral cortex of multiple sclerosis patients. J Neuropathol Exp Neurol 2003;62(7):723–32. Available from: https://doi.org/10.1093/jnen/62.7.723.

[206] Peterson JW, Bö L, Mörk S, Chang A, Trapp BD. Transected neurites, apoptotic neurons, and reduced inflammation in cortical multiple sclerosis lesions. Ann Neurol 2001;50(3):389–400. Available from: https://doi.org/10.1002/ana.1123.

[207] Saez-Calveras N, Stuve O. The role of the complement system in multiple sclerosis: a review. Front Immunol 2022;13:970486. Available from: https://doi.org/10.3389/fimmu.2022.970486.

[208] The validity of animal models to explore the pathogenic role of the complement system in multiple sclerosis: a review | PubMed. https://pubmed-ncbi-nlm-nih-gov.foyer.swmed.edu/36311030/; 2023 [Accessed September 5, 2023].

[209] Targeted complement inhibition at synapses prevents microglial synaptic engulfment and synapse loss in demyelinating disease | PubMed. https://pubmed-ncbi-nlm-nih-gov.foyer.swmed.edu/31883839/; 2023 [Accessed September 5, 2023].

[210] Yates RL, Esiri MM, Palace J, Jacobs B, Perera R, DeLuca GC. Fibrin(ogen) and neurodegeneration in the progressive multiple sclerosis cortex. Ann Neurol 2017;82(2):259–70. Available from: https://doi.org/10.1002/ana.24997.

[211] Evidence of early cortical atrophy in MS: relevance to white matter changes and disability | PubMed. https://pubmed-ncbi-nlm-nih-gov.foyer.swmed.edu/12682324/; 2023 [Accessed September 4, 2023].

[212] Androdias G, Reynolds R, Chanal M, Ritleng C, Confavreux C, Nataf S. Meningeal T cells associate with diffuse axonal loss in multiple sclerosis spinal cords. Ann Neurol 2010;68(4):465–76. Available from: https://doi.org/10.1002/ana.22054.

[213] Pathogenic T cell cytokines in multiple sclerosis | PubMed. https://pubmed-ncbi-nlm-nih-gov.foyer.swmed.edu/31611252/; 2023 [Accessed September 5, 2023].

[214] Huynh JL, Garg P, Thin TH, et al. Epigenome-wide differences in pathology-free regions of multiple sclerosis-affected brains. Nat Neurosci 2014;17(1):121–30. Available from: https://doi.org/10.1038/nn.3588.
[215] Roosendaal SD, Bendfeldt K, Vrenken H, et al. Grey matter volume in a large cohort of MS patients: relation to MRI parameters and disability. Mult Scler Houndmills Basingstoke Engl 2011;17(9):1098–106. Available from: https://doi.org/10.1177/1352458511404916.
[216] LaCroix MS, Mirbaha H, Shang P, et al. Tau seeding in cases of multiple sclerosis. Acta Neuropathol Commun 2022;10(1):146. Available from: https://doi.org/10.1186/s40478-022-01444-2.
[217] Anderson JM, Patani R, Reynolds R, et al. Abnormal tau phosphorylation in primary progressive multiple sclerosis. Acta Neuropathol (Berl) 2010;119(5):591–600. Available from: https://doi.org/10.1007/s00401-010-0671-4.
[218] Abnormally phosphorylated tau is associated with neuronal and axonal loss in experimental autoimmune encephalomyelitis and multiple sclerosis | PubMed. https://pubmed-ncbi-nlm-nih-gov.foyer.swmed.edu/18567922/; 2023 [Accessed September 5, 2023].
[219] Levy H, Assaf Y, Frenkel D. Characterization of brain lesions in a mouse model of progressive multiple sclerosis. Exp Neurol 2010;226(1):148–58. Available from: https://doi.org/10.1016/j.expneurol.2010.08.017.
[220] Popescu BFGh PI, Lucchinetti CF. Pathology of multiple sclerosis: where do we stand? Contin Lifelong Learn Neurol 2013;19(4 Multiple Sclerosis):901–21. Available from: https://doi.org/10.1212/01.CON.0000433291.23091.65.
[221] Patrikios P, Stadelmann C, Kutzelnigg A, et al. Remyelination is extensive in a subset of multiple sclerosis patients. Brain J Neurol 2006;129(Pt 12):3165–72. Available from: https://doi.org/10.1093/brain/awl217.
[222] Chang A, Staugaitis SM, Dutta R, et al. Cortical remyelination: a new target for repair therapies in multiple sclerosis. Ann Neurol 2012;72(6):918–26. Available from: https://doi.org/10.1002/ana.23693.
[223] Bramow S, Frischer JM, Lassmann H, et al. Demyelination versus remyelination in progressive multiple sclerosis. Brain J Neurol 2010;133(10):2983–98. Available from: https://doi.org/10.1093/brain/awq250.
[224] Harlow DE, Honce JM, Miravalle AA. Remyelination therapy in multiple sclerosis. Front Neurol 2015;6:257. Available from: https://doi.org/10.3389/fneur.2015.00257.

CHAPTER 3

Diagnosis and clinical features of multiple sclerosis

Svetlana Eckert, Channa Kolb and Bianca Weinstock-Guttman

Department of Neurology, Jacobs Comprehensive Ms Treatment and Research Center, Jacobs School of Medicine and Biomedical Sciences, University at Buffalo, State University of New York, Buffalo, NY, United States

OUTLINE

What is multiple sclerosis	38
Clinical symptoms of multiple sclerosis	38
Development and mechanisms of multiple sclerosis	38
Ethnic and racial disparities in multiple sclerosis	39
Clinical course of multiple sclerosis and multiple sclerosis types	40
Linking multiple sclerosis pathophysiology and in vivo imaging	41
Multiple sclerosis diagnostic criteria	43
Precursors to multiple sclerosis: clinically isolated syndrome and radiologically isolated syndrome	45
Recommendations on the timing of imaging for monitoring of multiple sclerosis disease activity and progression	45
Recommendations on the use of newer magnetic resonance imaging techniques for evaluation of multiple sclerosis	46
Summary	47
References	47

What is multiple sclerosis

Multiple sclerosis (MS) is an unpredictable chronic demyelinating inflammatory and neurodegenerative condition that is thought to be caused by a maladaptive autoreactive immune system and may present with a relapsing or progressive disease course.

Clinical symptoms of multiple sclerosis

Clinical presentations of MS can involve any location on the whole CNS axis, which can result in highly variable disease presentation and can make the diagnosis difficult. Various iterations of the diagnostic criteria for MS have been agreed upon over time, defining MS as a chronic disease process entailing proof of dissemination in space and time. Typical presentations of MS supporting a demyelinating event may include unilateral (and more rarely bilateral) optic neuritis, limb weakness, numbness, bladder/bowel dysfunction, fatigue, ataxia, diplopia, and cognitive dysfunction depending on the location of the inflammatory lesions. Cognitive worsening indeed is being studied as a possible primary relapse symptom throughout the MS disease course [1,2]. It should be noted that cognitive dysfunction, fatigue, and mood changes alone should not be considered as an initial presentation of MS, though frequently those symptoms can accompany a relapse [3]. Cognitive change has not been established as the sole symptom which would meet the criteria as a new onset symptom in MS as of yet [4]. However, cognitive impairment can be identified early on in radiologically isolated syndrome (RIS) and clinically isolated syndrome (CIS) patients and may be an indicator of early activity of MS pathology [5].

Development and mechanisms of multiple sclerosis

The inflammatory processes in MS are thought to be mediated by both the adaptive and innate immune systems. The adaptive immune system culprits for MS involve autoreactive T and B cells, whereas the innate immune system contributes to the damage through the action of activated macrophages and dendritic cells. Abnormal autoreactive inflammation caused by peripheral immune cells (including T and B cells) and some of their products called cytokines (which include interleukins and interferons) ultimately leads to the breakdown of the blood—brain barrier [6]. Loss of the blood—brain barrier integrity leads to the unwelcome transmigration of additional peripheral inflammatory cells that promote further inflammation within the central nervous system (CNS), leading to demyelination of axons, a hallmark of MS pathogenesis. In addition to the undesirable inflammation from circulating immune cells, the CNS generated autoinflammation response from activated microglia and astrocyte promote the chronicity of the disease and lead to the superimposed neurodegeneration process [7]. Furthermore, an earlier, prodromal phase of MS is becoming more widely recognized and explored, in which patients may develop nonspecific symptoms such as fatigue, depression, or self-described "mental fog" before having specific neurologic symptoms but significant enough to cause them to seek medical care [8]. With increasing knowledge regarding MS pathophysiology, it is also becoming evident that MS is a continuum process of inflammatory, demyelinating and neurodegenerative superimposed processes, rather than separate disease phases, although the inflammatory process predominate early in the disease while neurodegeneration in the later phases [9].

The diagnosis of MS has a broad age distribution on presentation, with the age at time of first demyelinating event mainly seen between 20 and 50 years of age. Pediatric presentations of MS (before age 18) and diagnosis of MS above the age of 50 also occur, although not as commonly (4%—5% of MS diagnoses) [10]. About 5% of patients are diagnosed with MS above the age of 50, considered as late onset MS (LOMS). Most LOMS patients present with motor dysfunction and spinal cord involvement (65% of this patient population). A systemic review and meta-analysis of late onset MS has shown that this patient population tends to have a more progressive form of Ms which is often misdiagnosed due to the patients age and often superimposed comorbidities [11]. Furthermore, the incidence of LOMS seems to be increasing as well [12]. MS has a female predominance, 2.5—3:1 female-to-male ratio [13], which becomes more prominent after puberty.

The prevalence of MS has increased over time worldwide from 2.3 million people in 2013 to 2.9 million people in 2023 [14], possibly due to better and more widely available diagnostic imaging and more inclusive diagnostic criteria as well as improved access to care. Other factors that may increase risk of MS to consider include lifestyle, diet, environmental exposures, and vitamin D levels [15].

Northern latitudes and areas farther from the equator have the highest prevalence of MS, which may be due to a combination of factors such as sun exposure, vitamin D levels as well as viral milieu. Exposure to Epstein Barr virus, especially later exposure presenting as clinical mononucleosis, has been closely linked with the risk of developing MS [16,17].

Ethnic and racial disparities in multiple sclerosis

Ethnic and racial disparities should also be considered when evaluating a patient for MS. MS was previously thought to predominantly affect White people of Northern European descent. Recent research has shown that there is a 47% higher risk of MS in Black women compared to White women in the United States [18]. Black patients also tend to have an earlier more aggressive disease course compared to White patients [18] (insert Fig. 3.1A and B). Genetic analysis has shown that Black patients of European descent have a similar risk of developing MS compared to White European patients. Black patients of African descent can have a variable or lower risk to develop MS [19]. Hispanic patients tend to be diagnosed with MS at a younger age and may have a more aggressive disease course compared to White patients. Compared to White patients, Hispanic patients will more often present with optic neuritis and transverse myelitis [20]. Importantly, the female predominance of MS is similar across races [21].

Overall genetic factors (primarily specific major histocompatibility complex as HLA 1501) [22] are known to contribute significantly to the risk of development of MS, with familial MS representing as high as 12.6% of all MS cases [23]. More than 200 single-nucleotide polymorphisms have been linked to the risk for development of MS [24]. Twin studies show that monozygotic twins exhibit much higher concordance rates for the development of MS that dizygotic twins (25%—30% vs 3%—7%, respectively) [25]. Family history of MS in general has been

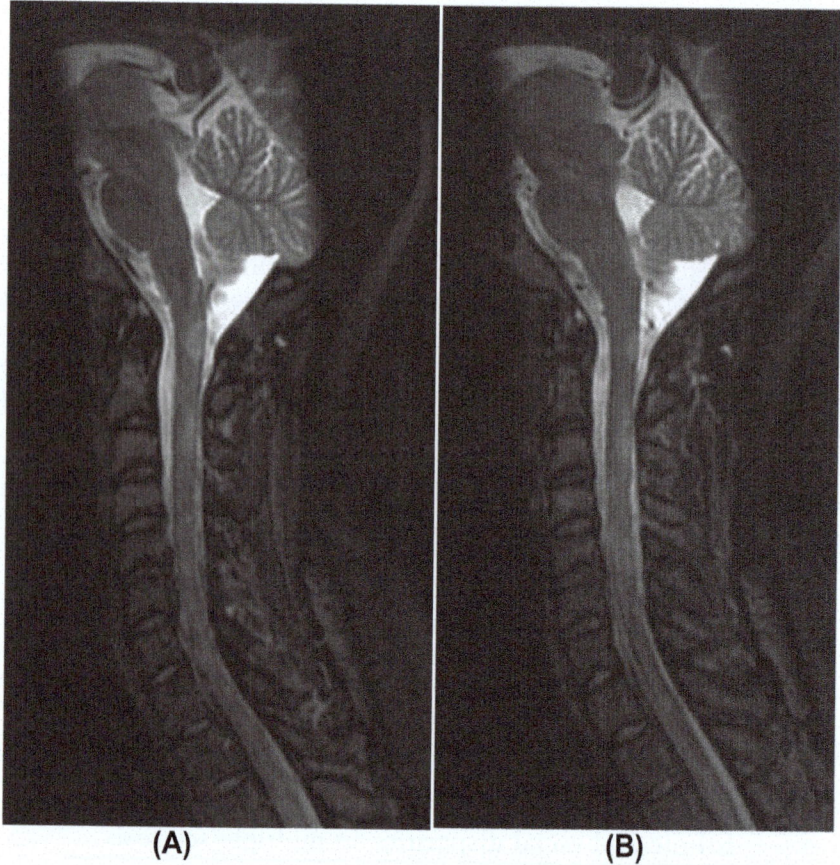

FIGURE 3.1 (A-B) Sagittal STIR images demonstrating multiple areas of T1 hyperintensity on a scan at diagnosis of a 26 year old female black MS patient with relatively few symptoms of hand numbness only.

reported to be much more frequent in MS patients, present in as many as 15%—20% of MS patients [23,26]. Furthermore, the estimates of lifetime risk of Ms in first-degree relatives of MS patients are approximately 3% (4% for siblings, 2% for parents, and 2% for children), and 10- to 30-fold greater than the age-adjusted risk in the general population (0.1%—0.3%) [23,26,27]. The variability in genetic risk suggests a multifactorial risk contribution, which includes a combination of genetic variations and various environmental and lifestyle exposures, such as viruses, smoking, obesity, sun exposure, and vitamin D levels.

Clinical course of multiple sclerosis and multiple sclerosis types

The main traditional clinical disease course of MS features relapses and continuous progression of disease at different stages. Relapses and progression in MS can be viewed as the inflammatory and the neurodegenerative components of MS, respectively. MS relapses are a salient feature in relapsing-remitting MS (RRMS), where patients present with periods of alternating new or worsening deficits (relapses) and clinical disease silence (remission). MS clinical relapses present as a new or an acute worsening of a previous neurological symptom which persists for more than 24 hours and is not due to infection or any other explanation. A patient may not recover completely back to baseline after every relapse, and with greater number of relapses, greater disability accumulation may occur. A prodromal phase is also believed to exist and can have a variable course toward progression to MS [28].

Traditional descriptions of MS disease course include the RRMS stage seen in 80% of cases, presenting with acute periods of inflammation and relative disease inactivity in periods of "remission." All RRMS patients eventually are believed to go on to the next stage of secondary progressive MS with symptoms worsening in the absence of relapses and/or new magnetic resonance imaging (MRI) lesions. Primary progressive MS patients are felt to have a different disease presentation, with a progressive clinical course at the onset of disease [13]. MS disease activity has also been described as active versus inactive, with activity indicating new/active MRI lesions and/or clinical relapses. Progressive MS patients can be classified as having stable or progressive disease, depending on change in disability overtime. However, in a newly emerging view of the disease, the processes of periods of inflammation with disease activity and progressive neurodegeneration are likely coexistent with a continuum of neurodegenerative and inflammatory activity rather than distinct disease phases [9].

Linking multiple sclerosis pathophysiology and in vivo imaging

Acute inflammation and compromise of the blood–brain barrier is thought to be the main process behind the acute neurological dysfunction seen in relapsing MS involving both the innate and adaptive immune systems.

The disruption of the blood–brain barrier integrity can be visualized on MRI after contrast administration as a gadolinium enhancing focal lesion (Fig. 3.2A–C). Such active MS lesions may present as round newly enhancing or open-ring enhancement on T1 pre- and postcontrast imaging and may reveal a central vein sign on gradient echo (GRE) or susceptibility-weighted imaging (SWI).

Inflammatory infiltration with ongoing or prior local inflammatory response results in vasogenic edema which on MRI is seen on T2 hyperintense sequences and is best visualized on T2 fluid-attenuated inversion recovery (FLAIR) sequences (Fig. 3.3A–C). Gliosis and loss of axons results in T1 hypointense lesions, with some of those termed "black holes" if a certain threshold of tissue death/destruction has been reached with significant axonal loss in those areas. As noted above, a central vein sign typical of MS lesions may also be noted in acute or chronic lesions on GRE and SWI images, indicating the perivascular location of demyelination and breakdown of the blood–brain barrier [29].

Additional MRI indicators of progressive MS pathophysiology may include, but are not exclusive to meningeal inflammation and chronic activation of microglia. The latter process is considered primarily CNS immune driven rather than due to the leakage across the blood–brain barrier [30]. MRI features of progression/neurodegeneration in MS can be

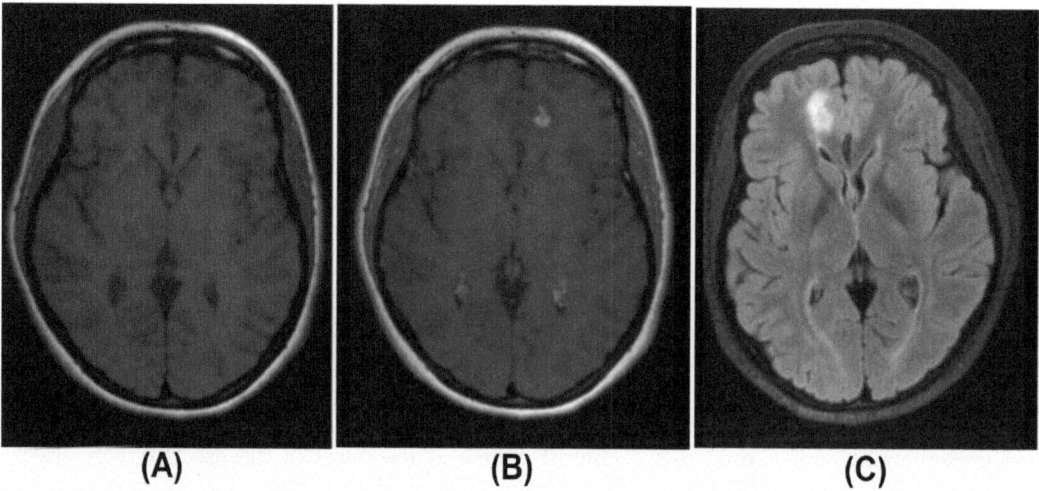

FIGURE 3.2 Pre-(A) and post-contrast (B) images demonstrating break down of the blood brain barrier with enhancement of the T1 lesion following administration of IV gadolinium-based contrast. Also included is a T2 FLAIR sequence demonstrating edema within the same lesion (C).

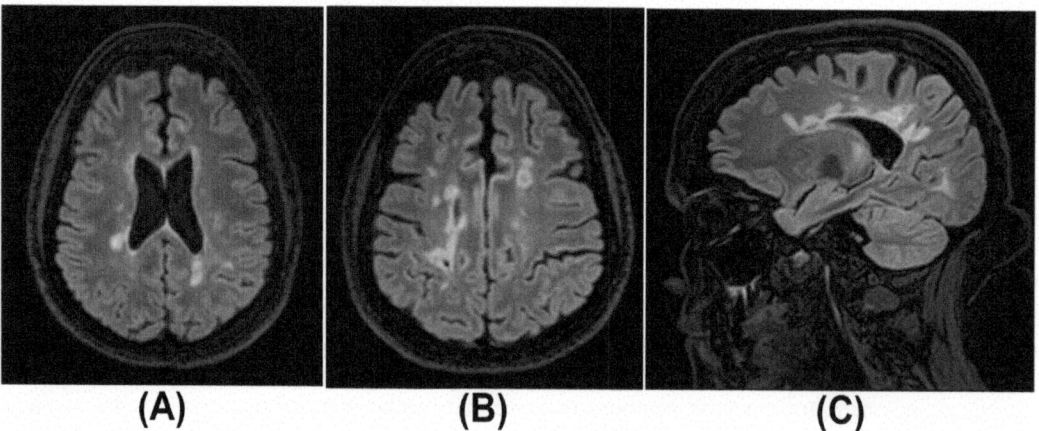

FIGURE 3.3 Typical Dawson fingers of MS lesions on axial (A-B) and sagittal (C) T2 FLAIR images. Some lesions can become confluent, resulting in difficulty evaluating for new lesions.

distinguished as cortical lesions, slowly expanding T2 lesions seen in the brain and some lesions that demonstrate remyelination, with the remyelination process being usually incomplete. Brain tissue loss (atrophy) is the primary MRI marker of neurodegeneration and progression of disease [31] (Fig. 3.4A−B). Brain-specific tissue atrophy within cortical, deep gray, or white matter may differ with disease course and age. Analytical techniques using specialized software, such as SIENA/SIENAX, MS matrix, and NeuroStream, to

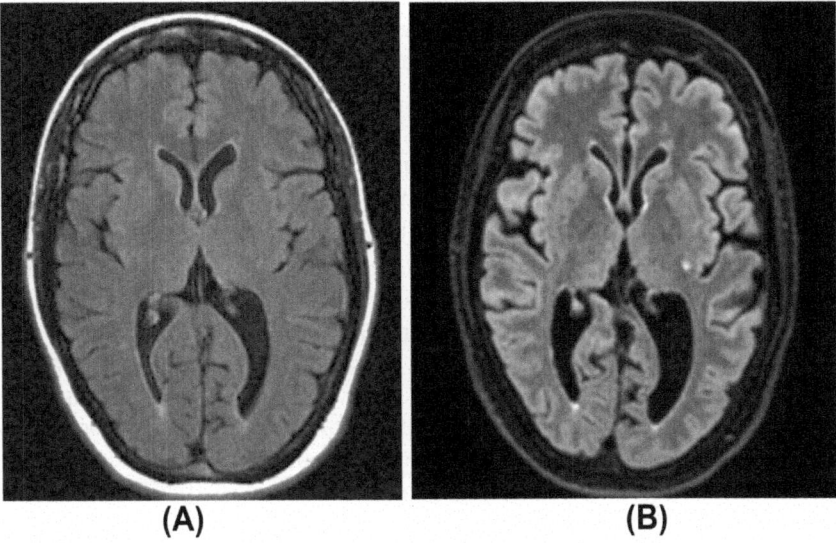

FIGURE 3.4 Cortical and central atrophy in a patient can be easily visualized on scans 18 years apart (initial scan (A) and most recent scan (B).

name a few, are used for measuring atrophy of the whole brain, deep nuclei, cortex, midbrain, and cerebellum [32,33].

Multiple sclerosis diagnostic criteria

The unpredictable nature of when and where inflammation will occur within the CNS axis, in part, explains why the clinical presentation of MS is so variable. Neuroimaging of MS patients plays a crucial role in the diagnosis and monitoring of MS disease activity, as MS patients will more commonly have T2 hyperintense lesions in certain anatomical regions (such as periventricular, corpus callosum, middle cerebellar peduncle, brainstem, optic nerves (ONs), and spinal cord).

One of the earlier MS diagnostic criteria, the Poser criteria, placed more weight on clinical exam and history. With the emergence of MRI as a valuable diagnostic tool, the McDonald criteria were established in 2001. That diagnostic criteria allowed paraclinical data, such as MRI and cerebral spinal fluid (CSF) to place more weight in fulfilling the diagnostic features of space and time dissemination with MS. The McDonald MS criteria were revised in 2005, 2010, and most recently in 2017 [3]. The 2017 McDonald criteria changes resulted in improved sensitivity in the diagnosis of MS, unfortunately also resulting in lower specificity for the disease [3]. It is important to mention that with lower sensitivity for MS diagnosis, there is an increased risk for misdiagnosis. Therefore one should be cautious when applying the most recent MS criteria and use it only when there is a high clinical suspicion for MS and include a rigorous exclusion of other "MS mimickers" [3]. On the other hand, a delay in the proper diagnosis of an MS patient can lead to irreversible disability. A 25-year retrospective analysis using data from the New York State Consortium confirms that using the updated diagnostic criteria does help decrease the time to diagnosis [34].

1. A patient with a history of ≥ 2 typical neurological attacks with objective evidence of ≥ 2 clinically evident lesions in a location typical for Ms will need no additional supporting evidence for diagnosis of clinically definite MS (CDMS).
2. A patient with a history of ≥ 2 clinical attacks and one clear historically objective evidence of a clinically relevant lesion on MRI will also fit the diagnostic criteria.
3. The presence of ≥ 2 clinical attacks with evidence of one clinical lesion not in a distinct anatomical location may require addition paraclinical evidence with MRI or CSF-specific oligoclonal bands to fulfill the dissemination in time criteria.
4. Patients with only one clinical attack and ≥ 2 clinically evident lesions on examination may require one additional clinical attack or MRI evidence of a demyelinating lesion or CSF-specific oligoclonal bands to fulfill the criteria for MS by dissemination in time criteria.
5. Patients with one clinical attack and one objective lesion on examination will require additional paraclinical evidence by MRI for dissemination in space or another clinical attack in another anatomical location.
6. Importantly, in contrast to the 2010 criteria, patients could be diagnosed with MS earlier than with the 2010 criteria, with the 2017 criteria allowing (1) oligoclonal bands to be used for dissemination in time, (2) both symptomatic and asymptomatic lesions to be counted toward dissemination in space and time, and (3) cortical lesions to be used to demonstrate dissemination in space [3].
7. In turn, primary progressive MS was defined as patients presenting with at least 1 year or of disability progression independent of clinical relapses and two of the following three characteristics [3]:
    a. One or more T2 hyperintense lesions characteristic of MS in one or more of the following brain regions: periventricular, cortical or juxtacortical, or infratentorial
    b. Two or more T2 hyperintense lesions in the spinal cord
    c. Presence of CSF-specific oligoclonal bands

The current (2017) MS diagnostic criteria include the following:

Clearly, MRI is essential to deciphering the nature of the disease on presentation and to improve diagnosis and monitoring, several iterations of recommendations on MRI use in MS have also been written, with the most recent one being the 2021 MAGNIMS-CMSC-NAIMS consensus recommendations [35], which provide a blueprint for ideal MRI sequences, slice thickness and techniques that will provide the best diagnostic value. For brain imaging, 3 T, 3D T2 FLAIR with 1 mm isotropic slice thickness images are preferred, with either 1.5 or 3 T strength and ≤ 5 mm axial and ≤ 3 mm sagittal slice thickness considered optimal for spinal cord imaging [35].

Previously, routine use of gadolinium-based contrast agents for MS disease monitoring and diagnosis was widespread. Due to relatively recent recognition of gadolinium deposition in the CNS after multiple contrast administrations [36], the most recent MRI recommendations advise against the routine use of gadolinium in MS scans. In contrast, gadolinium contrast may play an important role in the initial diagnosis to help establish the dissemination in time and space criteria [35].

Precursors to multiple sclerosis: clinically isolated syndrome and radiologically isolated syndrome

In some cases, patients develop isolated acute demyelination events in the brain hemispheres, cerebellum, brainstem, ON, or spinal cord. Such events are classified as CIS if there is no indication of dissemination in time and space that would meet the McDonald criteria for MS (i.e., negative oligoclonal bands and no MRI evidence of MS). As many as 85% of CIS patients will go on to develop either another clinical attack or new MRI lesions that result in a change of their diagnosis to CDMS [37]. CIS patients are, therefore, followed regularly both clinically and with regular MRIs.

Past and ongoing research is trying to identify those CIS patients who are at high risk for conversion for Ms to help guide earlier therapy initiation and prevention of relapses and disability accumulation in these patients. It is apparent that CIS patients with younger age at symptom onset, abnormal MRI findings, those with supratentorial lesions on presentation with CIS, and those with polysymptomatic and motor presentations may be more likely to convert to MS [37,38]. Serum neurofilament light chains (NfL) levels may also be a biomarker for higher risk of conversion to CIS [39].

Alternatively, MRI studies of the CNS are done in some patients for reasons other than symptoms suspicious for MS such as trauma or headaches that occasionally demonstrate typical locations of likely MS lesions in the absence of clinical symptoms. Such presentations are termed RIS and anywhere between 24% and 59% of patients with RIS may convert to CDMS over time [40,41]. Independent risk factors for conversion to CDMS may include high NfL levels and the presence of the central vein sign [41,42].

Recommendations on the timing of imaging for monitoring of multiple sclerosis disease activity and progression

Brain imaging: Most of the lesions in MS occur within the brain, and less frequently in the cervical and thoracic spine. It is recommended that the clinician obtain a baseline MRI in every patient before starting or changing the disease modifying therapy (DMT). The new baseline MRI after initiation of a DMT does not require the use of gadolinium, unless there is indication of highly active disease at baseline, or if the patient demonstrates unexpected clinical activity (relapse). For monitoring the response to DMT, a yearly brain MRI without contrast can be used. This interval can be increased further if a patient's disease has been stable after a few years on the same DMT [35].

Spinal cord imaging: Routine regular monitoring for new lesions in the spinal cord is not recommended, as lesions in the spinal cord tend to be clinically eloquent. There is, therefore, no need for C-spine or T-spine MRI unless clinical disease activity, a relapse, or significant disk disease that localizes to the spinal cord is suspected [35] per the recommendations. Nonetheless, clinical judgment should be used when ordering cervical spine imaging, as clinically silent or minimally symptomatic spinal cord lesions can also occur (Fig. 3.1A and B).

Recommendations on the use of newer magnetic resonance imaging techniques for evaluation of multiple sclerosis

As mentioned earlier, MS is characterized not only by ongoing inflammation indicated by new T2 hyperintense and T1 gadolinium-enhancing lesions but also by progressive atrophy, with sometimes enlarging or disappearing T2 hyperintense lesions. Various MRI automated analysis tools are currently in development for the use of monitoring progressive MS disease activity. The most recent imaging criteria indicate the following:

1. *Leptomeningeal enhancement (LME)*: While the presence of LME may be associated with higher EDSS, higher cortical volumes in some studies [43], its current use in MS is not recommended, as it may also be seen in NMOSD, MOG, and other neuroinflammatory diseases [35].
2. *Central vein sign (CVS)*: The CVS may be a promising technique for predicting progression of RIS to MS and may have significant prognostic value [42]. However, its' routine clinically use is not recommended as the necessary 3D T2*-weighted segmented echo-planar images are not yet widely available on clinical scanners (Fig. 3.5).
3. *Paramagnetic rim*: A paramagnetic rim on MRI imaging may correspond to activated microglia and can be seen in slowly expanding progressive Ms lesions, even in the absence of relapsing MS disease activity. Susceptibility-weighted sequences at 3 T can identify such paramagnetic rim lesions in around 50% of patient with Ms. Using this MRI technique to assess progression in MS patients has not been validated yet for clinical use [42].
4. *Quantitative/volumetric MRI techniques*: Multiple MRI volumetric protocols for various brain and spinal cord/brainstem regions are being studied. The total brain volume measurements, however, are notoriously variable from scanner to scanner and may also be affected even just by the hydration status of the patient [32]. The imaging recommendations, therefore, state that there is currently insufficient evidence for the use of quantitative MRI techniques in the clinical setting [33,42].

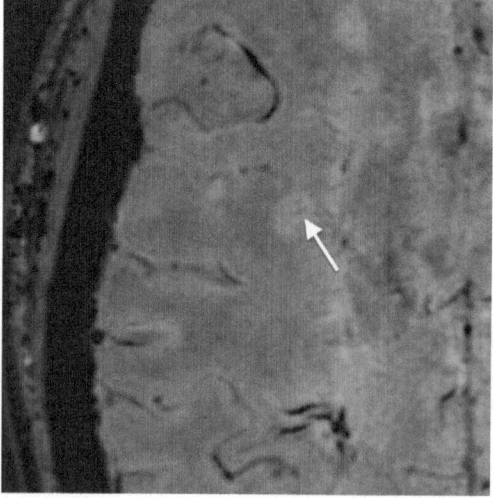

FIGURE 3.5 A clinical MRI demonstrating a faint central vein sign within a demyelinating lesion.

Summary

MS is an autoimmune neurologic disease which affects a wide age range of patients and can have variable clinical presentations, necessitating the use of ancillary testing to support or disprove the diagnosis of MS. A number of prodrome symptoms may be present years before the diagnosis. At this time prodromal symptoms for MS are recognized but are pleiomorphic and may apply to a myriad of other diseases. Genetic and environmental factors contribute to the risk of developing MS. The clinical course of MS can also differ in patients of different races and ethnicities. One must consider the typical age range for patients newly diagnosed with MS without excluding the outliers that are present, the pediatric and late onset MS patients. The updated 2017 McDonald Criteria helps with diagnostic sensitivity and does decrease the time to diagnosis. Accurate and timely diagnosis of MS is critical in preserving neurological function and staving off disability. One must exercise caution in using the updated McDonald as its' increased sensitivity may lead to misdiagnosis if misused and other MS mimickers are not considered.

MRI imaging has become an indispensable tool in the diagnosis and monitoring of MS disease. With further technological advances, various MRI techniques, such as brain and lesion volume measurements as well as SWI imaging for central vein sign, may eventually become incorporated into clinical use for MS as well. Serum biomarkers such as NfL are not specific enough to help with the diagnostic criteria for MS and can be also found in other MS mimickers. Cognitive dysfunction may be a sign of a prodrome, clinical relapse, or progression of disease. Other factors can contribute to a patient's cognitive dysfunction as well, so cognition alone cannot be solely relied upon. Due to the lack of a singular clinical tool to help aide in the diagnosis and monitoring of MS, advanced neuroimaging techniques play a critical role in patient care.

References

[1] Morrow SA, et al. Detecting isolated cognitive relapses in persons with MS. Mult Scler 2023;29(14):1786–94.
[2] Benedict RH, et al. Characterizing cognitive function during relapse in multiple sclerosis. Mult Scler 2014;20(13):1745–52.
[3] Thompson AJ, et al. Diagnosis of multiple sclerosis: 2017 revisions of the McDonald criteria. Lancet Neurol 2018;17(2):162–73.
[4] Benedict RHB, et al. Cognitive impairment in multiple sclerosis: clinical management, MRI, and therapeutic avenues. Lancet Neurol 2020;19(10):860–71.
[5] Kalb R, et al. Recommendations for cognitive screening and management in multiple sclerosis care. Mult Scler 2018;24(13):1665–80.
[6] Ward M, Goldman MD. Epidemiology and pathophysiology of multiple sclerosis. Contin (Minneap Minn) 2022;28(4):988–1005.
[7] Baecher-Allan C, Kaskow BJ, Weiner HL. Multiple sclerosis: mechanisms and immunotherapy. Neuron 2018;97(4):742–68.
[8] Marrie RA, et al. From the prodromal stage of multiple sclerosis to disease prevention. Nat Rev Neurol 2022;18(9):559–72.
[9] Vollmer TL, et al. Multiple sclerosis phenotypes as a continuum: the role of neurologic reserve. Neurol Clin Pract 2021;11(4):342–51.
[10] Lotti CBC, et al. Late onset multiple sclerosis: concerns in aging patients. Arq Neuropsiquiatr 2017;75(7):451–6.

[11] Naseri A, et al. Clinical features of late-onset multiple sclerosis: a systematic review and meta-analysis. Mult Scler Relat Disord 2021;50:102816.
[12] Vaughn CB, et al. Epidemiology and treatment of multiple sclerosis in elderly populations. Nat Rev Neurol 2019;15(6):329—42.
[13] Lublin FD, et al. Defining the clinical course of multiple sclerosis: the 2013 revisions. Neurology 2014;83 (3):278—86.
[14] MS International Federation, A.o.M. https://www.atlasofms.org/map/global/epidemiology/number-of-people-with-ms#about; 2023 [accessed May, 2023].
[15] Wallin MT, et al. The prevalence of MS in the United States: A population-based estimate using health claims data. Neurology 2019;92(10):e1029—40.
[16] Bjornevik K, et al. Longitudinal analysis reveals high prevalence of Epstein-Barr virus associated with multiple sclerosis. Science 2022;375(6578):296—301.
[17] Lanz TV, et al. Clonally expanded B cells in multiple sclerosis bind EBV EBNA1 and GlialCAM. Nature 2022;603(7900):321—7.
[18] Langer-Gould AM, et al. Racial and ethnic disparities in multiple sclerosis prevalence. Neurology 2022;98 (18):e1818—27.
[19] Goodin DS, et al. Genetic susceptibility to multiple sclerosis in African Americans. PLoS One 2021;16(8): e0254945.
[20] Ventura RE, et al. Hispanic Americans and African Americans with multiple sclerosis have more severe disease course than Caucasian Americans. Mult Scler 2017;23(11):1554—7.
[21] Amezcua L, McCauley JL. Race and ethnicity on MS presentation and disease course. Mult Scler 2020;26 (5):561—7.
[22] Ramagopalan SV, et al. Expression of the multiple sclerosis-associated MHC class II Allele HLA-DRB1*1501 is regulated by vitamin D. PLoS Genet 2009;5(2):e1000369.
[23] Patsopoulos NA. Genetics of multiple sclerosis: an overview and new directions. Cold Spring Harb Perspect Med 2018;8(7).
[24] Gresle MM, et al. Multiple sclerosis risk variants regulate gene expression in innate and adaptive immune cells. Life Sci Alliance 2020;3(7).
[25] Dyment DA, Sadovnick AD, Ebers GC. Genetics of multiple sclerosis. Hum Mol Genet 1997;6(10):1693—8.
[26] Compston A, Coles A. Multiple sclerosis. Lancet 2002;359(9313):1221—31.
[27] Sawcer S, Franklin RJ, Ban M. Multiple sclerosis genetics. Lancet Neurol 2014;13(7):700—9.
[28] Kuhlmann T, et al. Multiple sclerosis progression: time for a new mechanism-driven framework. Lancet Neurol 2023;22(1):78—88.
[29] Ontaneda D, Cohen JA, Sati P. Incorporating the central vein sign into the diagnostic criteria for multiple sclerosis. JAMA Neurol 2023;.
[30] Weiner HL. A shift from adaptive to innate immunity: a potential mechanism of disease progression in multiple sclerosis. J Neurol 2008;255(Suppl 1):3—11.
[31] Lassmann H, van Horssen J, Mahad D. Progressive multiple sclerosis: pathology and pathogenesis. Nat Rev Neurol 2012;8(11):647—56.
[32] Jakimovski D, et al. Late onset multiple sclerosis is associated with more severe ventricle expansion. Mult Scler Relat Disord 2020;46:102588.
[33] Matthews PM, D.G., Mittal D, Bai W, Scalfari A, Pollock KG, et al. The association between brain volume loss and disability in multiple sclerosis: a systematic review. Multiple Scler Relat Disord 2023;74.
[34] Jakimovski D, et al. Improvement in time to multiple sclerosis diagnosis: 25-year retrospective analysis from New York State MS Consortium (NYSMSC). Mult Scler 2023;29(6):753—6.
[35] Wattjes MP, et al. 2021 MAGNIMS-CMSC-NAIMS consensus recommendations on the use of MRI in patients with multiple sclerosis. Lancet Neurol 2021;20(8):653—70.
[36] Gulani V, et al. Gadolinium deposition in the brain: summary of evidence and recommendations. Lancet Neurol 2017;16(7):564—70.
[37] Çinar BP, Özakbaş S. Prediction of conversion from clinically isolated syndrome to multiple sclerosis according to baseline characteristics: a prospective study. Noro Psikiyatr Ars 2018;55(1):15—21.
[38] Novakova L, et al. Clinically isolated syndromes with no further disease activity suggestive of multiple sclerosis at the age of population life expectancy. Mult Scler 2014;20(4):496—500.

[39] Arrambide G, et al. Neurofilament light chain level is a weak risk factor for the development of MS. Neurology 2016;87(11):1076–84.
[40] Etemadifar M, et al. Conversion from radiologically isolated syndrome to multiple sclerosis. Int J Prev Med 2014;5(11):1379–86.
[41] Rival M, et al. Neurofilament light chain levels are predictive of clinical conversion in radiologically isolated syndrome. Neurol Neuroimmunol Neuroinflamm 2023;10(1).
[42] Suthiphosuwan S, et al. The central vein sign in radiologically isolated syndrome. AJNR Am J Neuroradiol 2019;40(5):776–83.
[43] Ineichen BV, et al. Leptomeningeal enhancement in multiple sclerosis and other neurological diseases: a systematic review and meta-analysis. Neuroimage Clin 2022;33:102939.

CHAPTER 4

Cognition in multiple sclerosis

Zachary L. Weinstock[1] and Ralph H.B. Benedict[2]

[1]Buffalo Neuroimaging Analysis Center (BNAC), Department of Neurology, Jacobs School of Medicine and Biomedical Sciences, University at Buffalo, State University of New York, Buffalo, NY, United States [2]Department of Neurology, Jacobs MS Center for Treatment and Research, Jacobs School of Medicine and Biomedical Sciences, University at Buffalo, State University of New York, Buffalo, NY, United States

OUTLINE

Introduction	51
A historical review of cognition in multiple sclerosis	52
Neuropsychological assessment	52
Cognitive dysfunction in multiple sclerosis	54
Acute cognitive decline as a marker of active disease in multiple sclerosis	58
Isolated cognitive relapse	59
Treatment of cognitive impairment in multiple sclerosis	61
Open questions and future directions	61
How can we optimally define and recognize cognitive change in people with multiple sclerosis?	61
What are the most useful magnetic resonance imaging markers for predicting progression of cognitive dysfunction and patient outcomes?	62
What determines the degree of cognitive change during and after multiple sclerosis relapse?	63
How should we translate lessons from the literature to clinical practice?	64
Conclusion	64
References	65

Introduction

Multiple sclerosis (MS) is a chronic autoimmune disorder affecting the central nervous system (CNS) characterized by inflammatory demyelination and neurodegeneration. Cognitive impairment (CI), although frequently unrecognized, is a common sign and symptom of MS that can markedly impact patients' independence, vocational status, and quality of life (QoL) [1]. While the clinical course of MS is different for each individual, an overwhelming majority (~83%) of people with MS (PwMS) are diagnosed with the relapsing-remitting MS (RRMS) phenotype, which is characterized by transient episodes of neurologic dysfunction (i.e., "relapses") followed by varying degrees of recovery (i.e., "remissions"). Given time and disease progression, RRMS may transition to a secondary-progressive MS (SPMS) marked by incomplete remissions as well as a steady decline in neurological functioning independent of relapse activity [2]. Therefore, CI may derive from incomplete recovery after relapse, gradual neurodegeneration, or some combination of the two. The former process, termed "cognitive relapse," is newly described in the literature and represents an area of active investigation [3–6]. Herein, we aim to provide an overview of MS-associated CI with a review of key neuropsychological tools and studies responsible for the now broad recognition of CI as a key facet of MS symptomatology. We also provide a summation of the literature describing cognitive relapse, and a discussion of how gaps in our knowledge might be addressed in future work.

A historical review of cognition in multiple sclerosis

For more than a century after Jean-Martin Charcot first noted "a marked enfeeblement of the memory" and "conceptions that are formed slowly" in his original description of MS in the late 19th century [7], little was known about MS-associated CI until researchers began to apply psychometric assessment in the 1980s. Using batteries of standardized neuropsychological tests, a few large-scale, controlled studies showed that CI in MS is far more common than previously thought [8,9]. In fact, we now know that CI impacts 45%–65% of patients, is only weakly correlated with physical disability, is more common in SPMS than in RRMS, and is frequently unrecognized in the absence of objective neuropsychological assessment [10,11]. Associations of CI with brain magnetic resonance imaging (MRI), as well as social, economic, and QoL problems are also widely recognized.

Neuropsychological assessment

In 1990, Peyser et al. made a first attempt at establishing guidelines for cognitive research in PwMS [12]. Amidst these guidelines was a recognition that comprehensive neuropsychological assessment lasted far too long (2–6 hours) to be compatible with routine clinical practice and therefore, the authors proposed an abbreviated, "core" battery of tests that would take no longer than 2 hours and facilitate comparison of findings across centers and studies. Soon after, a seminal study by Rao et al. would prove the wisdom of this goal when the authors demonstrated that just 4 indices from a similar battery of 23 tests offered comparable specificity and sensitivity in discriminating cognitively intact from impaired MS patients to that of the remaining measures combined [10]. Following minor modifications and the addition of the Symbol Digit Modalities Test (SDMT) [13] to augment assessment of cognitive processing speed, Rao devised the brief repeatable battery (BRB) for PwMS [14].

The BRB consists of the selective reminding test (SRT) [15], which measures verbal memory; the 10/36 spatial recall test (SPART) [14], which measures visuospatial memory; the SDMT and paced auditory serial addition test (PASAT) [16], which measure cognitive processing speed; and the controlled oral word association test (COWAT) [17], which measures verbal fluency. Taking less than 45 minutes to administer, the BRB was rapidly integrated into studies of cognition in MS and would soon become the most widely used neuropsychological assessment battery [18]. Although these tests were selected in part with assessment time in mind, the promise of the BRB became clearer as normative and preliminary data emerged. For example, Boringa et al. demonstrated weak-to-moderate correlations between the different tests, indicating that the BRB indeed measured different domains of cognition [19]. Furthermore, its validity was supported by studies demonstrating that performance on the BRB correlated with various measures derived from MRI [20,21]. Moreover, it revealed longitudinal cognitive change in MS [22], a finding which was reproduced across different languages and cultures [23,24]. Given these numerous advantages, the BRB was rapidly and widely adopted in MS research, and remains in use to this day [25].

Despite the many benefits afforded by the BRB, there were also several limitations that resulted in the proposal of alternative batteries. For example, Bever and colleagues were quick to recognize that longitudinal variability on the BRB could be confounded by practice effects as well as fatigue and affective state, as opposed to disease activity [26]. Furthermore, the BRB neglected key cognitive domains. As a result, in 2002 an expert panel was convened to identify a set of the most psychometrically sound measures that could be utilized for routine assessment of the principal features of MS-related CI in both the clinical and research settings. This consensus panel ultimately recommended a battery they named the minimal assessment of cognitive function in MS (MACFIMS) [27]. The MACFIMS builds upon the foundation of the BRB but with a few key changes. First, the 10/36 SPART and the SRT were replaced with the brief visuospatial memory test—revised (BVMT-R) [28] and the California verbal learning test—second edition (CVLT-II) [29], respectively. These substitutions were largely motivated by the new tests' relatively improved psychometric properties, especially with regard to alternate form equivalence and test-retest reliability. In addition to these substitutions, the MACFIMS also includes the Delis-Kaplan executive functioning system sorting test (DKEFS Sorting) [30], which measures executive functioning and the judgment of line orientation test (JLO) [31], which measures spatial processing. Subsequent research demonstrated a comparable sensitivity of the MACFIMS and BRB to overall cognitive status as well as an enhanced sensitivity of the BVMT-R relative to the 10/36 SPART [32]. However, despite comparing favorably to the BRB, the additional testing time (90 minutes) required by MACFIMS, along with a lack of non-English translations, impeded widespread adoption.

To address the limitations of MACFIMS, an international expert panel was convened in 2012 to develop a battery that would facilitate routine cognitive monitoring of MS patients around the world. The panel's objective was to recommend a brief clinical tool that could be administered by clinicians with minimal training in neuropsychology. Further, they sought to focus on tests with demonstrated reliability, validity, sensitivity, and specificity. Given these criteria, they ultimately recommended the brief international cognitive assessment for MS (BICAMS). With a test time of just 15 minutes, BICAMS measures processing speed, visual memory, and verbal memory using the SDMT, BVMT-R, and CVLT-II, respectively. The panel also produced a protocol for international adaptation and validation of the BICAMS tests [33]. In the years since, the BICAMS has been widely adapted (30 translation validation studies published or in progress, personal communication Dawn Langdon) and a meta-analysis of published national validation studies demonstrated that regardless of language or location, the BICAMS was capable of identifying cognitive dysfunction in each of the measured cognitive domains [34]. In fact, the tests chosen for the BICAMS are now widely regarded as the most sensitive tasks available for routine cognitive monitoring in MS [35]. In summary, the BICAMS represents a significant achievement in the quest for a broadly accessible, valid, and easily administered cognitive assessment tool for PwMS.

Among the various neuropsychological tests and batteries described above, it is worth highlighting the utility of the SDMT, which is the single test common to all three of the major cognitive batteries discussed here (see Table 4.1) [36]. Possessing superior psychometric properties (e.g., validity and reliability), as well as an unparalleled degree of sensitivity to cognitive change in PwMS, the SDMT is now widely recognized as the gold-standard assessment of cognitive function [37]. Additionally, several equivalent alternate forms are available to counter practice effects stemming from its increasingly routine use [38]. For these reasons, the SDMT is now used as an endpoint in clinical trials with work underway to incorporate it into the expanded disability status scale, the most common clinical measure of disability in MS [37,39–41].

Cognitive dysfunction in multiple sclerosis

As a direct result of the various standardized neuropsychological assessment tools developed over the past three decades, much has been learned about the prevalence of CI across different phenotypes, common profiles of cognitive dysfunction, the natural history

TABLE 4.1 The neuropsychological tests included in the brief repeatable battery [14], minimal assessment of cognitive function in multiple sclerosis [27], and the brief international cognitive assessment for multiple sclerosis [42].[a,b]

Cognitive domain	BRB (1990)	MACFIMS (2002)	BICAMS (2012)
Visual processing speed and working memory	SDMT [13]	SDMT [13]	SDMT [13]
Auditory processing speed and working memory	PASAT [16]	PASAT [16]	–
Auditory/verbal memory	SRT [15]	CVLT-II [29]	CVLT-II [29]
Visual/spatial memory	10/36 SPART [14]	BVMT-R [28]	BVMT-R [28]
Language	COWAT [17]	COWAT [17]	–
Spatial processing	–	JLO [31]	–
Executive Function	–	DKEFS Sorting [30]	–

[a]Reprinted from Weinstock ZL, Benedict RHB. Cognitive relapse in multiple sclerosis: new findings and directions for future research. Neuroscience 2022;3(3):513. Licensed under CC BY.
[b]In parenthesis next to each neuropsychological battery is the year in which it was published.

of cognitive dysfunction, and its real-world impact. Moreover, when combined with MRI technology, these tools have enabled scientists to better understand the neuropathological changes underlying observed behavioral and cognitive changes. Lastly, researchers are actively investigating factors which may protect against cognitive decline.

Defining cognitive impairment

How do we define CI? This key question must be addressed above all others, as it determines how neuropsychological data are interpreted in the clinic, as well as how subjects are grouped in cognitive studies (e.g., "cognitively impaired" vs "cognitively preserved"). The convention in neuropsychology is to consider an individual cognitively impaired if they perform worse than 1.5 standard deviations below demographic-adjusted normative expectations [1]. However, caveats abound. For example, alternative definitions of CI are not uncommon and likely contribute to differences in the reported prevalence of CI [43]. More conservative standards hold that an individual ought only be deemed "cognitively impaired" if they fall below this threshold on at least two different tests of the same cognitive domain. Conversely, a liberal approach might classify an individual as impaired if they perform worse than 1 standard deviation below normative expectations on just a single test. There are of course, costs and benefits to each of these different classification schemata. For example, the conservative approach enhances specificity at the cost of reduced sensitivity and an increased testing time. Meanwhile, a liberal definition of CI will likely capture numerous false positives.

Beyond mere differences regarding where a line is drawn in a z-distribution, there are inherent limitations to the use of z-scores based on normative data. In the clinical routine patients may be followed for years by a single provider. Here, change relative to baseline is most clinically relevant. Therefore there is a need for alternative means by which we might consider and interpret cognitive change over time. One such method was recently explored by Weinstock et. al. wherein the authors demonstrated the utility of a statistically derived reliable change threshold for the identification of longitudinal cognitive decline in MS patients [44]. Such an approach is unique in that it offers clinicians and researchers a means of evaluating an individual's evolving cognitive ability outside the somewhat rigid definition of "normal." However, it is also limited by its dependence on baseline—or at least longitudinal—data that is often unavailable.

In short, CI is not as easily defined as one might think. Although the neuropsychological convention is certainly the most common approach, the discerning reader should be aware of the method by which an investigator classifies cognitive status and further, how the different definitions might impact the reported results and conclusions. Further, careful review of longitudinal data should account for not only the level of performance compared to demographically matched controls, but also the degree of change from baseline using reliable change statistics that account for test reliability.

Prevalence and different phenotypes

As mentioned previously, the explosion of interest in MS-associated cognitive dysfunction that began in the 1990s stemmed from earlier observations that the prevalence of CI was substantially greater than previously expected, with an early review article placing the estimate at between 45% and 65% [8–10,45]. Although these early studies generally lacked the well validated testing batteries that exist today, modern investigations making use of such tools have supported this estimate [46]. Additional work has revealed more nuanced differences regarding the prevalence of CI in MS with respect to timing and the different phenotypes of MS. For example, although it is well established that cognitive dysfunction can occur in any stage of MS [47,48], it is now estimated that CI affects 20%–25% of patients with clinically or radiologically isolated syndrome, 30%–45% of patients with RRMS, and 50%–75% of patients with SPMS [1]. Although CI is understood to poorly correlate with physical disability, these estimates do lend support to the notion that more advanced brain pathology is associated with increasing odds of cognitive dysfunction [10]. While the prevalence of cognitive dysfunction in cases of pediatric-onset MS is less well studied, available literature reports cognitive decline in up to 56% of tested patients [49].

Common cognitive profiles

As with the motor and sensory symptoms of MS, CI is extremely heterogenous with highly variable presentations between patients. Taken as a whole, however, PwMS most frequently demonstrate deficits in the domains of cognitive processing speed as well as learning/memory [10,35]. On the MACFIMS test battery impaired processing speed was found in up to 51% of participants, compromised verbal memory in up to 34% of participants, and dysfunctional visual memory in up to 56% of participants [46]. Deficits in executive function and visuospatial processing were also reported at lesser frequencies of up to 28% and 22%, respectively [10,46]. Interestingly, these estimates of the prevalence of impaired cognitive domains roughly parallel the findings of a recent study by Wojcik and colleagues which used an event-based staging approach to suggest the following temporal sequencing of impairments: processing speed, visual learning, verbal learning, working memory/attention, and executive functioning [50]. The proposed order of CI may therefore explain in part the different patterns of CI observed in PwMS, and aligns with previous work showing that memory and executive function are rarely impaired independently of processing speed [51].

The consequences of cognitive impairment in multiple sclerosis

Early research revealed that up to 80% of PwMS were unemployed and that many patients lost their jobs within 5 years of diagnosis [52]. Although these data were initially assumed to result from patients' physical disturbances and/or age, LaRocca and colleagues demonstrated that such considerations accounted for less than 14% of observed variation in employment status among a large cohort of PwMS [53]. Shortly thereafter, Rao et al. showed that cognitively impaired PwMS were more likely to be unemployed than their cognitively preserved peers [54]. Linking these two studies, a group led by Beatty subsequently demonstrated that cognitive functioning variables explained 49% of observed variance in vocational status in PwMS [55]. Further, Morrow and colleagues found that longitudinal declines in processing speed and verbal memory were the most consistent predictors of negative employment changes [56]. Taken together these data indicate that CI is a primary driver of work disability in MS.

However, the impact of CI in MS extends beyond just work status. For example, related work has demonstrated that CI accounts for variance in activities of daily living [57], and correlates with poor QoL indicators, such as social activity and emotional well-being

[58,59], as well as other indicators of functional status [60,61]. Against this background and supported by the fact that 90% of MS diagnoses occur between the ages of 15 and 50 [55]—peak years for education, career development, and family building—the life-altering potential of CI in PwMS cannot be overstated.

Magnetic resonance imaging correlates of cognitive impairment in multiple sclerosis

MS is most commonly conceptualized as an autoimmune disorder targeting the myelin-rich white matter (WM) of the CNS [2,62]. As evidence to this fact, WM lesions visible on common MRI scan sequences are a primary component of the modern standards used to diagnose MS [63]. Early work revealed a moderate correlation of WM lesion burden with cognitive functioning [64], a finding that has often been replicated in the years since [65−67]. However, these studies also consistently signaled that lesion burden alone did not completely explain observed variance in cognitive performance among PwMS. More advanced MRI techniques which capture insidious changes to WM microstructure [68−70] as well as lesions affecting cortical myelin [67,71,72] succeeded in further explaining cognitive variability observed in PwMS, but these too paint an incomplete picture.

Approaching the problem from a different perspective, groups focusing on volumetric changes to the brain in MS have revealed that neurodegeneration is apparent in both cortical [66,67,73,74] and deep gray matter (GM) regions [69,75,76], and correlates with cognitive status independent of inflammatory WM pathology [77]. Of particular note is the cognitive relevance of thalamic atrophy, which has consistently demonstrated robust correlations with CI in both adult and pediatric MS populations [76,78−80]. In fact, thalamic atrophy may more strongly correlate with cognitive status than traditional measures like whole-brain atrophy or total lesion burden [76,80].

More recently, studies have increasingly focused on changes to the functional connectivity among GM structures in PwMS. Although such studies have noted altered connectivity patterns in conjunction with cognitive dysfunction, results are inconsistent with some regions demonstrating decreased functional connectivity whereas others demonstrate the opposite [81,82]. One possible explanation for this discrepancy suggests that timing and the degree of MS pathology may be the culprits [1]. Put simply, increased connectivity may indicate active compensation by neuronal resources early in disease whereas decreased connectivity may indicate depletion of neuronal reserves as the disease progresses. Taken as a whole, these data have ultimately led to the theory that pathological destabilization of brain networks is responsible for cognitive dysfunction in MS [83−85]. While these findings indicate that the field is likely closing in on the mechanisms responsible for CI in MS, an optimal set of clinically translatable measures of network integrity remains elusive. Furthermore, whether network destabilization can be managed and/or restored is an area of active research [1].

Factors protecting against cognitive impairment in multiple sclerosis

The concept of reserve originated in the field of Alzheimer's disease research which revealed that much like in MS [86,87], there were often discrepancies between observed pathology and cognitive performance [88]. This ultimately led to the twin theories of "brain reserve (BR)" [89] and "cognitive reserve (CR)" [90] which together propose that there are innate, heritable traits (e.g., maximal lifetime brain volume) as well as lifetime environmental factors (e.g., years of education), respectively which confer cognitive resilience against cerebral pathology. Although BR is considered almost entirely genetically predetermined and can therefore only assess relative risk, CR is believed to be a product of intellectually enriching experiences and therefore serves as a candidate for lifestyle modification to prevent or slow cognitive decline [35]. Notably, although evidence suggests a robust correlation of BR and CR [91], Sumowski and colleagues demonstrated independent effects of CR in both cross-sectional and longitudinal cohorts, strongly supporting the notion that cognitive decline in MS is influenced by modifiable lifestyle factors [92,93].

Personality and in particular, trait Conscientiousness is a similar protective factor. Conscientiousness, reflecting a proclivity to be well-organized and engage in goal-directed behavior, is associated with better cognitive performance [94,95] and employment [96,97]. Moreover, neuroimaging studies indicate that high conscientiousness correlates with reduced structural disruption [98–100], and through its synergy with CR [101], reduced functional disruption [102]. However, whether conscientiousness can be modified and/or targeted for clinical benefit remains an area of active investigation [103].

Acute cognitive decline as a marker of active disease in multiple sclerosis

The majority of the above studies assessed cognitive dysfunction in PwMS at times of clinical stability and comparing their performance with healthy controls. Next we consider how cognition may be affected during relapses in MS.

Before 2011, the concept of a "cognitive relapse" existed in only a handful of case reports and observational studies [104–107] lacking baseline cognitive data or control groups. The first controlled observation of cognitive change in relapse stemmed from the safety of Tysabri re-dosing and treatment (STRATA) [108]. The STRATA study was designed to evaluate the safety of natalizumab by following a cohort of former clinical trial participants for a year following reinitiation of treatment. One particular concern was progressive multifocal leukoencephalopathy (PML), a catastrophic demyelinating disorder caused by John Cunningham (JC) virus infection of an immunocompromised host [109]. Accordingly, patients were screened for cognitive and neurological changes before each monthly infusion. These examinations included the SDMT as well as the MS neuropsychological screening questionnaire (MSNQ) [110], which screens for self-reported cognitive difficulties. While no cases of PML were observed, 53 patients experienced MS relapse over the course of the study. Deeming these individuals "cases," Morrow et al. compared their raw SDMT scores from before, during, and after relapse with those from a selection of 115 clinically stable "controls" matched on demographics, baseline cognitive performance, and time from study initiation. Notably, the mean change in SDMT score between the visits immediately preceding and following an identified relapse was observed to differ significantly between cases (−1.2) and controls (+1.3). Further, performance recovered following relapse, signaling that the observed cognitive changes were transient and linked to disease activity status.

Morrow's work was limited by its post hoc and retrospective nature [108]. For example, the monthly visits were designed to coincide with drug administration rather than relapse onset and therefore, cases of relapse were often evaluated days if not weeks after a relapse had begun. Further, the design of the STRATA study neglected to account for other neurological and psychological measures that could influence measured cognitive performance such as fatigue and optic neuritis, wherein vision is affected by demyelination of the optic nerve.

Three years later, Benedict and colleagues addressed these limitations by studying relapsing MS patients prospectively [5]. Cognitive performance was measured at a clinically stable baseline, again on the day of relapse, and at 3-month follow-up. Furthermore, patients with neurological signs that could compromise cognitive performance such as optic neuritis were excluded from the study. When compared with a control group composed of matched clinically stable MS patients, only the relapsing group exhibited significant changes in SDMT performance between study timepoints. More specifically, the relapsing group's SDMT performance worsened by an average of nearly 4 points at relapse followed by a recovery of the same magnitude at 3-month follow-up. Meanwhile, the stable group trended towards modest improvement likely due to a practice effect. Therefore this work stands as the first demonstration of acute cognitive decline in a prospective study of relapse-adjacent cognitive change in PwMS.

Cases of cognitive relapse have been increasingly documented by other research groups (see Fig. 4.1). Most notable among these is a landmark study by McKay and colleagues which analyzed cognitive data from the Swedish MS Registry, a national database including nearly 4000 RRMS patients across a mean follow-up of over 10 years [111]. Because these data were sourced from a national registry rather than a controlled study with discrete timepoints, the authors broadly categorized each subject's available cognitive data as belonging to one of two periods based on disease activity status: "relapse" and "remission." The relapse period encompassed the time between 90 days prerelapse and 730 days (2 years) postrelapse, and was further divided into 10 discrete subperiods (90−61, 60−31, and 30−1 days prerelapse; 0−30, 31−60, 61−90, 91−180, 181−365, 366−550, and 551−730 days postrelapse). The remission period included all follow-up data outside this window. Using linear mixed models, the authors compared aggregate pre- and postrelapse SDMT data from the relapse periods to those from the remission period, which served as within-subjects reference points. This analysis revealed that cognitive performance didn't return to baseline levels until 18 months after relapse, suggesting that the cognitive effects of relapse last much longer than previously reported. Furthermore, consideration of available data from timepoints immediately preceding relapse (0−90 days prerelapse) revealed significant declines in SDMT performance as early as 30 days before the clinical recognition of relapse. The authors suggested that this observation may indicate a period of subclinical brain inflammation which could herald an impending relapse. Taken in the context of the literature, these data cement the status of acute cognitive decline as an important facet of active MS.

Isolated cognitive relapse

Shortly after the initial recognition of cognitive relapse by Morrow and Benedict, Pardini et al. reported that clinically meaningful declines of at least four points on the SDMT can occur in the absence of other neurological signs or symptoms [6]. Referring to these events as "isolated cognitive relapses" (ICRs), Pardini established a strict definition

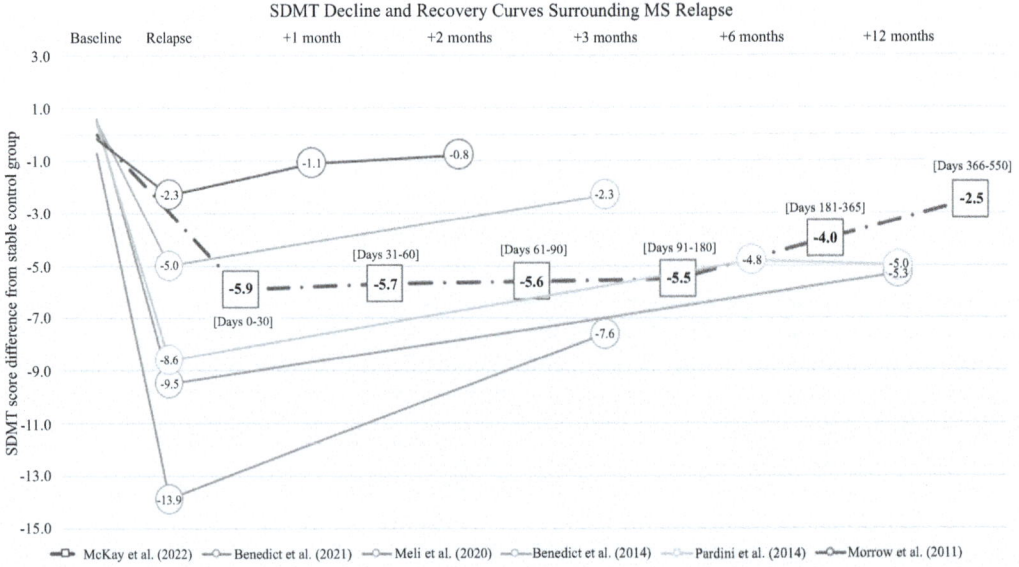

FIGURE 4.1 The SDMT decline and recovery curves surrounding MS relapse. The difference in raw SDMT scores between relapsing and stable MS patients across a variety of patient populations and timescales is shown. For each study, the mean score of the stable control group is subtracted from the mean score of the relapsing group. McKay et al. (bold text, red dashed line) represents the latest and most comprehensive study to date of cognitive change around MS relapse [111]. Because McKay used a national clinical database rather than a controlled study with discrete timepoints, each datapoint represents aggregate data from one of six distinct periods following relapse onset, indicated in brackets as the number of days postrelapse. Relapsing and stable groups were well matched at baseline across all studies with difference scores ranging from −0.7 to 0.6. Included for reference are data from studies of cognitive relapse described elsewhere in this paper such as those led by Benedict [5,112], Meli [3], Pardini [6], and Morrow [108]. Cognitive recovery after relapse exhibits marked variation across studies but is generally considered incomplete. *Credit: Reprinted from Weinstock ZL, Benedict RHB. Cognitive relapse in multiple sclerosis: new findings and directions for future research. Neuroscience 2022;3(3):513. Licensed under CC BY.*

requiring the following: (1) a clinically meaningful decline of at least four points on the SDMT; (2) the presence of gadolinium-enhancement on MRI indicating acute disease activity; and (3) an absence of clinical or subjective evidence of new neurological signs or symptoms. By these criteria, a retrospective analysis of 99 RRMS patients identified 17 cases of ICR. Notably, despite matching the cognitively stable subcohort on the basis of baseline SDMT performance and CR variables, the ICR group remained significantly worse at both 6- and 12-month follow-up. This finding of delayed/incomplete recovery has since been corroborated by other studies of cognitive relapse which taken together suggest that incomplete cognitive recovery after relapse may contribute to progressive cognitive decline in MS [1].

Interestingly, Pardini's original study of this phenomenon did not a reveal a significant decline by the ICR group on a self-report version of the MSNQ, which quantifies subjective impressions of cognitive decline [6]. This apparent discrepancy has since been used to question the purported existence of ICR by its critics [113]. It is worth noting however that in PwMS, self-reports of cognitive status are notoriously uncorrelated with brain MRI and cognitive testing, being driven more by mood than actual performance [114]. Indeed, a subsequent study by the same group reproduced Pardini's findings while also demonstrating a significant increase in cognitive difficulties as measured by an informant-report version of the MSNQ. Therefore, for the moment, the status of ICR as a true phenomenon remains a contested issue, awaiting further replication [113,115,116]. If confirmed, however, it may represent a fundamental shift in how we define active disease.

Treatment of cognitive impairment in multiple sclerosis

As discussed previously, CI in MS can profoundly impact the direction of patients' lives by disrupting educational pursuits, employment, and independence. However, there are as yet, no FDA approved medications for the treatment of cognitive dysfunction in MS. While a recent meta-analysis comparing the efficacy of first-line and disease-modifying therapies (DMTs) revealed small to moderate gains on the SDMT in groups treated with DMTs, the effects were so minor that cognitive considerations were not recommended for clinical decision-making [117].

There is, however, stronger evidence supporting the efficacy of nonpharmacological interventions such as cognitive training [1,118]. Broadly speaking, cognitive training can be classified as either restorative or compensatory. Restorative training targets specific cognitive domains with repetitive training exercises whereas compensatory approaches focus on the development of strategies to enhance cognitive functioning. Notably, both methods demonstrate medium-to-large effect sizes [119,120], with observed cognitive improvements related to improved functional connectivity in relevant brain networks [121,122]. Because the response to such therapies varies considerably between individuals [123], and interventions require a substantial time investment by both patients and providers, it is worth considering whether a patient is a good candidate for treatment. To this end, preliminary work suggests that measures of reserve may predict the patient response to cognitive rehabilitation, but further investigation is warranted [124]. If reserve does indeed influence treatment response, it would underscore a need for early cognitive intervention, when levels of reserve are intact.

Open questions and future directions

How can we optimally define and recognize cognitive change in people with multiple sclerosis?

CI is a complex, multidimensional phenomenon that eludes simple description. One approach is to diagnose neurocognitive disorder by comparing test scores to normative expectations. Another approach is to compare current performance to a previously acquired baseline. In the clinic setting, these approaches are viewed as complimentary. Weinstock and colleagues provide a threshold marking statistically reliable decline on SDMT [44]. However, no such cut-offs exist for other tests. Future research should therefore seek to define optimal thresholds for change across all of the BICAMS measures that are at once, statistically reliable and clinically meaningful as recommended by Benedict and Walton [125]. In support of this notion, a recent report by Benedict et al. focusing on cognitive recovery after relapse revealed a trend towards significant improvement on the CVLT-II by a group of relapsing MS subjects when compared to their clinically stable counterparts [112]. Similarly, a group led by Giedraitiene—while lacking baseline performance data—reported significant increases in mean performance across all three BICAMS tests by a cohort of MS patients assessed at relapse and 1-month follow-up [4]. These data emphasize the continued importance of testing batteries which touch upon a number of cognitive domains to more richly characterize cognitive relapse and perhaps, improve our understanding of how specific brain insults impact specific cognitive functions.

What are the most useful magnetic resonance imaging markers for predicting progression of cognitive dysfunction and patient outcomes?

As discussed previously, there are a number of different MRI techniques capable of measuring chronic and acute MS neuropathology affecting brain structure and function. However, the prognostic value of these measures, especially with respect to the cognitive component of MS, is limited. For example, although Eijlers et al. recently demonstrated that whole-brain cortical GM volume and regional MRI measures such as temporal atrophy were significant predictors of cognitive decline at follow-up, their best model explained just 35% of observed variance among subjects [126]. Therefore further research is clearly needed in order to define an optimal set of MRI markers which might indicate a patient's probable cognitive trajectory, quantify the degree of reserve that remains, and/or assess a patient's candidacy for interventions such as cognitive rehabilitation.

One promising avenue which may provide additional prognostic value is that of paramagnetic rim lesions (PRLs) seen using a specialized MRI technique known as quantitative susceptibility mapping. PRLs, which histopathologically correlate with thin layers of activated microglia at lesions edges, are generally understood to highlight areas of chronic, "smoldering" demyelination [127]. Evidence suggests that an increased number of PRLs on MRI indicates cases of more severe disease; correlating with greater lesion burden and brain atrophy, as well as a reduction in the age at which physical and cognitive disability is reached [128].

Furthermore, Marcille et al. recently demonstrated that the presence of at least one PRL may modify the impact of overall lesion burden on disability [129]. Additional prospective studies are needed to verify whether and to what degree PRLs may predict future cognitive deterioration and if so, whether such information is clinically actionable.

Another area of active research concerns neurologic reserve, which conceptualizes the brain's ability to buffer new or evolving pathology [130]. Both cognitive and BR metrics are traditionally conceptualized in terms of their maximal potential, with the former commonly estimated by an individual's total years of education whereas the latter is often estimated by intracranial volume which serves as a proxy for maximal lifetime brain volume [92]. Although these indices have proven useful in explaining some of the observed cognitive variance in PwMS, especially in its early stages when levels of reserve are largely intact, they fail to disclose how much reserve remains [93,131]. Quantifying remaining reserve is an important and clinically meaningful goal for future research as it may inform treatment decisions (e.g., aggressiveness of DMT) or perhaps, indicate whether a patient is likely to respond to cognitive rehabilitation [132]. Although there is as yet, no consensus on whether and how real-time levels of reserve might be measured, Weinstock et al. recently demonstrated that skull-size-normalized thalamic volume at the time of relapse significantly predicted cognitive recovery after relapse [133]. Further research should seek to understand exactly how reserve might be measured in "real-time" and whether reserve-based clinical decisions improve patient outcomes.

What determines the degree of cognitive change during and after multiple sclerosis relapse?

As noted previously, differences in cognitive performance between relapsing and stable MS cohorts appears to vary widely between studies. For example, in their initial recognition of cognitive relapse, Morrow reported that just 2.3 raw score points on the SDMT separated the relapsing and stable participants [108]. Meanwhile, Pardini noted that an impressive 13.9 raw score points separated cases of ICR from their clinically stable peers [6]. This raises an important question—what influences the severity of cognitive decline during relapse? Perhaps the clearest explanation derives from contrast-enhancing (CE) WM lesions, which highlight areas of acute inflammatory activity and blood–brain barrier breakdown. Several reports demonstrate moderate correlations of contrast-enhancement with CI, even in the absence of other neurological signs or symptoms (i.e., ICR) [134–136]. Although the exact mechanism underlying these observations is unclear, one hypothesis suggests that CE lesions disconnect specific GM regions from one another, precipitating dysfunction across associated brain networks [137]. Further data are needed, however, to determine whether the cognitive impact of such acute pathologies might be exacerbated by the relative burden of chronic pathologies, such as T1 hypointense lesions and GM atrophy. Additionally, we must consider how the impacts of both acute and chronic pathologies might be attenuated by protective factors such as a neurologic reserve [92]. For example, although CR is thought to reflect a gradually adaptive process [130], it is as yet unknown if reserve moderates acute cognitive decline during relapse.

Cognitive recovery also varies greatly between studies, with some cohorts returning to baseline function within a few months [5,108] whereas others demonstrate sustained CI well over a year after relapse [111]. However, it remains unclear which factors drive cognitive recovery after relapse. A role for neurologic reserve is suggested by Benedict et al. in a recent study which found that of many clinical, demographic, and MRI measures, only years of education—a proxy of CR—significantly predicted SDMT change after relapse [112]. It is worth noting however, that even when including trending variables, the authors could account for only 23% of observed variance. Therefore there is a need for further investigation of reserve-associated variables such as structural and functional connectivity, brain atrophy, and a more nuanced estimate of CR as offered by the cognitive reserve index questionnaire [138]. Although the clinical relevance of more accurately predicting cognitive recovery after relapse is yet to be explored, the scientific import of this goal is clear. Briefly, if the factors responsible for cognitive recovery can be identified, we will better understand what is necessary for proper cognitive functioning as well as how the brain adapts to acute and chronic injury. Such knowledge could perhaps, extend beyond the scope of MS and be applied towards similar conditions such as Alzheimer's, traumatic brain injury, and stroke.

How should we translate lessons from the literature to clinical practice?

Reports continue to advocate for a role of cognitive assessment in defining MS disease activity and progression. For example, one recent study revealed worsening in at least two cognitive domains by nearly 60% of purportedly stable MS patients who met criteria for no evidence of disease activity status [139,140]. This finding clearly demonstrates the inadequacy of current clinical standards to recognize all dimensions of disease progression which may in turn, delay offer of crucial interventions such as cognitive rehabilitation [1]. In fact, a consensus panel recently recommended a minimum of baseline SDMT screening and annual re-assessment for all MS patients older than 8 years of age [141]. These recommendations coincide with those from the MS Outcomes Assessment Consortium which advocate for a role of the SDMT as an endpoint in clinical trials to judge the cognitive impact of emerging DMTs [37]. In brief, the translation of routine neuropsychological assessment to clinical practice is sorely needed if we are to better detect acute disease activity, more accurately track disease progression, and fully appreciate treatment response.

Conclusion

In conclusion, cognition is clearly a common and impactful facet of MS symptomatology. Following decades of work by clinicians and scientists, baseline and routine cognitive assessments are now recommended for clinical uses [141]. Further, there exists a rapidly expanding body of evidence supporting the existence of cognitive relapse which may occur independently of other neurological signs or symptoms. Taken as a whole, available data suggest that cognition is an independent, often unrecognized component of the MS clinical picture and that objective evidence of cognitive decline indicates active disease, regardless of conventional neurological signs. This presents a strong case for the rapid and widespread translation of cognitive screening to clinical routine as it will empower clinicians to more adequately track disease activity and progression as well as treatment response. There are, however, glaringly large gaps in our understanding of cognition in MS that must be addressed. Notable among these are the lack of knowledge concerning how to best define and recognize cognitive change, the neuropathology driving cognitive relapse and recovery, and the effect of relapse-adjacent cognitive change on long-term cognitive trajectories. Against this background, additional large-scale prospective studies of cognitive relapse utilizing MRI and evidence-supported cognitive assessment are indicated.

References

[1] Benedict RHB, Amato MP, DeLuca J, Geurts JJG. Cognitive impairment in multiple sclerosis: clinical management, MRI, and therapeutic avenues. Lancet Neurol 2020;19:860–71.

[2] Lublin FD, Reingold SC, Cohen JA, et al. Defining the clinical course of multiple sclerosis: the 2013 revisions. Neurology 2014;83:278–86.

[3] Meli R, Roccatagliata L, Capello E, et al. Ecological impact of isolated cognitive relapses in MS. Mult Scler 2020;26:114–17.

[4] Giedraitiene N, Kaubrys G, Kizlaitiene R. Cognition during and after multiple sclerosis relapse as assessed with the brief international cognitive assessment for multiple sclerosis. Sci Rep 2018;8:8169.

[5] Benedict RH, Morrow S, Rodgers J, et al. Characterizing cognitive function during relapse in multiple sclerosis. Mult Scler 2014;20:1745–52.

[6] Pardini M, Uccelli A, Grafman J, Yaldizli O, Mancardi G, Roccatagliata L. Isolated cognitive relapses in multiple sclerosis. J Neurol Neurosurg Psychiatry 2014;85:1035–7.

[7] Charcot JM. Lectures on the diseases of the nervous system: delivered at La Salpetriere. Philadelphia, PA, USA: H. C. Lea; 1879.

[8] Rao SM. Neuropsychology of multiple sclerosis. Curr Opin Neurol 1995;8:216–20.

[9] McIntosh-Michaelis SA, Roberts MH, Wilkinson SM, et al. The prevalence of cognitive impairment in a community survey of multiple sclerosis. Br J Clin Psychol 1991;30:333–48.

[10] Rao SM, Leo GJ, Bernardin L, Unverzagt F. Cognitive dysfunction in multiple sclerosis. I. Frequency, patterns, and prediction. Neurology 1991;41:685–91.

[11] Potagas C, Giogkaraki E, Koutsis G, et al. Cognitive impairment in different MS subtypes and clinically isolated syndromes. J Neurol Sci 2008;267:100–6.

[12] Peyser JM, Rao SM, LaRocca NG, Kaplan E. Guidelines for neuropsychological research in multiple sclerosis. Arch Neurol 1990;47:94–7.

[13] Smith A. Symbol digit modalities test (SDMT) manual (revised). Los Angeles, CA: Western Psychological Services; 1982.

[14] Rao SMthe Cognitive Function Study Group of the National Multiple Sclerosis Society. A manual brief repeatable battery neuropsychological tests in multiple sclerosis. Milwaukee, WI: Medical College of Wisconsin; 1990.

[15] Buschke H, Fuld PA. Evaluating storage, retention, and retrieval in disordered memory and learning. Neurology 1974;24:1019–25.

[16] Gronwall DM. Paced auditory serial-addition task: a measure of recovery from concussion. Percept Mot Skills 1977;44:367–73.

[17] Benton AL, Hamsher K, Sivan AB. Multilingual aphasia examination. Iowa City, IA: AJA associates; 1994.

[18] Amato MP, Portaccio E, Goretti B, et al. The Rao's brief repeatable battery and stroop test: normative values with age, education and gender corrections in an Italian population. Mult Scler 2006;12:787–93.

[19] Boringa JB, Lazeron RH, Reuling IE, et al. The brief repeatable battery of neuropsychological tests: normative values allow application in multiple sclerosis clinical practice. Mult Scler 2001;7:263–7.

[20] Amato MP, Bartolozzi ML, Zipoli V, et al. Neocortical volume decrease in relapsing-remitting MS patients with mild cognitive impairment. Neurology 2004;63:89–93.

[21] Christodoulou C, Krupp LB, Liang Z, et al. Cognitive performance and MR markers of cerebral injury in cognitively impaired MS patients. Neurology 2003;60:1793–8.

[22] Amato MP, Ponziani G, Siracusa G, Sorbi S. Cognitive dysfunction in early-onset multiple sclerosis: a reappraisal after 10 years. Arch Neurol 2001;58:1602–6.

[23] Camp SJ, Stevenson VL, Thompson AJ, et al. A longitudinal study of cognition in primary progressive multiple sclerosis. Brain 2005;128:2891–8.

[24] Sepulcre J, Vanotti S, Hernandez R, et al. Cognitive impairment in patients with multiple sclerosis using the brief repeatable battery-neuropsychology test. Mult Scler 2006;12:187–95.

[25] Jandric D, Parker GJM, Haroon H, Tomassini V, Muhlert N, Lipp I. A tractometry principal component analysis of white matter tract network structure and relationships with cognitive function in relapsing-remitting multiple sclerosis. Neuroimage Clin 2022;34:102995.

[26] Bever Jr. CT, Grattan L, Panitch HS, Johnson KP. The brief repeatable battery of neuropsychological tests for multiple sclerosis: a preliminary serial study. Mult Scler 1995;1:165–9.

[27] Benedict RH, Fischer JS, Archibald CJ, et al. Minimal neuropsychological assessment of MS patients: a consensus approach. Clin Neuropsychol 2002;16:381−97.
[28] Benedict RH. The brief visuospatial memotry test revised (BVMT-R). Lutz, FL: Psychosocial Assessment Resources Inc.; 1997.
[29] Delis D, Kramer J, Kaplan E, Ober B. California verbal learning test, second edition (CVLT-II). San Antonio, TX: Psychological Corporation; 2000.
[30] Delis DC, Kaplan E, Kramer JH. Delis-Kaplan executive function system (D-KEFS): examiner's manual. San Antonio, TX: The Psychological Corporation; 2001.
[31] Benton AL, Sivan AB, Hamsher K, Varney NR, Spreen O. Contributions to neuropsychological assessment: a clinical manual. New York, NY: Oxford University Press; 1994.
[32] Strober L, Englert J, Munschauer F, Weinstock-Guttman B, Rao S, Benedict RH. Sensitivity of conventional memory tests in multiple sclerosis: comparing the Rao brief repeatable neuropsychological battery and the minimal assessment of cognitive function in MS. Mult Scler 2009;15:1077−84.
[33] Benedict RH, Amato MP, Boringa J, et al. Brief international cognitive assessment for MS (BICAMS): international standards for validation. BMC Neurol 2012;12:55.
[34] Corfield F, Langdon D. A systematic review and meta-analysis of the brief cognitive assessment for multiple sclerosis (BICAMS). Neurol Ther 2018;7:287−306.
[35] Sumowski JF, Benedict R, Enzinger C, et al. Cognition in multiple sclerosis: state of the field and priorities for the future. Neurology 2018;90:278−88.
[36] Weinstock ZL, Benedict RHB. Cognitive relapse in multiple sclerosis: new findings and directions for future research. Neuroscience 2022;3:510−20.
[37] Benedict RH, DeLuca J, Phillips G, et al. Validity of the symbol digit modalities test as a cognition performance outcome measure for multiple sclerosis. Mult Scler 2017;23:721−33.
[38] Benedict RH, Smerbeck A, Parikh R, Rodgers J, Cadavid D, Erlanger D. Reliability and equivalence of alternate forms for the symbol digit modalities test: implications for multiple sclerosis clinical trials. Mult Scler 2012;18:1320−5.
[39] Morrow SA, Conway D, Fuchs T, et al. Quantifying cognition and fatigue to enhance the sensitivity of the EDSS during relapses. Mult Scler 2021;27:1077−87.
[40] Strober L, DeLuca J, Benedict RH, et al. Symbol digit modalities test: a valid clinical trial endpoint for measuring cognition in multiple sclerosis. Mult Scler 2019;25:1781−90.
[41] Sacca F, Costabile T, Carotenuto A, et al. The EDSS integration with the brief international cognitive assessment for multiple sclerosis and orientation tests. Mult Scler 2017;23:1289−96.
[42] Langdon DW, Amato MP, Boringa J, et al. Recommendations for a brief international cognitive assessment for multiple sclerosis (BICAMS). Mult Scler 2012;18:891−8.
[43] Jak AJ, Bondi MW, Delano-Wood L, et al. Quantification of five neuropsychological approaches to defining mild cognitive impairment. Am J Geriatr Psychiatry 2009;17:368−75.
[44] Weinstock Z, Morrow S, Conway D, et al. Interpreting change on the symbol digit modalities test in people with relapsing multiple sclerosis using the reliable change methodology. Mult Scler 2022;28:1101−11.
[45] Peyser JM, Edwards KR, Poser CM, Filskov SB. Cognitive function in patients with multiple sclerosis. Arch Neurol 1980;37:577−9.
[46] Benedict RH, Cookfair D, Gavett R, et al. Validity of the minimal assessment of cognitive function in multiple sclerosis (MACFIMS). J Int Neuropsychol Soc 2006;12:549−58.
[47] Glanz BI, Holland CM, Gauthier SA, et al. Cognitive dysfunction in patients with clinically isolated syndromes or newly diagnosed multiple sclerosis. Mult Scler 2007;13:1004−10.
[48] Benedict RHB, DeLuca J, Enzinger C, Geurts JJG, Krupp LB, Rao SM. Neuropsychology of multiple sclerosis: looking back and moving forward. J Int Neuropsychol Soc 2017;23:832−42.
[49] Amato MP, Goretti B, Ghezzi A, et al. Neuropsychological features in childhood and juvenile multiple sclerosis: five-year follow-up. Neurology 2014;83:1432−8.
[50] Wojcik C, Fuchs TA, Tran H, et al. Staging and stratifying cognitive dysfunction in multiple sclerosis. Mult Scler 2022;28:463−71.
[51] Leavitt VM, Wylie G, Krch D, Chiaravalloti N, DeLuca J, Sumowski JF. Does slowed processing speed account for executive deficits in multiple sclerosis? Evidence from neuropsychological performance and structural neuroimaging. Rehabil Psychol 2014;59:422−8.

[52] Kornblith AB, La Rocca NG, Baum HM. Employment in individuals with multiple sclerosis. Int J Rehabil Res 1986;9:155−65.
[53] Larocca N, Kalb R, Scheinberg L, Kendall P. Factors associated with unemployment of patients with multiple sclerosis. J Chronic Dis 1985;38:203−10.
[54] Rao SM, Leo GJ, Ellington L, Nauertz T, Bernardin L, Unverzagt F. Cognitive dysfunction in multiple sclerosis. II. Impact on employment and social functioning. Neurology 1991;41:692−6.
[55] Beatty WW, Blanco CR, Wilbanks SL, Paul RH, Hames KA. Demographic, clinical, and cognitive characteristics of multiple sclerosis patients who continue to work. Neurorehabil Neural Repair 1995;9:167−73.
[56] Morrow SA, Drake A, Zivadinov R, Munschauer F, Weinstock-Guttman B, Benedict RH. Predicting loss of employment over three years in multiple sclerosis: clinically meaningful cognitive decline. Clin Neuropsychol 2010;24:1131−45.
[57] Kessler HR, Cohen RA, Lauer K, Kausch DF. The relationship between disability and memory dysfunction in multiple sclerosis. Int J Neurosci 1992;62:17−34.
[58] Benito-Leon J, Morales JM, Rivera-Navarro J. Health-related quality of life and its relationship to cognitive and emotional functioning in multiple sclerosis patients. Eur J Neurol 2002;9:497−502.
[59] Cutajar R, Ferriani E, Scandellari C, et al. Cognitive function and quality of life in multiple sclerosis patients. J Neurovirol 2000;6(Suppl 2):S186−90.
[60] Goverover Y, Chiaravalloti N, DeLuca J. Brief international cognitive assessment for multiple sclerosis (BICAMS) and performance of everyday life tasks: actual reality. Mult Scler 2016;22:544−50.
[61] Goverover Y, Strober L, Chiaravalloti N, DeLuca J. Factors that moderate activity limitation and participation restriction in people with multiple sclerosis. Am J Occup Ther 2015;69 6902260020p-9.
[62] Murray TJ. Diagnosis and treatment of multiple sclerosis. BMJ 2006;332:525−7.
[63] Thompson AJ, Banwell BL, Barkhof F, et al. Diagnosis of multiple sclerosis: 2017 revisions of the McDonald criteria. Lancet Neurol 2018;17:162−73.
[64] Rao SM, Leo GJ, Haughton VM, St Aubin-Faubert P, Bernardin L. Correlation of magnetic resonance imaging with neuropsychological testing in multiple sclerosis. Neurology 1989;39:161−6.
[65] Amato MP, Hakiki B, Goretti B, et al. Association of MRI metrics and cognitive impairment in radiologically isolated syndromes. Neurology 2012;78:309−14.
[66] Benedict RH, Bruce JM, Dwyer MG, et al. Neocortical atrophy, third ventricular width, and cognitive dysfunction in multiple sclerosis. Arch Neurol 2006;63:1301−6.
[67] Calabrese M, Agosta F, Rinaldi F, et al. Cortical lesions and atrophy associated with cognitive impairment in relapsing-remitting multiple sclerosis. Arch Neurol 2009;66:1144−50.
[68] Benedict RH, Bruce J, Dwyer MG, et al. Diffusion-weighted imaging predicts cognitive impairment in multiple sclerosis. Mult Scler 2007;13:722−30.
[69] Preziosa P, Rocca MA, Pagani E, et al. Structural MRI correlates of cognitive impairment in patients with multiple sclerosis: a multicenter study. Hum Brain Mapp 2016;37:1627−44.
[70] Rocca MA, Iannucci G, Rovaris M, Comi G, Filippi M. Occult tissue damage in patients with primary progressive multiple sclerosis is independent of T2-visible lesions − a diffusion tensor MR study. J Neurol 2003;250:456−60.
[71] Calabrese M, De Stefano N, Atzori M, et al. Detection of cortical inflammatory lesions by double inversion recovery magnetic resonance imaging in patients with multiple sclerosis. Arch Neurol 2007;64:1416−22.
[72] Calabrese M, Favaretto A, Martini V, Gallo P. Grey matter lesions in MS: from histology to clinical implications. Prion 2013;7:20−7.
[73] Amato MP, Razzolini L, Goretti B, et al. Cognitive reserve and cortical atrophy in multiple sclerosis: a longitudinal study. Neurology 2013;80:1728−33.
[74] Calabrese M, Rinaldi F, Mattisi I, et al. Widespread cortical thinning characterizes patients with MS with mild cognitive impairment. Neurology 2010;74:321−8.
[75] Bisecco A, Rocca MA, Pagani E, et al. Connectivity-based parcellation of the thalamus in multiple sclerosis and its implications for cognitive impairment: a multicenter study. Hum Brain Mapp 2015;36:2809−25.
[76] Houtchens MK, Benedict RH, Killiany R, et al. Thalamic atrophy and cognition in multiple sclerosis. Neurology 2007;69:1213−23.
[77] Benedict RH, Hulst HE, Bergsland N, et al. Clinical significance of atrophy and white matter mean diffusivity within the thalamus of multiple sclerosis patients. Mult Scler 2013;19:1478−84.

[78] Minagar A, Barnett MH, Benedict RH, et al. The thalamus and multiple sclerosis: modern views on pathologic, imaging, and clinical aspects. Neurology 2013;80:210—19.
[79] Batista S, Zivadinov R, Hoogs M, et al. Basal ganglia, thalamus and neocortical atrophy predicting slowed cognitive processing in multiple sclerosis. J Neurol 2012;259:139—46.
[80] Till C, Ghassemi R, Aubert-Broche B, et al. MRI correlates of cognitive impairment in childhood-onset multiple sclerosis. Neuropsychology 2011;25:319—32.
[81] Rocca MA, Valsasina P, Absinta M, et al. Default-mode network dysfunction and cognitive impairment in progressive MS. Neurology 2010;74:1252—9.
[82] Tona F, Petsas N, Sbardella E, et al. Multiple sclerosis: altered thalamic resting-state functional connectivity and its effect on cognitive function. Radiology 2014;271:814—21.
[83] Schoonheim MM, Meijer KA, Geurts JJ. Network collapse and cognitive impairment in multiple sclerosis. Front Neurol 2015;6:82.
[84] Tewarie P, Steenwijk MD, Tijms BM, et al. Disruption of structural and functional networks in long-standing multiple sclerosis. Hum Brain Mapp 2014;35:5946—61.
[85] Welton T, Constantinescu CS, Auer DP, Dineen RA. Graph theoretic analysis of brain connectomics in multiple sclerosis: reliability and relationship with cognition. Brain Connect 2020;10:95—104.
[86] Chiaravalloti ND, DeLuca J. Cognitive impairment in multiple sclerosis. Lancet Neurol 2008;7:1139—51.
[87] Filippi M, Rocca MA, Benedict RH, et al. The contribution of MRI in assessing cognitive impairment in multiple sclerosis. Neurology 2010;75:2121—8.
[88] Bennett DA, Schneider JA, Arvanitakis Z, et al. Neuropathology of older persons without cognitive impairment from two community-based studies. Neurology 2006;66:1837—44.
[89] Satz P. Brain reserve capacity on symptom onset after brain injury: a formulation and review of evidence for threshold theory. Neuropsychology 1993;7:273—95.
[90] Stern Y. What is cognitive reserve? Theory and research application of the reserve concept. J Int Neuropsychol Soc 2002;8:448—60.
[91] Deary IJ, Penke L, Johnson W. The neuroscience of human intelligence differences. Nat Rev Neurosci 2010;11:201—11.
[92] Sumowski JF, Rocca MA, Leavitt VM, et al. Brain reserve and cognitive reserve in multiple sclerosis: what you've got and how you use it. Neurology 2013;80:2186—93.
[93] Sumowski JF, Rocca MA, Leavitt VM, et al. Brain reserve and cognitive reserve protect against cognitive decline over 4.5 years in MS. Neurology 2014;82:1776—83.
[94] Fuchs TA, Wojcik C, Wilding GE, et al. Trait conscientiousness predicts rate of longitudinal SDMT decline in multiple sclerosis. Mult Scler 2020;26:245—52.
[95] Roy S, Drake AS, Eizaguirre MB, et al. Trait neuroticism, extraversion, and conscientiousness in multiple sclerosis: link to cognitive impairment? Mult Scler 2018;24:205—13.
[96] Jaworski 3rd MG, Fuchs TA, Dwyer MG, et al. Conscientiousness and deterioration in employment status in multiple sclerosis over 3 years. Mult Scler 2021;27:1125—35.
[97] Strober LB, Christodoulou C, Benedict RH, et al. Unemployment in multiple sclerosis: the contribution of personality and disease. Mult Scler 2012;18:647—53.
[98] Benedict RH, Hussein S, Englert J, et al. Cortical atrophy and personality in multiple sclerosis. Neuropsychology 2008;22:432—41.
[99] Fuchs TA, Benedict RH, Wilding G, et al. Trait conscientiousness predicts rate of brain atrophy in multiple sclerosis. Mult Scler 2020;26:1433—6.
[100] Fuchs TA, Dwyer MG, Kuceyeski A, et al. White matter tract network disruption explains reduced conscientiousness in multiple sclerosis. Hum Brain Mapp 2018;39:3682—90.
[101] Roy S, Schwartz CE, Duberstein P, et al. Synergistic effects of reserve and adaptive personality in multiple sclerosis. J Int Neuropsychol Soc 2016;22:920—7.
[102] Fuchs TA, Benedict RHB, Bartnik A, et al. Preserved network functional connectivity underlies cognitive reserve in multiple sclerosis. Hum Brain Mapp 2019;40:5231—41.
[103] Fuchs TA, Jaworski 3rd MG, Youngs M, et al. Preliminary support of a behavioral intervention for trait conscientiousness in multiple sclerosis. Int J MS Care 2022;24:45—53.

[104] Patzold T, Schwengelbeck M, Ossege LM, Malin JP, Sindern E. Changes of the MS functional composite and EDSS during and after treatment of relapses with methylprednisolone in patients with multiple sclerosis. Acta Neurol Scand 2002;105:164–8.

[105] Ozakbas S, Cagiran I, Ormeci B, Idiman E. Correlations between multiple sclerosis functional composite, expanded disability status scale and health-related quality of life during and after treatment of relapses in patients with multiple sclerosis. J Neurol Sci 2004;218:3–7.

[106] Foong J, Rozewicz L, Quaghebeur G, Thompson AJ, Miller DH, Ron MA. Neuropsychological deficits in multiple sclerosis after acute relapse. J Neurol Neurosurg Psychiatry 1998;64:529–32.

[107] Franklin GM, Nelson LM, Filley CM, Heaton RK. Cognitive loss in multiple sclerosis. Case reports and review of the literature. Arch Neurol 1989;46:162–7.

[108] Morrow SA, Jurgensen S, Forrestal F, Munchauer FE, Benedict RH. Effects of acute relapses on neuropsychological status in multiple sclerosis patients. J Neurol 2011;258:1603–8.

[109] Weissert R. Progressive multifocal leukoencephalopathy. J Neuroimmunol 2011;231:73–7.

[110] Benedict RH, Munschauer F, Linn R, et al. Screening for multiple sclerosis cognitive impairment using a self-administered 15-item questionnaire. Mult Scler 2003;9:95–101.

[111] McKay KA, Bedri SK, Manouchehrinia A, et al. Reduction in cognitive processing speed surrounding multiple sclerosis relapse. Ann Neurol 2022;91:417–23.

[112] Benedict RH, Pol J, Yasin F, et al. Recovery of cognitive function after relapse in multiple sclerosis. Mult Scler 2021;27:71–8.

[113] Baldwin C, Morrow SA. Do isolated cognitive relapses exist? No. Mult Scler 2021;27:1488–9.

[114] Benedict RH, Cox D, Thompson LL, Foley F, Weinstock-Guttman B, Munschauer F. Reliable screening for neuropsychological impairment in multiple sclerosis. Mult Scler 2004;10:675–8.

[115] Pardini M. Do isolated cognitive relapses exist? Yes. Mult Scler 2021;27:1486–7.

[116] Ruet A. Do isolated cognitive relapses exist? Commentary. Mult Scler 2021;27:1489–90.

[117] Landmeyer NC, Burkner PC, Wiendl H, et al. Disease-modifying treatments and cognition in relapsing-remitting multiple sclerosis: a meta-analysis. Neurology 2020;94:e2373–83.

[118] Bossa M, Manocchio N, Argento O. Non-pharmacological treatments of cognitive impairment in multiple sclerosis: a review. Neuroscience 2022;3:476–94.

[119] Goverover Y, Chiaravalloti N, Genova H, DeLuca J. A randomized controlled trial to treat impaired learning and memory in multiple sclerosis: the self-GEN trial. Mult Scler 2018;24:1096–104.

[120] Messinis L, Nasios G, Kosmidis MH, et al. Efficacy of a computer-assisted cognitive rehabilitation intervention in relapsing-remitting multiple sclerosis patients: a multicenter randomized controlled trial. Behav Neurol 2017;2017:5919841.

[121] Chiaravalloti ND, Moore NB, DeLuca J. The efficacy of the modified story memory technique in progressive MS. Mult Scler 2020;26:354–62.

[122] Filippi M, Riccitelli G, Mattioli F, et al. Multiple sclerosis: effects of cognitive rehabilitation on structural and functional MR imaging measures – an explorative study. Radiology 2012;262:932–40.

[123] Fuchs TA, Ziccardi S, Dwyer MG, et al. Response heterogeneity to home-based restorative cognitive rehabilitation in multiple sclerosis: an exploratory study. Mult Scler Relat Disord 2019;34:103–11.

[124] Fuchs TA, Ziccardi S, Benedict RHB, et al. Functional connectivity and structural disruption in the default-mode network predicts cognitive rehabilitation outcomes in multiple sclerosis. J Neuroimaging 2020;30:523–30.

[125] Benedict RH, Walton MK. Evaluating cognitive outcome measures for MS clinical trials: what is a clinically meaningful change? Mult Scler 2012;18:1673–9.

[126] Eijlers AJC, van Geest Q, Dekker I, et al. Predicting cognitive decline in multiple sclerosis: a 5-year follow-up study. Brain 2018;141:2605–18.

[127] Absinta M, Sati P, Schindler M, et al. Persistent 7-tesla phase rim predicts poor outcome in new multiple sclerosis patient lesions. J Clin Invest 2016;126:2597–609.

[128] Absinta M, Sati P, Masuzzo F, et al. Association of chronic active multiple sclerosis lesions with disability in vivo. JAMA Neurol 2019;76:1474–83.

[129] Marcille M, Hurtado Rua S, Tyshkov C, et al. Disease correlates of rim lesions on quantitative susceptibility mapping in multiple sclerosis. Sci Rep 2022;12:4411.

[130] Sumowski JF, Leavitt VM. Cognitive reserve in multiple sclerosis. Mult Scler 2013;19:1122−7.
[131] Sumowski JF, Chiaravalloti N, DeLuca J. Cognitive reserve protects against cognitive dysfunction in multiple sclerosis. J Clin Exp Neuropsychol 2009;31:913−26.
[132] Amato MP. "Brain reserve" and "cognitive reserve" should always be taken into account when studying neurodegeneration − commentary. Mult Scler 2018;24:577−8.
[133] Weinstock Z, Dwyer MG, Bergsland N, et al. Thalamic volume predicts recovery from cognitive relapse (P14-4.010). Neurology 2022;98:3340.
[134] Bellmann-Strobl J, Wuerfel J, Aktas O, et al. Poor PASAT performance correlates with MRI contrast enhancement in multiple sclerosis. Neurology 2009;73:1624−7.
[135] Damasceno A, Damasceno BP, Cendes F. Subclinical MRI disease activity influences cognitive performance in MS patients. Mult Scler Relat Disord 2015;4:137−43.
[136] Fenu G, Arru M, Lorefice L, et al. Does focal inflammation have an impact on cognition in multiple sclerosis? An MRI study. Mult Scler Relat Disord 2018;23:83−7.
[137] Dineen RA, Vilisaar J, Hlinka J, et al. Disconnection as a mechanism for cognitive dysfunction in multiple sclerosis. Brain 2009;132:239−49.
[138] Nucci M, Mapelli D, Mondini S. Cognitive reserve index questionnaire (CRIq): a new instrument for measuring cognitive reserve. Aging Clin Exp Res 2012;24:218−26.
[139] Lublin FD. Disease activity free status in MS. Mult Scler Relat Disord 2012;1:6−7.
[140] Damasceno A, Damasceno BP, Cendes F. No evidence of disease activity in multiple sclerosis: Implications on cognition and brain atrophy. Mult Scler 2016;22:64−72.
[141] Kalb R, Beier M, Benedict RH, et al. Recommendations for cognitive screening and management in multiple sclerosis care. Mult Scler 2018;24:1665−80.

CHAPTER 5

Multiple sclerosis treatment

Hernan Inojosa and Tjalf Ziemssen

Department of Neurology, Center of Clinical Neuroscience, University Hospital Carl Gustav Carus, Technical University of Dresden, Dresden, Saxony, Germany

OUTLINE

Introduction	71	Sphingosine 1-phosphate—receptor modulators (ozanimod, ponesimod, fingolimod, siponimod)	80
Escalation versus induction approach?	72		
Personalized decision	73	B-cell depletion	82
Interferon beta 1-a and interferon beta 1-b	74	Natalizumab	84
Glatiramer acetate	77	Alemtuzumab	85
Dimethyl fumarate and diroximel fumarate (monomethyl fumarate)	78	*Cladribine*	86
		Conflict of interest	88
Teriflunomide	79	References	88

Introduction

The treatment of multiple sclerosis (MS) has experienced a drastic revolution in the last several years. In the mid-1990s, the first disease-modifying treatments (DMTs) were approved after analysis of favorable data for interferon beta-1b in the course of MS. Since then and specially in the last decade, due to the research advances regarding the pathophysiology, diagnosis, and treatment of MS, newer and more specific DMTs have been approved [1]. Therapeutic options are changing from unspecific agents with not fully understood pharmacodynamics to more specific treatments. Different mechanisms of action and administration are approved to be used already at early stages of the disease and recently also on progressive courses.

> **BOX 5.1**
>
> **Several DMTs are available for MS.**
>
> More than 20 DMTs have been approved in MS. These can positively alter the course of patients with both relapsing and progressive MS forms.

There is more than enough evidence to support the use of DMTs already in very early disease stages [2]. A conversion from clinically isolated syndrome (CIS) to MS, relapses, disability accumulation, brain magnetic resonance imaging (MRI) activity, and time to diagnosis of a secondary progressive course can be positively modified through current MS treatments. Data show that patients with a MS diagnosis established in the more recent years develop comparable neurological deficits (measured, e.g., with the expanded disability status score [EDSS]) later in their disease course than those with a first diagnosis in the 1990s and also have a lower conversion rate to secondary-progressive MS (SPMS) [3].

Currently, both United States Food and Drug Administration (FDA) and European Medicines Agency (EMA) have approved over 20 DMTs for the treatment of MS. Several further options are in advanced stages of clinical trials promising novel alternatives for MS patients. Drugs may be applied at subcutaneous, intramuscular, oral or at intravenous ways. While most of them have been approved for use in relapsing-remitting MS (RRMS), first therapeutic options for primary-progressive MS (PPMS) and SPMS were made available only in recent years (Box 5.1).

As more treatment options are available, complex strategies of initial treatment, treatment sequences, induction or escalation may be necessary to take advantage of the available resources [4]. An optimal treatment goal would be the stabilization or slowing of the disease activity and progression, as a cure or complete recovery of symptoms in progressive MS patients are not possible to date.

Escalation versus induction approach?

There is enough clinical evidence to support the use of DMTs already in early phases of the disease and these should be recommended to every patient with a MS diagnosis [5,6]. Even in patients with CIS, DMTs reduce the probability of conversion to confirmed MS at follow-up. However, the strategy regarding which therapy to use is still on discussion.

An "escalation" approach has been widely used in the treatment of MS, where drugs from a first-line spectrum are initially used and successively escalated to more effective alternatives depending on the disease activity over time [7]. Alternatively, an "induction" approach has emerged as an alternative, where high-efficacy DMTs are used already on early disease stages, though rather in patients with a more highly active disease course (Box 5.2) [8]. For the latter, the possibility of an initial immune reconstitution therapy (e.g., through autologous stem cell transplantation, alemtuzumab, cladribine, or even B-cell depletion) may provide a clinical and subclinical stability over several years without continuous immune therapy and

BOX 5.2
Highly active MS.

The definition of highly active MS may also be challenging, as there is currently no established standard. Frequently, demographic and clinical factors (e.g., relapse frequency, severity and recovery disability accumulation) and brain imaging aspects (e.g., gadolinium-enhanced lesions), and response to previous DMTs are considered.

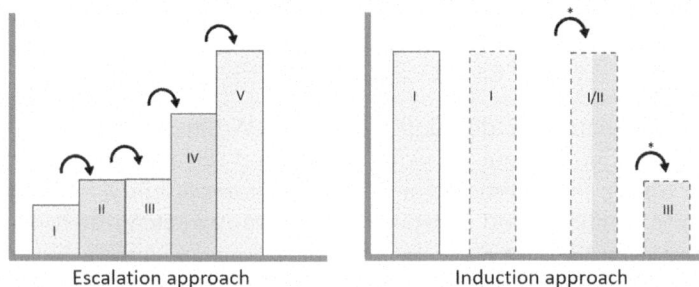

FIGURE 5.1 Graphical presentation of both escalation and induction strategies in the treatment of multiple sclerosis. In the escalation approach, disease-modifying therapies are successively switched beginning with drugs from the basic spectrum toward higher-efficacy ones as disease activity occurs. In the induction strategy, a high-efficacy treatment is used at a first point, ideally early on the disease course and in certain cases as an immune reconstitution. This may be repeated if necessary in a further time point. In certain cases, a continuous maintenance therapy with high effective drugs may be considered after an initial stabilization.

could be considered in highly active patients for a long-term remission; for other patients, a continuous high-efficacy treatment may be preferred [9] (Fig. 5.1).

Safety profiles and wide experience with first-line DMTs have supported to date their use in milder forms of MS. However, the induction strategy ("hit early and hit hard") offers a better approach especially in those with highly active forms of the disease. Large cohort studies have demonstrated that patients receiving high-efficacy DMTs early have lower disability in the long run compared to patients who receive initially first-line treatments [10]. Risk of conversion to SPMS may be decreased in up to 44% after 5.8 years follow-up in patients treated at early disease stages with high-efficacy DMTs [2]. Also from a financial aspect, even though usually newer and more effective drugs may have a higher price, the health system related costs seem to be more effective on the long term, also considering that secondary costs related to SPMS may be decreased.

Personalized decision

To date, several risk factors or clinical hallmarks have been described for a high disease activity or higher risk of progression to SPMS. However, these are not the only aspects to

> **BOX 5.3**
>
> **A single disease with a several faces.**
>
> First DMT may vary drastically between:
>
> - 29-year-old woman with desire to have children and a recent RRMS diagnosis after sensitive manifestations on the left upper limb with a fully recovery after corticosteroids, low cerebral MRI-lesions burden and no spinal lesions.
>
> - 43-year-old man with new diagnosis of RRMS after paraparesis with incomplete recovery after corticosteroids and two relapses in the previous year. Currently >15 lesions on brain MRI and 2 cervical and thoracic lesions.

be considered in the treatment decision (Box 5.3). We recommend a patient-centered approach and a direct involvement in a shared decision setting. Family plans, preferences of administration forms, sociodemographic aspects, compatibility with social and work life, possible adverse reactions, and perspective of therapy switch due to factors other than MS-activity are some of the factors to be considered in the choice. Not only the DMTs (main focus in this chapter) are important in the MS management, but also symptomatic treatments (e.g., pain, spasticity, urinary, or bowel symptoms), treatment of acute relapses (eventually with, e.g., steroids, plasmapheresis), and a proper rehabilitation (e.g., physiotherapie, ergotherapy) are necessary.

The greatest barrier against the use of high-efficacy therapies are the still not fully described safety profiles, especially regarding long-term side effects. Further studies are ongoing to determine the benefit of this strategy, which may support the clinical decisions, such as the TRaditional versus Early Aggressive Therapy for MS (TREAT-MS) or the Determining the Effectiveness of earLy Intensive Versus Escalation Approaches for the Treatment of Relapsing-Remitting Multiple Sclerosis (DELIVER-MS) open trials [11,12].

A baseline MRI is recommended before starting DMTs for further monitoring. We also recommend in case of therapy switch or start a "rebaseline" MRI at 3–6 months to detect lesions that may have appeared before therapeutic onset of the newer drug [13]. A regular clinical and radiological monitoring is required to promptly detect disease activity (Fig. 5.2 and Table 5.1).

Interferon beta 1-a and interferon beta 1-b

Presentation: Interferon beta 1-a: Avonex 30 μg, Plegridy 125 μg (Peginterferon), Rebif 22 and 44 μg. Inteferon-beta 1-b: Betaferon 250 μg, Extavia 250 μg.

Administration: subcutaneous (Plegridy once every 2 weeks, Rebif three times per week, Betaferon or Extavia every 2 days) or intramuscular (Avonex once per week) injections.

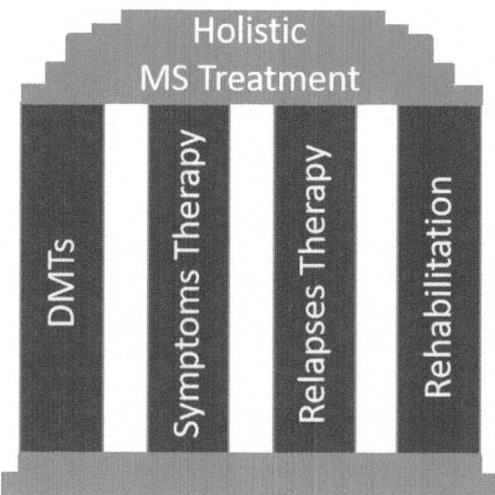

FIGURE 5.2 Not only are disease-modifying therapies important in the management of multiple sclerosis patients, but a complete and holistic treatment also requires controlling symptoms and relapses, as well as implementing targeted rehabilitation strategies.

Mechanism of action: Interferons are cytokines with immunomodulatory effects that play an important role, especially in the native immunity. Its specific role in MS is not clearly understood, as it may has a complex pro- and antiinflammatory activity. Interferon beta may suppress autoinflammatory responses after biding interferon-receptors in MS patients with regulation of cytokines secretion through JAK-STAT family of signal transducers and a further point suppressing T cell activation, which according to animal models have a central importance in the pathogenesis of MS.

Efficacy: Interferone was the first DMTs in the treatment of MS. Already in 1986, a small randomized, double-blind, placebo-controlled trial demonstrated reduced relapse rates in a group of patients treated with interferon [14]. A further study in 1993 could confirm significant lower exacerbation rates and MRI activity with a dosage effect [15]. A larger randomized double-blinded, placebo-controlled, multicenter phase III trial could demonstrate also a reduction of clinical disability. The use of interferon beta-1a delayed the proportion of patients with progressive disease after 104 weeks as well. Additionally, there is also evidence of a positive effect reducing the rate of brain atrophy in up to 68% at the first-year according to the European interferon beta-1a dose comparison study [16].

Since then, several trials have demonstrated the efficacy and effectivity of interferon on the treatment of MS, including among others the BENEFIT, PRISM, ADVAMCE, EVIDENCE, or INCOMIN studies.

Interferon was also tested in patients with SPMS with positive results in from a European trial against placebo including 718 SPMS patients [17]. These results were, however, not confirmed in further randomized trials and in several meta-analyses [18]. Before the approval of more recent drugs, it was more frequently used in patients with active SPMS and simultaneously relapse activity.

Safety: Because of long-term data and described safety profiles regarding long-term side-effects, interferons are still an attractive option, especially in patients with mild MS activity. Flu-like symptoms (chills, myalgia, and fever) or local reactions to subcutan interferons are the most frequent reactions. Special care should be taken in patients with

TABLE 5.1 Summary of multiple sclerosis therapies modified from multiple sclerosis therapy consensus group [1].

	Relapsing MS		Progressive MS		
				Secondary-progressive MS	Primary-progressive MS
CIS	Relapsing-remitting MS		Secondary-progressive MS	Without relapses, with radiological activity	With clinical/radiological activity
	(High) active MS first and second lines		With relapses		
Interferone beta 1-a i.m.	*Induction therapies*				
Interferone beta 1-a s.c.	Alemtuzumab		Cladribine*	Siponimod	Ocrelizumab
Interferon beta 1-b s.c.	Cladribine		Interferon beta 1-b s.c.		
	Ocrelizumab		Ocrelizumab		
			Ofatumumab		
			Ponesimod		
	Continous therapies		Siponimod		
	Natalizumab				
	Ofatumumab				
	S1P-modulation (fingolimod, ozanimod, ponesimod)				
	Mild/moderate MS				
	Dimethyl fumarate				
	Diroximel fumarate				
	Interferone				
	Teriflunomide				

To date, DMTs in relapsing-remitting MS have been classically separated in first-line and high-efficacy drugs. The choice of disease-modifying therapy should base on the MS subtype and clinical and radiological marks of disease activity. An escalation approach is, especially in patients with highly active MS not recommended. A distinction should be made between secondary-progressive MS with and without additional relapse activity. *DMT*, disease-modifying therapies; *MS*, multiple sclerosis; *CIS*, clinically isolated syndrome.

depression or anxiety. Menstrual disorders with dysmenorrhea and metrorrhagia—especially between menstrual periods—are also described, as well as impotence.

Blood counts and monitoring of liver and renal parameters is recommended after 1 and 3 months of therapy. Depending on these values, controls may be performed every 6 or 12 months.

In real-world experience after over 1000 pregnancies, there was no increased risk of severe malformations after intake of interferone beta before conception or during the first trimester. Few experiences have been collected with take of interferone beta in the second and third trimesters. Therapy with interferons can be considered in women with potential childbearing or during the pregnancy as an individual choice.

Glatiramer acetate

Presentation: Copaxone, Clift (20 and 40 mg/mL injection solution or pen with autoinjector for Copaxone).

Administration: Subcutaneous injections (20 mg every day or 40 mg three times per week).

Mechanism of action: The mechanism of action for its positive effect in RRMS is still not completely understood. Probably, immune modulation of components of innate immunity (such as monocytes and dendritic cells) occurs more at an antigen-specific level (e.g., competitive binding with myelin antigens), with a further effects on the presentation and function of B- and T-cells. This may have an effect on the secretion of regulatory and auto-inflammatory immunoglobulins.

Efficacy: Five randomized pivotal studies supported the efficacy of this drug in MS. A placebo-controlled study with 1404 MS patients randomized to receive glatiramer acetate or placebo [19]. This study demonstrated a significant effectivity of a dose of 40 mg/mL three times per week for the reduction of relapses and MRI activity. This was also demonstrated in a further study using 20 mg/day. No direct comparison has been performed between both doses.

A study involving 943 patients with PPMS demonstrated no significant delay in time to sustained accumulation of disability in patients with Glatiramer acetate compared to placebo, although an interestinging trend was seen. Additionally, a positive effect on SPMS patients was not demonstrated.

The daily administration of glatiramer acetate has also shown a positive effect in patients with CIS in the PreCISe study and may be considered in a dose of 20 mg/d in these patients as alternative to interferone [20].

Safety: Local reactions at the injection site may be frequent (including erythema, pain, pruritus, inflammation, local atrophy), these may be mitigated with alternation of injection site and application of temperature measures (cold) before the injection.

No toxicity, teratogenicity, or relevant epidemiological data are known to recommend against the use of glatiramer acetate in women with potential desire of pregnancy. It should be discontinued at evidence of pregnancy. In individual cases, treatment may be continued even after it begins.

Dimethyl fumarate and diroximel fumarate (monomethyl fumarate)

Presentation: Tecfidera (120 and 240 mg capsules) and Vumerity (231 mg capsules).

Administration: Oral, twice per day. A gradual starting dosing scheme is recommended at begin of therapy for a better toleration und compliance.

Mechanism of action: Both dimethyl fumarate (DMF) and diroximel fumarate (DOF) are fumaric esters that are in vivo degraded to the active metabolite monomethyl fumarate (MMF). Esterases metabolize both agents ubiquitous in gastrointestinal tract after duodenal absorbtion. The median time to peak drug concentration for MMF is similar for both drugs after 2.5 to 3 hours with comparable peak plasma concentrations.

The exact mechanisms and effect in MS are still not fully understood. MMF exerts an antiinflammatory and immunomodulatory effect through activation of nuclear factor (erythroid-derived 2)-like 2 (Nrf2) transcriptional pathways with a pleiotropic effect. This may downregulate proinflammatory cytokines and shift Th1/Th17 responses and produce a lymphocytopenia predominantly of CD8 + T-cells and CD4 +.

Efficacy: Efficacy studies were initial performed for DMF. After the development of DOF, which aims for better tolerance, the efficacy profile is expected to be similar to that of MMF, as the concentrations of the active metabolite are similar.

The Phase 3 studies of DMF, DEFINE and CONFIRM, could demonstrate after 2 years of treatment a clinically meaningful reduction of relapses and annualized relapse rates in comparison to placebo [21,22]. An open 8-year extension study ENDORSE could confirm these results as well as performed an assessment of long-term safety [23]. MRI outcomes also demonstrated fewer Gd-enhancing lesions, new or enlarging T2 and T1 lesions in the treatment groups.

Less disease progression was observed in long-term follow-ups for patients under these therapies compared to placebo, supporting a preventive effect of SPMS. However, not enough data are available to support its use already on progressive MS forms. Small studies have demonstrated a positive trend with inconclusive impact on EDSS.

Safety: At beginning of therapy, gastrointestinal tract adverse reactions (diarrhea, nausea, vomiting, and abdominal pain), which may continue to occur intermittently in the course of therapy. Flushing (mostly secondary to prostaglandin-mediated vasodilatation, rash, pruritus, erythema, hot feeling) is another frequent reaction that may lead to discontinuation of treatment.

A gradually dosing scheme as well as the intake of these medicaments with food reduce these adverse events. In certain patients, a premedication with acetylsalicylic acid 30 minutes prior medication (e.g., 325 mg) could reduce flushing but potential risks of NSAIDs should be considered.

A regular control of leukocytes and especially lymphocytes cell counts is recommended as the risk of opportunistic infections (e.g., progressive multifocal leukoencephalopathy or herpes zoster) could be increased. A treatment switch should be considered if repeated lymphopenia $<0.7 \times 10^9$/L.

Gastrointestinal tolerability was tested in the EVOLVE-MS-2 trial and DOF had better profiles. However, these patients may also present these adverse reactions [24].

A temporary reduction of DMF to 120 mg or of DOF to 231 mg twice a day may be possible to reduce adverse reactions, but the recommended maintenance dose should be resumed within 1 month.

Both DMF and DOF are contraindicated during pregnancy. Data from approval and animal studies reflected, however, no increased risk of teratogen effects. Therapy with these drugs should be discontinued by known pregnancy desire or at latest if pregnancy is suspected.

Teriflunomide

Presentation: Aubagio 7 mg and 14 mg tablets
Administration: Oral daily
Mechanism of action: Teriflunomide is the active metabolized of leflunomide, an approved drug for the treatment of rheumatoid arthritis. It has an immunomodulatory and antiinflammatory effect with reversible inhibition of the mitochondrial enzyme dihydroorotate dehydrogenase; however, the precise mechanisms are not fully understood. This enzyme is involved in the synthesis of rapid proliferating cells (e.g., T and B lymphocytes) dependent of pyrimidine. A mild lymphocyte count reduction was frequent but not mandatory in placebo-controlled studies.

Efficacy: Approval of teriflunomide was granted with the TEMSO and TOWER studies, which randomized 1088 and 1169 patients to receive teriflunomide or placebo [25,26]. Annualized relapse rate, 3-month disability progression, MRI burden and Gd-enhancing lesions were significantly reduced in both studies compared to placebo.

An efficacy in patients with high disease activity was also observed. However, no superiority to interferon beta-1a has been described in the TENERE trial [27]. In this trial, fewer permanent discontinuations (due to clinical outcomes and adverse events) were observed in the teriflunomide arm.

Safety: Laboratory screening, including blood counts, liver profiles, and pancreas enzymes, is mandatory. By known or suspected abdominal diseases, an abdomen sonography is recommended. Measure of blood pressure and screen for latent tuberculosis are also recommended by the FDA. Teriflunomide is contraindicated in case of liver disease, chronic active infections, immundeficiency, or significant blood alterations. Renal insufficiency, pregnancy, and breastfeeding are other important contraindications.

Teriflunomide shows generally a good tolerability. Most common reported adverse events were increased blood pressure (mostly reversible), mild liver parameters increase, alopecia, nausea, diarrhea, and arthralgia. Severe hepatic dysfunction has also described.

A low grade lymphopenia may be tolerated in the first 6 weeks of treatment. Less frequent is the onset of rhabdomyolysis or peripheral neuropathy.

On treatment with teriflunomide, monthly liver function tests are recommended for the first 6 months. Afterwards, blood controls may be performed every 3, 6, or 12 months depending on the patient's profiles. A sonography of the abdomen and gastroenterological consultation may be considered in case of suspicion of abdominal adverse events (Fig. 5.3).

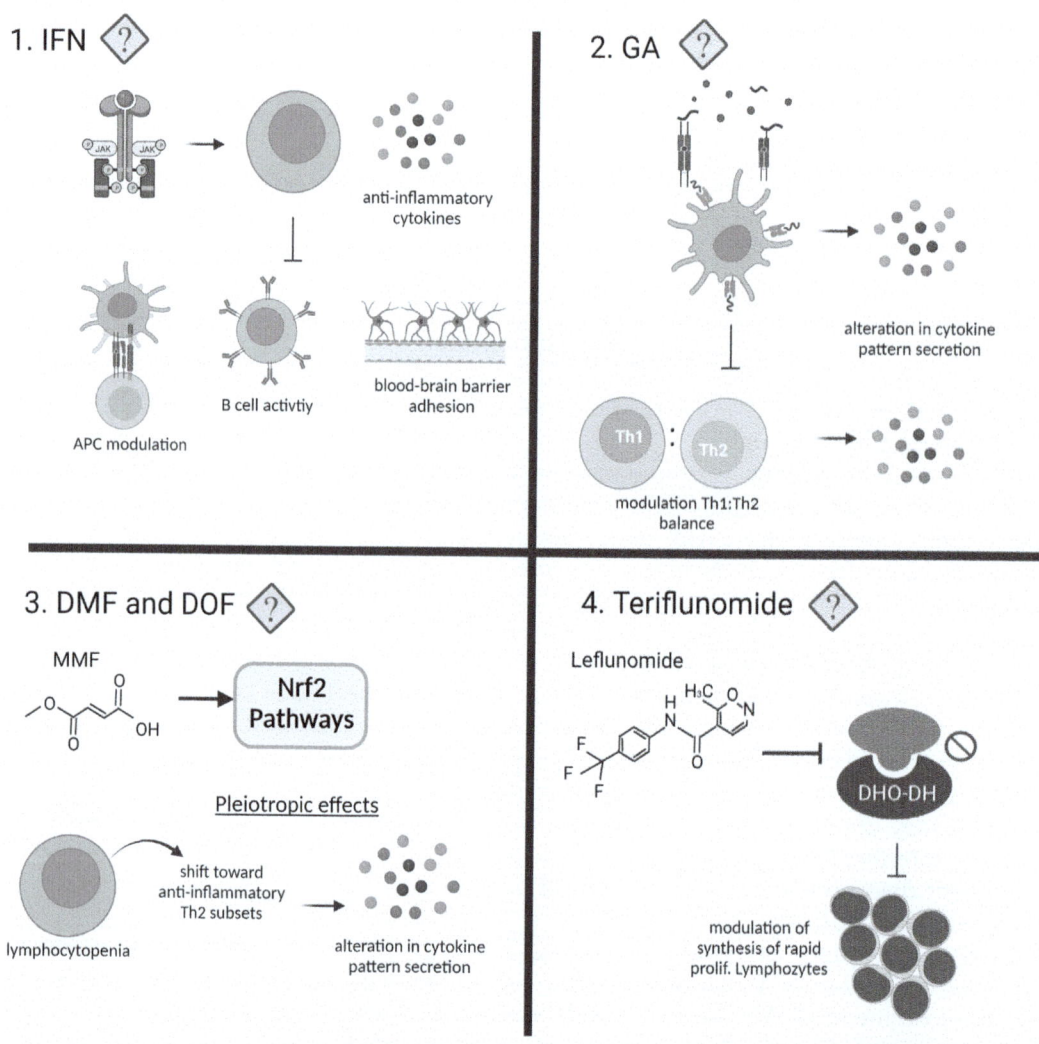

FIGURE 5.3 Mechanism of action of selected disease-modifying therapies. Interferon, glatiramer acetate, dimethyl and diroximel fumarate and teriflunomide are used in clinically isolated syndrome and mild MS forms. These have been classically described as first-line MS therapies. Their mechanism of action is not completely understood. A modulation of JAK receptors (interferone), direct antigen-ligand effect (glatiramer acetate), activation of Nrf2 pathways through monomethyl fumarate (DMF and DOF), or inhibition of dihydro-orotate dehydrogenase through leflunomide (teriflunomide) may result in diverse, mostly pleiotropic effects on the immune system.

Sphingosine 1-phosphate—receptor modulators (ozanimod, ponesimod, fingolimod, siponimod)

Presentation: The presentation of MS-approved drugs and their current indications are listed in Table 5.2.

TABLE 5.2 Sphingosine 1-phosphate–receptor modulators approved for multiple sclerosis.

Family of S1P-receptor modulators					
Drug	Commercial name	FDA approval	Label	Composition	Titration regimen
Fingolimod	Gilenya	2010	Highly active RRMS[a]	0.25 and 0.5 mg capsules	No
Ozanimod	Zeposia	2020	Active RRMS Ulcerative colitis	0.23, 0.46, and 0.92 mg capsules	Yes
Ponesimod	Ponvory	2021	Active RRMS	2, 3, 4, 5, 6, 7, 8, 9, 10, and 20 mg tablets (contain lactose)	Yes
Siponimod	Mayzent	2019	Active SPMS	0.25, 1, and 2 mg tablets	Yes

[a]Despite Despite a full and adequate course of treatment with one disease-modifying therapy, some patients may experience two or more disabling relapses in a year, along with one or more Gd-enhancing lesions and/or a significant increase in T2 burden.

Administration: Oral, once daily. A gradually dose escalation is required for initiation of treatments with ozanimod, ponesimod, and siponimod. Treatment initiation packs are available to facilitate the intake.

Before treatment begin with siponimod, the genotype of the enzyme CYP2C9 has to be mandatory assessed to determine the metabolizer status. Siponimod is contraindicated in case of a CYP2C9*3*3 genotype.

Mechanism of action: Fingolimod, ozanimod, ponesimod, and siponimod are modulators of the sphingosine 1-phosphate (S1P) receptor. Fingolimod is metabolized to the active metabolite Fingolimod phosphate.

Fingolimod binds nonselectively to the S1P receptor located on lymphocytes and neural cells in the central nervous systems and causes a redistribution of lymphocytes, as it blocks their trafficking and capacity to egress from lymphoid tissues. As this receptor is also present in various sites (e.g., cardiovascular, gastrointestinal, or nervous system). Its effects in those systems are responsible of several adverse effects. In vitro investigations suggest a significant effect on neurodegeneration or remyelination by effects on S1P1, S1P3, and S1P5 receptors.

Ponesimod, ozanimod, and siponimod have a similar mechanism of action. However, their effects are more selective in the S1P receptor from subtype 1 located and lymphocytes, reducing adverse events. Ozanimod and siponimod bind with high affinity to S1P receptors 1 and 5 and have minimal or no activity on S1P2, S1P3, and S1P4.

Efficacy: Two Phase 3 trials (FREEDOMS and FREEDOMS 2) demonstrated the efficacy of Fingolimod reducing relapse rates and new or enlarging lesions on MRI compared to placebo [28,29]. The follow-up of these trials also demonstrated a lesser risk of disability progression. The TRANSFORMS trial showed additionally a benefit of Fingolimod against interferon beta in these outcomes.

Further approval trials could demonstrate the efficacy of ozanimod (SUNBEAM and RADIANCE) and ponesimod (OPTIMUN) in patients with active RRMS also in a direct comparison with interferon beta 1-a and teriflunomide, respectively [30,31].

Siponimod was approved after the EXPAND trial, which demonstrated efficacy in SPMS. Thus it was the first drug approved for this disease form [32].

To date, no head-to-head trials have compared the efficacy of S1P receptor modulators.

Safety: Due to the ubiquitous distribution of S1P receptors with several subtypes, certain adverse events may occur as a result of their effects on other organs. Effects on the S1P4 receptor in cardial tissue may contribute to bradycardia and vasoconstriction.

A laboratory screening, including cell count, liver enzymes, viral profiles (such as Varizela Zoster Virus, hepatitis B and C, and HIV), and an electrocardiogram, should be performed before therapy begin. In case of previous tuberculosis, a chest X-ray is recommended. A regular check-up for prevention or early detection of skin cancer and macular edema is necessary under these DMTs.

These drugs are contraindicated in patients with known immunodeficiency and risk for opportunistic infections, severe active infections, active malignancies, and sever liver impairment. Cardiovascular disease must be assessed as among other myocardial infarction, transient ischemic attack/stroke, severe heart failure, or arrhythmias are also contraindication.

For fingolimod, a cardiac monitoring is mandatory for 6 hours after the first dose. In the case of ozanimod or siponimod, it may be shorter depending on the cardiovascular risk factors.

Pregnancy and breastfeeding are not advised under S1P modulation, as these drugs are teratogenic. Time of onset of 3- and 6-month confirmed disease progression was significantly delayed by this therapy. Absolute risk of progression was also reduced compared to placebo. This effect was more marked in the subgroup of patients with active disease (relapses in the last 2 years or Gd-enhancing lesions at baseline).

A therapy monitoring is recommended every 3 months under these immune therapies. Specially a monitoring of liver enzymes and lymphocytes counts is mandatory. These therapies should be discontinued if a lymphopenia of <0.2 GPt/L or if liver enzymes are elevated upper 5 upper limit of normal is confirmed after repeated controls.

B-cell depletion

Presentation: Ofatumumab 20 mg solution (Kesimpta) and ocrelizumab 300 mg solution (Ocrevus). Rituximab was previously frequently used before approval of newer drugs. B-cell-depletive DMTs are listed in Table 5.3.

Administration: Ofatumumab is administered subcutaneously once a month after the initial dosing, and premedication is not required under the guidance of healthcare professionals. Ocrelizumab is administered intravenously every 6 months after the initial dosing, in two separate infusions.

Mechanism of action: Both ofatumumab and ocrelizumab are monoclonal antibodies anti-CD20, a transmembrane phosphoprotein expressed on the surface of B-cells (from pre-B to mature B lymphocytes) in bone marrow and periphery. CD20 is not expressed on stem cells, pro-B cells, or plasma cells. Through these drugs, B-cells are depleted from peripheral blood, principally through Kupffer cells mediated antibody-dependent phagocytosis in the liver.

TABLE 5.3 General characteristics of anti-CD20 drugs oftaumumab, ocrelizumab, and rituximab.

	Ofatumumab	Ocrelizumab	Rituximab
Label	Active relapsing MS	Active relapsing MS Early PPMS and MRI activity	Off-label
Commercial name	Kesimpta	Ocrevus	MabThera, Rituxan
Dosis	20 mg monthly	600 mg every 6 months	Varying, e.g., 375 mg/m^2 weekly, four doses
Initial dosis scheme	20 mg at weeks 0, 1, and 2	300 mg at weeks 0 and 2	
Composition	Fully human	Recombinant humanized	Chimeric mouse-human
Antibody	Monoclonal anti-CD 20 immunoglobulin G1	Monoclonal anti-CD 20 immunoglobulin G1	
Anti-drug antibodies	Not identified	Very low incidence (~1%), rare neutralizing	Frequent (in 11% to >50%)

Initially, the role of B cells in MS was unknown, as it was postulated though based on mice models (e.g., experimental autoimmune encephalomyelitis) that MS was a T-cell mediated disease. Recent data suggest a key function of B-cells in the activation of T-cells beyond production of autoantibodies. Intrathecal production of antibodies (demonstrated through oligoclonal bands) is the most consistent laboratory abnormality in MS. Certain B-cells subtypes (e.g., memory B-cells) may play a central role on regulation of pro- and antiinflammatory regulation and antigen presentation. A reduction of cerebrospinal fluid concentration not only of B-cells but also of T-cells implies a regulation of trafficking as consequence of B-cell elimination.

Efficacy: Initial studies with rituximab (e.g., HERMES trial) led to increasing interest on targeting B-cells on patients with MS [33]. Ocrelizumab was approved after two randomized trials against interferons on patients with relapsing MS and disease activity in the previous 2 years (OPERA I and II) [34]. Patients treated with ocrelizumab had better outcomes in both clinical and radiological endpoints, with significant reduction in relapse rates and confirmed disability progression (measured up to 96 weeks) and MRI activity as well.

The ORATORIO study evaluated its efficacy in patients with PPMS at early disease stages (such as age <55 years, EDSS 3.0–6.5, and disease duration <15 years) and MRI evidence of activity [35]. This study showed a significant reduction of disease progression and deterioration of walking speed compared to placebo. An additional study (ORATORIO-HAND) demonstrated a benefit also in the function of the upper extremities reducing disability progression [36]. Recent preliminary data from the CONSONANCE study reports also a benefit in several aspects in patients with SPMS, including cognition. A current study is evaluating higher doses of ocrelizumab on progressive MS patients.

Ofatumumab's efficacy was evaluated compared to teriflunomide on a double-dummy design in the ASCLEPIOS I and II studies on active relapsing MS patients [37]. Relapse rates, disability progression, and MRI activity were significant lower on ofatumumab patients compared to placebo.

Safety: Injection site reactions such as erythema, swelling, itching, and pain were observed especially after the first ofatumumab injections.

Infusion-related reactions have been described under ocrelizumab, including headache, hypotension, flushing, fever, or dyspnea. To reduce these reactions, intravenous steroids and antihistamine are recommended before ocrelizumab infusions. Infusions may last at least 3.5 hours or alternatively 2 hours in patients with gut toleration.

Upper respiratory tract infections and urinary tract reactions were frequently reported for both treatments. Initial screening for hepatitis B virus (including HBsAg and HBcAb) is important, as hepatitis B reactivation with consequent fulminant hepatitis and death has been reported with anti-CD20 medications.

All immunizations should be completed at least 4–6 weeks before treatment initiation. Both drugs may interfere with the effectiveness of inactivated vaccines, and live attenuated vaccines are contraindicated.

progressive multifocal leukoencephalopathy (PML) has been observed in patients treated with certain anti-CD20 medications. Vigilance for symptoms and immediate suspension of treatment are recommended if a PML is suspected. An incidence of malignancies has not been confirmed compared to rates expected for an MS population.

There is still insufficient data regarding the safety of these drugs during pregnancy. Lymphopenia and B-cell depletion may occur in infants born to mother treated with anti-CD20 drugs. Women of childbearing potential should use effective contraception methods while being treated with ofatumumab or ocrelizumab and 6 months after the end of therapy.

Natalizumab

Presentation: Tysabri 300 mg vials or 150 mg in prefilled syringes.

Administration: Subcutaneous or intravenous injection every 4 weeks.

Mechanism of action: Natalizumab is a humanized monoclonal antibody that binds the $\alpha 4\beta 1$ subunit of human integrins on leukocytes and thus, blocks its interaction with the receptor vascular cell adhesion molecule-1 (VCAM-1), the mucosal addressin cell adhesion molecule-1 (MadCAM-1), osteopontin, and a domain of fibronectin. This action impedes the migration of leukocytes into the central nervous system across the blood–brain barrier, as the VCAM-1 is upregulated in the presence of proinflammatory cytokines in the context of inflammation in MS. Consequently, inflammation is reduced in the central nervous system.

Efficacy: The clinical efficacy of natalizumab was evaluated for the intravenous infusion initially in the AFFIRM trial and has been considered similar for the subcutaneous injection based on pharmacokinetic similarities [38]. This trial randomized relapsing MS patients with clinically active disease in the last year to receive natalizumab or placebo.

Data from a phase 4 study on the Tysabri observational program confirmed a relative safety profile and a persistent low relapse rate after over 4 years follow-up [39]. Recently, the NOVA trial aimed to evaluate the efficacy and safety profile of switching from a 4-weekly to a 6-weekly regimen. Increased rates of new or enlarging lesions were described in the once every 6 weeks compared to the once every 4 weeks group, but due to an extreme lower activity in the once every 4 weeks groups, statistical differences were not interpretable [40]. Safety profiles were similar between both regimes.

The further trials REFINE and DELIVER assessed the subcutaneous formulations on patients without or with previous natalizumab treatment with positive results, supporting the possibility of this practical administration form [41]. Patient's preferences for an optimal compliance should be considered in the decision of the administration method.

Safety: The most frequent adverse reactions associated with natalizumab were headache, upper respiratory and urinary tract infections, fatigue, nausea, arthralgia, and dizziness.

An increased infection risk has been described for natalizumab, especially for opportunistic infections with certain cases of fatal encephalitis or meningitis due to herpes infections or JC-virus infections. Risk of PML due to JC-virus, a severe complication that may frequently be fatal, increases at longer treatment duration. Patients should be active informed about this risk and a laboratory monitoring should be performed before and under therapy. A cohort analysis including over 37,000 patients treated with natalizumab revealed an estimated probability of PML after 6 years of up to 2.7% in patients with previous immunosuppressant and positive antibodies anti-JC-virus [42]. The risk of PML increases over treatment duration and according to the antibody index [42]. An extended interval dosing should be individually considered as an extension already of 1 or 2 weeks could result in a large reduction of risk of PML. However, the risk is not completely eliminated.

In 2-year controlled trials, approximately 6% of patients developed persistent antibodies against natalizumab, which may reduce the efficacy and increase hypersensitivity reactions.

Data from prospective pregnancy registries suggest a relative safety profile of natalizumab, as the rate of defects after birth does not seem to be increased compared to other MS populations. In case of highly active disease, natalizumab may be continued if needed. A monitoring of blood counts (especially platelet) is recommended in neonates.

Alemtuzumab

Presentation: Lemtrada 12 mg.

Administration: Intravenous, initially two treatment courses (5 days and 3 days) separated by 12 months. Additional treatment courses should be determined on an individual basis. Pretreatment with corticosteroids required for the first 3 days of each course. Antihistaminergics and antipyretics may be considered.

Mechanism of action: Alemtuzumab is an IgG1 humanized monoclonal antibody directed against the glycoprotein CD52, present on T and B cells. It has a human and murine (rat) components. It produces a cellular cytolysis and complement-mediated lysis with consequent depletion and repopulation of lymphocytes.

Efficacy: This drug was approved for patients with highly active MS forms including those with activity despite treatment with a previous DMT or patients with two or more disabling relapses in 1 year, and MRI significant activity (new Gd-enhancing lesions or increase in T2 lesion burden).

Three studies evaluated the efficacy and safety of alemtuzumab (CARE-MS I, CARE-MS II, and CAMMS 223) [43,44]. In these trials, a decrease on relapse rates and relapse severity was reported compared to patients treated with interferons. Patients treated with alemtuzumab had also less disability accumulation and a lower MRI disease activity. Data from 5- and 9-year follow-ups showed consistent effects considering relapse rates or new MRI activity (including brain volume loss) [45,46].

A third or fourth treatment course was indicated in a follow-up study in 26%−31% of patients after 4 years [47]. In the Cambridge cohort, up to 48% of patients after 12 years needed this treatment. Retreatment may be offered for patients with new relapses or more than two new, enlarging, or Gd-enhancing lesions on MRI monitoring. A lymphocyte-repopulation did not appear to be related to onset of disease activity. At increased repetition of alemtuzumab courses, rates of development of antibodies antialemtuzumab and patients with insufficient lymphocyte depletion increase. Neutralizing antibodies was associated with an inhibition of the therapeutic effect in 18.8% of patients treated with a third course in a cohort of 32 patients. Measurement of antidrug antibodies titers and neutralization can be considered before retreatment and support therapeutic decisions (e.g., switch to another DMT).

Safety: Alemtuzumab is contraindicated in patients with severe infection, HIV infection, uncontrolled hypertension, history of stroke, angina pectoris, myocardial infarction, known coagulopathy, or other autoimmune disease. An electrocardiogram should be performed previous treatment with alemtuzumab and a laboratory screening (blood count, serum transaminases, creatinine, thyroid function, and urine analysis).

Infusion-associated reactions have been frequently described, probably due to a cytokine release during infusion. These may be variable and include headache, rash, pyrexia, flushing, fatigue, dyspnea, chest discomfort, so on. Due to immunosuppression, infections may occur in more than 70% of patients treated with alemtuzumab, although they are often not severe. Rare cases of PML have been described. Prophylaxis for herpes necessary for 1 month following treatment (e.g., acyclovir 200 mg twise a day).

Clinical and laboratory follow-up for at least 48 months after the last infusion should be performed on every patient after treatment with alemtuzumab as autoimmune mediated conditions may occur. Thyroid disorders (in up to 36.8% of patients after a median of 6.1 years follow-up), thrombocytopenic purpura, acquired hemophilia A, nephropathies, hepatitis, or sarcoidosis are described. The importance of safety monitoring after treatment courses must be informed to the patients.

It is recommended to proof and complete immunization at least 6 weeks before treatment courses.

A safe contraception is required during alemtuzumab and up to 4 months after administration. A potential risk to the fetus because of crossing of the placenta barrier is possible (Fig. 5.4).

Cladribine

Presentation: Mavenclad 10 mg.

Administration: Oral, adapted to body weight (3.5 mg/kg) over 2 years, two treatment courses of 1.75 mg/kg per year. Each course includes two treatment weeks (4−5 days with 10 or 20 mg per day) separated by 1 month.

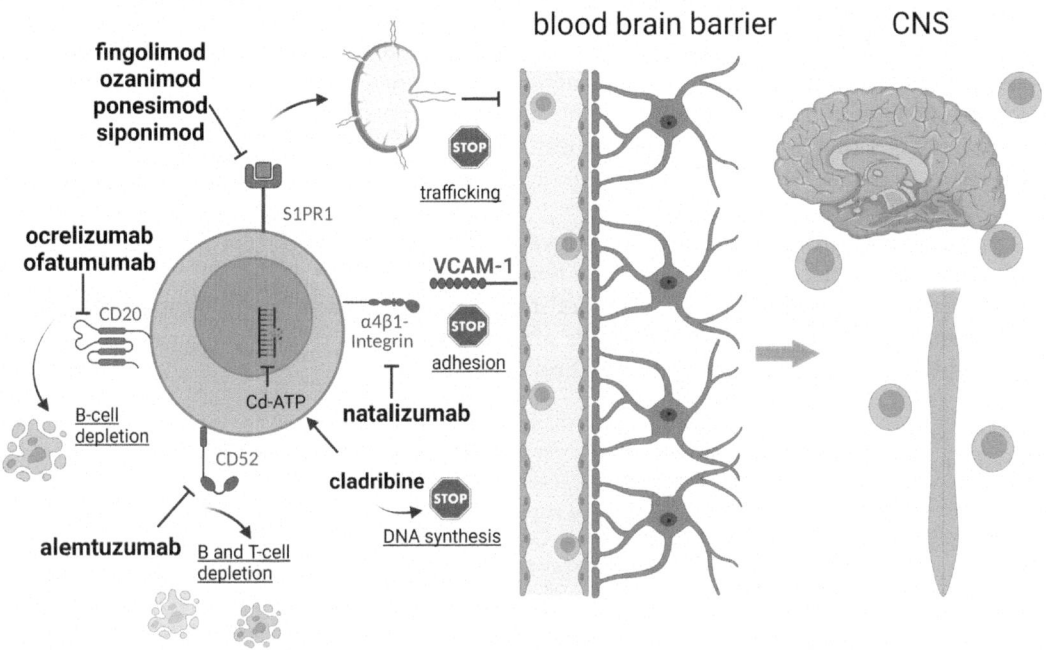

FIGURE 5.4 Mechanism of action of selected disease-modifying therapies. S1P-receptor modulators (e.g., fingolimod, ozanimod, ponesimod, and siponimod) regulate the trafficking from lymphocytes in the lymph nodes into the blood stream. Natalizumab binds to α4β1-integrin and impedes its interaction with the adhesion molecule VCAM-1, which is more expressed on cells at lumen of blood vessels close to site of MS active lesions. The monoclonal antibodies ocrelizumab and ofatumumab bind the CD20 molecule present at B-cells with a further depletion. Alemtuzumab acts at the CD52 molecule and effects a B- and T-cell depletion.

Mechanism of action: Cladribine is phosphorylated to its active triphosphate form, 2-chlorodeoxyadenosine triphosphate. Lymphocytes (T- and B-cells) are especially susceptible to apoptosis due to high levels of deoxycytidine kinase. This is mediated primary through interference in the DNA synthesis through inhibition of ribonucleotide reductase and incorporation of deoxyadenosine triphosphate into DNA. It also causes DNA single-strand breaks and possibly apoptosis via release of cytochrome c (caspase-dependent and −independent) and apoptosis-inducing factor.

Depletion of both T- and B-cells may reduce the autoimmune reactions in central nervous system.

Efficacy: Cladribine is approved for the treatment of highly active RRMS patients. Its efficacy was tested in the CLARITY trial, a randomised, placebo-controlled study [48]. This trail demonstrated a significant decrease in the annualized relapse rates, T1 Gd-enhancing and active T2 lesions, as well as an increase in the proportion of relapse- and disability-free patients over 96 weeks. A consistent similar effect was seen in patients with high disease activity.

An additional study evaluated the effects on patients with SPMS. In those patients with additional relapse activity, a positive effect with reduction of annualized relapse rate was observed. However, no effects on disability progression measured through the EDSS was found [49].

Safety: A normal lymphocyte count is required before treatment begin and at least 800 cells/mm^3 before the second treatment year. Cladribine is contraindicated in cases of active chronic infection, HIV infection, active malignancy, renal impairment, pregnancy, and breastfeeding. Appropriate screening should be performed. Treatment in immunocompromised patients is also contraindicated. A close hematological monitoring is also required. Malignancies were more frequent in patients treated with cladribine in clinical studies.

Lymphopenia is the most clinically relevant adverse reaction. Herpes zoster may occur with higher incidences depending on the lymphopenia grade.

Effective contraception is recommended for both women and men before starting each treatment cycle and for at least 6 months after the last dose, due to potential interference with DNA synthesis, teratogenic effects, and toxicity.

Conflict of interest

HI received speaker feed from Roche and financial support for research activities from Biogen and Alexion. TZ received personal compensation from Biogen, Bayer, Celgene, Novartis, Roche, Sanofi, and Teva for the consulting services. Ziemssen received additional financial support for the research activities from Bayer, Biogen, Novartis, Teva, and Sanofi.

References

[1] Wiendl H, Gold R, Berger T, Derfuss T, Linker R, Mäurer Mdie Multiple Sklerose Therapie Konsensus, G. Multiple Sklerose Therapie Konsensus Gruppe (MSTKG): Positionspapier zur verlaufsmodifizierenden Therapie der Multiplen Sklerose 2021 (white paper). Nervenarzt 2021;92(8):773–801. Available from: https://doi.org/10.1007/s00115-021-01157-2.

[2] Brown JWL, Coles A, Horakova D, Havrdova E, Izquierdo G, Prat A, et al. Association of initial disease-modifying therapy with later conversion to secondary progressive multiple sclerosis. JAMA 2019;321(2):175–87. Available from: https://doi.org/10.1001/jama.2018.20588.

[3] Cree BA, Gourraud PA, Oksenberg JR, Bevan C, Crabtree-Hartman E, Gelfand JM, et al. Long-term evolution of multiple sclerosis disability in the treatment era. Ann Neurol 2016;80(4):499–510. Available from: https://doi.org/10.1002/ana.24747.

[4] Inojosa H, Proschmann U, Akgün K, Ziemssen T. The need for a strategic therapeutic approach: multiple sclerosis in check Therap Adv Chronic Dis 2022;13:20406223211063032. Available from: https://doi.org/10.1177/20406223211063032.

[5] Tintoré M. Early MS treatment. Int MS J 2007;14(1):5–10.

[6] Comi G, Filippi M, Barkhof F, Durelli L, Edan G, Fernández O, et al. Effect of early interferon treatment on conversion to definite multiple sclerosis: a randomised study. Lancet 2001;357(9268):1576–82. Available from: https://doi.org/10.1016/s0140-6736(00)04725-5.

[7] Rieckmann P, Traboulsee A, Devonshire V, Oger J. Escalating immunotherapy of multiple sclerosis. Therap Adv Neurol Disord 2008;1(3):181–92. Available from: https://doi.org/10.1177/1756285608098359.

[8] Gajofatto A, Benedetti MD. Treatment strategies for multiple sclerosis: when to start, when to change, when to stop? World J Clin Case 2015;3(7):545–55. Available from: https://doi.org/10.12998/wjcc.v3.i7.545.

[9] AlSharoqi IA, Aljumah M, Bohlega S, Boz C, Daif A, El-Koussa S, et al. Immune reconstitution therapy or continuous immunosuppression for the management of active relapsing-remitting multiple sclerosis patients? A narrative review. Neurol Ther 2020;9(1):55–66. Available from: https://doi.org/10.1007/s40120-020-00187-3.

[10] Harding K, Williams O, Willis M, Hrastelj J, Rimmer A, Joseph F, et al. Clinical outcomes of escalation vs early intensive disease-modifying therapy in patients with multiple sclerosis. JAMA Neurol 2019;76(5):536–41. Available from: https://doi.org/10.1001/jamaneurol.2018.4905.

[11] Ontaneda D, Tallantyre EC, Raza PC, Planchon SM, Nakamura K, Miller D, et al. Determining the effectiveness of early intensive versus escalation approaches for the treatment of relapsing-remitting multiple sclerosis: the DELIVER-MS study protocol. Contemp Clin Trials 2020;95:106009. Available from: https://doi.org/10.1016/j.cct.2020.106009.

[12] Ontaneda D, Tallantyre E, Kalincik T, Planchon SM, Evangelou N. Early highly effective versus escalation treatment approaches in relapsing multiple sclerosis. Lancet Neurol 2019;18(10):973–80. Available from: https://doi.org/10.1016/s1474-4422(19)30151-6.

[13] Wattjes MP, Ciccarelli O, Reich DS, Banwell B, de Stefano N, Enzinger C, et al. 2021 MAGNIMS-CMSC-NAIMS consensus recommendations on the use of MRI in patients with multiple sclerosis. Lancet Neurol 2021;20(8):653–70. Available from: https://doi.org/10.1016/s1474-4422(21)00095-8.

[14] Hojati Z, Kay M, Dehghanian F. Mechanism of action of interferon beta in treatment of multiple sclerosis. In: Minagar A, editor. Multiple sclerosis. San Diego, CA: Academic Press; 2016, p. 365–92.

[15] Interferon beta-1b is effective in relapsing-remitting multiple sclerosis. I. Clinical results of a multicenter, randomized, double-blind, placebo-controlled trial. The IFNB Multiple Sclerosis Study Group. (1993). Neurology, 1993;43(4):655. Available from: https://doi.org/10.1212/wnl.43.4.655.

[16] Hardmeier M, Wagenpfeil S, Freitag P, Fisher E, Rudick RA, Kooijmans M, et al. Rate of brain atrophy in relapsing MS decreases during treatment with IFNbeta-1a. Neurology 2005;64(2):236–40. Available from: https://doi.org/10.1212/01.Wnl.0000149516.30155.B8.

[17] Placebo-controlled multicentre randomised trial of interferon beta-1b in treatment of secondary progressive multiple sclerosis. European Study Group on interferon beta-1b in secondary progressive MS. Lancet, 1998;352(9139):1491–1497.

[18] La Mantia L, Vacchi L, Di Pietrantonj C, Ebers G, Rovaris M, Fredrikson S, et al. Interferon beta for secondary progressive multiple sclerosis. Cochrane Database Syst Rev 2012;1:Cd005181. Available from: https://doi.org/10.1002/14651858.CD005181.pub3.

[19] Khan O, Rieckmann P, Boyko A, Selmaj K, Zivadinov R. Three times weekly glatiramer acetate in relapsing-remitting multiple sclerosis. Ann Neurol 2013;73(6):705–13. Available from: https://doi.org/10.1002/ana.23938.

[20] Comi G, Martinelli V, Rodegher M, Moiola L, Bajenaru O, Carra A, et al. Effect of glatiramer acetate on conversion to clinically definite multiple sclerosis in patients with clinically isolated syndrome (PreCISe study): a randomised, double-blind, placebo-controlled trial. Lancet 2009;374(9700):1503–11. Available from: https://doi.org/10.1016/s0140-6736(09)61259-9.

[21] Fox RJ, Miller DH, Phillips JT, Hutchinson M, Havrdova E, Kita M, et al. Placebo-controlled Phase 3 study of oral BG-12 or glatiramer in multiple sclerosis. N Engl J Med 2012;367(12):1087–97. Available from: https://doi.org/10.1056/NEJMoa1206328.

[22] Gold R, Kappos L, Arnold DL, Bar-Or A, Giovannoni G, Selmaj K, et al. Placebo-controlled Phase 3 study of oral BG-12 for relapsing multiple sclerosis. N Engl J Med 2012;367(12):1098–107. Available from: https://doi.org/10.1056/NEJMoa1114287.

[23] Gold R, Arnold DL, Bar-Or A, Hutchinson M, Kappos L, Havrdova E, et al. Long-term effects of delayed-release dimethyl fumarate in multiple sclerosis: interim analysis of ENDORSE, a randomized extension study. Mult Scler 2017;23(2):253–65. Available from: https://doi.org/10.1177/1352458516649037.

[24] Naismith RT, Wundes A, Ziemssen T, Jasinska E, Freedman MS, Lembo AJ, et al. Diroximel fumarate demonstrates an improved gastrointestinal tolerability profile compared with dimethyl fumarate in patients with relapsing-remitting multiple sclerosis: results from the randomized, double-blind, Phase III EVOLVE-MS-2 study. CNS Drugs 2020;34(2):185–96. Available from: https://doi.org/10.1007/s40263-020-00700-0.

[25] Confavreux C, O'Connor P, Comi G, Freedman MS, Miller AE, Olsson TP, et al. Oral teriflunomide for patients with relapsing multiple sclerosis (TOWER): a randomised, double-blind, placebo-controlled, phase 3 trial. Lancet Neurol 2014;13(3):247–56. Available from: https://doi.org/10.1016/s1474-4422(13)70308-9.

[26] O'Connor P, Wolinsky JS, Confavreux C, Comi G, Kappos L, Olsson TP, et al. Randomized trial of oral teriflunomide for relapsing multiple sclerosis. N Engl J Med 2011;365(14):1293–303. Available from: https://doi.org/10.1056/NEJMoa1014656.

[27] Vermersch P, Czlonkowska A, Grimaldi LM, Confavreux C, Comi G, Kappos L, et al. Teriflunomide versus subcutaneous interferon beta-1a in patients with relapsing multiple sclerosis: a randomised, controlled phase 3 trial. Mult Scler 2014;20(6):705–16. Available from: https://doi.org/10.1177/1352458513507821.

[28] Kappos L, Radue E-W, O'Connor P, Polman C, Hohlfeld R, Calabresi P, et al. A placebo-controlled trial of oral fingolimod in relapsing multiple sclerosis. N Engl J Med 2010;362(5):387–401. Available from: https://doi.org/10.1056/NEJMoa0909494.

[29] Calabresi PA, Radue E-W, Goodin D, Jeffery D, Rammohan KW, Reder AT, et al. Safety and efficacy of fingolimod in patients with relapsing-remitting multiple sclerosis (FREEDOMS II): a double-blind, randomised, placebo-controlled, phase 3 trial. Lancet Neurol 2014;13(6):545–56. Available from: https://doi.org/10.1016/S1474-4422(14)70049-3.

[30] Cohen JA, Comi G, Selmaj KW, Bar-Or A, Arnold DL, Steinman L, et al. Safety and efficacy of ozanimod versus interferon beta-1a in relapsing multiple sclerosis (RADIANCE): a multicentre, randomised, 24-month, phase 3 trial. Lancet Neurol 2019;18(11):1021–33. Available from: https://doi.org/10.1016/S1474-4422(19)30238-8.

[31] Kappos L, Fox RJ, Burcklen M, Freedman MS, Havrdová EK, Hennessy B, et al. Ponesimod compared with teriflunomide in patients with relapsing multiple sclerosis in the active-comparator Phase 3 OPTIMUM study: a randomized clinical trial. JAMA Neurol 2021;78(5):558–67. Available from: https://doi.org/10.1001/jamaneurol.2021.0405.

[32] Kappos L, Bar-Or A, Cree BAC, Fox RJ, Giovannoni G, Gold R, et al. Siponimod versus placebo in secondary progressive multiple sclerosis (EXPAND): a double-blind, randomised, phase 3 study. Lancet 2018;391 (10127):1263–73. Available from: https://doi.org/10.1016/S0140-6736(18)30475-6.

[33] Hauser SL, Waubant E, Arnold DL, Vollmer T, Antel J, Fox RJ, et al. B-cell depletion with rituximab in relapsing-remitting multiple sclerosis. N Engl J Med 2008;358(7):676–88. Available from: https://doi.org/10.1056/NEJMoa0706383.

[34] Hauser SL, Bar-Or A, Comi G, Giovannoni G, Hartung H-P, Hemmer B, et al. Ocrelizumab versus interferon beta-1a in relapsing multiple sclerosis. N Engl J Med 2016;376(3):221–34. Available from: https://doi.org/10.1056/NEJMoa1601277.

[35] Montalban X, Hauser SL, Kappos L, Arnold DL, Bar-Or A, Comi G, et al. Ocrelizumab versus placebo in primary progressive multiple sclerosis. N Engl J Med 2016;376(3):209–20. Available from: https://doi.org/10.1056/NEJMoa1606468.

[36] Fox EJ, Markowitz C, Applebee A, Montalban X, Wolinsky JS, Belachew S, et al. Ocrelizumab reduces progression of upper extremity impairment in patients with primary progressive multiple sclerosis: Findings from the phase III randomized ORATORIO trial. Mult Scler 2018;24(14):1862–70. Available from: https://doi.org/10.1177/1352458518808189.

[37] Hauser SL, Bar-Or A, Cohen JA, Comi G, Correale J, Coyle PK, et al. Ofatumumab versus teriflunomide in multiple sclerosis. N Engl J Med 2020;383(6):546–57. Available from: https://doi.org/10.1056/NEJMoa1917246.

[38] Polman CH, O'Connor PW, Havrdova E, Hutchinson M, Kappos L, Miller DH, et al. A randomized, placebo-controlled trial of natalizumab for relapsing multiple sclerosis. N Engl J Med 2006;354(9):899–910. Available from: https://doi.org/10.1056/NEJMoa044397.

[39] Butzkueven H, Kappos L, Wiendl H, Trojano M, Spelman T, Chang I, et al. Long-term safety and effectiveness of natalizumab treatment in clinical practice: 10 years of real-world data from the Tysabri observational program (TOP). J Neurol Neurosurg Psychiatry 2020;91(6):660–8. Available from: https://doi.org/10.1136/jnnp-2019-322326.

[40] Foley JF, Defer G, Ryerson LZ, Cohen JA, Arnold DL, Butzkueven H, et al. Comparison of switching to 6-week dosing of natalizumab versus continuing with 4-week dosing in patients with relapsing-remitting multiple sclerosis (NOVA): a randomised, controlled, open-label, phase 3b trial. Lancet Neurol 2022;21:608–19. Available from: https://doi.org/10.1016/S1474-4422(22)00143-0.

[41] Trojano M, Ramió-Torrentà L, Grimaldi LM, Lubetzki C, Schippling S, Evans KC, et al. A randomized study of natalizumab dosing regimens for relapsing-remitting multiple sclerosis. Mult Scler J 2021;27(14):2240–53. Available from: https://doi.org/10.1177/13524585211003020.

[42] Ho PR, Koendgen H, Campbell N, Haddock B, Richman S, Chang I. Risk of natalizumab-associated progressive multifocal leukoencephalopathy in patients with multiple sclerosis: a retrospective analysis of data from four clinical studies. Lancet Neurol 2017;16(11):925–33. Available from: https://doi.org/10.1016/s1474-4422(17)30282-x.

[43] Alemtuzumab vs. interferon beta-1a in early multiple sclerosis. (2008). N Engl J Med, 2008;359(17):1786–1801. Available from: https://doi.org/10.1056/NEJMoa0802670.
[44] Cohen JA, Coles AJ, Arnold DL, Confavreux C, Fox EJ, Hartung HP, et al. Alemtuzumab versus interferon beta 1a as first-line treatment for patients with relapsing-remitting multiple sclerosis: a randomised controlled phase 3 trial. Lancet 2012;380(9856):1819–28. Available from: https://doi.org/10.1016/s0140-6736(12)61769-3.
[45] Havrdova E, Arnold DL, Cohen JA, Hartung H-P, Fox EJ, Giovannoni G, et al. Alemtuzumab CARE-MS I 5-year follow-up: durable efficacy in the absence of continuous MS therapy. Neurology 2017;89(11):1107–16. Available from: https://doi.org/10.1212/WNL.0000000000004313.
[46] Ziemssen T, Bass AD, Berkovich R, Comi G, Eichau S, Hobart JTOPAZ Investigators, T. Efficacy and safety of alemtuzumab through 9 years of follow-up in patients with highly active disease: post hoc analysis of CARE-MS I and II patients in the TOPAZ extension study. CNS Drugs 2020;34(9):973–88. Available from: https://doi.org/10.1007/s40263-020-00749-x.
[47] Havrdova E, Horakova D, Kovarova I. Alemtuzumab in the treatment of multiple sclerosis: key clinical trial results and considerations for use. Therap Adv Neurol Disord 2015;8(1):31–45. Available from: https://doi.org/10.1177/1756285614563522.
[48] Giovannoni G, Comi G, Cook S, Rammohan K, Rieckmann P, Sørensen PS, et al. A placebo-controlled trial of oral cladribine for relapsing multiple sclerosis. N Engl J Med 2010;362(5):416–26. Available from: https://doi.org/10.1056/NEJMoa0902533.
[49] Rice GPA, Filippi M, Comi Gfor the Cladribine Clinical Study Group and for the Cladribine MRI Study Group. Cladribine and progressive MS. Clinical and MRI outcomes of a multicenter controlled trial. Neurology 2000;54(5):1145–55. Available from: https://doi.org/10.1212/wnl.54.5.1145.

CHAPTER 6

Use of magnetic resonance imaging and quantitative imaging reports in clinical care of multiple sclerosis

Tomas Uher[1] and Manuela Vaneckova[2]

[1]Department of Neurology and Center of Clinical Neuroscience, First Faculty of Medicine and General University Hospital in Prague, Charles University, Prague, Czech Republic
[2]Department of Radiology, First Faculty of Medicine and General University Hospital in Prague, Charles University, Prague, Czech Republic

OUTLINE

Introduction	94
Magnetic resonance imaging in multiple sclerosis diagnosis	94
Magnetic resonance imaging in differential diagnosis of multiple sclerosis	98
Magnetic resonance imaging in the prediction of disease activity	100
Magnetic resonance imaging for monitoring disease activity	100
Magnetic resonance imaging in drug safety surveillance	106
Quantitative magnetic resonance imaging reports in clinical care	109
Disclosures	113
References	114

Introduction

The clinical phenotypes of multiple sclerosis (MS) include a wide range of neurological signs and symptoms that originate from damage to the brain and spinal cord. In MS, disease presentation is nonspecific and often subclinical; therefore, diagnosing and monitoring MS is challenging. This has prompted an investigation of specific and sensitive biomarkers for the disease. Among different paraclinical measures, magnetic resonance imaging (MRI) is currently the most established tool for diagnosing, predicting, and monitoring disease activity and drug safety surveillance in MS [1,2].

Freedom from clinical and radiological disease activity has become a new goal in the management of MS [3]. The demand for early diagnosis and early detection of treatment failure has also increased [4]. All these factors increase the importance of MRI in the management of MS. In this chapter, we present an overview of the use of MRI in the clinical care of MS.

Magnetic resonance imaging in multiple sclerosis diagnosis

The diagnostic criteria for MS include clinical, MRI, and laboratory evidence. Given that the radiological presentation may be more specific for a diagnosis of MS than the clinical or laboratory findings, MRI plays an essential role in diagnosis. The imaging hallmark of MS is focal demyelinated lesions within the brain and spinal cord. Although central nervous system (CNS) lesions in MS represent a heterogeneous and dynamic group of focal brain pathologies, ranging from edema and inflammation to demyelination, gliosis, and axonal loss, it is well-accepted that they are neuroinflammatory in origin [5,6]. The diagnosis of MS requires the dissemination of disease activity in space (DIS) and dissemination of disease activity in time (DIT) in patients with typical clinical symptomatology of MS originating from the CNS. MRI can be used to determine the DIS and DIT. To determine the DIS, brain or spinal cord lesions must be identified at different locations that are characteristic of MS (Table 6.1); however, to determine the DIT, lesion formation must be shown to be progressing over time (Table 6.2) [1,2].

Assessing lesions on conventional MRI involves an evaluation of their morphology, size, location, and borders. Brain lesions are hyperintense on T2-weighed images (T2 lesions), usually have a round to ovoid shape, are distributed asymmetrically between hemispheres, show no mass effect, and usually have either no or very little vasogenic edema. The long axis of focal lesions should be at least 3 mm; thus, small punctate lesions should not be used to meet the MRI criteria. MS lesions are most often localized in the periventricular and juxtacortical or cortical regions. Other typical locations include the infratentorial region and spinal cord, especially the cervical spine. Periventricular lesions are typically perpendicular to the lateral ventricle and form wedge-shaped hyperintensities with a broad base to the lateral ventricle ("Dawson's fingers") [1,7]. In the periventricular region, lesions are also typical in the corpus callosum, usually with a calloso-striatal orientation. Other typical locations of MS lesions include the juxtacortical region and cerebral cortex [1,7]. However, cortical lesion imaging is challenging, as even the most sensitive sequences (e.g., double inversion recovery [DIR], phase-sensitive inversion recovery

TABLE 6.1 Evidence of at least one T2 hyperintense lesion in at least two (out of four) distinct anatomical locations within the CNS is necessary to determine DIS on MRI.

Location	Evaluation
a. Periventricular	— T2-hyperintense WM lesion which abuts lateral ventricles. Includes lesions in the corpus callosum, but excludes lesions in the deep GM (amygdala, caudate, putamen, hippocampus, hypothalamus, nucleus accumbens, globus pallidus, thalamus). — Although they are most often seen in patients with MS, they are not specific for MS. They are often present in older patients (aged >50 years) or patients with vascular risk factors or other comorbidities, such as migraines or small vessel disease. In these cases, identifying ≥3 mm periventricular lesions is recommended. — "Capping" at the ventral and dorsal horns of the lateral ventricles occurs with normal aging. — Lesions touching the third ventricle or the cerebral aqueduct are not considered periventricular and are not typical for MS. — Lesions near the fourth ventricle are considered infratentorial lesions. — Linear plate-like lesions parallel to the lateral ventricles ("periventricular banding" or "halo") are not typical for MS. — Lesions need to be distinguished from enlarged perivascular (Virchow-Robin spaces), which has signal intensity as CSF (hypersignal on T2WI and hyposignal on FLAIR).
b. Juxtacortical or cortical	— T2-hyperintense WM lesion touching the cortex or intracortical lesions. — Mostly includes leukocortical lesions. Intracortical and subpial type lesions are rarely detected by conventional imaging. — Cortical lesions are often located along the subpial surface (but are rarely detected by conventional imaging) and are associated with inflammation in the adjacent meninges. — Consideration of cortical lesions has been added to the McDonald 2017 criteria.
c. Infratentorial	— T2-hyperintense WM lesion in the brainstem (mesencephalon, pons, oblongata, cerebellar peduncles) or cerebellum. — Usually adjacent to inner or outer CSF borders. — Lesions of the thalamus, subthalamus, hypothalamus, and epithalamus are supratentorial, and are thus not considered infratentorial. — Asymptomatic and symptomatic infratentorial lesions could be included as MRI evidence of DIS and DIT in the McDonald 2017 criteria.

d. Spinal cord	— T2-hyperintense lesion in the cervical, thoracic, or lumbar spinal cord.
	— Asymptomatic and symptomatic spinal cord lesions could be included as MRI evidence of DIS and DIT in the McDonald 2017 criteria.
	— If 2 or more spinal cord lesions are detected in patients with primary progressive MS, no brain lesions on MRI are needed for evidence of DIS (additionally, no distinction between symptomatic and asymptomatic spinal cord lesions is needed).
	— Spinal MRI (whole spine or at least between C1 and T4–5) is recommended especially
	  a) when symptoms originate from the spinal cord.
	  b) if the phenotype is primary progressive.
	  c) in atypical populations; or
	  d) to increase diagnostic confidence.
	— Spinal MRI seems to be less useful for diagnosing MS in children than in adults.

TABLE 6.2 Evidence of new CNS lesion formation over time is necessary to determine DIT on MRI. There are two possible scenarios.

Scenario	MRI sequence and evaluation
Simultaneous presence of contrast-enhancing and nonenhancing CNS lesions at any time.	– Patients with possible MS should have a diagnostic MRI that includes postcontrast T1WI.
A new T2 and/or contrast-enhancing lesion on follow-up MRI (with reference to a baseline scan), irrespective of the timing of the baseline MRI.	– Protocols with 3D, isotropic voxel, and automatic subtraction software increase the sensitivity to detect new T2 lesions. – Detecting only enlarged (not new) T2 lesions is not considered for determining dissemination in time

[PSIR], or magnetization-prepared 2 rapid gradient-echo (MP2RAGE)) detect only a minor fraction of the cortical lesions that are detected by histopathology and cortical lesions can be easily confused with artifacts [1,7–9]. Finally, infratentorial lesions are usually adjacent to the fourth ventricle or reach the outer surface of the brainstem (e.g., adjacent to the cerebrospinal fluid [CSF] borders). In the midbrain, lesions are typically located in the cerebral peduncles and close to the periaqueductal gray matter (GM); in the pons, they are located at the periphery of the pons; while in the medulla oblongata, they are typically in a paramedian location either uni- or bilaterally. Lesions may also be located in the cerebellar hemispheres, medial longitudinal fasciculus, or trigeminal nerve root entry zones [1,7].

Some hyperintense T2 lesions display as hypointense lesions on spin echo T1-weighted imaging (T1WI). There are two types of T1 hypointense lesions. The first type involves new lesions that disappear within 6–12 months, probably due to remyelination or resolution of edema. The second type involves T1 hypointense lesions that persist after 12 months. These lesions are called "black holes" and represent areas of permanent focal axonal loss. The most accurate sequence for their detection is spin echo because gradient-echo sequences tend to overestimate their presence [10,11]. Although T1 hypointense lesions are not considered in the diagnostic criteria of MS, their presence may have negative prognostic value [12].

Except for the very early stages of lesion formation, MS lesions usually do not show diffusion restriction on diffusion-weighted imaging (DWI), which distinguishes them from stroke, CNS lymphoma or brain abscesses [13].

The presence of contrast-enhanced T1 lesions indicates acute inflammation. These lesions are associated with blood–brain barrier breakdown due to active inflammation and persist usually for only 2–8 weeks (typically between 3 and 4 weeks). Similar to T2 lesions, the long axis should be at least 3 mm. Occasionally, persistent contrast-enhanced lesions (>3 months), in which active inflammation is questionable and may indicate either chronic focal disruption of the blood–brain barrier or an alternative diagnosis, such as a vascular abnormality (developmental venous abnormality, capillary telangiectasia) sarcoidosis, high-grade glioma, or CNS lymphoma are observed. Contrast enhancement have usually either a nodular (smaller lesions), ring, or horseshoe (larger lesions) shape. Although lesions in MS have "open-ring" enhancement, some large lesions may show less typical "closed-ring" enhancement [7]. The leptomeningeal enhancement may be visible on post-contrast FLAIR with a 15-minute delay after contrast administration. Although more pronounced in chronic stages, this feature is not specific to MS [14–16].

Spinal cord lesions are hyperintense on T2-weighted imaging (T2WI) and affect both the white matter (WM) and GM. They are usually small (though still at least 3 mm), circumscribed (well-demarcated), cigarette shaped on sagittal images and wedge-shaped on axial images, short (one to two spinal segments), involve <50% of the axial cord area, are located mostly in the peripheral dorsal and/or lateral spinal cord, and usually reach the outer surface of the spinal cord. However, spinal cord lesions can also be located in the anterior WM and central GM, although lesions strictly located in the GM are not usual [7]. In the acute stage, spinal cord lesions may be associated with mild edema and increased regional spinal cord volume, making the evaluation of spinal cord atrophy over time challenging. Spinal cord lesions tend to be aligned with the long axis of the cord and most often affect the cervical and upper thoracic spine [7,17,18]. On the other hand, lesions in the central area that affect more than 50% of cross-sectional area are typical for neuromyelitis optica spectrum disorders (NMOSD) or myelin oligodendrocyte glycoprotein antibody-associated disease (MOGAD) and are typically longer than three spinal segments. MOGAD also typically involves the conus medullaris [17,19,20]. In a proportion of patients, not only focal but also diffuse spinal cord lesions are observed, making it difficult for them to be distinguished from artifacts [17,18,21]. The origin of diffuse spinal cord lesions is still unclear, and different hypotheses have been proposed, including the transition of focal lesions to diffuse lesions through disease progression. Diffuse spinal cord lesions are not included in the current MS diagnostic criteria because of their low specificity and limited reliability in lesion detection [2,7]. Spinal cord MRIs are often associated with major imaging artifacts. Therefore, to distinguish MS spinal cord lesions from artifacts, at least two sequences should be combined, such as T2WI along with proton density- weighted images (PDWI) or short tau inversion recovery (STIR). If only one sequence is available, the lesion should be detected in two planes. If a contrast agent is administered for postcontrast brain imaging, postcontrast sagittal T1WI of the spinal cord is recommended. Spinal cord lesions are typically isointense on T1WI; however, acute contrast-enhancing spinal cord lesions may occasionally occur. If possible, MRI should cover the whole spinal cord; however, if this is not feasible, at least the cervical and upper thoracic spine up to T4—5 should be covered. Transversal images allow for the location of lesions (lateral, latero-dorsal, or central) to be better defined and provide helpful information for the differential diagnosis of MS [1,7,20].

Radiologically isolated syndrome (RIS) is diagnosed in individuals that meet the MRI criteria for DIS who have no clinical manifestations of MS or other clear explanations of CNS lesions. For RIS to be diagnosed, a larger number of typically focal T2 lesions located in MS-specific regions is required [22]. However, according to the revised 2023 diagnostic criteria, RIS can be also diagnosed in patients with at least one lesion in two of the four typical MS locations, along with two of the following three features: a) the presence of CSF-restricted oligoclonal bands (OCB), b) at least one spinal cord lesion, and c) active lesion (new T2 or contrast-enhancing lesion). Currently, clinical symptomatology is required to diagnose MS and initiate immunomodulatory treatment [23].

Solitary sclerosis is a rare condition diagnosed in patients with a combination of solitary inflammatory lesions (of the brain WM, cervicomedullary junction, or spinal cord) and clinical phenotypes suggestive of primary progressive MS. CSF-specific oligoclonal bands might be present, and there is no clinical or radiological evidence of new lesion activity. These patients do not fulfill the McDonald 2017 criteria for MS because they do not have DIS. The pathophysiology of solitary sclerosis and its potential relationship with MS are unclear [24].

In general, the use of 3 T scanners, which have a higher sensitivity for detecting MS lesions in the brain, is preferred. The basic MRI protocol involves a three-dimensional (3D)-FLAIR sequence (a key sequence for diagnosis) and T2WI in the transverse plane. When 3D-FLAIR is not available, it can be substituted with 2D-FLAIR in the transverse and sagittal planes. However, 3D-FLAIR images have better sensitivity for detecting brainstem lesions than 2D-FLAIR or 2D-T2WI [1].

Another important sequence includes postcontrast T1WI (3D or 2D in the transverse plane). Here, it is important to wait at least 5 min after applying the contrast agent to perform the postcontrast scan. Assessment of brain volume changes requires high-resolution 3D-T1WI with isotropic voxels. Although DWI is not necessary, it may help with the differential diagnosis in some cases. The standard MRI has limited sensitivity for the detection of cortical lesions. Hence, the DIR sequence has been developed to improve the detection of cortical lesions, which are already included in the McDonald 2017 diagnostic criteria. Thus adding this sequence to the diagnostic protocol may be considered.

Magnetic resonance imaging in differential diagnosis of multiple sclerosis

Although MRI plays a central role in the diagnosis of MS, misinterpretation and an overreliance on MRI pathology, especially in patients with nonspecific neurological symptoms, are frequent contributors to the misdiagnosis of MS [25–27]. However, MRI scans that do not meet the imaging diagnostic criteria in patients with a neurological presentation suggestive of MS do not exclude a potential future diagnosis of MS and should indicate imaging follow-up to determine the presence of DIS and DIT. The added value of repeated spinal cord MR images will likely be significantly lower compared to brain MR images [1]. It is important to emphasize that the McDonald criteria were developed to identify patients with clinically isolated syndrome who are at the highest risk of developing clinically definitive MS; they were not intended to differentiate MS from other medical conditions [1,2]. Indeed, the number of patients with other diseases affecting the CNS may meet the diagnostic criteria for MS, especially when DIS is considered (Fig. 6.1). Therefore clinical symptomatology and laboratory findings should be considered in the diagnosis of MS to decrease the likelihood of a potential misdiagnosis. Currently, significant concern has been expressed regarding misdiagnosis, which can lead to serious consequences. Misdiagnosis may be associated not only with inappropriate treatment of the underlying disease but also with considerable psychological burden, risk of immunomodulatory drug adverse events, financial burden and increased risk of disability [25,26,28,29]. Taken together, when a diagnosis of MS is unclear, maintaining the threshold for additional testing, including spinal MRI, is recommended. Some additional MRI sequences with the potential to improve differential diagnosis in patients with MS have been proposed, such as susceptibility weighted images (SWI), DIR, or MRI of the optic nerve [1].

DIR, particularly in 3D acquisition, or PSIR sequences can improve the detection of cortical MS lesions; however, interpretation of these sequences is associated with high interrater variability and a risk of mistaking cortical lesions for neuroimaging artifacts, especially by inexperienced neuroradiologists. Hence, these sequences should be restricted to centers with sufficient expertise and image acquisition standardization. Cortical lesions are more specific for MS than WM lesions and are not present in the majority of patients with

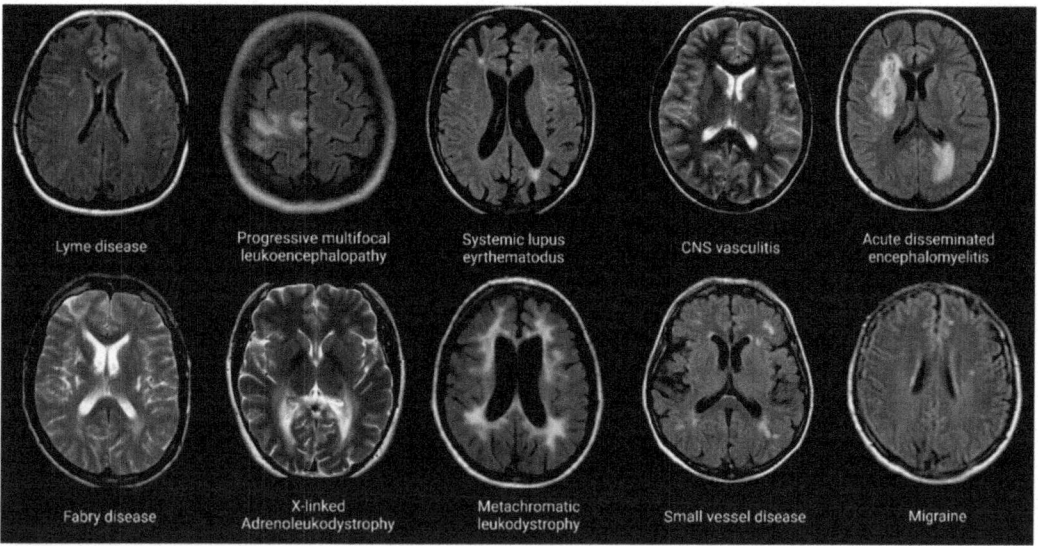

FIGURE 6.1 Axial brain MR images of patients with diagnoses other than MS that meet the MRI criteria for DIS.

NMOSD (only present in 3%), migraines, or in healthy subjects, but can be present in patients with vasculitis [1,7,9,30].

The use of gradient sequences (T2*-weighted imaging [T2*-WI] or SWI; 3 T scanner recommended) can show the so-called central vein sign (CVS). The combination of gradient and FLAIR sequences produces FLAIR* images, which are more sensitive for CVS evaluation. To differentiate MS from other diagnoses, a cutoff of 40%–50% of lesions with a CVS has been suggested [31,32]. However, the optimal gradient sequences for detecting the CVS are not yet available for all clinical scanners, and the cutoff may depend on the sequence and individual experience of neuroradiologist. In patients with a very high lesion burden or large confluent lesions, counting and evaluating all lesions would be challenging. Additionally, determining the number of lesions with the CVS may not be possible in patients with very few lesions (often in early disease stages, when differential diagnosis is most important). Therefore the CVS should be used for the differential diagnosis with caution and in MRI centers that have experience with this method [1].

SWI or quantitative susceptibility mapping using 3 T scanners can be used to identify paramagnetic rim lesions occurring in approximately 50% of patients. Their presence increases with disease progression; thus, they are mostly observed in advanced disease stages. The paramagnetic rim reflecting iron deposition in the microglia and macrophages at the edge of chronic active lesions is rarely observed in other brain diseases. Therefore detecting the paramagnetic rim could potentially help with the differential diagnosis. However, future research is needed to provide recommendations for its use in practice [1,7,33].

Imaging of optic nerve lesions on MRI requires additional coronal T2 sequences with fat suppression. Optic nerve lesions are usually hyperintense on T2WI or enhanced on postcontrast T1WI (perioptic nerve sheath enhancement). Although a clinically silent optic nerve lesion could provide evidence of DIS, this is not included in the McDonald 2017 criteria because of its limited add-on value for improving the accuracy of MS diagnosis [1,2,7].

In some patients with MS, leptomeningeal enhancement is observed (small foci or thin lines of contrast enhancement when using delayed gadolinium-enhanced 3D-FLAIR sequences). However, leptomeningeal enhancement on MRI is nonspecific and can be present across many other pathologies, including neuroinfections, neoplasmas, and other neuroinflammatory diseases, such as MOGAD or sarcoidosis. Although the presence of leptomeningeal enhancement in patients with MS may be associated with worse clinical outcomes, its routine use in patients with MS is not currently recommended in clinical practice [14–16].

Below, we present an overview of MRI red flags that, when identified, should raise suspicion for an alternative diagnosis to MS (Table 6.3) [1,7,25,26,34–37].

Magnetic resonance imaging in the prediction of disease activity

The heterogeneous presentation of MS phenotypes results in limited predictability of the disease course and treatment response. Compared to clinical or biochemical measures, abnormal MRI findings are the most informative prognostic markers of future disease activity. However, predicting the disease course using MRI in clinical practice has limitations. First, it is widely accepted that conventional MRI measures significantly underestimate ongoing CNS damage. This can be illustrated by the very low detection rate of cortical lesions [7,9] and the inability of conventional MRI to detect pathological changes in normal-appearing WM or GM [38–40]. On the other hand, the extent of pathology visualized by conventional MRI methods may represent the "tip of the iceberg," which correlates relatively well with the overall CNS disease burden. Second, research studies have shown an association between early MRI measures and future disease activity; however, predicting the MS course in individual patients is fraught due to the relatively low accuracy of MRI predictors [41]. Even though predictions using MRI at the individual patient level is imperfect, it provides certain information about the increased risk of future disease activity. The predictive role of MRI is important, especially in early disease stages before treatment is initiated, because it may help stratify patients based on their risk of disease activity and identify those who should start with more efficacious immunomodulatory treatments. The conventional MRI measures used for the prediction of disease activity include the T1 and T2 lesion number, lesion load (volume), and presence of contrast-enhancing lesions. However, the location of lesions in the infratentorial region or spinal cord may also provide valuable information on individual risk assessment. Further details of the most commonly used MRI predictors are described in Table 6.4 [42–47].

Magnetic resonance imaging for monitoring disease activity

Traditional clinical predictors are not sufficiently sensitive to monitor disease activity. Only a very small proportion of brain or spinal cord lesions detected on MRI are symptomatic. Indeed, new lesions on brain MRI are approximately 5–10 times more frequent than clinical relapses, especially in relapsing–remitting patients [1,48]. Lesion MRI activity has been shown to be the most informative and sensitive surrogate of subclinical MS

TABLE 6.3 Some MRI red flags that should increase suspicion of a possible alternative diagnosis.

MRI "red flags"	Possible alternative diagnoses
Brain	
Calcifications	Mitochondrial disease, parasitic infections, brain tumors, celiac disease, Cogan's syndrome, some inherited metabolic diseases, sarcoidosis, systemic lupus erythematodes, systemic sclerosis or scleroderma
Dilation of the Virchow-Robin spaces	Mucopolysaccharidosis, primary CNS angiitis, cerebral small vessel disease
Edematous lesions with a typical marbled or "arch bridge" appearance	Neuromyelitis optica spectrum disorders (NMOSD)
Hypothalamic and/or pituitary stalk infiltration	Granulomatosis with polyangiitis, Histiocytosis disorders, IgG4-related disease, neuroinfections, sarcoidosis, drug-induced
Large lesions without mass effect and enhancement	Progressive multifocal leukoencephalopathy (PML)
Lesions across GM/WM boundaries	Stroke, cerebral autosomal dominant arteriopathy with subcortical infarcts and leukoencephalopathy (CADASIL), lupus erythematosus, CNS vasculitis, PML
Lesions in the basal ganglia, thalamus, or hypothalamus	Acute disseminated encephalomyelitis (ADEM), Behçet's disease, mitochondrial disease, Sussac's syndrome
Multifocal "snowball-like" lesions in the central corpus callosum	Sussac's syndrome
Petechial hemorrhage, pial and subarachnoid hemorrhage	CNS vasculitis
Predominance of lesions at the cortical or subcortical junction	Embolic infarction, PML, CNS vasculitis
Restriction on DWI	Abscess, acute ischemia, lymphoma, hyperacute MS lesion
Selective or predominant involvement of the anterior temporal lobe (can also be lesions of the inferior frontal lobe or u-fibers at the vertex, external capsule, and insula)	Cerebral autosomal recessive arteriopathy with subcortical infarcts and leukoencephalopathy (CADASIL), cathepsin A-related arteriopathy with strokes and leukoencephalopathy (CARASAL)
Symmetrically distributed confluent WM lesions	Leukodystrophies and other inherited metabolic disorders
T1 hyperintensity of the pulvinar	Hepatic encephalopathy, manganese and calcium deposits, toxicity
T2 hypersignal lesions with microhemorrhages	Stroke with hemorrhagic transformation, CNS vasculitis, vascular malformation, CADASIL
Brainstem	
Central lesions along the transverse pontine fibers	Stroke, ischemic small-vessel disease, central pontine myelinolysis (MS lesions are usually located at the periphery of the pons)
Fluffy lesions in the brain parenchyma, ADEM-like lesions, large lesions involving the brainstem in particular areas adjacent to the fourth ventricle and the cerebellar peduncles	MOGAD

(Continued)

TABLE 6.3 (Continued)

MRI "red flags"	Possible alternative diagnoses
Large and infiltrating brainstem lesions	Behçet's disease, Bickerstaff's brainstem encephalitis, CARASAL, o central pontinne myelopathy, mitochondrial disorders, NMOSD, pontine glioma, other brain tumors
Lesions in the periaqueductal area, area postrema (often as paired and discrete), and solitary tract	NMOSD
Regional atrophy of the brainstem	Behçet's disease, adult-onset Alexander's disease
Spinal cord	
Central spinal cord lesions that involve >50% of the cross-sectional spinal cord area	NMOSD, MOGAD
Complete transverse myelopathy	Wide range of alternative diagnoses other than MS
Diffuse abnormalities in the posterior columns of the spinal cord	B12 or copper deficiency, paraneoplastic disorder, neurosyphilis
Involvement of the caudal spinal cord	MOGAD
Leptomeningeal or nerve root involvement	Sarcoidosis, infectious disease, malignancy
Lesions in the anterior two-thirds of the spinal cord ("snake eye" or "owl's eye" sign), bilateral hyperintensities of the anterior GM horns	Ischemia or spinal cord compression occurring in spondylotic myelopathy
Long and selective involvement of the WM columns	Metabolic diseases (vitamin B12 or copper deficiency)
Longitudinally extensive myelitis, lesions extending over three or more vertebral segments	Autoimmune disorders (NMOSD, antiphospholipid syndrome, MOGAD, Behçet's disease, lupus erythematosus, sarcoidosis, Sjögren's syndrome), infectious or para-infectious myelitis, paraneoplastic conditions, vascular diseases (spinal cord infarct, arteriovenous malformations) and others
Micro or macro-bleeds in the spinal cord	Arteriovenous fistula
Prominent involvement of the central GM and spinal cord swelling	NMOSD, MOGAD
Optic nerve	
Absence of an acute T2-hyperintense optic nerve lesion	Ischemic and toxic optic neuropathies; Leber's hereditary optic neuropathy do not show T2-hyperintensity
Posterior optic nerve involvement including the chiasm	NMOSD
Simultaneous bilateral optic nerve involvement and a long optic nerve lesion	NMOSD, MOGAD
Soft tissue enhancement extrinsic to the optic nerve affecting the orbit, orbital apex, or cavernous sinus	Granulomatous disease, tumor, infection, MOGAD
Contrast enhancement	
Band-like brain lesion enhancement	Baló's concentric sclerosis
Cloud-like brain lesion enhancement	NMOSD, MOGAD

(Continued)

TABLE 6.3 (Continued)

MRI "red flags"	Possible alternative diagnoses
Contrast enhancement of lesion >3 months	Brain tumors, chronic neuroinfections, histiocytosis disorders, lymphoma, sarcoidosis, venous anomaly, capillary telangiectasia
Dorsal subpial enhancement combined with enhancement of the central canal of the spinal cord ("trident sign")	Sarcoidosis, rarely B12 deficiency
Meningeal enhancement	Aseptic meningitis, lymphomatosis or meningeal carcinomatosis, sarcoidosis, neurosyphilis, vasculitis
Brain cortical lesion enhancement only	Subacute ischemia
Patchy and persistent brain lesion enhancement	Capillary telangiectasia
Patchy/punctate or large ring enhancement in the spinal cord	NMOSD
Punctiform parenchymal contrast enhancement	Sarcoidosis, CNS vasculitis
Punctiform parenchymal contrast enhancement confined to the pons (can also be the peduncles, hemispheres, and cervical spinal cord)	Chronic lymphocytic inflammation with pontine perivascular enhancement responsive to steroids (CLIPPERS)
Simultaneous contrast enhancement of all lesions	Lymphoma, sarcoidosis, CNS vasculitis, neoplasm, neuroinfections
Transverse pancake-like band of enhancement at the spinal cord just below the site of maximal stenosis	Cervical spondylosis with cord compression

TABLE 6.4 Conventional MRI predictors of future disease activity used in clinical practice.

MRI marker	Comments
T2 lesion number and volume	— There is no clear cutoff value, but more lesions are associated with increased future disease activity — Greater lesion volume (cumulative size) is also associated with a higher risk of disease activity
Presence of contrast-enhancing lesion	— A greater number of enhancing lesions is associated with a greater risk of disease activity — Persistently enhancing lesions have not a clear predictive value
New T2 lesion New contrast-enhancing lesion	— A cutoff for the number of new T2 lesions that would indicate treatment failure has not been established. In most studies, it ranges from 2 to 3 — Detection rate of new T2 lesions is highly influenced by the MRI protocol, changes to the scanner (different scanner or updates), use of subtraction software, and radiologist's expertise — Clinicians must consider not only the number but also the size and location of new lesions for treatment decision making (mainly infratentorial and spinal cord lesions) — A new contrast-enhancing lesion is associated with a greater risk of future disease activity compared with a new T2 lesion
T1 lesion number and volume	— A greater number and larger size (volume) is associated with greater risk — Presence of "black holes" is a risk factor for future disease activity
Presence of infratentorial, cortical, or spinal cord lesion	— All are associated with a higher risk of future disease activity
Increased rate of global and regional brain volume loss	— A pathological cutoff of whole brain volume loss ≤0.4% per year has been proposed; however, it is difficult to apply to individual patients due to the high technical and biological variability of MRI volumetric measurements

activity [45,49]. Therefore standard monitoring of MS patients includes not only regular assessments of new neurologic signs and symptoms, but also regular MRI follow-ups.

Monitoring radiological disease activity includes searching for new active lesions on follow-up MRIs (sometimes called combined unique active lesions). Active lesions are defined as current new contrast-enhancing lesions on T1WI or new or enlarged lesions on T2WI compared with the most-recent MRI. A new lesion on T2WI is usually defined as a rounded or oval lesion (≥3 mm in size) arising from an area previously considered normal-appearing brain tissue. Enlarged T2 lesions show an identifiable increase in size from a previously stable-appearing lesion. There is no standard definition of an enlarged lesion. In the case of small lesions, an increase in volume of at least 50% is usually considered a sign of lesion activity. In cases of large or confluent lesions, the establishment of lesion activity is more ambiguous. To obtain information about ongoing disease activity in patients, brain MRIs should be performed at least every 12 months. In selected cases, a greater frequency is recommended [1]. Conversely, slightly longer intervals (2–3 years) could be considered in clinically stable patients with a long disease duration who are not receiving high-efficacy immunomodulatory treatments. Very low serum neurofilament levels may help in the identification of patients with a very low probability of recent radiologic disease activity during the preceding year, thus replacing the need for annual brain MRI monitoring in some proportion of MRI visits in clinically stable patients [50]. However, this approach needs to be confirmed in future studies.

The occurrence of new lesions indicates ongoing disease activity and may provide information about treatment efficacy. Radiological disease activity itself, or in association with clinical disease activity, is usually a reason to change the treatment plan or escalate treatment with the aim of decreasing ongoing disease activity and the risk of disability progression in the future. Although the basic principle of monitoring radiological disease activity in MS is simple, real-world practice is challenging. Serial MRI follow-up has its limitations, including the need for standardized MRI scanning protocols, high cost, and unsolved challenges associated with accessible measures of the neurodegenerative component of MS. In addition, asymptomatic spinal cord lesions are not accompanied by active asymptomatic brain lesions in approximately 10% of clinically stable patients [51]. However, in routine settings, MR images frequently do not cover the entire spinal cord, and the detection of active spinal cord lesions is challenging due to frequent imaging artifacts and relatively low sensitivity [1,18,52]. More details are provided in Table 6.5.

In recent years, automated segmentation-based tools for the coregistration of consecutive MR images have been introduced. Some of these tools are commercially available and approved by regulatory authorities (LesionQuant from NeuroQuant [53]; Icobrain/MSmetrix from Icometrix [54]; open-source software such as ITK-SNAP [55]; and FLAIR fusion with syngo.via from Siemens [56]).

Segmentation-based software can increase the detection rate of active T2 lesions. This is highly valuable, especially in patients with a very high lesion burden or in those with new or newly enlarged MRI lesions that are very small and are easily missed even by experienced neuroradiologists. Although these methods enhance the sensitivity of MRI monitoring, they are not yet widely used because many neuroradiologists and neurologists are still unfamiliar with their use and their application requires standardized image acquisition (the same MRI scanner and protocol) (Fig. 6.2) [1].

TABLE 6.5 Recommendations and challenges of MRI monitoring of MS disease activity in practice.

	Recommendation	Challenge in practice
MRI scanner	– Identical MRI scanner, preferably 3 T (minimal 1.5 T)	– Use of different MRI scanners make detection of active lesions difficult
Sequence for brain MRI	– 3D-FLAIR (if high quality, T2WI is not needed) – 3D-FLAIR outperforms 2D-FLAIR in the detection of active lesions – Postcontrast T1WI only if needed – Use identical MRI protocol	– Changing the MRI protocol makes detection of active lesions difficult – 2D or thick slice thickness decreases the detection rate of active lesions
Head repositioning	– Use automatic repositioning techniques	– Repositioning errors may affect assessment of lesion activity and measurement of lesions and brain volume
Frequency	– Every 12 months (or 3–6 months after new treatment initiation)	– Often postponed due to organizational issues
Evaluation of lesion activity	– Detect all active T2 lesions – Consider using automated segmentation-based software for coregistration of consecutive MRI (3D-FLAIR) images – Experienced neuroradiologist – MRI reports describing the number, size, and location of active lesions – Review by neurologist to appreciate lesion characteristics (size, location, morphology, borders)	– In patients with a high lesion burden, detection of small active lesions may be challenging. In these cases, segmentation-based tools for coregistration of consecutive MRI scans can be helpful in increasing the detection rate of new lesion activity – Evaluation by inexperienced radiologist – Unstandardized and incomplete MRI reports – MR images are often not seen by clinicians who only review MRI reports
Evaluation of neurodegeneration	– Volumetric software and scanning on single MRI scanner	– With a few exceptions, this is not yet recommended for use in individual patients

One of the main limitations of MRI in MS is the histopathological heterogeneity of MS lesions and challenges of MRI monitoring of chronic inflammation in progressive MS [5,6,57]. To address this issue, a new MRI measure of chronic active lesions (e.g., smoldering lesions or slowly expanding lesions) characterized by chronic inflammation and ongoing tissue loss has been described. These are defined as concentric regions of existing lesions with slow local expansion and often progressive hypointensity on T1WI. This type of lesion is seen mostly in patients with progressive MS phenotypes but can also occur in relapsing–remitting patients. MS lesions with a paramagnetic rim on SWI sequences (e.g., phase-rim lesions) are characterized by ongoing chronic inflammatory demyelination, iron accumulation in microglia and macrophages, and slow expansion [1,7,58]. Slowly expanding lesions show only partial correspondence with iron-rim lesions, suggesting that both types of chronic active lesions represent different aspects or stages of MS pathology [33]. Importantly, both lesion types are associated with disability progression [12,59]; however, since their identification is challenging and not standardized, their use in clinical practice remains limited [1,33].

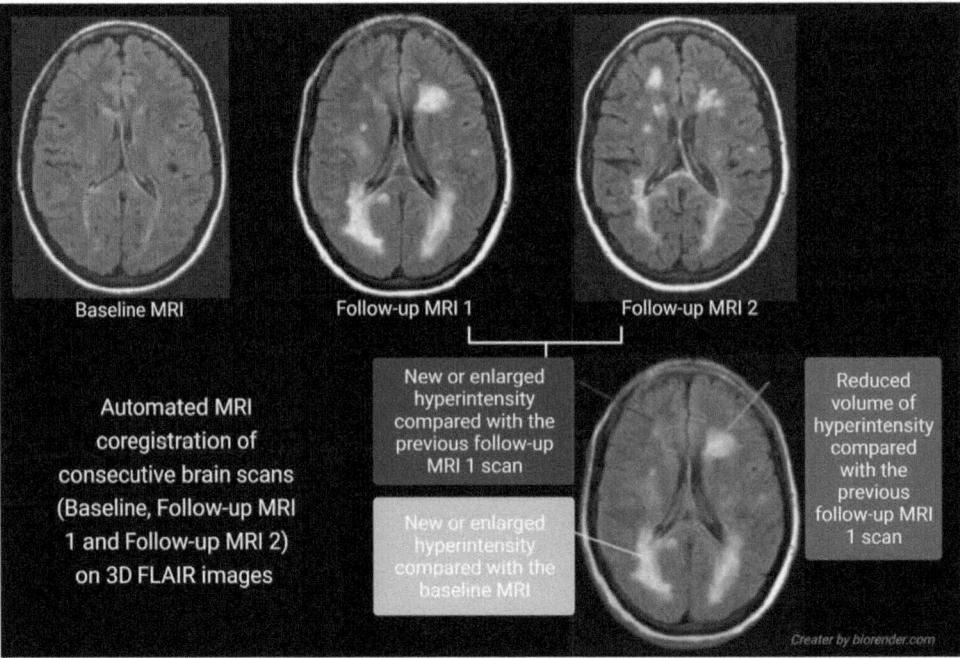

FIGURE 6.2 Example of automated segmentation-based tool for coregistration of consecutive MR images (3D-FLAIR) using ScanView software.

Many years ago, the use of gadolinium-based contrast agents was recommended for routine monitoring of radiological disease activity. With improvements in MRI scanners and sequences (high-field MRI, lower slice thickness, 3D isotropic imaging, and longitudinal subtraction techniques), this recommendation has changed, and the use of contrast agents is recommended only in specific clinical situations and at the lowest possible dose [1]. This recommendation is also supported by evidence of contrast agent deposition in the brain, especially in patients who receive linear rather than macrocyclic chelates. Given that the average duration of contrast enhancement is only approximately 2–8 weeks [7,60] and almost all new T1 enhancing lesions have their imaging correlate in T2WI, monitoring new or enlarged T2 lesions is considered a reliable tool for the detection of radiological disease activity in most cases (Table 6.6) [1,61].

Magnetic resonance imaging in drug safety surveillance

MRI drug safety monitoring is usually performed along with regular (yearly) imaging, which is indicated for the standard monitoring of radiological disease activity. Fortunately, adverse events associated with MS treatment that are detected on MRI are relatively rare. These may include vascular events including strokes, opportunistic neuroinfections, or posterior reversible encephalopathy syndrome [62–66]. Neoplastic processes of the CNS are

TABLE 6.6 Clinical scenarios in which contrast agent administration (postcontrast T1WI) should be considered.

Indication	Explanation
Diagnostic MRI	– The morphology of contrast-enhancing lesions is helpful in the differential diagnosis of MS (for details see Table 6.3) – Simultaneous presence of contrast-enhancing and nonenhancing lesions provides evidence of DIT
Required evidence of actual radiological disease activity	– Usually not needed but can be helpful 3–6 months after treatment initiation to differentiate between acute lesions, lesions that occurred soon after treatment initiation (a delayed onset of treatment effect), and those that occurred before treatment initiation – When assessing lesion activity based on T2 lesions (confluent lesions, comorbidities, etc.) is difficult – When recent baseline MRI is missing, and it is not possible to assess whether new lesions are acute/recent or not – If required by healthcare provider to show disease activity
Suspected new comorbidity, including PML infection (based on clinical examination or safety MRI monitoring)	– For the early detection and monitoring of PML and PML-immune reconstitution inflammatory syndrome (IRIS) – The morphology of contrast-enhancing lesions is helpful to distinguish MS lesions from other brain pathologies
Atypical spinal cord lesion	– The morphology of enhancing lesions is helpful to distinguish MS lesions from other spinal cord pathologies

also rare, usually accidental, and their association with immunomodulatory treatment has not yet been established [67,68]. When treatment-related adverse events involving the CNS are suspected, additional MRI is often required. In these cases, the MRI protocols depend on clinical suspicion. For example, in cases of suspected stroke following alemtuzumab administration, sequences including DWI, gradient sequences, venography, or angiography may be required for the detection of hemorrhages and vascular pathology. In cases of suspected neuroinfection or neoplasm, administration of a contrast agent may be required to differentiate between atypical MS lesions and non-MS pathology.

The most important indication for additional surveillance brain MRI scans is John Cunningham virus (JCV) positivity in patients treated with natalizumab [69,70] (rarely observed in patients on other therapies) [71,72] who are at higher risk of PML infection (Fig. 6.3). This indication also temporarily applies to JCV-positive patients with a higher risk of PML infection who were recently switched from natalizumab to other immunomodulatory therapies. In these patients, PML can still occur during the following 9–12 months after changing treatment (carryover cases) [1].

PML infections may occur in clinically asymptomatic patients; thus, early detection of PML using regular safety MRI monitoring is crucial for early diagnosis and intervention. Given that a lower volume of PML lesions at diagnosis is associated with better outcomes, early detection of PML is critical so that immunomodulatory treatment can be discontinued as soon as possible. For early detection of PML, the safety MRI protocol is

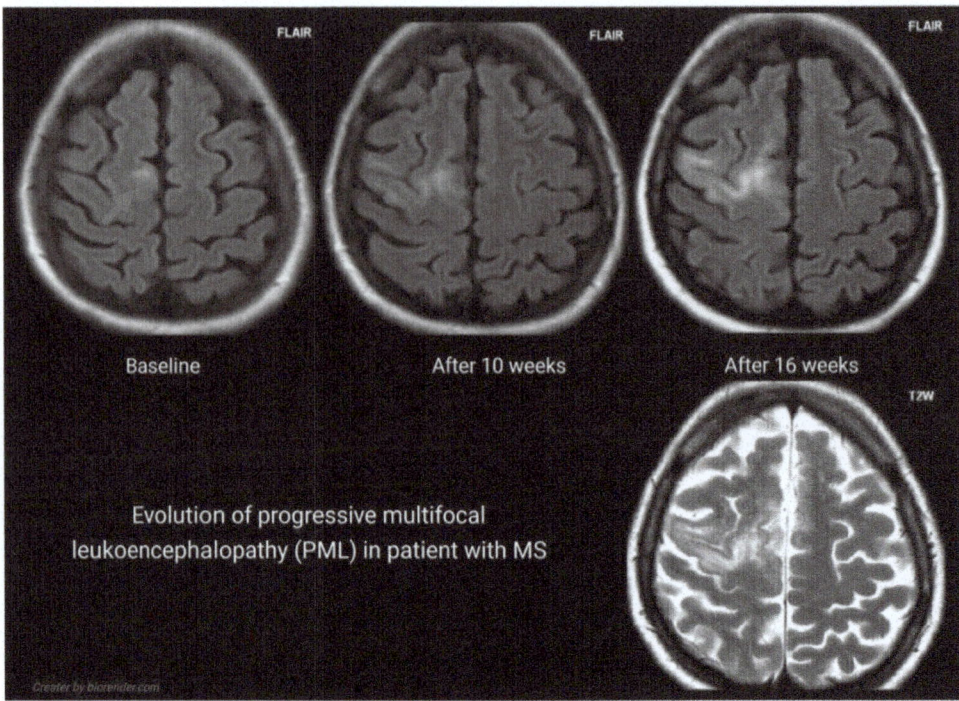

FIGURE 6.3 Progressive multifocal leukoencephalopathy (PML): 48-year-old relapsing–remitting patient with MS diagnosed with PML. The brain MRI shows a discrete subcortical lesion in the right frontal lobe. The first follow-up MRI after 10 weeks shows lesion enlargement (time of PML diagnosis). The second follow-up MRI, 16 weeks from baseline, shows further lesion enlargement on FLAIR imaging and small intralesional focal microlesions on T2WI.

recommended every 3–6 months (frequency depends on the JCV titer and other individual risk factors). The safety MRI protocol includes the most sensitive FLAIR (3D-FLAIR preferred, the same as for routine yearly MRI monitoring of disease activity), T2WI, and DWI sequences. Contrast-enhanced T1WI is recommended either in patients with new lesions suspected to be PML on safety MRI monitoring or to monitor PML infection and detect PML-immune reconstitution inflammatory syndrome (IRIS). PML is characterized by subcortical lesions with well-defined borders. T2WI often shows small hypersignal lesions inside PML lesions or at the borders resembling the milky way, which is typical for PML but not for MS. T1 hypointense lesions can occur in later stages of the infection. Usually, lesions are sharply delineated at the cortical border. The mass effect is absent in classic PML, and fine irregular peripheral contrast enhancement of lesions is observed less frequently in PML than in MS. DWI can sometimes show restriction, typically at the border. For IRIS, edema, mass effect, contrast enhancement of lesions, or small focal perivascular enhancement are typical. Lesions usually do not show restrictions on DWI [1,69].

In cases of suspected PML lesions (especially if the volume of the PML lesion is small and the infection is in an early stage) but with negative JC virus DNA in the CSF, a spinal tap should be repeated to confirm the diagnosis [1,73].

Quantitative magnetic resonance imaging reports in clinical care

Not only inflammatory lesion activity but also neurodegeneration are important components of disease progression associated with the development of physical [44,74–79] and cognitive disability [80–83]. One of the most investigated approaches to quantify neurodegeneration in patients with MS is to evaluate brain volume loss. Brain volume loss is driven by brain atrophy caused by the loss of myelin, glial cells, neurons, and axons due to inflammatory demyelination and neurodegeneration. However, brain volume loss may also be influenced by volume changes in the brain's nontissue components, such as fluid shifts due to inflammation, hydration, endocrine influences, or environmental factors [75,77,78,84–86]. For example, high-dose corticosteroids or initiation of immunomodulatory treatment is often associated with a reduction in neuroinflammation, resulting in partial or complete resolution of brain edema and a reduction of brain volume, which cannot be attributed to brain tissue atrophy. This phenomenon, called pseudoatrophy, can be observed within 6–12 months of antiinflammatory treatment initiation, and may occur in WM as well as in GM [87–89]. All these transient changes in brain nontissue components confound the estimation of real brain volume loss caused by real neurodegeneration and complicated interpretation of volumetric data in practice [78,86,90,91].

Previous studies have shown that patients with higher rates of brain volume loss are at higher risk of disease activity [75,77,78,89,92]. On the other hand, patients with stable or improving clinical disability develop less brain atrophy over time [93,94]. Therefore assessment of the evolution of brain volume loss within individual patients could provide additional markers of disease activity and facilitate the identification of patients at the highest risk of accumulating permanent disability in the future [4]. Although most of the associations between brain volume loss and clinical measures are observed at the group level, significant effort has been made to bring measurements of brain volume into clinical practice for use at the individual-patient level [90,95–99]. Unfortunately, longitudinal MRI measures currently have high intraindividual variability due to a number of biological and technical biases that hamper the confident evaluation of brain volume loss in clinical practice [78,86,89,91]. Despite this, a recent survey conducted by the British Society for Neuroradiologists showed that almost 50% of centers routinely acquire volumetric sequences [100]. The most important challenges associated with the assessment of brain volume loss in clinical practice for clinical decision making in individual patients include the following issues:

1. One of the most important assumptions for reliable estimation of brain volume loss over time is that all consecutive MRI scans are performed using the same MRI scanner (without major scanner upgrades) and with the same MRI protocol (same parameters of sequences, optimally in 3D with isotropic voxels). Volumetric comparisons of two longitudinal brain MR images performed on different scanners or with different protocols (differences in image characteristics, field homogeneity, sequence parameters, and postprocessing routines) can have an enormous impact on measurement errors (i.e., decreased precision), making it almost impossible to assess the precise change in brain volume of a single patient [101–103]. Some approaches have been attempted to minimize this problem. For example, the so-called "similarity index" is a normalized value that aims to quantify the similarity of two follow-up MR images of the same patient with respect to feasibility and reliability of

volumetric comparisons. According to this concept, a pair of follow-up MR images with a high similarity index may be suitable for reliable volumetric comparisons. On the other hand, MRI pairs with low similarity index scores are not suitable for volumetric comparison [104]. Another approach to address nonstandardized MRI acquisition is to measure the lateral ventricular volume as a simple proxy for whole-brain volume on conventional FLAIR imaging. This technique has been validated using a tool called the Neurologic Software Tool for REliable Atrophy Measurement (NeuroSTREAM) on lower resolution images and has shown good agreement with whole brain volume [98]. Nevertheless, dealing with nonstandardized MRI acquisition is always imperfect and represents an additional bias to biological variability. In this context, highly standardized MRI data acquisition is required. However, this assumption is often difficult to satisfy in clinical practice, particularly in MRI centers with multiple scanners or frequent scanner upgrades. Further efforts by responsible authorities, including neurologists and neuroradiologists, are required to improve image acquisition standardization.

2. The proposed cutoff value for the pathological rate of annualized brain volume loss is approximately −0.40%, depending on the specificity and sensitivity of the threshold. This cutoff, which attempts to discriminate between physiological and pathological brain volume loss, has unsatisfactory accuracy [96,97]. However, this is not the only limitation. The technical measurement errors (e.g., scan−rescan error) of the most accurate volumetric software (e.g., SIENA) are around 0.15%−0.20%; however, the range of this error for individual patients is between 0% and 0.60% [105,106]. Although the mean measurement error might appear reasonable for group level analyses, the measurement error range (up to 0.60%) exceeds the proposed cutoff value (−0.40%) [86,90]. The similarity of the absolute value of the cutoff for pathological brain volume loss and the potential measurement error limits its use in individual patients because clinicians may not know whether measured brain volume loss in patients is pathological or the result of measurement error. Importantly, reported data on the accuracy of volumetric analyses are mostly based on the results of clinical trials. Thus biological variability and measurement errors of brain volume assessment could be further magnified in real-world settings where the standardization of MRI acquisition and external conditions is typically lower. This implies that the presence of pathological brain atrophy between two MRI scans can only be reliably ascertained at the individual level when the rate of brain volume loss exceeds a very high level; for example, in the case of the SIENA method [105,106], this is approximately ≤1% (cutoff value: around 0.40% plus upper range of potential measurement error around 0.60%). However, even this approach cannot exclude false positive classifications of patients with a normal rate of brain volume loss but a temporal decrease in brain volume due to brain fluid shifts, resulting from, for example, dehydration or transient hormonal changes. Considering that the majority of patients have lower rates of brain volume loss, this approach could be used to identify only a very small proportion of patients with accelerated brain atrophy. In other words, this approach can be used to identify only the tip of the iceberg of patients with accelerated brain atrophy [86,90]. Finally, given that in elderly patients with MS, the aging process and comorbidities are likely to contribute more to overall brain atrophy than in younger patients, the proportion of physiological and pathological brain volume loss may be age dependent [107−110]. Thus specific cutoff

values of pathological brain volume loss must be established for different age ranges. In addition, sex needs to be taken into account, as men have larger brain volumes and different patterns of age-related brain volume loss than women [89,108,111].
3. A number of biological factors, such as fluid shifts due to inflammation [112], hydration status [85], time of day (larger brain volume in the morning) [113], pregnancy [84], other endocrine influences [114], metabolic and cardiovascular factors [115–118], and various lifestyle and risk factors [75], are well-known modifiers of brain volume. Some of these factors are dynamic (change over time) and might be more relevant, especially in cases when the time between two longitudinal MRI scans is short or in patients with a relatively low rate of brain volume loss [119]. The potential neuroprotective effects of highly effective treatment leading to brain volume stabilization or pseudoatrophy following the initiation of immunomodulatory treatments must also be taken into account [74,87,88,120]. To avoid misinterpretation of accelerated brain volume loss due to pseudoatrophy, another baseline MRI at 6–12 months after treatment initiation is recommended [89]. In contrast, in some of these subjects, paradoxical brain volume increases may be observed as a result of neuroinflammation with associated tissue edema [112,119]. We also need to emphasize the differential effect of various immunomodulatory treatments on the evolution of brain volume loss [89]. Unfortunately, except for treatment interventions and time of day, many of these important biological confounders are difficult to control in routine practice.
4. Currently, there are a number of high-quality manual, semiautomated, and automated softwares [75,77,78] available for the assessment of global and regional brain volume measures, including Structural Image Evaluation using Normalization of Atrophy Cross-sectional (SIENAX) [106], FreeSurfer [121,122], NeuroQuant [53,123], Icobrain/MSmetrix [54,99,124], the model-based segmentation/registration tool FMRIB Integrated Registration and Segmentation Tool (FIRST) [125,126], and others [89]. Longitudinal registration-based methods, such as SIENA [105,106], longitudinal Freesurfer [127], Icobrain/MSmetric [99,104], and others [89] are available, which directly measure relative brain volume changes over time.

Volumetric analyses can be performed in a local MRI unit (or a radiological department). This option requires dedicated personnel (physicists or neuroradiologists) that are able to perform volumetric analyses, including lesion filling. It is extremely important to not only perform volumetric analyses, but also provide volumetric results to clinicians in a fast and digestible fashion to increase their applicability in clinical practice and research.

Alternatively, MRI data can be sent externally to central reading centers such as Icobrain/MSmetrix [54,99,124], NeuroQuant (CorTech Labs) [53,123], and Biometrica MS (Jung Diagnostics) [128], among others [89]. The advantage of this approach is the high quality and standardization of volumetric analyses, complexity of volumetric reports, and simple transfer of MRI scans via clouds. Except for a summary of lesion activity, lesion volume, and lesion distribution, quantitative MRI reports usually provide cross-sectional and longitudinal whole and regional brain volume change metrics and comparisons of brain volumes to age- and sex-matched normative reference populations. However, dedicated personnel must be able to send anonymized imaging data outside the MRI unit, and thus the question of data safety may be a concern. If a large amount of data are analyzed, the cost of external services must also be considered.

Finally, attempts have been made to include volumetric software which used normative referent data from healthy controls, such as AI-Rad Companion (previous MorphoBox prototype) from SIEMENS [129–131] or NeuroQuant (CorTech Labs) [53,123]. Some advantages of this approach include the automatic supervision of data quality control, immediate availability of volumetric data, no need for additional personnel for MRI analysis, and access to normative volumetric data through comparison with large samples of healthy controls. Indeed, it is currently possible to buy these additional software packages, which provide normalized cross-sectional global and regional brain volume data. This approach may thus represent a promising future alternative.

In summary, currently available platforms provide quantitative MRI reports for clinical practice. Some of these platforms are even approved by regulators (FDA clearance and CE mark received) [89]. A number of medical facilities treating MS patients use these reports as part of their standard clinical care and share them with the patients. Although providing detailed reports about lesion activity may also be an advantage for centers where these services are not provided, there are some potential pitfalls associated with the clinical interpretation of the quantitative MRI reports, especially when considering the assessment of brain volume changes. The main challenges arise from the fact that the analyzed MRI scans are from real-world clinical practice, and thus are often performed on different MRI scanners or using different protocols. This may dramatically increase the measurement error of volumetric assessments. Additionally, even if technical variability were minimal, there are important issues with biological variability, which is very difficult to control. Finally, we do not have clear data indicating that the rate of brain volume loss is abnormal at a certain age. These drawbacks make the interpretation of changes in brain volume loss over a short follow-up period (12 months) almost impossible. Despite these limitations, volumetric MRI assessment of the brain and spinal cord for clinical decision-making in individual patients may be feasible and relatively reliable in the following scenarios:

1. The accuracy of brain volume loss assessment may be improved in individual patients who undergo a series of consecutive MRI scans over a mid- or long-term follow-up instead of a single longitudinal scan. This approach could represent an option to partially minimize the high intra-subject variability of a single image and provide measurement accuracy, enabling its applicability in individual patients. The minimal number of MRI scans needed, and the shortest follow-up duration required to obtain a reasonable measurement error substantially lower than the suggested pathological cutoffs of brain atrophy remains to be confirmed in further research. Preliminary findings have shown that a longer follow-up (≥ 2 years) rather than a greater number of MRI scans over a short-term follow-up may provide better precision for measuring brain volume loss in individual patients [95]. Further research is needed to investigate the clinical relevance of this approach.
2. Cross-sectional measures, such as absolute T2 lesion volume or normalized brain volume (e.g., brain parenchymal fraction) have substantially lower relative measurement errors and higher intersubject variability than longitudinal outcomes. Therefore these cross-sectional MRI markers could be applied for risk stratification of future disability and estimation of brain reserve [44,80,92,132–136]. However, because of the cross-sectional nature of these measures, they are not suitable for frequent monitoring of brain volume loss (Table 6.7).

TABLE 6.7 Modifiers of brain volume measures.

Modifier		Comment
Technical variability	Different MRI scanner	— Nonstandardized MRI acquisition prevents comparison of MRI volumetric measurements over time
	MRI scanner hardware/software upgrade	— For volumetric analysis, only artifact-free MR images should be included
	Change in MRI protocol	
	Impact of head positioning and movements	
	Change/upgrade in volumetric software	
Biological variability	Hydration status	— Very difficult to evaluate and control
	Daytime scanning	— Organizationally difficult to maintain uniform time of day for scanning individual patients
	Hormonal changes (puberty, menstrual cycle, pregnancy)	— Further research is needed to determine the extent of their effect on brain volume estimation
	Recent high-dose corticosteroids or initiation of immunomodulatory treatment	— Often associated with the pseudoatrophy effect
	Comorbidities, medical interventions, smoking, alcohol, and drug use	— Their presence and their effect on brain volume in individual patients is difficult to control and estimate
Interpretation of volumetric measurement	Unclear pathological cutoff value of annualized whole and regional brain volume loss	— Cutoffs for whole brain volume changes differ according to age, and patterns of age-related brain volume loss trajectories differ according to sex — Cutoffs for pathological regional brain volume loss have not been established yet — Suggested cutoff values of pathological brain volume loss are similar to measurement error and have relatively low sensitivity and/or specificity

The spinal cord is often affected in patients with MS, and its involvement is associated with disease progression. The rate of spinal cord volume loss appears to be greater than the rate of brain volume loss and is also associated with the risk of disability progression [17,18,21,137]. However, assessment of spinal cord atrophy (often quantified as percent change of mean upper cervical cord area) is even more technically challenging due to anatomical and imaging features and is affected by the number of technical measurement errors, and thus not yet ready for use in clinical practice [89].

Disclosures

This work was supported by the project National Institute for Neurological Research (Program EXCELES, ID project No. LX22NPO5107)—funded by the European Union-Next Generation EU; the Czech Ministry of Health project—grant NU22-04-00193; the Czech Ministry of Education—project Cooperatio LF1, research area Neuroscience; and institutional support of the hospital research project MH CZ-DRO-VFN64165.

References

[1] Wattjes MP, Ciccarelli O, Reich DS, Banwell B, de Stefano N, Enzinger C, et al. 2021 MAGNIMS-CMSC-NAIMS consensus recommendations on the use of MRI in patients with multiple sclerosis. Lancet Neurol 2021;20(8):653−70. Available from: https://doi.org/10.1016/S1474-4422(21)00095-8.

[2] Thompson AJ, Banwell BL, Barkhof F, Carroll WM, Coetzee T, Comi G, et al. Diagnosis of multiple sclerosis: 2017 revisions of the McDonald criteria. Lancet Neurol 2018;17(2):162−73. Available from: https://doi.org/10.1016/S1474-4422(17)30470-2.

[3] Havrdova E, Galetta S, Stefoski D, Comi G. Freedom from disease activity in multiple sclerosis. Neurology 2010;74(17):S3−7. Available from: https://doi.org/10.1212/WNL.0b013e3181dbb51c.

[4] Giovannoni G, Butzkueven H, Dhib-Jalbut S, Hobart J, Kobelt G, Pepper G, et al. Brain health: time matters in multiple sclerosis. Mult Scler Relat Disord 2016;9(Suppl 1):S5−48. Available from: https://doi.org/10.1016/j.msard.2016.07.003.

[5] Tobin WO, Kalinowska-Lyszczarz A, Weigand SD, Guo Y, Tosakulwong N, Parisi JE, et al. Clinical correlation of multiple sclerosis immunopathologic subtypes. Neurology 2021;97(19):e1906−13. Available from: https://doi.org/10.1212/WNL.0000000000012782.

[6] Frischer JM, Weigand SD, Guo Y, Kale N, Parisi JE, Pirko I, et al. Clinical and pathological insights into the dynamic nature of the white matter multiple sclerosis plaque. Ann Neurol 2015;78(5):710−21. Available from: https://doi.org/10.1002/ana.24497.

[7] Filippi M, Preziosa P, Banwell BL, Barkhof F, Ciccarelli O, De Stefano N, et al. Assessment of lesions on magnetic resonance imaging in multiple sclerosis: practical guidelines. Brain 2019;142(7):1858−75. Available from: https://doi.org/10.1093/brain/awz144.

[8] Beck ES, Sati P, Sethi V, Kober T, Dewey B, Bhargava P, et al. Improved visualization of cortical lesions in multiple sclerosis using 7T MP2RAGE. AJNR Am J Neuroradiol 2018;39(3):459−66. Available from: https://doi.org/10.3174/ajnr.A5534.

[9] Bouman PM, Steenwijk MD, Pouwels PJW, Schoonheim MM, Barkhof F, Jonkman LE, et al. Histopathology-validated recommendations for cortical lesion imaging in multiple sclerosis. Brain 2020;143(10):2988−97. Available from: https://doi.org/10.1093/brain/awaa233.

[10] Trapp BD, Peterson J, Ransohoff RM, Rudick R, Mork S, Bo L. Axonal transection in the lesions of multiple sclerosis. N Engl J Med 1998;338(5):278−85. Available from: https://doi.org/10.1056/NEJM199801293380502.

[11] Barkhof F, McGowan JC, van Waesberghe JH, Grossman RI. Hypointense multiple sclerosis lesions on T1-weighted spin echo magnetic resonance images: their contribution in understanding multiple sclerosis evolution. J Neurol Neurosurg Psychiatry 1998;64(Suppl 1):S77−9.

[12] Elliott C, Belachew S, Wolinsky JS, Hauser SL, Kappos L, Barkhof F, et al. Chronic white matter lesion activity predicts clinical progression in primary progressive multiple sclerosis. Brain 2019;142(9):2787−99. Available from: https://doi.org/10.1093/brain/awz212.

[13] Eisele P, Szabo K, Griebe M, Rossmanith C, Forster A, Hennerici M, et al. Reduced diffusion in a subset of acute MS lesions: a serial multiparametric MRI study. AJNR Am J Neuroradiol 2012;33(7):1369−73. Available from: https://doi.org/10.3174/ajnr.A2975.

[14] Zivadinov R, Ramasamy DP, Hagemeier J, Kolb C, Bergsland N, Schweser F, et al. Evaluation of leptomeningeal contrast enhancement using pre- and postcontrast subtraction 3D-FLAIR imaging in multiple sclerosis. AJNR Am J Neuroradiol 2018;39(4):642−7. Available from: https://doi.org/10.3174/ajnr.A5541.

[15] Zivadinov R, Ramasamy DP, Vaneckova M, Gandhi S, Chandra A, Hagemeier J, et al. Leptomeningeal contrast enhancement is associated with progression of cortical atrophy in MS: a retrospective, pilot, observational longitudinal study. Mult Scler J 2017;23(10):1336−45. Available from: https://doi.org/10.1177/1352458516678083.

[16] Absinta M, Vuolo L, Rao A, Nair G, Sati P, Cortese ICM, et al. Gadolinium-based MRI characterization of leptomeningeal inflammation in multiple sclerosis. Neurology 2015;85(1):18−28. Available from: https://doi.org/10.1212/Wnl.0000000000001587.

[17] Ciccarelli O, Cohen JA, Reingold SC, Weinshenker BG, International Conference on Spinal Cord I, Imaging in Multiple S, et al. Spinal cord involvement in multiple sclerosis and neuromyelitis optica spectrum disorders. Lancet Neurol 2019;18(2):185−97. Available from: https://doi.org/10.1016/S1474-4422(18)30460-5.

[18] Andelova M, Uher T, Krasensky J, Sobisek L, Kusova E, Srpova B, et al. Additive effect of spinal cord volume, diffuse and focal cord pathology on disability in multiple sclerosis. Front Neurol 2019;10:820. Available from: https://doi.org/10.3389/fneur.2019.00820.
[19] Jurynczyk M, Craner M, Palace J. Overlapping CNS inflammatory diseases: differentiating features of NMO and MS. J Neurol Neurosurg Psychiatry 2015;86(1):20–5. Available from: https://doi.org/10.1136/jnnp-2014-308984.
[20] Mariano R, Flanagan EP, Weinshenker BG, Palace J. A practical approach to the diagnosis of spinal cord lesions. Pract Neurol 2018;18(3):187–200. Available from: https://doi.org/10.1136/practneurol-2017-001845.
[21] Andelova M, Vodehnalova K, Krasensky J, Hardubejova E, Hrnciarova T, Srpova B, et al. Brainstem lesions are associated with diffuse spinal cord involvement in early multiple sclerosis. BMC Neurol 2022;22(1):270. Available from: https://doi.org/10.1186/s12883-022-02778-z.
[22] Okuda DT, Mowry EM, Beheshtian A, Waubant E, Baranzini SE, Goodin DS, et al. Incidental MRI anomalies suggestive of multiple sclerosis: the radiologically isolated syndrome. Neurology 2009;72(9):800–5. Available from: https://doi.org/10.1212/01.wnl.0000335764.14513.1a.
[23] Lebrun-Frénay C, Okuda DT, Siva A, Landes-Chateau C, Azevedo CJ, Mondot L, et al. The radiologically isolated syndrome: revised diagnostic criteria. Brain 2023;146(8):3431–43. Available from: https://doi.org/10.1093/brain/awad073.
[24] Keegan BM, Kaufmann TJ, Weinshenker BG, Kantarci OH, Schmalstieg WF, Paz Soldan MM, et al. Progressive solitary sclerosis: gradual motor impairment from a single CNS demyelinating lesion. Neurology 2016;87(16):1713–19. Available from: https://doi.org/10.1212/WNL.0000000000003235.
[25] Solomon AJ, Bourdette DN, Cross AH, Applebee A, Skidd PM, Howard DB, et al. The contemporary spectrum of multiple sclerosis misdiagnosis: a multicenter study. Neurology 2016;87(13):1393–9. Available from: https://doi.org/10.1212/WNL.0000000000003152.
[26] Solomon AJ, Corboy JR. The tension between early diagnosis and misdiagnosis of multiple sclerosis. Nat Rev Neurol 2017;13(9):567–72. Available from: https://doi.org/10.1038/nrneurol.2017.106.
[27] Arrambide G, Brownlee WJ, Flanagan EP, Amato MP, Amezcua L, Banwell BL, et al. Differential diagnosis of suspected multiple sclerosis: an updated consensus approach. Lancet Neurol 2023;22(8):750–68. Available from: https://doi.org/10.1016/S1474-4422(23)00148-5.
[28] Uher T, Adzima A, Srpova B, Noskova L, Maréchal B, Maceski AM, et al. Diagnostic delay of multiple sclerosis: prevalence, determinants and consequences. Mult Scler 2023;29(11–12):1437–51. Available from: https://doi.org/10.1177/13524585231197076.
[29] Cobo-Calvo A, Tur C, Otero-Romero S, Carbonell-Mirabent P, Ruiz M, Pappolla A, et al. Association of Very Early Treatment Initiation With the Risk of Long-term Disability in Patients With a First Demyelinating Event. Neurology 2023;101(13)e1280-e1292. Available from: https://doi.org/10.1212/WNL.0000000000207664.
[30] Kim W, Lee JE, Kim SH, Huh SY, Hyun JW, Jeong IH, et al. Cerebral cortex involvement in neuromyelitis optica spectrum disorder. J Clin Neurol 2016;12(2):188–93. Available from: https://doi.org/10.3988/jcn.2016.12.2.188.
[31] Maggi P, Absinta M, Grammatico M, Vuolo L, Emmi G, Carlucci G, et al. Central vein sign differentiates multiple sclerosis from central nervous system inflammatory vasculopathies. Ann Neurol 2018;83(2):283–94. Available from: https://doi.org/10.1002/ana.25146.
[32] Sinnecker T, Clarke MA, Meier D, Enzinger C, Calabrese M, De Stefano N, et al. Evaluation of the central vein sign as a diagnostic imaging biomarker in multiple sclerosis. JAMA Neurol 2019;76(12):1446–56. Available from: https://doi.org/10.1001/jamaneurol.2019.2478.
[33] Arnold DL, Belachew S, Gafson AR, Gaetano L, Bernasconi C, Elliott C. Slowly expanding lesions are a marker of progressive MS – No. Mult Scler 2021;27(11):1681–3. Available from: https://doi.org/10.1177/13524585211017020.
[34] Kister I. The multiple sclerosis lesion checklist. Practic Neurol 2018; July/August.
[35] Miller DH, Weinshenker BG, Filippi M, Banwell BL, Cohen JA, Freedman MS, et al. Differential diagnosis of suspected multiple sclerosis: a consensus approach. Mult Scler J 2008;14(9):1157–74. Available from: https://doi.org/10.1177/1352458508096878.
[36] Patel J, Pires A, Derman A, Fatterpekar G, Charlson RE, Oh C, et al. Development and validation of a simple and practical method for differentiating MS from other neuroinflammatory disorders based on

lesion distribution on brain MRI. J Clin Neurosci 2022;101:32–6. Available from: https://doi.org/10.1016/j.jocn.2022.04.035.

[37] Solomon AJ, Arrambide G, Brownlee WJ, Flanagan EP, Amato MP, Amezcua L, et al. Differential diagnosis of suspected multiple sclerosis: an updated consensus approach. Lancet Neurol 2023;22(8):750–68. Available from: https://doi.org/10.1016/S1474-4422(23)00148-5.

[38] Vaneckova M, Piredda GF, Andelova M, Krasensky J, Uher T, Srpova B, et al. Periventricular gradient of T1 tissue alterations in multiple sclerosis. Neuroimage Clin 2022;34103009. Available from: https://doi.org/10.1016/j.nicl.2022.103009.

[39] Giorgio A, De Stefano N. Advanced structural and functional brain MRI in multiple sclerosis. Semin Neurol 2016;36(2):163–76. Available from: https://doi.org/10.1055/s-0036-1579737.

[40] Miller DH, Thompson AJ, Filippi M. Magnetic resonance studies of abnormalities in the normal appearing white matter and grey matter in multiple sclerosis. J Neurol 2003;250(12):1407–19. Available from: https://doi.org/10.1007/s00415-003-0243-9.

[41] Stankiewicz JM, Weiner HL. An argument for broad use of high efficacy treatments in early multiple sclerosis. Neurol Neuroimmunol Neuroinflamm 2020;7(1). Available from: https://doi.org/10.1212/NXI.0000000000000636.

[42] Tintore M, Rovira A, Rio J, Otero-Romero S, Arrambide G, Tur C, et al. Defining high, medium and low impact prognostic factors for developing multiple sclerosis. Brain 2015;138(Pt 7):1863–74. Available from: https://doi.org/10.1093/brain/awv105.

[43] Uher T, Horakova D, Kalincik T, Bergsland N, Tyblova M, Ramasamy DP, et al. Early magnetic resonance imaging predictors of clinical progression after 48 months in clinically isolated syndrome patients treated with intramuscular interferon beta-1a. Eur J Neurol 2015;22(7):1113–23. Available from: https://doi.org/10.1111/ene.12716.

[44] Uher T, Vaneckova M, Sobisek L, Tyblova M, Seidl Z, Krasensky J, et al. Combining clinical and magnetic resonance imaging markers enhances prediction of 12-year disability in multiple sclerosis. Mult Scler 2017;23(1):51–61. Available from: https://doi.org/10.1177/1352458516642314.

[45] Wattjes MP, Rovira A, Miller D, Yousry TA, Sormani MP, de Stefano MP, et al. Evidence-based guidelines: MAGNIMS consensus guidelines on the use of MRI in multiple sclerosis—establishing disease prognosis and monitoring patients. Nat Rev Neurol 2015;11(10):597–606. Available from: https://doi.org/10.1038/nrneurol.2015.157.

[46] Cortese R, Giorgio A, Severa G, De Stefano N. MRI prognostic factors in multiple sclerosis, neuromyelitis optica spectrum disorder, and myelin oligodendrocyte antibody disease. Front Neurol 2021;12679881. Available from: https://doi.org/10.3389/fneur.2021.679881.

[47] Allen CM, Mowry E, Tintore M, Evangelou N. Prognostication and contemporary management of clinically isolated syndrome. J Neurol Neurosurg Psychiatry 2020;. Available from: https://doi.org/10.1136/jnnp-2020-323087.

[48] Uher T, Havrdova E, Sobisek L, Krasensky J, Vaneckova M, Seidl Z, et al. Is no evidence of disease activity an achievable goal in MS patients on intramuscular interferon beta-1a treatment over long-term follow-up? Mult Scler 2017;23(2):242–52. Available from: https://doi.org/10.1177/1352458516650525.

[49] Rovira A, Wattjes MP, Tintore M, Tur C, Yousry TA, Sormani MP, et al. Evidence-based guidelines: MAGNIMS consensus guidelines on the use of MRI in multiple sclerosis-clinical implementation in the diagnostic process. Nat Rev Neurol 2015;11(8):471–82. Available from: https://doi.org/10.1038/nrneurol.2015.106.

[50] Uher T, Schaedelin S, Srpova B, Barro C, Bergsland N, Dwyer M, et al. Monitoring of radiologic disease activity by serum neurofilaments in MS. Neurol Neuroimmunol Neuroinflamm 2020;7(4). Available from: https://doi.org/10.1212/NXI.0000000000000714.

[51] Zecca C, Disanto G, Sormani MP, Riccitelli GC, Cianfoni A, Del Grande F, et al. Relevance of asymptomatic spinal MRI lesions in patients with multiple sclerosis. Mult Scler 2016;22(6):782–91. Available from: https://doi.org/10.1177/1352458515599246.

[52] Gass A, Rocca MA, Agosta F, Ciccarelli O, Chard D, Valsasina P, et al. MRI monitoring of pathological changes in the spinal cord in patients with multiple sclerosis. Lancet Neurol 2015;14(4):443–54. Available from: https://doi.org/10.1016/S1474-4422(14)70294-7.

[53] Brune S, Hogestol EA, Cengija V, Berg-Hansen P, Sowa P, Nygaard GO, et al. LesionQuant for assessment of MRI in multiple sclerosis-a promising supplement to the visual scan inspection. Front Neurol 2020;11546744. Available from: https://doi.org/10.3389/fneur.2020.546744.

[54] Jain S, Ribbens A, Sima DM, Cambron M, De Keyser J, Wang C, et al. Two time point MS lesion segmentation in brain MRI: an expectation-maximization framework. Front Neurosci 2016;10:576. Available from: https://doi.org/10.3389/fnins.2016.00576.

[55] Yushkevich PA, Piven J, Hazlett HC, Smith RG, Ho S, Gee JC, et al. User-guided 3D active contour segmentation of anatomical structures: significantly improved efficiency and reliability. Neuroimage 2006;31(3):1116–28. Available from: https://doi.org/10.1016/j.neuroimage.2006.01.015.

[56] Cantin STT, Lamain E, Bakir M. FLAIR fusion in multiple sclerosis follow-up: an indispensable tool in clinical routine. MAGNETOM Flash 2017;3:100–3.

[57] Absinta M, Sati P, Reich DS. Advanced MRI and staging of multiple sclerosis lesions. Nat Rev Neurol 2016;12(6):358–68. Available from: https://doi.org/10.1038/nrneurol.2016.59.

[58] Hametner S, Dal Bianco A, Trattnig S, Lassmann H. Iron related changes in MS lesions and their validity to characterize MS lesion types and dynamics with ultra-high field magnetic resonance imaging. Brain Pathol 2018;28(5):743–9. Available from: https://doi.org/10.1111/bpa.12643.

[59] Treaba CA, Conti A, Klawiter EC, Barletta VT, Herranz E, Mehndiratta A, et al. Cortical and phase rim lesions on 7 T MRI as markers of multiple sclerosis disease progression. Brain Commun 2021;3(3)fcab134. Available from: https://doi.org/10.1093/braincomms/fcab134.

[60] Cotton F, Weiner HL, Jolesz FA, Guttmann CR. MRI contrast uptake in new lesions in relapsing-remitting MS followed at weekly intervals. Neurology 2003;60(4):640–6. Available from: https://doi.org/10.1212/01.wnl.0000046587.83503.1e.

[61] Eichinger P, Schon S, Pongratz V, Wiestler H, Zhang H, Bussas M, et al. Accuracy of unenhanced MRI in the detection of new brain lesions in multiple sclerosis. Radiology 2019;291(2):429–35. Available from: https://doi.org/10.1148/radiol.2019181568.

[62] Thormann A, Magyari M, Koch-Henriksen N, Laursen B, Sorensen PS. Vascular comorbidities in multiple sclerosis: a nationwide study from Denmark. J Neurol 2016;263(12):2484–93. Available from: https://doi.org/10.1007/s00415-016-8295-9.

[63] Fine AJ, Sorbello A, Kortepeter C, Scarazzini L. Central nervous system herpes simplex and varicella zoster virus infections in natalizumab-treated patients. Clin Infect Dis 2013;57(6):849–52. Available from: https://doi.org/10.1093/cid/cit376.

[64] Holmoy T, Fevang B, Olsen DB, Spigset O, Bo L. Adverse events with fatal outcome associated with alemtuzumab treatment in multiple sclerosis. BMC Res Notes 2019;12(1):497. Available from: https://doi.org/10.1186/s13104-019-4507-6.

[65] Hong Y, Tang HR, Ma M, Chen N, Xie X, He L. Multiple sclerosis and stroke: a systematic review and meta-analysis. BMC Neurol 2019;19(1):139. Available from: https://doi.org/10.1186/s12883-019-1366-7.

[66] Linda H, von Heijne A. A case of posterior reversible encephalopathy syndrome associated with gilenya((R)) (fingolimod) treatment for multiple sclerosis. Front Neurol 2015;6:39. Available from: https://doi.org/10.3389/fneur.2015.00039.

[67] Plantone D, Renna R, Sbardella E, Koudriavtseva T. Concurrence of multiple sclerosis and brain tumors. Front Neurol 2015;6:40. Available from: https://doi.org/10.3389/fneur.2015.00040.

[68] Bahmanyar S, Montgomery SM, Hillert J, Ekbom A, Olsson T. Cancer risk among patients with multiple sclerosis and their parents. Neurology 2009;72(13):1170–7. Available from: https://doi.org/10.1212/01.wnl.0000345366.10455.62.

[69] Wattjes MP, Wijburg MT, van Eijk J, Frequin S, Uitdehaag BMJ, Barkhof F, et al. Inflammatory natalizumab-associated PML: baseline characteristics, lesion evolution and relation with PML-IRIS. J Neurol Neurosurg Psychiatry 2018;89(5):535–41. Available from: https://doi.org/10.1136/jnnp-2017-316886.

[70] Schwab N, Schneider-Hohendorf T, Melzer N, Cutter G, Wiendl H. Natalizumab-associated PML: challenges with incidence, resulting risk, and risk stratification. Neurology 2017;88(12):1197–205. Available from: https://doi.org/10.1212/WNL.0000000000003739.

[71] Berger JR, Cree BA, Greenberg B, Hemmer B, Ward BJ, Dong VM, et al. Progressive multifocal leukoencephalopathy after fingolimod treatment. Neurology 2018;90(20):e1815–21. Available from: https://doi.org/10.1212/WNL.0000000000005529.

[72] Diebold M, Altersberger V, Decard BF, Kappos L, Derfuss T, Lorscheider J. A case of progressive multifocal leukoencephalopathy under dimethyl fumarate treatment without severe lymphopenia or immunosenescence. Mult Scler 2019;25(12):1682–5. Available from: https://doi.org/10.1177/1352458519852100.

[73] Kuhle J, Gosert R, Buhler R, Derfuss T, Sutter R, Yaldizli O, et al. Management and outcome of CSF-JC virus PCR-negative PML in a natalizumab-treated patient with MS. Neurology 2011;77(23):2010−16. Available from: https://doi.org/10.1212/WNL.0b013e31823b9b27.
[74] Uher T, Krasensky J, Malpas C, Bergsland N, Dwyer MG, Kubala Havrdova E, et al. Evolution of brain volume loss rates in early stages of multiple sclerosis. Neurol Neuroimmunol Neuroinflamm 2021;8(3). Available from: https://doi.org/10.1212/NXI.0000000000000979.
[75] Zivadinov R, Jakimovski D, Gandhi S, Ahmed R, Dwyer MG, Horakova D, et al. Clinical relevance of brain atrophy assessment in multiple sclerosis. Implications for its use in a clinical routine. Expert Rev Neurother 2016;16(7):777−93. Available from: https://doi.org/10.1080/14737175.2016.1181543.
[76] Zivadinov R, Uher T, Hagemeier J, Vaneckova M, Ramasamy DP, Tyblova M, et al. A serial 10-year follow-up study of brain atrophy and disability progression in RRMS patients. Mult Scler 2016;22(13):1709−18. Available from: https://doi.org/10.1177/1352458516629769.
[77] De Stefano N, Airas L, Grigoriadis N, Mattle HP, O'Riordan J, Oreja-Guevara C, et al. Clinical relevance of brain volume measures in multiple sclerosis. CNS Drugs 2014;28(2):147−56. Available from: https://doi.org/10.1007/s40263-014-0140-z.
[78] Azevedo CJ, Pelletier D. Whole-brain atrophy: ready for implementation into clinical decision-making in multiple sclerosis? Curr Opin Neurol 2016;29(3):237−42. Available from: https://doi.org/10.1097/WCO.0000000000000322.
[79] Sormani MP, Arnold DL, De Stefano N. Treatment effect on brain atrophy correlates with treatment effect on disability in multiple sclerosis. Ann Neurol 2014;75(1):43−9. Available from: https://doi.org/10.1002/ana.24018.
[80] Uher T, Vaneckova M, Sormani MP, Krasensky J, Sobisek L, Dusankova JB, et al. Identification of multiple sclerosis patients at highest risk of cognitive impairment using an integrated brain magnetic resonance imaging assessment approach. Eur J Neurol 2017;24(2):292−301. Available from: https://doi.org/10.1111/ene.13200.
[81] Benedict RH, Zivadinov R. Risk factors for and management of cognitive dysfunction in multiple sclerosis. Nat Rev Neurol 2011;7(6):332−42. Available from: https://doi.org/10.1038/nrneurol.2011.61.
[82] Motyl J, Friedova L, Vaneckova M, Krasensky J, Lorincz B, Blahova Dusankova J, et al. Isolated cognitive decline in neurologically stable patients with multiple sclerosis. Diagnostics (Basel) 2021;11(3). Available from: https://doi.org/10.3390/diagnostics11030464.
[83] Rocca MA, Amato MP, De Stefano N, Enzinger C, Geurts JJ, Penner IK, et al. Clinical and imaging assessment of cognitive dysfunction in multiple sclerosis. Lancet Neurol 2015;14(3):302−17. Available from: https://doi.org/10.1016/S1474-4422(14)70250-9.
[84] Uher T, Kubala Havrdova E, Vodehnalova K, Krasensky J, Capek V, Vaneckova M, et al. Pregnancy-induced brain magnetic resonance imaging changes in women with multiple sclerosis. Eur J Neurol 2022;29(5):1446−56. Available from: https://doi.org/10.1111/ene.15245.
[85] Nakamura K, Brown RA, Araujo D, Narayanan S, Arnold DL. Correlation between brain volume change and T2 relaxation time induced by dehydration and rehydration: implications for monitoring atrophy in clinical studies. Neuroimage Clin 2014;6:166−70. Available from: https://doi.org/10.1016/j.nicl.2014.08.014.
[86] Narayanan S, Nakamura K, Fonov VS, Maranzano J, Caramanos Z, Giacomini PS, et al. Brain volume loss in individuals over time: source of variance and limits of detectability. Neuroimage 2020;214116737. Available from: https://doi.org/10.1016/j.neuroimage.2020.116737.
[87] Zivadinov R, Reder AT, Filippi M, Minagar A, Stuve O, Lassmann H, et al. Mechanisms of action of disease-modifying agents and brain volume changes in multiple sclerosis. Neurology 2008;71(2):136−44. Available from: https://doi.org/10.1212/01.wnl.0000316810.01120.05.
[88] De Stefano N, Giorgio A, Gentile G, Stromillo ML, Cortese R, Gasperini C, et al. Dynamics of pseudo-atrophy in RRMS reveals predominant gray matter compartmentalization. Ann Clin Transl Neurol 2021;8(3):623−30. Available from: https://doi.org/10.1002/acn3.51302.
[89] Sastre-Garriga J, Pareto D, Battaglini M, Rocca MA, Ciccarelli O, Enzinger C, et al. MAGNIMS consensus recommendations on the use of brain and spinal cord atrophy measures in clinical practice. Nat Rev Neurol 2020;16(3):171−82. Available from: https://doi.org/10.1038/s41582-020-0314-x.
[90] Opfer R, Ostwaldt AC, Walker-Egger C, Manogaran P, Sormani MP, De Stefano N, et al. Within-patient fluctuation of brain volume estimates from short-term repeated MRI measurements using SIENA/FSL. J Neurol 2018;265(5):1158−65. Available from: https://doi.org/10.1007/s00415-018-8825-8.

References

[91] Rocca MA, Battaglini M, Benedict RH, De Stefano N, Geurts JJ, Henry RG, et al. Brain MRI atrophy quantification in MS: from methods to clinical application. Neurology 2017;88(4):403—13. Available from: https://doi.org/10.1212/WNL.0000000000003542.

[92] Sormani MP, Kappos L, Radue EW, Cohen J, Barkhof F, Sprenger T, et al. Defining brain volume cutoffs to identify clinically relevant atrophy in RRMS. Mult Scler 2017;23(5):656—64. Available from: https://doi.org/10.1177/1352458516659550.

[93] Uher T, Horakova D, Bergsland N, Tyblova M, Ramasamy DP, Seidl Z, et al. MRI correlates of disability progression in patients with CIS over 48 months. Neuroimage Clin 2014;6:312—19. Available from: https://doi.org/10.1016/j.nicl.2014.09.015.

[94] Ghione E, Bergsland N, Dwyer MG, Hagemeier J, Jakimovski D, Ramasamy DP, et al. Disability improvement is associated with less brain atrophy development in multiple sclerosis. AJNR Am J Neuroradiol 2020;41(9):1577—83. Available from: https://doi.org/10.3174/ajnr.A6684.

[95] Uher T, Krasensky J, Sobisek L, Seidl Z, Bergsland N, Dwyer MG, et al. The role of high-frequency MRI monitoring in the detection of brain atrophy in multiple sclerosis. J Neuroimaging 2018;28(3):328—37. Available from: https://doi.org/10.1111/jon.12505.

[96] Uher T, Vaneckova M, Krasensky J, Sobisek L, Tyblova M, Volna J, et al. Pathological cut-offs of global and regional brain volume loss in multiple sclerosis. Mult Scler 2017;25(4):541—53. Available from: https://doi.org/10.1177/1352458517742739.

[97] De Stefano N, Stromillo ML, Giorgio A, Bartolozzi ML, Battaglini M, Baldini M, et al. Establishing pathological cut-offs of brain atrophy rates in multiple sclerosis. J Neurol Neurosurg Psychiatry 2016;87(1):93—9. Available from: https://doi.org/10.1136/jnnp-2014-309903.

[98] Dwyer MG, Silva D, Bergsland N, Horakova D, Ramasamy D, Durfee J, et al. Neurological software tool for reliable atrophy measurement (NeuroSTREAM) of the lateral ventricles on clinical-quality T2-FLAIR MRI scans in multiple sclerosis. Neuroimage Clin 2017;15:769—79. Available from: https://doi.org/10.1016/j.nicl.2017.06.022.

[99] Smeets D, Ribbens A, Sima DM, Cambron M, Horakova D, Jain S, et al. Reliable measurements of brain atrophy in individual patients with multiple sclerosis. Brain Behav 2016;6(9)e00518. Available from: https://doi.org/10.1002/brb3.518.

[100] Schmierer K, Campion T, Sinclair A, van Hecke W, Matthews PM, Wattjes MP. Towards a standard MRI protocol for multiple sclerosis across the UK. Br J Radiol 2019;92(1101)20180926. Available from: https://doi.org/10.1259/bjr.20180926.

[101] Biberacher V, Schmidt P, Keshavan A, Boucard CC, Righart R, Samann P, et al. Intra- and interscanner variability of magnetic resonance imaging based volumetry in multiple sclerosis. Neuroimage 2016;142:188—97. Available from: https://doi.org/10.1016/j.neuroimage.2016.07.035.

[102] De Stefano N, Battaglini M, Pareto D, Cortese R, Zhang J, Oesingmann N, et al. MAGNIMS recommendations for harmonization of MRI data in MS multicenter studies. Neuroimage Clin 2022;34102972. Available from: https://doi.org/10.1016/j.nicl.2022.102972.

[103] Storelli L, Rocca MA, Pagani E, Van Hecke W, Horsfield MA, De Stefano N, et al. Measurement of whole-brain and gray matter atrophy in multiple sclerosis: assessment with MR imaging. Radiology 2018;288(2):554—64. Available from: https://doi.org/10.1148/radiol.2018172468.

[104] Sima, D.M., Horakova, D., Nguyen, A.L., Van Hecke, W., Kalincik, T., Barnett, M.H., et al. (2019). Assessing the reliability of longitudinal MRI examinations in multiple sclerosis follow-up. Paper presented at the ECTRIMS, Stockholm.

[105] Smith SM, De Stefano N, Jenkinson M, Matthews PM. Normalized accurate measurement of longitudinal brain change. J Comput Assist Tomogr 2001;25(3):466—75. Available from: https://doi.org/10.1097/00004728-200105000-00022.

[106] Smith SM, Zhang Y, Jenkinson M, Chen J, Matthews PM, Federico A, et al. Accurate, robust, and automated longitudinal and cross-sectional brain change analysis. Neuroimage 2002;17(1):479—89. Available from: https://doi.org/10.1006/nimg.2002.1040.

[107] Battaglini M, Gentile G, Luchetti L, Giorgio A, Vrenken H, Barkhof F, et al. Lifespan normative data on rates of brain volume changes. Neurobiol Aging 2019;81:30—7. Available from: https://doi.org/10.1016/j.neurobiolaging.2019.05.010.

[108] Kiraly A, Szabo N, Toth E, Csete G, Farago P, Kocsis K, et al. Male brain ages faster: the age and gender dependence of subcortical volumes. Brain Imaging Behav 2016;10(3):901–10. Available from: https://doi.org/10.1007/s11682-015-9468-3.

[109] Azevedo CJ, Cen SY, Jaberzadeh A, Zheng L, Hauser SL, Pelletier D. Contribution of normal aging to brain atrophy in MS. Neurol Neuroimmunol Neuroinflamm 2019;6(6):e616. Available from: https://doi.org/10.1212/NXI.0000000000000616.

[110] Uher T, Krasensky J, Malpas C, Bergsland N, Dwyer MG, Kubala Havrdova E, et al. Evolution of Brain Volume Loss Rates in Early Stages of Multiple Sclerosis. Neurol Neuroimmunol Neuroinflamm 2021;8(3):e979. Available from: https://doi.org/10.1212/NXI.0000000000000979.

[111] Ruigrok AN, Salimi-Khorshidi G, Lai MC, Baron-Cohen S, Lombardo MV, Tait RJ, et al. A meta-analysis of sex differences in human brain structure. Neurosci Biobehav Rev 2014;39:34–50. Available from: https://doi.org/10.1016/j.neubiorev.2013.12.004.

[112] Warntjes JBM, Tisell A, Hakansson I, Lundberg P, Ernerudh J. Improved precision of automatic brain volume measurements in patients with clinically isolated syndrome and multiple sclerosis using edema correction. AJNR Am J Neuroradiol 2018;39(2):296–302. Available from: https://doi.org/10.3174/ajnr.A5476.

[113] Trefler A, Sadeghi N, Thomas AG, Pierpaoli C, Baker CI, Thomas C. Impact of time-of-day on brain morphometric measures derived from T1-weighted magnetic resonance imaging. Neuroimage 2016;133:41–52. Available from: https://doi.org/10.1016/j.neuroimage.2016.02.034.

[114] Hagemann G, Ugur T, Schleussner E, Mentzel HJ, Fitzek C, Witte OW, et al. Changes in brain size during the menstrual cycle. PLoS One 2011;6(2):e14655. Available from: https://doi.org/10.1371/journal.pone.0014655.

[115] Graetz C, Groger A, Luessi F, Salmen A, Zoller D, Schultz J, et al. Association of smoking but not HLA-DRB1*15:01, APOE or body mass index with brain atrophy in early multiple sclerosis. Mult Scler 2018;25(5):661–8. Available from: https://doi.org/10.1177/1352458518763541.

[116] Jakimovski D, Gandhi S, Paunkoski I, Bergsland N, Hagemeier J, Ramasamy DP, et al. Hypertension and heart disease are associated with development of brain atrophy in multiple sclerosis: a 5-year longitudinal study. Eur J Neurol 2019;26(1):87–e88. Available from: https://doi.org/10.1111/ene.13769.

[117] Kappus N, Weinstock-Guttman B, Hagemeier J, Kennedy C, Melia R, Carl E, et al. Cardiovascular risk factors are associated with increased lesion burden and brain atrophy in multiple sclerosis. J Neurol Neurosurg Psychiatry 2016;87(2):181–7. Available from: https://doi.org/10.1136/jnnp-2014-310051.

[118] Lorincz B, Jury EC, Vrablik M, Ramanathan M, Uher T. The role of cholesterol metabolism in multiple sclerosis: from molecular pathophysiology to radiological and clinical disease activity. Autoimmun Rev 2022;21(6)103088. Available from: https://doi.org/10.1016/j.autrev.2022.103088.

[119] Uher T, Bergsland N, Krasensky J, Dwyer MG, Andelova M, Sobisek L, et al. Interpretation of brain volume increase in multiple sclerosis. J Neuroimaging 2021;31(2):401–7. Available from: https://doi.org/10.1111/jon.12816.

[120] Investigators CT, Coles AJ, Compston DA, Selmaj KW, Lake SL, Moran S, et al. Alemtuzumab vs. interferon beta-1a in early multiple sclerosis. N Engl J Med 2008;359(17):1786–801. Available from: https://doi.org/10.1056/NEJMoa0802670.

[121] Dale AM, Fischl B, Sereno MI. Cortical surface-based analysis. I. Segmentation and surface reconstruction. Neuroimage 1999;9(2):179–94. Available from: https://doi.org/10.1006/nimg.1998.0395.

[122] Fischl B, Sereno MI, Dale AM. Cortical surface-based analysis. II: inflation, flattening, and a surface-based coordinate system. Neuroimage 1999;9(2):195–207. Available from: https://doi.org/10.1006/nimg.1998.0396.

[123] Pareto D, Sastre-Garriga J, Alberich M, Auger C, Tintore M, Montalban X, et al. Brain regional volume estimations with NeuroQuant and FIRST: a study in patients with a clinically isolated syndrome. Neuroradiology 2019;61(6):667–74. Available from: https://doi.org/10.1007/s00234-019-02191-3.

[124] Steenwijk MD, Amiri H, Schoonheim MM, de Sitter A, Barkhof F, Pouwels PJW, et al. Agreement of MSmetrix with established methods for measuring cross-sectional and longitudinal brain atrophy. Neuroimage Clin 2017;15:843–53. Available from: https://doi.org/10.1016/j.nicl.2017.06.034.

[125] Zhang QX, Ling YF, Sun Z, Zhang L, Yu HX, Kamau SM, et al. Protective effect of whey protein hydrolysates against hydrogen peroxide-induced oxidative stress on PC12 cells. Biotechnol Lett 2012;34(11):2001–6. Available from: https://doi.org/10.1007/s10529-012-1017-1.

[126] Jain S, Sima DM, Ribbens A, Cambron M, Maertens A, Van Hecke W, et al. Automatic segmentation and volumetry of multiple sclerosis brain lesions from MR images. Neuroimage Clin 2015;8:367−75. Available from: https://doi.org/10.1016/j.nicl.2015.05.003.

[127] Reuter M, Schmansky NJ, Rosas HD, Fischl B. Within-subject template estimation for unbiased longitudinal image analysis. Neuroimage 2012;61(4):1402−18. Available from: https://doi.org/10.1016/j.neuroimage.2012.02.084.

[128] Opfer R, Suppa P, Kepp T, Spies L, Schippling S, Huppertz HJ, et al. Atlas based brain volumetry: how to distinguish regional volume changes due to biological or physiological effects from inherent noise of the methodology. Magn Reson Imaging 2016;34(4):455−61. Available from: https://doi.org/10.1016/j.mri.2015.12.031.

[129] Tsang, A., Fartaria, M.J., Perea, R.D., Corredor-Jerez, R., Liao, S., Benzinger, T.L. S., et al. (2020). Quantitative MRI metrics in routine clinical practice: a validation study from a large heterogeneous cohort of multiple sclerosis patients. Paper presented at the ISMRM & SMRT virtual conference & exhibition, 8−14 August.

[130] Mietchen D, Gaser C. Computational morphometry for detecting changes in brain structure due to development, aging, learning, disease and evolution. Front Neuroinform 2009;3:25. Available from: https://doi.org/10.3389/neuro.11.025.2009.

[131] Roche A, Bénédicte M, Kober T, Krueger G, Hagman P, Maeder P, et al. Assessing brain volumes using morphobox prototype. MAGNETOM Flash 2017;68:33−7.

[132] Bakshi R, Healy BC, Dupuy SL, Kirkish G, Khalid F, Gundel T, et al. Brain MRI predicts worsening multiple sclerosis disability over 5 years in the SUMMIT study. J Neuroimaging 2020;30(2):212−18. Available from: https://doi.org/10.1111/jon.12688.

[133] Vonk JMJ, Ghaznawi R, Zwartbol MHT, Stern Y, Geerlings MI, Group UC- S-S. The role of cognitive and brain reserve in memory decline and atrophy rate in mid and late-life: the SMART-MR study. Cortex 2022;148:204−14. Available from: https://doi.org/10.1016/j.cortex.2021.11.022.

[134] Bermel RA, Sharma J, Tjoa CW, Puli SR, Bakshi R. A semiautomated measure of whole-brain atrophy in multiple sclerosis. J Neurol Sci 2003;208(1-2):57−65. Available from: https://doi.org/10.1016/s0022-510x(02)00425-2.

[135] Bose G, Healy BC, Lokhande HA, Sotiropoulos MG, Polgar-Turcsanyi M, Anderson M, et al. Early predictors of clinical and MRI outcomes using least absolute shrinkage and selection operator (LASSO) in multiple sclerosis. Ann Neurol 2022;92(1):87−96. Available from: https://doi.org/10.1002/ana.26370.

[136] Sumowski JF, Rocca MA, Leavitt VM, Meani A, Mesaros S, Drulovic J, et al. Brain reserve against physical disability progression over 5 years in multiple sclerosis. Neurology 2016;86(21):2006−9. Available from: https://doi.org/10.1212/WNL.0000000000002702.

[137] Casserly C, Seyman EE, Alcaide-Leon P, Guenette M, Lyons C, Sankar S, et al. Spinal cord atrophy in multiple sclerosis: a systematic review and meta-analysis. J Neuroimaging 2018;28(6):556−86. Available from: https://doi.org/10.1111/jon.12553.

CHAPTER 7

Magnetic resonance imaging markers in multiple sclerosis clinical trials and emerging imaging biomarkers

Eleonora Tavazzi[1] and Niels Bergsland[2]

[1]Multiple Sclerosis Centre, IRCCS Mondino Foundation, Pavia, Italy [2]Buffalo Neuroimaging Analysis Center (BNAC), Department of Neurology, Jacobs School of Medicine and Biomedical Sciences, University at Buffalo, State University of New York, Buffalo, NY, United States

OUTLINE

Introduction	123	Emerging imaging biomarkers	128
Conventional magnetic resonance imaging markers in multiple sclerosis	124	Chronic active lesions	128
		Cortical lesions	129
T2 lesions	124	Atrophied T2-lesion volume	129
T1 lesions	125	Conclusions	130
Gadolinium-enhancing lesions	126	References	130
Atrophy measurements	127		

Introduction

After defining/characterizing multiple sclerosis (MS) as a specific nosological entity [1], two of the main milestones in the process of unraveling all the unknowns on the topic have been the advent of magnetic resonance imaging (MRI) [2] and the approval of steroids and interferon-ß, as the first pharmacological treatments [3,4]. The former revolutionized the knowledge on MS-related pathogenic mechanisms. Before the introduction of MRI, it was necessary to rely on anatomopathological studies, which typically involved patients in the later stages of the disease, and findings were inherently cross-sectional. As such, a proper understanding of the longitudinal evolution of the underlying pathology was for the most part lacking. Evidence of disease activity, as shown by MRI, was quickly incorporated in the diagnostic criteria, facilitating the differential diagnostic process and allowing for a timely diagnosis.

Since the approval of the first disease modifying therapy (DMT) along with the many newer compounds that have been added to the pharmaceutical armament, the disease course for many patients has changed dramatically, from an irreversible process of progressive clinical worsening leading almost inevitably to the loss of autonomy in 15–20 years, to a very satisfactory control on clinical relapses and a partial slowing of the progressive component. In this regard, MRI has become an essential component of randomized controlled trials (RCTs) assessing the efficacy and safety of essentially all of the now approved DMTs [5].

This chapter presents data on the main imaging markers currently used in RCTs and it will discuss emerging imaging biomarkers already used or in the development phase.

Conventional magnetic resonance imaging markers in multiple sclerosis

The main conventional MRI markers used in RCTs are the so-called T2- and T1-lesion loads, which represent the lesion burden in terms of both the number and volume of lesions within the brain, as well as the number and volume of gadolinium-enhancing lesions.

T2 lesions

T2 lesion load, observed as hyperintense areas on T2-weighted MRI scans, reflects the total burden of lesions accumulated over time and is one of the most widely used imaging biomarkers in MS research. T2 lesions are highly sensitive indicators of disease activity, although they do not specifically differentiate between the various underlying pathological processes such as edema, inflammation, demyelination, or axonal loss. RCTs employing T2 lesion load as a primary or secondary endpoint have substantially contributed to the understanding of MS. For example, the pivotal trials leading to the approval of interferon-ß compounds and glatiramer acetate demonstrated a reduction in T2 lesion volume, supporting their role in modifying the disease course [4,6,7]. These trials set a precedent for using T2 lesion load as a surrogate marker for disease activity in MS.

The use of T2 lesion load as a metric in clinical trials, however, is not without its limitations. While the presence of lesions along with their volume correlate with disease activity, they do not always parallel the clinical outcomes or the progression of disability. This discrepancy is often referred to as the "clinico-radiological paradox," where patients with a high lesion load may have a relatively mild disability and vice versa [8]. This paradox challenges the reliability of T2 lesion load as the sole endpoint in clinical trials and underlines the necessity for additional measures of disease activity and progression.

Furthermore, the total number/volume of T2 lesions are cumulative markers, incorporating both new and preexisting damage, and do not provide information about the current disease activity. As such, T2 lesion activity (i.e., development of new or enlarging T2 lesions) at follow-up visits is also important in monitoring overall disease activity. Nevertheless, such activity does not necessarily inform on when the activity actually happened, which is particularly relevant since many months often pass before the next visit.

Despite these challenges, T2 lesion load remains a valuable and widely accepted biomarker in MS clinical trials. Its utility is enhanced when combined with other imaging techniques such as brain atrophy measures and advanced MRI techniques that can assess the integrity of neural tissue, such as magnetization transfer ratio and diffusion tensor imaging.

The future of Ms treatment relies on a multifaceted approach to drug evaluation, where T2 lesion load will continue to be an important, albeit not singular, component of comprehensive trial design.

T1 lesions

T1 lesions, particularly T1 hypointense lesions, commonly referred to as "black holes," are indicative of severe and mostly irreversible damage such as axonal loss and brain tissue destruction. In clinical trials, T1-lesion load is measured as the volume or number of hypointense lesions on T1-weighted MRI scans that do not enhance simultaneously on a T1-weighted scans acquired after administration of gadolinium. The accumulation of T1 black holes is associated with the progression of disability in MS [9]. Unlike T2-hyperintense lesions, which can reflect a range of pathological changes including inflammation and edema, T1 black holes are considered to be more specific markers of neurodegeneration and permanent neurological impairment. RCTs incorporating T1 lesion load as an endpoint can offer better insights into the neuroprotective effects of treatments. A study might demonstrate a reduction in the accumulation of new T1 black holes over the course of the trial, suggesting that a given treatment has a clearer stabilizing effect on the disease process compared to a drug that does not. For instance, in trials for therapies like natalizumab and fingolimod, a decrease in the accumulation of T1 lesions was correlated with a reduction in disability progression [10,11]. These findings support the use of T1 lesion load as an important imaging endpoint in Ms RCTs. However, the evaluation of T1 lesion load is not without its challenges. The presence of T1 black holes can be a lagging indicator of disease progression, as they represent tissue loss that has already occurred. Thus changes in T1 lesion load may not be apparent until months or years after the underlying damage has begun. This delayed appearance can complicate the interpretation of short-term clinical trials, where changes in T1 lesion load may not fully capture the immediate impact of a therapy. Moreover, the quantification of T1 lesion load can be influenced by MRI scanner settings and the method of lesion measurement, which can lead to variability in results across studies. Standardizing MRI protocols and lesion measurement techniques is crucial for ensuring the reliability and comparability of T1 lesion load data across different clinical trials. Moreover, it has been shown that the degree of hypointensity also reflects differences in the underlying pathology, with a lesion that is more hypointense showing more damage compared to one that is less hypointense [12]. Despite these limitations, T1 lesion load remains an essential biomarker for RCTs in MS. It provides valuable information about the extent of irreversible central nervous system (CNS) damage and helps to predict long-term disability outcomes. As treatments for MS continue to evolve, the role of T1 lesion load in clinical trials is likely to expand, particularly as researchers seek to understand the neurodegenerative aspects of MS and develop therapies aimed at protecting and repairing the nervous system.

Gadolinium-enhancing lesions

Gadolinium-enhancing lesions, detected by MRI, are a key biomarker for active inflammation within the CNS, providing real-time insights into the disease's activity [13]. Gadolinium is a paramagnetic contrast agent used in MRI to visualize the breakdown of the blood–brain barrier (BBB), which is a characteristic feature of active MS lesions. When gadolinium is administered intravenously, it does not usually cross an intact BBB. However, in areas where the BBB is compromised, which occurs during active inflammation in Ms, gadolinium leaks into the CNS, resulting in a hyperintense signal on T1-weighted MRI scans. In RCTs, the presence and number of gadolinium-enhancing lesions can serve as a primary or secondary endpoint. They provide a measure of acute disease activity, allowing one to assess the immediate effect of a therapeutic agent on the inflammatory process of MS. A significant reduction in the number of these lesions compared to placebo or a comparator drug indicates a positive impact of the treatment on controlling disease activity. For instance, the use of natalizumab, a monoclonal antibody targeting the cell adhesion molecule α4-integrin, has been shown to result in a marked reduction in gadolinium-enhancing lesions in patients with relapsing-remitting MS (RRMS) [10]. This finding was instrumental in demonstrating the drug's efficacy and supporting its approval for clinical use. Similarly, other DMTs, such as fingolimod and ocrelizumab, have shown a decrease in gadolinium-enhancing lesions in RCTs, reinforcing their role in reducing the inflammatory activity of Ms [11,14]. The use of gadolinium-enhancing lesions in RCTs also aids in the stratification of patients. Patients with a high number of enhancing lesions at baseline may have a more aggressive disease course and may respond differently to treatment [15]. This stratification can optimize the design of clinical trials by ensuring that treatment effects are not masked by variations in disease activity among participants. However, the use of gadolinium-enhancing lesions as an endpoint has limitations. The transient nature of these lesions means (lasting typically 4 weeks, with a range of 2–8 weeks) [16] that their presence can vary greatly depending on the timing of the MRI scan relative to the onset of the lesion. Consequently, frequent MRI scans are required to accurately capture changes in lesion activity over time, which can be costly and burdensome for trial participants. Moreover, not all active lesions will enhance with gadolinium at the same time, and some may be missed if the timing of the scan does not coincide with the peak of BBB disruption. The role of gadolinium-enhancing lesions in RCTs also extends beyond mere measurement of treatment response [5]. It can provide valuable information regarding the pathophysiology of MS, help in the identification of rapid responders to therapy, and allow for the early detection of suboptimal treatment outcomes. This information is critical for the personalization of treatment, offering a tailored therapeutic approach based on individual disease activity. Despite the insights provided by gadolinium-enhancing lesions, the medical community remains aware of the concerns associated with repeated use of gadolinium-based contrast agents, such as the potential for gadolinium deposition in the brain [17,18]. Ongoing research aims to address these concerns by developing newer contrast agents with better safety profiles [19] and by refining noncontrast MRI techniques that can also detect active lesions [20].

Atrophy measurements

The use of atrophy measurements in RCTs for MS has become increasingly important as our understanding of the disease evolves. Ms is characterized not only by the presence of inflammatory lesions but also by the progressive loss of neural tissue (i.e., tissue atrophy) [21]. This atrophy is considered a key marker of neurodegeneration and is associated with long-term disability in Ms patients [22]. Atrophy can be measured using various MRI techniques, which quantify the loss of brain volume over time. Common assessments include whole-brain atrophy, cortical atrophy, and central atrophy, encompassing the deep gray matter structures such as the thalamus. In RCTs, brain atrophy measurements are often used as secondary or exploratory endpoints. As atrophy is a relatively slow process, changes in brain volume may be relatively difficult to detect over the short durations typical of many clinical trials. However, the importance of atrophy as a predictor of long-term disability [22] has prompted its inclusion in clinical trials as a crucial measure of a drug's potential neuroprotective effect. For instance, in trials of high-efficacy DMTs like natalizumab and alemtuzumab, reductions in brain volume loss have been observed compared to comparator therapies, indicating a potential for these drugs to slow the progression of atrophy [23,24]. Furthermore, studies have shown that even in the early stages of MS, atrophy measurements can predict the accumulation of disability, underscoring the importance of early intervention [25]. Despite its clear relevance, accurate atrophy measurement at the individual level can be challenging due to natural variations in brain volume, which can be influenced by factors such as hydration status [26], diurnal fluctuations [27], menstrual phase [28], among others. To account for these variables, standardized imaging protocols and analysis techniques are required. Moreover, atrophy must be differentiated from pseudoatrophy, a rapid reduction in brain volume that can occur due to the resolution of edema following the initiation of antiinflammatory treatments [29,30]. Despite these challenges, the inclusion of atrophy measurements in RCTs is essential for a comprehensive assessment of treatment effects. Atrophy is a more direct measure of the neurodegenerative component of MS than inflammatory lesion load. Therefore drugs that can demonstrate an impact on reducing the rate of atrophy offer hope for addressing the progressive aspect of MS, which is often the most debilitating. The focus on atrophy in RCTs also represents a shift in the MS treatment paradigm—from a sole focus on the inflammatory activity to a more balanced approach that also targets neurodegeneration. As such, the development of neuroprotective agents is likely to be guided by their effects on atrophy, with the aim of preserving neurological function and improving quality of life for MS patients.

Emerging imaging biomarkers

Chronic active lesions

The result of new DMTs being approved has dramatically changed MS prognosis, allowing for an excellent control of the inflammatory activity, clinically measurable as marked reduction of relapse rate and radiologically represented by significant improvements in all the above-described MRI markers. However, literature data concordantly describe the presence of progression independent of relapse activity (PIRA) [31–33] that parallels the relapse-associated worsening. PIRA develops from the early disease stages [33] and is associated with accelerated brain tissue loss in RRMS [34]. This phenomenon is likely sustained by so-called smoldering inflammation, a chronic inflammatory process resulting from several simultaneous pathological mechanisms. Among them there is the formation of meningeal lymphoid aggregates, releasing soluble inflammatory factors that contribute to the self-maintenance of compartmentalized inflammation [35–37]; the persistent and widespread activation of microglial cells throughout brain parenchyma [38–40]; the persistence of activated T lymphocytes scattered throughout the CNS [41]. Whereas the presence of smoldering inflammation has been defined histopathologically, its identification in vivo has only recently become possible via the aid of MRI. MRI markers related to all these processes are of prominent relevance as they capture and allow to visualize, and possibly quantify, progressive tissue damage in MS.

Chronic active lesions are an MRI marker that is thought to reflect to smoldering inflammation most closely. These lesions are represented by slowly-expanding lesions (SELs) and/or paramagnetic rim lesions (PRLs) [42–44]. SELs are detected through deformation-based volumetric MRI using serial T1- and T2-weighted MRI, as they reflect a phenomenon evolving over time [44]. They have been identified retrospectively on large clinical trials datasets of both RRMS and progressive MS, and are characterized by a local expansion of T2-hyperintense lesions, together with a gradual increase of T1-hypointensity, indicating marked tissue damage, mainly due to axonal loss. Although present in all MS phenotypes, they are more frequent in PMS and correlate with clinical disability and cognitive impairment [44–47].

PRLs can be identified applying susceptibility-weighted imaging (SWI), a technique sensitive to paramagnetic elements such as iron [48]. PRLs can be distinguished from non-iron rim containing lesions (n-IRLs) for the typical hypointense rim at the periphery of the lesions themselves [49], histopathologically characterized by microglia/macrophages enriched with iron. The hypointense rim colocalize with the initial contrast enhancement of newly-developed lesions, which usually fades over the 4–6 weeks following its appearance. Instead, in a small percentage of lesions, this edge becomes chronically inflamed reflecting failed tissue repair and irreversible tissue damage, with microglial cells containing iron derived from oligodendrocyte disruption or leaking from a disrupted BBB [49,50]. PRL dynamics are characterized by a slow, gradual enlargement for the first years, sometimes followed by a progressive disappearance of the iron-rim lesions in a timespan of 5–7 years [51]. They are more destructive than non-PRLs, as indicated by the greater T1-hypointensity, suggesting the presence of irreversible tissue damage that correlates with a worse disease course [52]. As PRLs can be detected early in the disease [53] and the association between their presence/persistence over time with a poor clinical outcome, they may potentially serve as an ideal MRI marker to identify those patients at higher risk for clinical disability accrual. These patients could then be treated with a more aggressive therapy. However, they have been rarely included is RCTs as of now [54], owing in part to the fact SWI is not necessarily available on all MRI systems. Moreover, consensus guidelines on how to identify PRLs have only recently become available [55] and there is still debate on what the best technique is for visualizing them [55,56].

As SELs and PRLs are both indicative of chronic inflammatory activity and they share some features, such as a higher frequency in progressive forms of MS and the association with greater clinical disability, some studies have aimed at analyzing their relationship [57,58]. Concordantly, the results of these studies reported a significantly higher number of SELs than PRLs, with a colocalization of the two types of lesions in 39.5% of lesions [58]; a linear association between the number of PRLs as well SELs and clinical disability, and a more severe clinical progression when PRLs and SELs occurred jointly [57]. Nevertheless, it seems evident that a single approach will not be sufficient to capture the full extent of chronic active lesions in MS.

Cortical lesions

The development of nonconventional sequences, such as double-inversion recovery and phase-sensitive inversion recovery, has allowed for a better visualization of cortical lesions (CLs) [59–61], an imaging marker of cortical demyelination significantly associated with clinical disability [62], clinical relapses [63], and cognitive impairment [64,65]. The use of 7 tesla (T) T2* gradient-echo sequences has given further insight on the topic, facilitating a more detailed quantification of CLs and the identification of different types of lesion according to the anatomical location (i.e., intracortical, leukocortical lesions, and subpial lesions [66]) strongly associated with disease stages and clinical disability [67–69]. Besides the association with markers of clinical and cognitive impairment, CLs are a potentially promising marker for their prognostic value, as demonstrated in different studies in which CLs were among the few MRI parameters able to predict long-term disease course and the development of cognitive dysfunction [64,70,71]. Some exploratory studies analyzed the effect of first-line and high-efficacy DMTs on CLs, showing that interferon ß-1a, glatiramer acetate, dimethyl fumarate, and natalizumab were able to prevent new CLs formation [72–74]. However, the limited visualization of cortical (especially subpial) lesions with 1.5 and 3 T MRI and several technical issues limit their widespread use in the context of RCTs, together with lack of standardized imaging approach to identify them limit, as of now, their potential use as MRI biomarkers in RCTs.

Atrophied T2-lesion volume

Chronic inflammatory activity leading to irreversible tissue damage, as illustrated above, manifests itself through lesions expansion. However, lesions can also shrink and be partially or completely substituted by cerebrospinal fluid, reflecting again neural tissue damage and axonal loss. A novel MRI marker reflecting this phenomenon is the so-called atrophied T2-lesion volume, reflecting the process of lesional tissue loss evolving over time [75,76].

The exact histopathological correlate of atrophied T2-lesion volume has not been identified. However, the association with clinical disability and the ability to predict evolution toward secondary progressive forms [77,78] indicate the usefulness of this measure as MRI marker of neurodegeneration in the context of RCTs [79].

Conclusions

The use of MRI has become a crucial component of all clinical trials evaluating the effects of DMTs in MS. As of now, RCTs typically rely on the measures that have been used extensively (i.e., lesion-based outcomes). However, as the acute inflammatory component of MS can be effectively managed with modern DMTs, future trials will likely rely on more advanced, emerging MRI markers that can offer better specificity regarding the underlying pathology, particularly with respect to neurodegenerative changes.

References

[1] Schumacher GA. Multiple sclerosis. Arch Neurol 1966;14:571–3.
[2] Young IR, Hall AS, Pallis CA, Legg NJ, Bydder GM, Steiner RE. Nuclear magnetic resonance imaging of the brain in multiple sclerosis. Lancet 1981;2:1063–6.
[3] Miller H, Newell DJ, Ridley A. Multiple sclerosis. Trials of maintenance treatment with prednisolone and soluble aspirin. Lancet 1961;1:127–9.
[4] Placebo-controlled multicentre randomised trial of interferon beta-1b in treatment of secondary progressive multiple sclerosis. European Study Group on interferon beta-1b in secondary progressive MS. Lancet 1998;352:1491–7.
[5] Sormani MP, Bruzzi P. MRI lesions as a surrogate for relapses in multiple sclerosis: a meta-analysis of randomised trials. Lancet Neurol 2013;12:669–76.
[6] Johnson KP, Brooks BR, Cohen JA, et al. Copolymer 1 reduces relapse rate and improves disability in relapsing-remitting multiple sclerosis: results of a phase III multicenter, double-blind placebo-controlled trial. The Copolymer 1 Multiple Sclerosis Study Group. Neurology 1995;45:1268–76.
[7] Randomised double-blind placebo-controlled study of interferon beta-1a in relapsing/remitting multiple sclerosis. PRISMS (Prevention of Relapses and Disability by Interferon beta-1a Subcutaneously in Multiple Sclerosis) Study Group. Lancet 1998;352:1498–504.
[8] Barkhof F, Bruck W, De, Groot CJ, et al. Remyelinated lesions in multiple sclerosis: magnetic resonance image appearance. Arch Neurol 2003;60:1073–81.
[9] Valizadeh A, Moassefi M, Barati E, Ali Sahraian M, Aghajani F, Fattahi MR. Correlation between the clinical disability and T1 hypointense lesions' volume in cerebral magnetic resonance imaging of multiple sclerosis patients: a systematic review and meta-analysis. CNS Neurosci Ther 2021;27:1268–80.
[10] Polman CH, O'Connor PW, Havrdova E, et al. A randomized, placebo-controlled trial of natalizumab for relapsing multiple sclerosis. N Engl J Med 2006;354:899–910.
[11] Kappos L, Radue EW, O'Connor P, et al. A placebo-controlled trial of oral fingolimod in relapsing multiple sclerosis. N Engl J Med 2010;362:387–401.
[12] Adusumilli G, Trinkaus K, Sun P, et al. Intensity ratio to improve black hole assessment in multiple sclerosis. Mult Scler Relat Disord 2018;19:140–7.
[13] Grossman RI, Gonzalez-Scarano F, Atlas SW, Galetta S, Silberberg DH. Multiple sclerosis: gadolinium enhancement in MR imaging. Radiology 1986;161:721–5.
[14] Hauser SL, Bar-Or A, Comi G, et al. Ocrelizumab versus Interferon Beta-1a in Relapsing Multiple Sclerosis. N Engl J Med 2017;376:221–34.

[15] Arrambide G, Iacobaeus E, Amato MP, et al. Aggressive multiple sclerosis (2): treatment. Mult Scler 2020;26 1352458520924595.
[16] Filippi M, Preziosa P, Banwell BL, et al. Assessment of lesions on magnetic resonance imaging in multiple sclerosis: practical guidelines. Brain 2019;142:1858−75.
[17] Gulani V, Calamante F, Shellock FG, Kanal E, Reeder SB. International Society for Magnetic Resonance in M. Gadolinium deposition in the brain: summary of evidence and recommendations. Lancet Neurol 2017;16:564−70.
[18] Zivadinov R, Bergsland N, Hagemeier J, et al. Cumulative gadodiamide administration leads to brain gadolinium deposition in early MS. Neurology 2019;93:e611−23.
[19] Boehm-Sturm P, Haeckel A, Hauptmann R, Mueller S, Kuhl CK, Schellenberger EA. Low-molecular-weight iron chelates may be an alternative to gadolinium-based contrast agents for T1-weighted contrast-enhanced MR imaging. Radiology 2018;286:537−46.
[20] Gupta A, Al-Dasuqi K, Xia F, et al. The use of noncontrast quantitative MRI to detect gadolinium-enhancing multiple sclerosis brain lesions: a systematic review and meta-analysis. AJNR Am J Neuroradiol 2017;38:1317−22.
[21] Sastre-Garriga J, Pareto D, Battaglini M, et al. MAGNIMS consensus recommendations on the use of brain and spinal cord atrophy measures in clinical practice. Nat Rev Neurol 2020;16:171−82.
[22] Zivadinov R, Jakimovski D, Gandhi S, et al. Clinical relevance of brain atrophy assessment in multiple sclerosis. Implications for its use in a clinical routine. Expert Rev Neurother 2016;16:777−93.
[23] Rudick RA, Fisher E, Lee JC, Simon J, Jacobs L. Use of the brain parenchymal fraction to measure whole brain atrophy in relapsing-remitting MS. Multiple Sclerosis Collaborative Research Group. Neurology 1999;53:1698−704.
[24] Coles AJ, Twyman CL, Arnold DL, et al. Alemtuzumab for patients with relapsing multiple sclerosis after disease-modifying therapy: a randomised controlled phase 3 trial. Lancet 2012;380:1829−39.
[25] De Stefano N, Airas L, Grigoriadis N, et al. Clinical relevance of brain volume measures in multiple sclerosis. CNS Drugs 2014;28:147−56.
[26] Nakamura K, Brown RA, Araujo D, Narayanan S, Arnold DL. Correlation between brain volume change and T2 relaxation time induced by dehydration and rehydration: implications for monitoring atrophy in clinical studies. Neuroimage Clin 2014;6:166−70.
[27] Nakamura K, Brown RA, Narayanan S, Collins DL, Arnold DL, Alzheimer's Disease Neuroimaging I. Diurnal fluctuations in brain volume: statistical analyses of MRI from large populations. Neuroimage 2015;118:126−32.
[28] Hagemann G, Ugur T, Schleussner E, et al. Changes in brain size during the menstrual cycle. PLoS One 2011;6:e14655.
[29] Sastre-Garriga J, Tur C, Pareto D, et al. Brain atrophy in natalizumab-treated patients: a 3-year follow-up. Mult Scler 2015;21:749−56.
[30] De Stefano N, Arnold DL. Towards a better understanding of pseudoatrophy in the brain of multiple sclerosis patients. Mult Scler 2015;21:675−6.
[31] Kappos L, Wolinsky JS, Giovannoni G, et al. Contribution of relapse-independent progression vs relapse-associated worsening to overall confirmed disability accumulation in typical relapsing multiple sclerosis in a pooled analysis of 2 randomized clinical trials. JAMA Neurol 2020;77:1132−40.
[32] Muller J, Cagol A, Lorscheider J, et al. Harmonizing definitions for progression independent of relapse activity in multiple sclerosis: a systematic review. JAMA Neurol 2023;80:1232−45.
[33] Portaccio E, Bellinvia A, Fonderico M, et al. Progression is independent of relapse activity in early multiple sclerosis: a real-life cohort study. Brain 2022;145:2796−805.
[34] Cagol A, Schaedelin S, Barakovic M, et al. Association of brain atrophy with disease progression independent of relapse activity in patients with relapsing multiple sclerosis. JAMA Neurol 2022;79:682−92.
[35] Serafini B, Rosicarelli B, Magliozzi R, Stigliano E, Aloisi F. Detection of ectopic B-cell follicles with germinal centers in the meninges of patients with secondary progressive multiple sclerosis. Brain Pathol 2004;14:164−74.
[36] Magliozzi R, Howell O, Vora A, et al. Meningeal B-cell follicles in secondary progressive multiple sclerosis associate with early onset of disease and severe cortical pathology. Brain 2007;130:1089−104.
[37] Ransohoff RM. Multiple sclerosis: role of meningeal lymphoid aggregates in progression independent of relapse activity. Trends Immunol 2023;44:266−75.

[38] Airas L, Yong VW. Microglia in multiple sclerosis — pathogenesis and imaging. Curr Opin Neurol 2022;35:299—306.
[39] Distefano-Gagne F, Bitarafan S, Lacroix S, Gosselin D. Roles and regulation of microglia activity in multiple sclerosis: insights from animal models. Nat Rev Neurosci 2023;24:397—415.
[40] Absinta M, Maric D, Gharagozloo M, et al. A lymphocyte-microglia-astrocyte axis in chronic active multiple sclerosis. Nature 2021;597:709—14.
[41] Viglietta V, Baecher-Allan C, Weiner HL, Hafler DA. Loss of functional suppression by CD4 + CD25 + regulatory T cells in patients with multiple sclerosis. J Exp Med 2004;199:971—9.
[42] Absinta M, Sati P, Fechner A, Schindler MK, Nair G, Reich DS. Identification of chronic active multiple sclerosis lesions on 3T MRI. AJNR Am J Neuroradiol 2018;39:1233—8.
[43] Maggi P, Sati P, Nair G, et al. Paramagnetic rim lesions are specific to multiple sclerosis: an international multicenter 3T MRI study. Ann Neurol 2020;88:1034—42.
[44] Elliott C, Wolinsky JS, Hauser SL, et al. Slowly expanding/evolving lesions as a magnetic resonance imaging marker of chronic active multiple sclerosis lesions. Mult Scler 2019;25:1915—25.
[45] Preziosa P, Pagani E, Meani A, et al. Slowly expanding lesions predict 9-year multiple sclerosis disease progression. Neurol Neuroimmunol Neuroinflamm 2022;9.
[46] Elliott C, Belachew S, Wolinsky JS, et al. Chronic white matter lesion activity predicts clinical progression in primary progressive multiple sclerosis. Brain 2019;142:2787—99.
[47] Calvi A, Carrasco FP, Tur C, et al. Association of slowly expanding lesions on MRI with disability in people with secondary progressive multiple sclerosis. Neurology 2022;98:e1783—93.
[48] Liu C, Li W, Tong KA, Yeom KW, Kuzminski S. Susceptibility-weighted imaging and quantitative susceptibility mapping in the brain. J Magn Reson Imaging 2015;42:23—41.
[49] Absinta M, Sati P, Schindler M, et al. Persistent 7-tesla phase rim predicts poor outcome in new multiple sclerosis patient lesions. J Clin Invest 2016;126:2597—609.
[50] Hochmeister S, Grundtner R, Bauer J, et al. Dysferlin is a new marker for leaky brain blood vessels in multiple sclerosis. J Neuropathol Exp Neurol 2006;65:855—65.
[51] Dal-Bianco A, Grabner G, Kronnerwetter C, et al. Long-term evolution of multiple sclerosis iron rim lesions in 7T MRI. Brain 2021;144:833—47.
[52] Dal-Bianco A, Schranzer R, Grabner G, et al. Iron rims in patients with multiple sclerosis as neurodegenerative marker? A 7-Tesla magnetic resonance study. Front Neurol 2021;12:632749.
[53] Ng Kee Kwong KC, Mollison D, Meijboom R, et al. Rim lesions are demonstrated in early relapsing-remitting multiple sclerosis using 3 T-based susceptibility-weighted imaging in a multi-institutional setting. Neuroradiology 2022;64:109—17.
[54] Reich DS, Arnold DL, Vermersch P, et al. Safety and efficacy of tolebrutinib, an oral brain-penetrant BTK inhibitor, in relapsing multiple sclerosis: a phase 2b, randomised, double-blind, placebo-controlled trial. Lancet Neurol 2021;20:729—38.
[55] Bagnato F, Sati P, Hemond CC, et al. Imaging chronic active lesions in multiple sclerosis: a consensus statement. Brain 2024;.
[56] Reeves JA, Mohebbi M, Zivadinov R, et al. Reliability of paramagnetic rim lesion classification on quantitative susceptibility mapping (QSM) in people with multiple sclerosis: single-site experience and systematic review. Mult Scler Relat Disord 2023;79:104968.
[57] Calvi A, Clarke MA, Prados F, et al. Relationship between paramagnetic rim lesions and slowly expanding lesions in multiple sclerosis. Mult Scler 2023;29:352—62.
[58] Elliott C, Rudko DA, Arnold DL, et al. Lesion-level correspondence and longitudinal properties of paramagnetic rim and slowly expanding lesions in multiple sclerosis. Mult Scler 2023;29:680—90.
[59] Calabrese M, Filippi M, Gallo P. Cortical lesions in multiple sclerosis. Nat Rev Neurol 2010;6:438—44.
[60] Ciccarelli O, Chen JT. MS cortical lesions on double inversion recovery MRI: few but true. Neurology 2012;78:296—7.
[61] Sethi V, Yousry TA, Muhlert N, et al. Improved detection of cortical MS lesions with phase-sensitive inversion recovery MRI. J Neurol Neurosurg Psychiatry 2012;83:877—82.
[62] Chase A. Multiple sclerosis: 7T MRI reveals cortical lesions associated with disability in MS. Nat Rev Neurol 2015;11:486.
[63] Puthenparampil M, Poggiali D, Causin F, et al. Cortical relapses in multiple sclerosis. Mult Scler 2016;22:1184—91.

[64] Ziccardi S, Pisani AI, Schiavi GM, et al. Cortical lesions at diagnosis predict long-term cognitive impairment in multiple sclerosis: a 20-year study. Eur J Neurol 2023;30:1378–88.

[65] Papadopoulou A, Muller-Lenke N, Naegelin Y, et al. Contribution of cortical and white matter lesions to cognitive impairment in multiple sclerosis. Mult Scler 2013;19:1290–6.

[66] Mainero C, Louapre C, Govindarajan ST, et al. A gradient in cortical pathology in multiple sclerosis by in vivo quantitative 7T imaging. Brain 2015;138:932–45.

[67] Harrison DM, Roy S, Oh J, et al. Association of cortical lesion burden on 7-T magnetic resonance imaging with cognition and disability in multiple sclerosis. JAMA Neurol 2015;72:1004–12.

[68] Beck ES, Maranzano J, Luciano NJ, et al. Cortical lesion hotspots and association of subpial lesions with disability in multiple sclerosis. Mult Scler 2022;28:1351–63.

[69] Nielsen AS, Kinkel RP, Madigan N, Tinelli E, Benner T, Mainero C. Contribution of cortical lesion subtypes at 7T MRI to physical and cognitive performance in MS. Neurology 2013;81:641–9.

[70] Haider L, Chung K, Birch G, et al. Linear brain atrophy measures in multiple sclerosis and clinically isolated syndromes: a 30-year follow-up. J Neurol Neurosurg Psychiatry 2021;.

[71] Pisani AI, Scalfari A, Crescenzo F, Romualdi C, Calabrese M. A novel prognostic score to assess the risk of progression in relapsing-remitting multiple sclerosis patients. Eur J Neurol 2021;28:2503–12.

[72] Marastoni D, Crescenzo F, Pisani AI, et al. Two years' effect of dimethyl fumarate on focal and diffuse gray matter pathology in multiple sclerosis. Mult Scler 2022;28:2090–8.

[73] Calabrese M, Bernardi V, Atzori M, et al. Effect of disease-modifying drugs on cortical lesions and atrophy in relapsing-remitting multiple sclerosis. Mult Scler 2012;18:418–24.

[74] Puthenparampil M, Cazzola C, Zywicki S, et al. NEDA-3 status including cortical lesions in the comparative evaluation of natalizumab versus fingolimod efficacy in multiple sclerosis. Ther Adv Neurol Disord 2018;11 1756286418805713.

[75] Dwyer MG, Bergsland N, Ramasamy DP, Jakimovski D, Weinstock-Guttman B, Zivadinov R. Atrophied brain lesion volume: a new imaging biomarker in multiple sclerosis. J Neuroimaging 2018;28:490–5.

[76] Zivadinov R, Bergsland N, Dwyer MG. Atrophied brain lesion volume, a magnetic resonance imaging biomarker for monitoring neurodegenerative changes in multiple sclerosis. Quant Imaging Med Surg 2018;8:979–83.

[77] Genovese AV, Hagemeier J, Bergsland N, et al. Atrophied brain T2 lesion volume at MRI is associated with disability progression and conversion to secondary progressive multiple sclerosis. Radiology 2019;293:424–33.

[78] Tavazzi E, Bergsland N, Kuhle J, et al. A multimodal approach to assess the validity of atrophied T2-lesion volume as an MRI marker of disease progression in multiple sclerosis. J Neurol 2020;267:802–11.

[79] Zivadinov R, Pei J, Clayton D, et al. Evolution of atrophied T2 lesion volume in primary-progressive multiple sclerosis: results from the phase 3 ORATORIO study. J Neurol Neurosurg Psychiatry 2023;.

SECTION I

Imaging of multiple sclerosis

PART 1

Conventional MRI use in multiple sclerosis

CHAPTER 8

Magnetic resonance imaging physics and image acquisition

Junghun Cho[1] and Alexey Dimov[2]

[1]Department of Biomedical Engineering, State University of New York at Buffalo, Buffalo, NY, United States [2]Department of Radiology, Weill Cornell Medicine, New York, NY, United States

OUTLINE

Introduction	139	Relationship between magnetic resonance imaging signal and tissue properties	144
Magnetic resonance imaging image formation	140	Relaxation: longitudinal (T_1) and transverse (T_2) relaxation	144
The main magnetic field (B_0): Larmor precession and magnetic resonance imaging signal detection	140	T_1-, T_2-, and proton density weighting in MR imaging	146
The gradient field (G): Larmor frequency modulation and image formation	141	Biological water motion: diffusion and perfusion	148
The radiofrequency field (B_1): spin dynamics and excitation	141	Tissue magnetism: T_2^* relaxation and quantitative susceptibility mapping	152
Basics of the magnetic resonance imaging pulse sequences	142	Summary	155
		References	155

Introduction

Magnetic resonance imaging (MRI) has become an essential tool in the diagnosis and management of multiple sclerosis (MS), providing noninvasive imaging of the brain and spinal cord to highlight areas of inflammation or scarring [1–4]. To investigate MRI abnormalities driven by pathophysiology in neurologic disorders, including Ms, it is crucial to understand basics concepts of how the MRI signal is generated and how it correlates with tissue properties such as relaxation, diffusion, perfusion, and magnetism. This chapter will elucidate how an MRI image is generated and how the MRI signal relates to tissue properties from a physics perspective.

Magnetic resonance imaging image formation

An MRI system consists of three magnetic fields that interact with the nuclear spins in the body: the main magnetic field (B_0), the gradient fields (G), and the radiofrequency (RF) fields (B_1). The main magnetic field (B_0) is essential for Larmor precession, which is the foundation of magnetic resonance. The gradient fields (G) play a critical role in image formation by setting different frequencies at different locations using a spatially linearly varying magnetic field. The RF field (B_1) is responsible for transmitting the RF field to excite nuclear spins and for receiving the MRI signal. Operating under the static main magnetic field, the MRI uses a sequence of RF pulses and gradient field waveforms, known as the pulse sequence, to generate image contrasts.

The main magnetic field (B_0): Larmor precession and magnetic resonance imaging signal detection

The main magnet in MRI generates a homogenous magnetic field (B_0), setting the stage for Larmor precession. The most commonly used main magnets are superconducting electromagnets, especially in clinical MRI scanners with field strengths higher than 1 T. The main magnet consists of a superconductive solenoid coil, cooled by helium liquid, which carries a constant direct current. According to the Biot-Savart Law, a current generates a magnetic field, and in this case, the constant current running through the solenoid can generate a uniform magnetic field inside.

When our body is placed in an external magnetic field B (mainly from B_0 in an MRI), magnetic moments (often referred to as "spin," m) of the protons in water molecules (~60% of our body weight) precess clockwise around B at an angular velocity ω. This phenomenon is known as Larmor precession:

$$\omega = -\gamma B \tag{8.1}$$

Here, γ is the gyromagnetic ratio, which is specific to the configuration of unpaired nucleons or electrons of the atoms. The value of γ varies by atomic species, for example, $\gamma = 2\pi \times 42.58 MHz/T$ for the protons of water molecules. The magnitude of ω is referred to as the Larmor frequency ($\omega = |\gamma|B$).

For a magnetic field in the z-direction (generally the direction set by the main magnet in MRI), the spin (m) rotates clockwise in the transverse plane (x, y) with Larmor frequency (ω): $m_\perp = me^{-i\omega t}$ where $m_\perp$ is the transverse component of the magnetic moment and m is its magnitude. Each spin generates its own magnetic field (b_s). Due to the oscillating transverse components of magnetic moment of the spin, its magnetic field b_s also oscillates at ω. If a conductor loop (i.e., receiver coil) is positioned near the processing spins, the magnetic field flux through the loop's area A ($\Phi_b = \int b_S \cdot dA$) oscillates at ω. According

1. Conventional MRI use in multiple sclerosis

to Faraday's law, this oscillating magnetic flux induces an electromotive force (i.e., a voltage signal $s = -\frac{d\Phi_b}{dt} = i\omega\Phi_b$) in the receiver coil. In this way, the signal from spin precession can be detected in MRI: $s = i\omega m e^{-i\omega t}$ where $\Phi_b = m_\perp$ assuming the coil sensitivity to be unity for simplicity [5]. As each spin at a different location (r) generates a magnetic flux, the detected signal in a receiver coil is the summation of magnetic flux from all the spins:

$$s(t) = \int i\omega(r)m(r)e^{-i\omega(r)t}d^3r \tag{8.2}$$

This is the general signal equation for a group of spins, integrating the spin magnetization $m(r)$ across the image volume. The key to image formation is related to how to obtain the spin magnetization's spatial distribution $m(r)$ from the detected signal by the receiver coil, $s(t)$.

The gradient field (G): Larmor frequency modulation and image formation

The main magnetic field B_0 does not allow the resolution of spins at different locations, as it is spatially homogeneous. The gradient system generates an additional magnetic field that varies linearly with the gradient (G) over the spatial location (r) of the spin. This is achieved by spatially arranged current loops (e.g., a Helmholtz coil), and the resulting field is much smaller than the main magnetic field ($G \cdot r \ll B_0$). Consequently, the total field varies linearly in space ($B(r) = B_0 + G \cdot r$), which modulates the Larmor frequency of protons to also vary linear with space: $\omega = \gamma(B_0 + G \cdot r) = \omega_0 + \gamma G \cdot r$. This allows the spatial encoding of the MRI signal. By filtering out the high-frequency constant factor ($i\omega_0 e^{-i\omega_0 t}$), the MR signal detected by the receiver coil can be expressed as $s(t) = \int m(r)e^{-i\gamma G \cdot rt}d^3r$.

By introducing the Fourier phase factor ($\phi(k,r) = 2\pi\bar{\gamma}G \cdot rt = 2\pi k \cdot r$, where $\bar{\gamma} = \gamma/2\pi$ and $k = \bar{\gamma}Gt$), the detected MR signal under the gradient field, $s(t)$, can be expressed as the Fourier transform ($\mathscr{F}$) of the object's magnetization $m(r)$: $s(t) = \int m(r)e^{-i2\pi k \cdot r}d^3r = \mathscr{F}[m(r)]$. Consequently, $m(r)$ can be retrieved from $s(t)$ via inverse Fourier transform:

$$m(r) = \mathscr{F}^{-1}[s(t(k))] \tag{8.3}$$

where k represents the spatial frequency vector (often referred to as k-space vector). Data acquired at time t corresponds to the data sampled at point k in k-space ($k = \bar{\gamma}Gt$). Once all data in k-space are obtained, an image of the spin distribution of the object can be generated using an inverse Fourier transform. Hence, the characteristics of the object's image are determined by the k-space data sampling. For instance, the field of view (FOV) and resolution (Δx) of an MRI image are determined by the k-space sampling interval (Δk) and maximum (k_{max}), respectively (FOV $= 1/\Delta k$ and $\Delta x = 1/2k_{max}$). This is the fundamental process of image formation in MRI.

The radiofrequency field (B_1): spin dynamics and excitation

The RF system plays a crucial role in spin excitation and signal reception in MRI. In the main magnetic field (B_0), the hydrogen proton spins in the body align with B_0, which means that the magnetization vector (m) points in the z-direction (the equilibrium state).

---

1. Conventional MRI use in multiple sclerosis

To acquire an MRI signal, these aligned spins need to be excited out of their equilibrium state to have a transverse component in the x−y plane, which then allows the gradient fields to generate spatial encoding. This can be accomplished by using the RF system. When the RF system is turned on, an RF magnetic pulse (B_1) that oscillates at the Larmor frequency (ω_0) is applied in the x−y plane to flip m into the transverse plane. The way magnetization changes due to an applied magnetic field (B) is governed by the following torque equation [6]:

$$\frac{dm}{dt} = -\gamma B \times m \qquad (8.4)$$

In a simple scenario where $B = B_0$ in the z-direction, the solution of the torque equation is $m_\perp = me^{-i\omega_0 t}$, that is, Larmor precession. To flip m from the z-direction into x−y plane, an additional RF field (B_1) needs to be applied in the x−y plane. Moreover, B_1 should oscillate at the Larmor frequency ($B_1 = b_1 e^{-i\omega_0 t}$ where b_1 is the magnitude of the RF field), that is, B_1 should rotate with m, to generate a consistent flip of the magnetization from the z-axis into the x−y plane (a process called "resonance"). The flip angle (α) of the magnetization depends on the RF pulse profiles during its duration (T): $\alpha = \gamma \int_0^T b_1(t)dt$. In this way, all the spins in the body that have the same resonance frequency (ω_0) can be excited by the RF system, enabling three-dimensional (3D) image acquisition. For 2D image acquisition, a gradient field (G) can be additionally applied to spread the resonance frequency along the z-direction, and an RF field consisting of a band of frequencies ($\Delta\omega$) can be applied to excite the spins within a specific range of locations (Δz) that have their resonance frequencies falling in $\Delta\omega$: $\Delta z = \Delta\omega/\gamma G$.

After the RF pulse is turned off, the perturbed magnetization returns to their equilibrium state (e.g., from the x−y plane to the z-direction), a process known as relaxation. The nature of this process will be discussed in latter sections of the present chapter.

Basics of the magnetic resonance imaging pulse sequences

After the initial RF excitation, the amplitude of the MR signal decays exponentially to zero within just a few milliseconds due to the rapid dephasing of spins relative to each other. As a result, this short-lived free induction decay (FID) signal is not directly measured. Instead, MRI relies on the sophisticated manipulation of the induced magnetization by a sequence of RF pulses and gradient waveforms, known as a pulse sequence. Depending on the mechanism of signal formation, two broad classes of pulse sequences can be identified: gradient echo (GRE) and spin echo (SE).

- In GRE (Fig. 8.1), a negative gradient lobe is turned on immediately after the RF pulse, causing the transverse magnetization $m_\perp$ to dephase much faster than it would under normal FID. After this, a positive gradient is applied. Consequently, spins that were precessing at lower frequencies due to their positions in the negative gradient will start precessing faster, and vice versa. Spins that were earlier dephasing will now start to realign, producing an increase in the generated RF signal, known as the echo. The time interval between the RF excitation and the echo, defined by the gradient waveforms, is referred to as the echo time (TE).
- In SE (Fig. 8.2), after the initial excitation RF pulse, the spins are allowed to dephase without interference for a certain period of time. Then, another RF pulse flips all the

Magnetic resonance imaging image formation 143

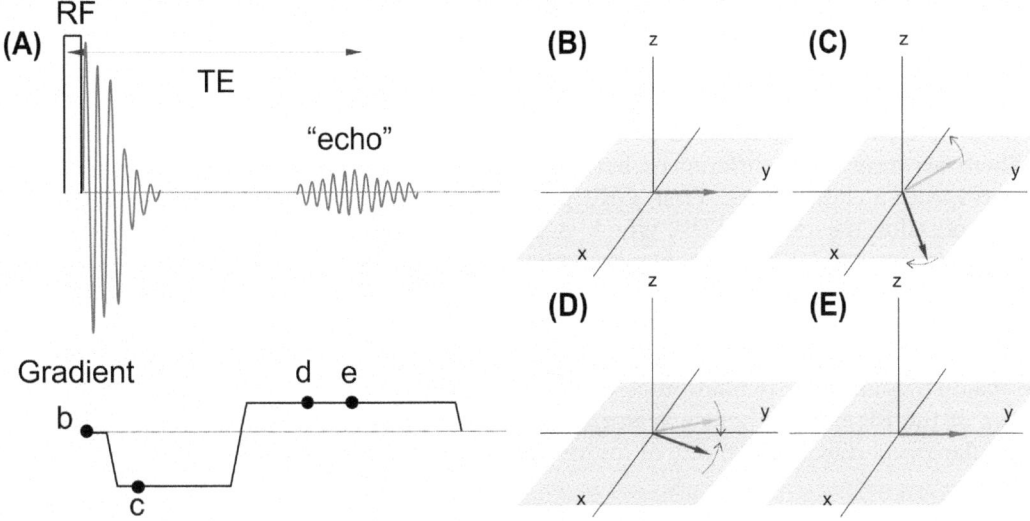

FIGURE 8.1 (A) Gradient echo pulse sequence. (B) In the rotating frame of reference, all spins are aligned immediately after RF excitation. (C) The applied negative gradient causes some spins to precess faster (red arrow), which rapidly increases spins dispersion. (D) The opposite gradient lobe begins to rephase spins, causing slower spins to precess faster and vice versa, eventually forming an echo (E).

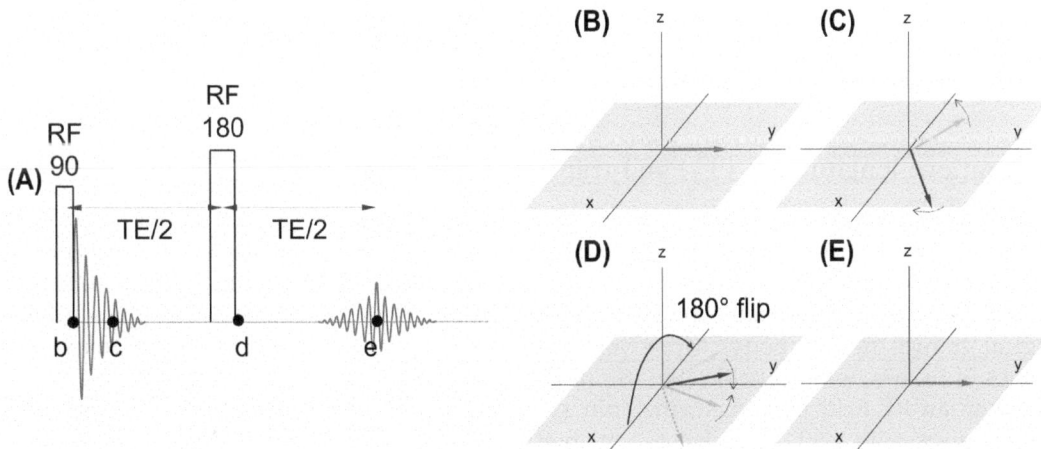

FIGURE 8.2 (A) Spin echo pulse sequence. (B) All spins are initially in phase and start dephasing over time (C) until the 180° pulse is applied, which reverses their phases (D). Then, the faster precessing spins (red arrow) start "catching up" with the slower spins, which reduces the phase dispersion and forms an echo (E).

spins by 180° about the y-axis, effectively reversing their phases. Consequently, the faster precessing spins appear to be "catching up" with the slower precessing ones. As a result, after a time period equal to the delay between the 90° and 180° RF pulses, all the spins realign, forming an echo.

1. Conventional MRI use in multiple sclerosis

The excitation process described above is repeated multiple times during the MRI acquisition. Typically, during each of the "RF-echo" blocks, only one k-space line is acquired, and then the process is repeated as required until the entire k-space is sampled adequately to provide the desired imaging resolution. The time interval between two consecutive RF excitations is referred to as the repetition time (TR).

There are important differences between the SE and GRE sequences that determine their respective clinical uses. First, GRE utilizes only one RF pulse, allowing for quicker echo generation (i.e., shorter TE), which is useful for rapid signal acquisition. Second, the gradient reversal in GRE only refocuses spins that have been dephased by the first gradient lobe. Unlike SE, other mechanisms contributing to signal decay (e.g., due to static magnetic field inhomogeneities) are not compensated for.

As one might expect, the measured MRI signal, as defined in Eq. 8.2, depends on how the magnetization $m(r)$ is manipulated, including the type of the sequence used and the timing of the RF and gradient waveforms. The selection of these parameters is guided by molecular properties of the tissue of interest and its pathology.

Relationship between magnetic resonance imaging signal and tissue properties

In an MRI voxel (~ 1 mm^3), millions of cells coexist, forming various types of tissue. Consequently, the MRI signal is influenced by tissue properties, including the relaxation of water proton spins in biological tissue (T_1 and T_2), biological water motions (diffusion and perfusion), and tissue magnetism (T_2^* and quantitative susceptibility mapping [QSM]). Understanding these concepts provides a basis for interpreting abnormalities in these biomarkers in neurologic disorders.

Relaxation: longitudinal (T_1) and transverse (T_2) relaxation

Relaxation refers to the process by which spins return to their thermal equilibrium state after being excited (i.e., perturbed) by the RF pulse. This process involves two key types of relaxation: longitudinal (T_1) and transverse (T_2) relaxation. Both T_1 and T_2 relaxation are critical for generating image contrast in MRI, differentiating between various types of biological tissues in our body. Moreover, pathological changes often alter the relaxation values of tissues, which enables the use of MRI for disease diagnosis and monitoring.

When an RF field is applied, the spin system deviates from thermal equilibrium state. For instance, with a 90° RF pulse, the magnetization flips from the z-direction into the x−y plane. After the applied 90° RF pulse is turned off, the magnetization returns to its equilibrium state.

- Longitudinal relaxation (T_1): In the z-direction, the longitudinal magnetization component (m_z) regrows toward its thermal equilibrium as the spins lose their energy to the thermal reservoir (also known as lattice). This process is referred to as longitudinal (T_1) relaxation.
- Transverse relaxation (T_2): While m_z returns to its thermal equilibrium, the transverse magnetization component ($m_\perp$) decays to zero due to thermal motions and interactions

among the spins. This process is known as transverse (T_2) relaxation. T_2 relaxation is caused by spin–spin interactions under irreversible microscopic molecular thermal fluctuations. T_2 is always shorter than T_1, as the latter originates from a part of these spin–spin interactions, the energy of which mainly corresponds to the Larmor frequency.

Although relaxation is a quantum mechanical phenomenon, a semiquantitative understanding of its cause, random thermal fluctuation, can be achieved through the random walk model [5]. A spin maintains its state (e.g., position and spin direction) for a certain average time (τ, defined as correlation time) before changing due to thermal collision. During τ, a spin coherently accumulates phase change from the field generated by other spins. Hence, the longer τ is, the faster spins dephase (i.e., shorter T_2). τ is proportional to the viscosity of the medium in which the spins are present. In biological tissue, this viscosity is primarily determined by cellular contents (cellularity) that surround the spins (e.g., biomolecules, membranes, and impurities). Consequently, T_2 may provide information about cellularity. For instance, in normal white matter, the spins adhere to cellular contents such as myelin lipid bilayers ("structured spins"), leading to a long correlation time. This results in large spin dephasing, that is, a short T_2. However, in demyelinated lesions in patients with MS, the nearly free spins have short correlation times, resulting in small spin dephasing and long T_2 values. This leads to a hyperintense demyelinated MS lesion in T_2-weighted images [5,7].

As T_1 relaxation occurs when spins lose energy to the thermal reservoir, which may be considered analogous to the reverse process of RF excitation, T_1 must originate from the part of spin–spin interaction that allows an $\hbar\omega$ energy release, which corresponds the energy at Larmor frequency. Consequently, T_1 may provide information about microscopic environmental restrictions on the movements of magnetic sources. If the movements of these sources induce large magnetic fields that fluctuate at Larmor frequency, T_1 relaxation would be rapid. For instance, when paramagnetic contrast agent sources are injected into patients with MS, if the blood–brain barrier (BBB) is compromised (e.g., in newly formed lesions), these parametric sources can move dynamically, leading to shortened T_1 times (which appears as hyperintensity in T_1-weighted images) [1].

By incorporating T_1 and T_2 relaxation into the magnetization torque equation (Eq. 8.4), the Bloch equation can be derived.

$$\frac{d\boldsymbol{m}}{dt} = -\gamma \boldsymbol{B} \times \boldsymbol{m} + \frac{1}{T_1}(m_0 - m_z)\hat{z} - \frac{1}{T_2}(m_x\hat{x} + m_y\hat{y}) \quad (8.5)$$

Here, m_0 is the magnetization value at thermal equilibrium. If no other magnetic field is present, the Bloch equation can be split into two decoupled equations, one describing the transverse component $\left(\frac{dm_\perp}{dt} = -\frac{m_\perp}{T_2}\right)$ and the other describing the longitudinal component $\left(\frac{dm_z}{dt} = -\frac{m_z - m_0}{T_1}\right)$. The solution to each equation shows the exponential behavior for both transverse magnetization ($m_\perp = m_{\perp 0} e^{-t/T_2}$ where $m_{\perp 0}$ is the initial $m_\perp$ value in the absence of any other magnetic field) and longitudinal magnetization ($m_z = m_{z0} e^{-t/T_1} + m_0(1 - e^{-t/T_1})$ where m_{z0} is the initial m_z value). The relaxation times (T_1 and T_2), which indicate how quickly the longitudinal and transverse magnetizations return to equilibrium, reflect the tissue's chemical environments. As such, they contribute to the MRI contrast between different tissue types, as well as between normal and abnormal tissue.

---

1. Conventional MRI use in multiple sclerosis

T_1-, T_2-, and proton density weighting in MR imaging

Using the solutions of the Bloch equation specified above, it is now possible to understand the relationship between the timing of the MR pulse sequences and the generated image contrast. For instance, in an SE experiment consisting of a repeated 90° RF – TR – 90° RF – TR –... sequence, the signal intensity S is proportional to the following expression [5]:

$$S \propto m_0 \left(1 - e^{-\frac{TR}{T_1}}\right) e^{-\frac{TE}{T_2}} \tag{8.6}$$

- When TE is set to be short compared to T_2, the exponential term $e^{-(TE/T_2)}$ approaches 1, and all T_2-related effects disappear from the equation. Furthermore, if TR $\gg T_1$, then T_1 effects become negligible as well, and the only remaining parameter that describes tissue contrast is m_0. Since m_0 is proportional to the number of spins within a voxel, this contrast is referred to as proton density weighting (PD).
- When both TE and TR are short, the term $\left(1 - e^{-(TR/T_1)}\right)$ becomes significantly different from 1. The differences in T_1 signal recovery between tissues then become more noticeable, producing T_1-weighting (T_1w).
- When both TE and TR are long, the difference in contrast is driven by the $e^{-(TE/T_2)}$ factor describing T_2 effects. Accordingly, such contrast is called T_2-weighting (T_2w).

Due to differences in molecular and cellular composition, brain tissues and their pathologies are characterized by distinct T_1 and T_2 values. By manipulating MRI pulse sequence parameters, it is possible to utilize this difference to highlight specific anatomic features (Fig. 8.3).

Notably, T_2-weighted sequences are an essential imaging modality for diagnosing MS in the brain and spinal cord [7,8]. The loss of myelin causes T_2 prolongation, and as a result, MS lesions appear as hyperintense regions with well-defined boundaries on T_2w images. However, sometimes the surrounding white matter may show a hyperintense

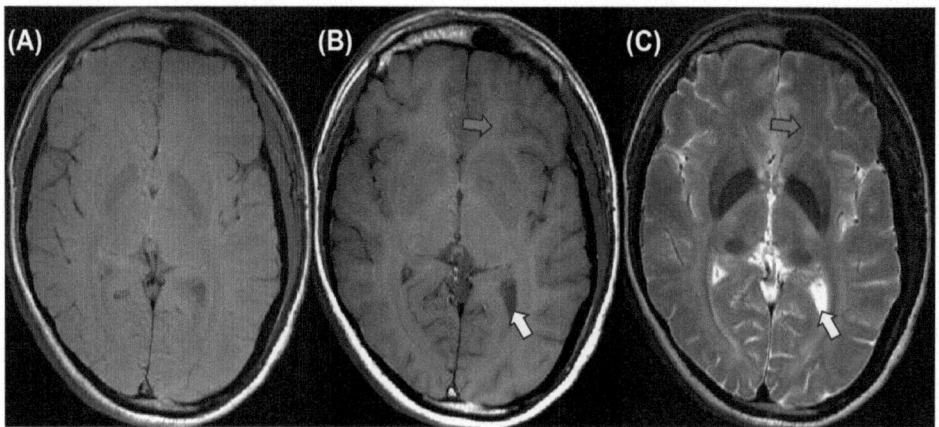

FIGURE 8.3 Examples of proton density weighting (PD) (A), T_1w (B) and T_2w (C) SE images of the brain in a healthy adult. Ventricular cerebrospinal fluid (yellow arrows) with long T_1 and T_2 appears dark in T_1w but bright in T_2w images. On the contrary, white matter (red arrows) with relatively short T_1 and T_2 appears bright on T_1w and dark on T_2w.

halo, likely indicating edema during the acute stage of inflammation. MS lesions are typically circular or oval in shape and vary in size, ranging from a few millimeters to over one centimeter in diameter. MS lesion formation tends to occur in certain brain regions [8], often involving periventricular white matter, subcortical, and infratentorial regions. As a result, pure T_2w contrast might be suboptimal for MS lesions adjacent to gray matter and cerebrospinal fluid (CSF), both of which have longer T_2 compared to white matter [9]. To improve lesion conspicuity while maintaining T_2 weighting, differences in T_1 relaxation times between the lesion and CSF matter can be exploited [10]. Compared to the traditional GRE and SE sequences discussed above, this approach, called fluid-attenuated inversion recovery (FLAIR), utilizes an extra 180° inversion RF pulse placed prior to the 90° excitation pulse. By adjusting the delay between the inversion and excitation pulses (inversion time, TI) such that the longitudinal component of CSF signal is zero at the time of excitation, it is possible to effectively null the signal from CSF (Fig. 8.4).

Consequently, compared to T_2w, T_2-FLAIR more effectively depicts MS lesions in periventricular and subcortical regions (Fig. 8.5), as the FLAIR sequence eliminates the partial volume effect from the CSF.

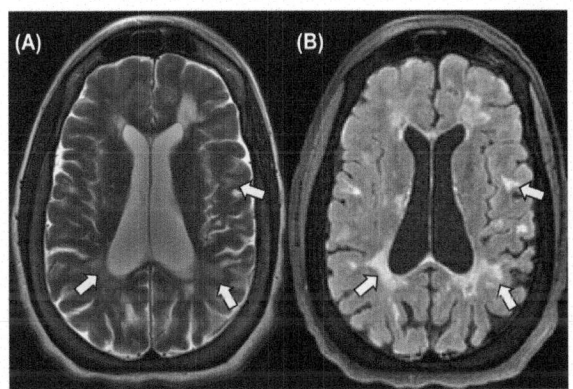

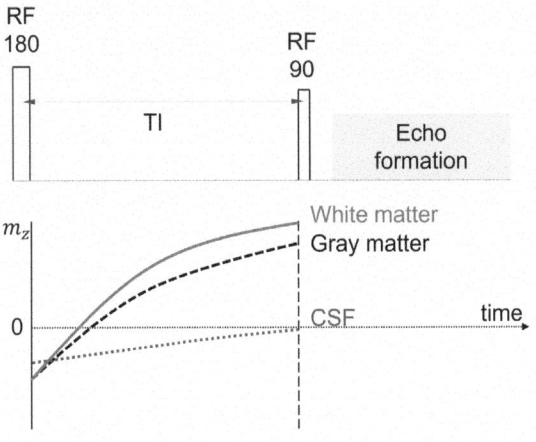

FIGURE 8.4 Schematic of the fluid-attenuated inversion recovery (FLAIR) pulse sequence and plots of white matter, gray matter, and CSF longitudinal magnetization evolution over time. Inversion time (TI) is strategically set so that CSF signal is nulled at the time of the excitation.

FIGURE 8.5 Comparison of appearance on T_2w (A) and T_2-FLAIR (B) images in a patient with Ms. Note the improved delineation of the white matter pathology in periventricular and subcortical lesions (yellow arrows) on FLAIR due to the elimination of the partial voxel effect from the CSF.

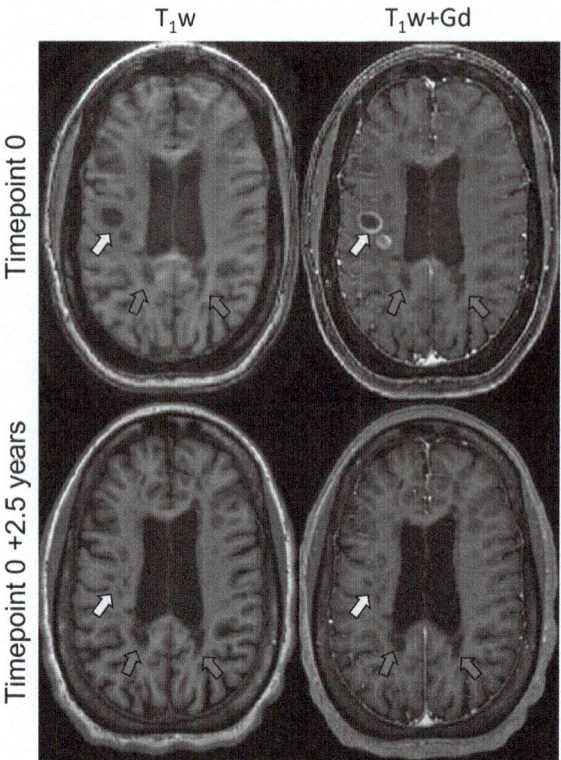

FIGURE 8.6 T_1w images in the same Ms patient at two timepoints. On precontrast images, lesions (arrows) appear as hypointense "black holes." Acute lesions (yellow arrows) become apparent on postcontrast images due to the distinct signal enhancement at the initial scan; on the followup examination, enhancement disappears, and lesions shrink due to the resolution of edema.

On T_1w images (Fig. 8.6), MS lesions appear as iso- or hypointense regions. The majority of the lesions are hypointense at the time of onset, likely due to edema, with or without concurrent myelin destruction [11]. As edema resolves and inflammatory processes cease, lesions may return to an isointense appearance. Chronic hypointense foci, often referred to as "black holes," indicate areas with more severe tissue damage. Inflammation in acute MS lesions increases BBB permeability, leading to contrast enhancement on postgadolinium T_1w imaging [12]. The highly paramagnetic gadolinium shortens the apparent T_1 of blood, resulting in a hyperintense appearance of acute lesions. This contrast enhancement persists for approximately 1 month, allowing the classification of MS lesions into acute and chronic categories.

Biological water motion: diffusion and perfusion

The body's water motion includes incoherent, microscopic, random thermal motion (diffusion), which is determined by cellular geometry, as well as coherent macroscopic circulatory motion (perfusion in microvascular spaces), which is determined by tissue vascularity. These types of water motion (diffusion and perfusion) can be measured by different imaging sequences in MRI, each of which is sensitive to a specific type of water motion.

1. Conventional MRI use in multiple sclerosis

Diffusion: Understanding water diffusion in brain tissue can provide important physiological information, as water diffusion can be restricted or altered in neurologic disorders. For instance, in an ischemic stroke, the cells in the ischemic core become swollen (cytotoxic edema), which restricts water movement in the extracellular space. MRI can detect the effects of this reduced water diffusion by applying additional diffusion-sensitive gradients, a technique known as diffusion weighted imaging (DWI) [13–15]. Also, in MS, the integrity of white matter tracts can be investigated by using diffusion tensor imaging (DTI), as the directionality of water diffusion can provide information about the orientation and integrity of these fiber tracts [16].

Diffusion is a random thermal motion that can be modeled as random walk [5]. If a water molecule begins at the origin at a specific time in a 1D space, it undergoes rapid collisions, for example, on average one collision in 10^{-12} seconds for free water, and the water molecule experiences millions of collisions from the RF excitation to the signal measurement time (t, on the order of 10 milliseconds) in MRI. This results in a random back-and-forth path of the water molecule (x). By averaging over many molecules, the average of random molecule positions $\langle x \rangle = 0$ and the variation (i.e., the spatial spread over time) increases linearly with time: $\langle x^2 \rangle = 2Dt$ where D is diffusion coefficient (generally in the unit of mm^2/s). Based on $\langle x^2 \rangle$, the probability of a molecule being located at x after experiencing numerous collisions during time t can be expressed as a Gaussian distribution (in a 1D case): $p(x,t) = \frac{1}{\sqrt{2\pi \langle x^2 \rangle}} e^{-\frac{x^2}{2\langle x^2 \rangle}} = \frac{1}{\sqrt{4\pi Dt}} e^{-\frac{x^2}{4Dt}}$. This probability equation is the solution to Fick's second law of diffusion: $\frac{\partial p}{\partial t} = \frac{\partial}{\partial x}\left(D \frac{\partial p}{\partial x}\right)$. This Fick's law can be generalized to 3D and incorporated into the Bloch equation, as magnetization m is linearly proportional to the particle density ($n = N \cdot p$ where N is the total number of particles), which leads to the Torrey–Bloch equation [17]:

$$\frac{d\mathbf{m}}{dt} = -\gamma \mathbf{B} \times \mathbf{m} + \frac{1}{T_1}(m_0 - m_z)\hat{z} - \frac{1}{T_2}(m_x \hat{x} + m_y \hat{y}) + \nabla^T D \nabla \mathbf{m} \quad (8.7)$$

The diffusion effect on the MRI signal can be measured by applying an additional diffusion-sensitive gradient. If a gradient field (G) is applied and the T_2 term is ignored (for simplicity), the transverse component of the Torrey–Bloch equation can be simplified as $\frac{dm_\perp}{dt} = -i\gamma \mathbf{G} \cdot \mathbf{r} m_\perp + D\nabla^2 m_\perp$ [17]. The solution to this equation shows an exponential behavior [18]:

$$m_\perp = m_{\perp 0} e^{-bD} \quad (8.8)$$

Here, $m_{\perp 0}$ is the transverse magnetization without the gradient (G). The b-factor (b) is a variable, determined by the applied gradient profile (e.g., the strength and duration), defined as $b(t) = \gamma^2 \int_0^t \left(\int_0^{t'} G(t'')dt''\right)^2 dt'$. In a rectangular bipolar gradient that consists of two strong pulses of magnitude (G) and duration (δ), separated by time interval (Δ), $b = (2\pi\gamma G)^2 \delta^2 (\Delta - \frac{\delta}{3})$ [19].

DWI typically employs a single b-value ($s \propto e^{-bD}$). To obtain the diffusion coefficient (D), at least two b-values are required: $D = \frac{\ln(m_{\perp 1}/m_{\perp 2})}{b_2 - b_1}$. Since the directions of the gradients can align along any combinations of the three axes (x, y, z) and signal decay occurs due to water diffusion along the diffusion gradient direction, the diffusion experiment can be repeated with different gradient directions to obtain different elements of the diffusion tensor. This forms the basis of DTI, which is useful for investigating white matter tractography [20,21]. DTI enables a quantitative analysis of the magnitude and directionality of water molecules' movement (see Chapter 2.2.1 for a comprehensive review of DTI in MS). The diffusion tensor can be diagonalized to obtain three eigenvalues or diffusivities ($\lambda_1, \lambda_2, \lambda_3$), which can be combined to define quantitative scalar indices such as mean diffusivity (MD) and fractional anisotropy (FA) [22]. MD, also referred to as the apparent diffusion coefficient (ADC), measures the degree of diffusion, independent of direction: $MD = (\lambda_1 + \lambda_2 + \lambda_3)/3$. Also, FA characterizes the extent to which the distribution of diffusion in a voxel is directional: $FA = \sqrt{\frac{1}{2}\frac{(\lambda_1-\lambda_2)^2 + (\lambda_1-\lambda_3)^2 + (\lambda_2-\lambda_3)^2}{\lambda_1^2 + \lambda_2^2 + \lambda_3^2}}$.

In MS, myelin sheaths, which restrict water diffusion, are destructed, leading to higher ADC values in the lesions compared to the normal appearing white mater (NAWM) (Fig. 8.7C). On the contrary, the ADC values might be lower in acute MS lesions, where the tissue damage is not yet complete, compared to chronic "black holes" that resemble free water [23]. Unlike MD, FA is lower in MS lesions compared to NAWM, reflecting demyelination, axonal injury, and accumulation of inflammatory cells [24] (Fig. 8.7D).

Perfusion: Perfusion refers to the regional blood flow in arterioles, capillaries, and venules, which is driven by blood pressure. This process is distinct from diffusion, which relates to the intrinsic thermal motion in tissues. Oxygen and essential nutrients are delivered via this perfusion process, making it crucial for maintaining the functionality of brain and other organs. For instance, a disruption in brain perfusion, such as that caused by artery clotting in an ischemic stroke, can rapidly lead to brain tissue damage, and ultimately, death.

Several MRI techniques have been proposed to quantify perfusion (cerebral blood flow [CBF]) based on tracer kinetic modeling using gadolinium chelate contrast agents (Gd). These include dynamic susceptibility contrast (DSC) MRI, which utilizes the T_2^* effect of

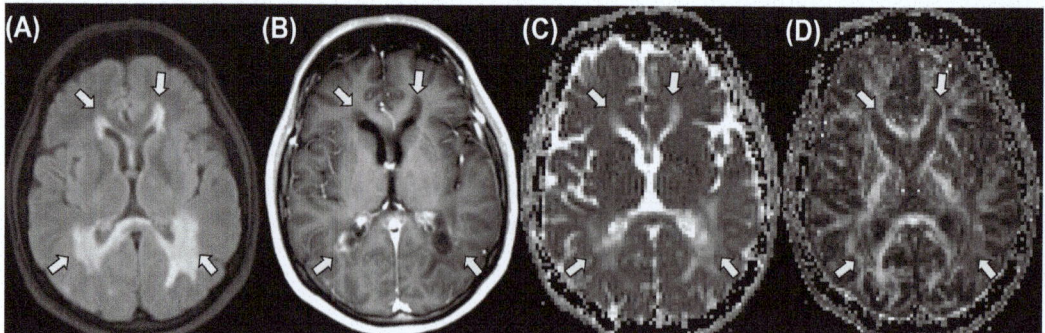

FIGURE 8.7 T_2FLAIR (A), postcontrast T_1w (B) image, apparent diffusion coefficient (ADC), (C) and fractional anisotropy (FA) (D) maps in a patient with Ms. Ms lesions (yellow arrows) appear as hyperintense regions in ADC map and hypointense on FA due to the destruction of the highly organized myelin sheaths which anisotropically restrict water diffusion.

Gd (see section "Tissue magnetism: T_2^* relaxation and quantitative susceptibility mapping") [25,26], and dynamic contrast enhanced (DCE) MRI, which considers T1 shortening due to Gd [27,28]. However, both DSC and DCE MRI require burdensome intravenous contrast agent injections, which can also potentially induce nephrogenic systemic fibrosis [29]. On the other hand, arterial spin labeling (ASL), a technique recently recognized for its effectiveness, eliminates the necessity of the intravenous contrast agent injection and even intravenous access [30–34]. ASL has shown good agreement with the reference standard positron emission tomography–based CBF [35], and is widely used in the study of neurologic disorders [36–38]. It is commercially available on all major MRI platforms.

To estimate CBF, ASL uses arterial blood water as an endogenous diffusible tracer which equilibrates with the surrounding tissue (e.g., the extravascular extracellular space). The magnetization of arterial blood is inverted (i.e., "labeled") by using an RF pulse (e.g., 180° RF pulse). After a predetermined delay known as "postlabeling delay" (T_{PLD}), which allows for the labeled blood to flow to the brain tissue, labeled images (m^{tag}) are obtained. These images contain signals from both labeled water and static tissue water. Separately, control images (m^{ctrl}) are also acquired without the prior labeling of arterial blood water. The signal difference between the labeled and control images ($\Delta m = m^{tag} - m^{ctrl}$) provides CBF estimation based on the Bloch equation, which takes water flow into account [34]:

$$\frac{d\boldsymbol{m}}{dt} = -\gamma \boldsymbol{B} \times \boldsymbol{m} + \frac{1}{T_1}(m_0 - m_z)\hat{z} - \frac{1}{T_2}(m_x\hat{x} + m_y\hat{y}) + CBF\left(\boldsymbol{m}_a - \frac{\boldsymbol{m}}{\lambda}\right) \quad (8.9)$$

Here, $\boldsymbol{m}_a$ is the labeled arterial magnetization. λ is blood/tissue water partition coefficient (~ 0.9 mL/g, reflecting freely diffusible water). The flow term in the Bloch equation (Eq. 8.9) can be derived from Kety's equation ($\frac{dC_t}{dt} = CBF(C_a - C_v)$, where C_t, C_a, C_v are tracer concentrations in tissue, artery, and vein, respectively), with the assumption of freely diffusible tracers ($C_t = \lambda C_v$) [5].

As the prepared $\boldsymbol{m}_a$ is typically aligned along the $\boldsymbol{B}_0$, Eq. (8.9) can be simplified into the longitudinal component: $\frac{dm_z}{dt} = \frac{1}{T_1}(m_0 - m_z) + CBF(m_a - \frac{m_z}{\lambda})$. By subtracting the control images from the labeled images (Δm), the relationship between CBF and Δm can be obtained: $\frac{d\Delta m}{dt} + \frac{\Delta m}{T_{1app}} = CBF \cdot \Delta m_a$ where $T_{1app} = \frac{1}{T_1} + \frac{CBF}{\lambda}$ and Δm_a are the arterial magnetization difference between the labeled and control images. Consequently, CBF can be obtained by solving this equation. For instance, in a continuous ASL where constant arterial magnetization is delivered with the time using RF field, $CBF = \frac{\lambda \cdot \Delta m}{2\alpha \cdot T_{1a} \cdot m^{ctrl}\left\{e^{-T_{PLD}/T_{1a}} - e^{-(T_{PLD}+\tau)/T_{1a}}\right\}}$ where α is the tagging efficiency, τ is the duration of the labeling pulse, and T_{1a} is the longitudinal relaxation time of blood [39].

In MS, the disease's involvement with vascular pathology has suggested that perfusion measurement can provide insights into its pathogenesis and lesion development [40,41]. Moreover, perfusion imaging has demonstrated its potential in identifying new lesion activity, and it might be useful for modulating therapy based on the microcirculation changes in lesions (Fig. 8.8). A comprehensive review of perfusion-weighted imaging is given in Chapter 2.2.6.

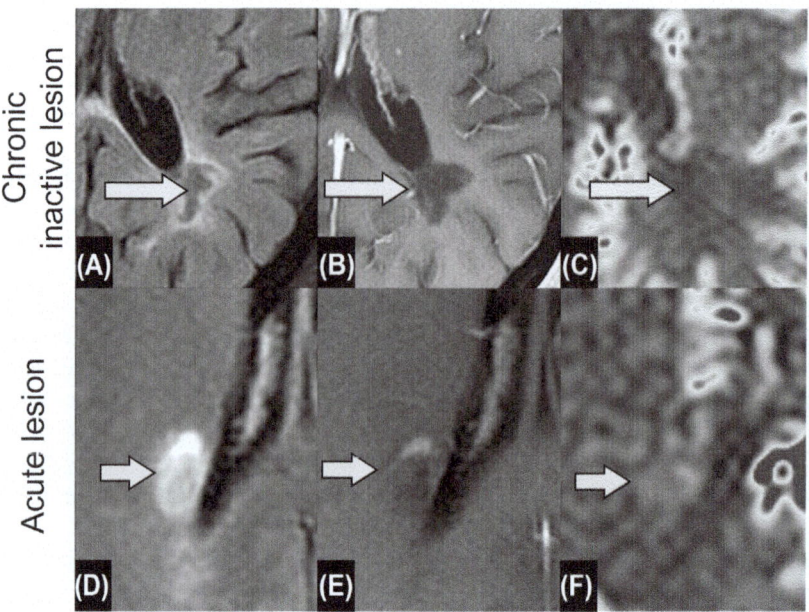

FIGURE 8.8 T$_2$FLAIR (A, D), postcontrast T$_1$w (B, E), and relative cerebral blood flow (CBF) (C, F) in chronic inactive and acute lesions (arrows). The acute lesion demonstrates a marked increase in perfusion, indicating vasodilation and increased blood flow due to Ms-related vascular inflammation. Nonenhancing chronic lesions are less likely to demonstrate perivenular inflammation and often have thickened vessel walls, resulting in tissue hypoperfusion. *Source: Adapted from Sheng H, Zhao B, Ge Y. Blood perfusion and cellular microstructural changes associated with iron deposition in multiple sclerosis lesions. Front Neurol 2019;10:747. https://doi.org/10.3389/fneur.2019.00747.*

Tissue magnetism: T$_2$* relaxation and quantitative susceptibility mapping

A static, inhomogeneous susceptibility field, which is distinct from the randomly fluctuating fields that induce the T$_1$ and T$_2$ relaxation, can affect MRI signal. This susceptibility field can be generated by the tissue itself (e.g., through iron accumulation in the tissue or venous deoxyhemoglobin) or by the susceptibility differences at the air-tissue interfaces. T$_2$* relaxation and QSM consider the effects of the susceptibility field on the MRI magnitude and phase signal, respectively.

T$_2$* relaxation: While T$_2$ relaxation accounts for the effects of randomly fluctuating field in the transverse plane, T$_2$* relaxation considers nonrandom, static field inhomogeneity, $\frac{b(r)\omega_0}{\gamma}$. Given the susceptibility field inhomogeneity, the spin frequency can be expressed as $\omega(r) = \omega_0 + \gamma G \cdot r + b(r)\omega_0$. Consequently, the MRI signal equation (Eq. 8.2) expands to $s = \int m(r) e^{-ib(r)\omega_0 t} \cdot e^{-\frac{t}{T2(r)}} \cdot e^{-2\pi i k \cdot r} d^3 r$, taking into account the consideration of the T$_2$ effect and assuming that the field inhomogeneity field is much smaller than the main magnetic field ($\omega_0 \gg \gamma G \cdot r + b(r)\omega_0$). By applying an inverse Fourier transform, this MRI signal equation at TE becomes the following equation [5]:

$$S(r) = \overline{m} \cdot \Delta V \cdot e^{-\frac{TE}{T2(r)}} \cdot \langle e^{-ib \cdot \omega_0 \cdot TE} \rangle \tag{8.10}$$

Here, $\bar{m}$ is the average spin distribution in a voxel ($\int_{r-\Delta r/2}^{r+\Delta r/2} \frac{d^3r' m(r')}{\Delta V}$). ΔV is a voxel volume. b is the inhomogeneity field scaled to ω_0. $\langle \cdot \rangle$ is the spatial average over a voxel. By taking the cumulant expansion up to the second order, the MR signal (Eq. 8.10) can be expressed as:

$$S(r) = \bar{m} \cdot \Delta V \cdot e^{-\frac{TE}{T_2^*(r)}} \cdot e^{-i\langle b \rangle \omega_0 TE} \quad (8.11)$$

Here,

$$\frac{1}{T_2^*} = \frac{1}{T_2} + \frac{1}{2}\omega_0^2 TE(\langle b^2 \rangle - \langle b \rangle^2) = \frac{1}{T_2} + \frac{1}{T_2'} \quad (8.12)$$

While T_2 arises from randomly fluctuating spin–spin interactions due to irreversible microscopic molecular thermal fluctuations, T_2' results from static mesoscopic or macroscopic magnetic field inhomogeneity (the field variation within a voxel). T_2^* relaxation is particularly important in functional magnetic resonance imaging (fMRI) due to its sensitivity to changes in blood oxygenation, that is, blood oxygen level dependent effect [42,43]. A detailed review of fMRI in the context of MS is provided in Chapter 2.2.5. Also, T_2^* relaxation has been used to detect and quantify iron overload in various body organs, including the liver [44,45].

Quantitative susceptibility mapping: The average inhomogeneity field, $\langle b \rangle$, which can be measured from the MRI phase signal (Eq. 8.11), can be expressed by the dipole field: $\langle b \rangle = d * \chi$ where d is dipole kernel, $*$ is convolution operator, and χ is voxel-wise magnetic susceptibility. This dipole field equation can be derived from taking the Taylor expansion of the Maxwell equations up to the first order [46]. QSM solves the inversion problem of the dipole field equation to estimate χ [47–51]. Given that χ is sensitive disease-related biometals such as iron and calcium, QSM has been widely used in clinical settings [52,53], including investigating iron accumulation in Ms [54–59].

In MS, QSM serves as a biomarker for inflammatory and neurodegenerative activities [58,60,61]. Reflecting the dynamic nature of inflammation and myelin destruction, the magnetic susceptibility of white matter MS lesions changes during the course of their development. In the acute enhancing stage, lesion susceptibility stars at the level of NAWM, followed by quick increase due to the influx of the iron-laden microglia and breakdown of the myelin sheath (Fig. 8.9). In the subsequent chronic nonenhancing stage, the susceptibility of the lesion first plateaus and then slowly decays back to levels similar to that of NAWM over an average of 6–8 years due to the clearance of macrophages [54].

QSM uniquely allows for the mapping of spatial distribution of iron, representative of iron-laden activated microglia and macrophages within the MS lesions. Histopathology of chronic inflammatory lesions reveals activated microglia at the lesion border [62], and iron has been identified in activated microglia within the lesion rim [63]. Iron rim lesions have been shown to be more likely to develop permanent tissue damage (T_1w "black holes") and gradually expand over time [64,65]. QSM allows identification of this clinically significant MS lesion subtype through direct visualization of the paramagnetic iron rims (Fig. 8.10). A more comprehensive review of QSM in MS is given in Chapter 2.2.4.

1. Conventional MRI use in multiple sclerosis

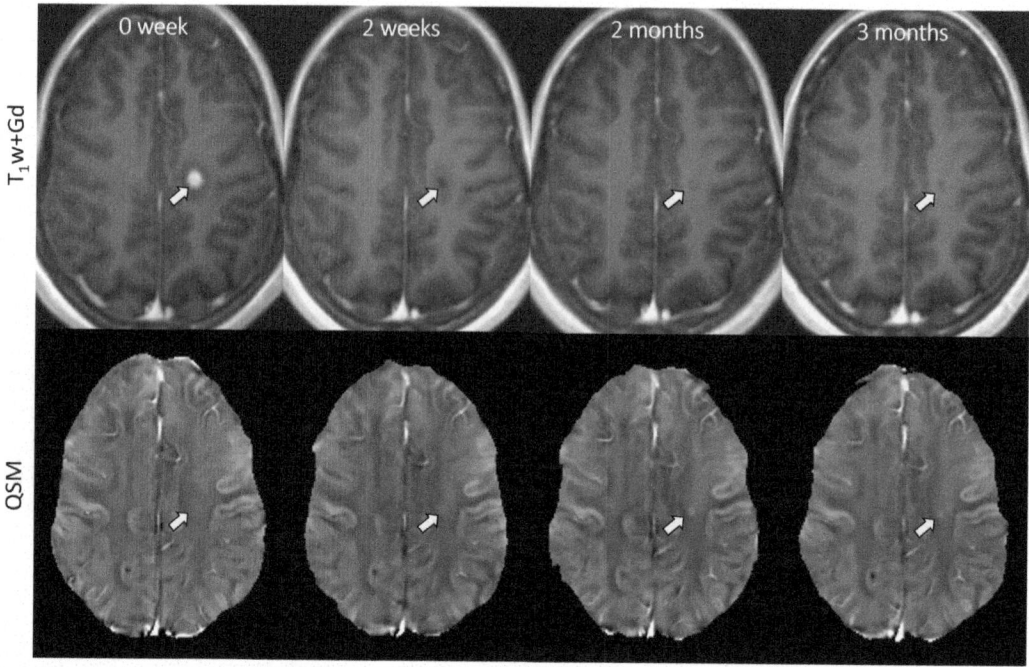

FIGURE 8.9 Gadolinium-enhanced T_1w images (top row) and corresponding quantitative susceptibility mapping (QSM) (bottom row) of a new acute Ms lesion (arrows) imaged over the course of 3 months. The observed rapid increase in lesion susceptibility behind the closed blood–brain barrier is the product of myelin loss and influx of iron-laden proinflammatory microglial cells.

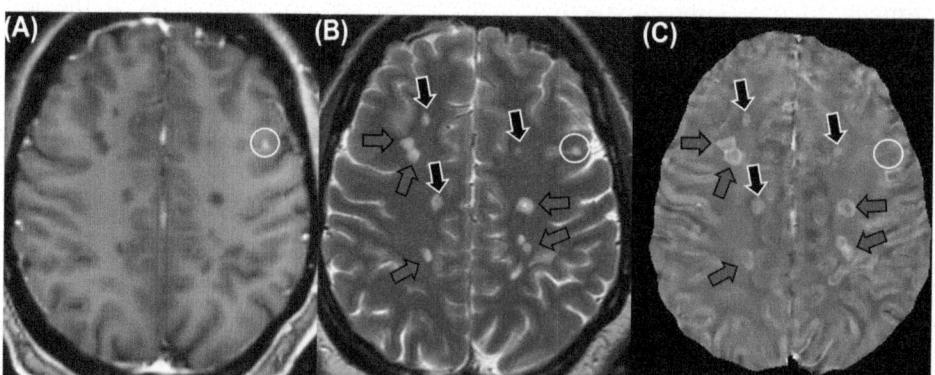

FIGURE 8.10 Gadolinium-enhanced T_1w (A), T_2w (B), and QSM images of an Ms brain. Two types of lesions are readily identifiable on QSM—lesions with clearly defined paramagnetic rim (red arrows) and homogeneously hyperintense lesions (black arrows). Such lesion classification is not possible on traditional anatomical contrasts. Also note Gd-enhancing acute Ms lesion appearing isointense on QSM (yellow circle).

1. Conventional MRI use in multiple sclerosis

Summary

Understanding how an MRI image is formed and how the MRI signal is associated with tissue properties is crucial for connecting abnormalities in MRI images to pathophysiological processes in neurologic disorders, including MS. The topics discussed in this chapter—including MRI image formation and the relationship between MRI signal and tissue properties in terms of relaxation, biological water motion, and tissue magnetism—contribute to a fundamental understating of MRI image interpretation in neurologic disorders.

References

[1] Hemond CC, Bakshi R. Magnetic resonance imaging in multiple sclerosis. Cold Spring Harb Perspect Med 2018;8(5). Available from: https://doi.org/10.1101/cshperspect.a028969.

[2] Kaunzner UW, Gauthier SA. MRI in the assessment and monitoring of multiple sclerosis: an update on best practice. Ther Adv Neurol Disord 2017;10(6):247–61. Available from: https://doi.org/10.1177/1756285617708911.

[3] Lövblad K-O, Anzalone N, Dörfler A, Essig M, Hurwitz B, Kappos L, et al. MR imaging in multiple sclerosis: review and recommendations for current practice. Am J Neuroradiol 2010;31(6):983–9. Available from: https://doi.org/10.3174/ajnr.A1906.

[4] Wattjes MP, Rovira À, Miller D, Yousry TA, Sormani MP, de Stefano MP, et al. Evidence-based guidelines: MAGNIMS consensus guidelines on the use of MRI in multiple sclerosis—establishing disease prognosis and monitoring patients. Nat Rev Neurol 2015;11(10):597–606. Available from: https://doi.org/10.1038/nrneurol.2015.157.

[5] Wang Y. Principles of magnetic resonance imaging: physics concepts, pulse sequences, & biomedical applications. CreateSpace Independent Publishing Platform; 2012. Available from: https://books.google.com/books?id=LzwOMwEACAAJ.

[6] Griffiths DJ. Introduction to electrodynamics. 4 ed. Cambridge University Press; 2017. Available from: https://doi.org/10.1017/9781108333511.

[7] Filippi M, Rocca MA, De Stefano N, Enzinger C, Fisher E, Horsfield MA, et al. Magnetic resonance techniques in multiple sclerosis: the present and the future. Arch Neurol 2011;68(12):1514–20. Available from: https://doi.org/10.1001/archneurol.2011.914.

[8] Pretorius PM, Quaghebeur G. The role of MRI in the diagnosis of MS. Clin Radiol 2003;58(6):434–48. Available from: https://doi.org/10.1016/s0009-9260(03)00089-8.

[9] Bakshi R, Ariyaratana S, Benedict RH, Jacobs L. Fluid-attenuated inversion recovery magnetic resonance imaging detects cortical and juxtacortical multiple sclerosis lesions. Arch Neurol 2001;58(5):742–8. Available from: https://doi.org/10.1001/archneur.58.5.742.

[10] Bydder GM, Young IR. MR imaging: clinical use of the inversion recovery sequence. J Comput Assist Tomogr 1985;9(4):659–75. Available from: https://www.ncbi.nlm.nih.gov/pubmed/2991345.

[11] van Waesberghe JH, van Walderveen MA, Castelijns JA, Scheltens P, Lyclama a Nijeholt GJ, Polman CH, et al. Patterns of lesion development in multiple sclerosis: longitudinal observations with T1-weighted spin-echo and magnetization transfer MR. AJNR Am J Neuroradiol 1998;19(4):675–83. Available from: https://www.ncbi.nlm.nih.gov/pubmed/9576653.

[12] Cotton F, Weiner HL, Jolesz FA, Guttmann CR. MRI contrast uptake in new lesions in relapsing-remitting MS followed at weekly intervals. Neurology 2003;60(4):640–6. Available from: https://doi.org/10.1212/01.wnl.0000046587.83503.1e.

[13] Everdingen KJv, Grond Jv d, Kappelle LJ, Ramos LMP, Mali WPTM. Diffusion-weighted magnetic resonance imaging in acute stroke. Stroke 1998;29(9):1783–90. Available from: https://doi.org/10.1161/01.STR.29.9.1783.

[14] González RG, Schaefer PW, Buonanno FS, Schwamm LH, Budzik RF, Rordorf G, et al. Diffusion-weighted MR imaging: diagnostic accuracy in patients imaged within 6 hours of stroke symptom onset. Radiology 1999;210(1):155–62. Available from: https://doi.org/10.1148/radiology.210.1.r99ja02155.

[15] Roberts TPL, Rowley HA. Diffusion weighted magnetic resonance imaging in stroke. Eur J Radiol 2003;45 (3):185–94. Available from: https://doi.org/10.1016/S0720-048X(02)00305-4.
[16] Filippi M, Pagani E, Preziosa P, Rocca MA. The role of DTI in multiple sclerosis and other demyelinating conditions. In: Van Hecke W, Emsell L, Sunaert S, editors. Diffusion tensor imaging: a practical handbook. New York: Springer; 2016. p. 331–41. Available from: https://doi.org/10.1007/978-1-4939-3118-7_16.
[17] Torrey HC. Bloch equations with diffusion terms. Phys Rev 1956;104(3):563–5. Available from: https://doi.org/10.1103/PhysRev.104.563.
[18] Jones DK. Diffusion MRI: theory, methods, and applications. Oxford University Press; 2010. Available from: https://doi.org/10.1093/med/9780195369779.001.0001.
[19] Stejskal EO, Tanner JE. Spin diffusion measurements: spin echoes in the presence of a time-dependent field gradient. J Chem Phys 2004;42(1):288–92. Available from: https://doi.org/10.1063/1.1695690.
[20] Alexander AL, Lee JE, Lazar M, Field AS. Diffusion tensor imaging of the brain. Neurotherapeutics 2007;4 (3):316–29. Available from: https://doi.org/10.1016/j.nurt.2007.05.011.
[21] Le BD, Mangin JF, Poupon C, Clark CA, Pappata S, Molko N, et al. Diffusion tensor imaging: concepts and applications. J Magn Reson Imaging 2001;13(4):534–46.
[22] Basser PJ. Inferring microstructural features and the physiological state of tissues from diffusion-weighted images. NMR Biomed 1995;8(7–8):333–44. Available from: https://doi.org/10.1002/nbm.1940080707.
[23] Roychowdhury S, Maldjian JA, Grossman RI. Multiple sclerosis: comparison of trace apparent diffusion coefficients with MR enhancement pattern of lesions. AJNR Am J Neuroradiol 2000;21(5):869–74. Available from: https://www.ncbi.nlm.nih.gov/pubmed/10815662.
[24] Filippi M, Cercignani M, Inglese M, Horsfield MA, Comi G. Diffusion tensor magnetic resonance imaging in multiple sclerosis. Neurology 2001;56(3):304–11. Available from: https://doi.org/10.1212/wnl.56.3.304.
[25] Rosen BR, Belliveau JW, Vevea JM, Brady TJ. Perfusion imaging with NMR contrast agents. Magn Reson Med 1990;14(2):249–65. Available from: https://doi.org/10.1002/mrm.1910140211.
[26] Boxerman JL, Quarles CC, Hu LS, Erickson BJ, Gerstner ER, Smits M, et al. Consensus recommendations for a dynamic susceptibility contrast MRI protocol for use in high-grade gliomas. Neuro-Oncology 2020;22 (9):1262–75. Available from: https://doi.org/10.1093/neuonc/noaa141.
[27] Tofts PS, Kermode AG. Measurement of the blood-brain barrier permeability and leakage space using dynamic MR imaging. 1. Fundamental concepts. Magn Reson Med 1991;17(2):357–67. Available from: https://doi.org/10.1002/mrm.1910170208.
[28] Brix G, Semmler W, Port R, Schad LR, Layer G, Lorenz WJ. Pharmacokinetic parameters in CNS Gd-DTPA enhanced MR imaging. J Comput Assist Tomogr 1991;15(4):621–8. Available from: https://doi.org/10.1097/00004728-199107000-00018.
[29] Sadowski EA, Bennett LK, Chan MR, Wentland AL, Garrett AL, Garrett RW, et al. Nephrogenic systemic fibrosis: risk factors and incidence estimation. Radiology 2007;243(1):148–57. Available from: https://doi.org/10.1148/radiol.2431062144.
[30] Detre JA, Leigh JS, Williams DS, Koretsky AP. Perfusion imaging. Magn Reson Med 1992;23(1):37–45. Available from: https://doi.org/10.1002/mrm.1910230106.
[31] Golay X, Petersen ET. Arterial spin labeling: benefits and pitfalls of high magnetic field. Neuroimaging Clin N Am 2006;16(2):259–68. Available from: https://doi.org/10.1016/j.nic.2006.02.003.
[32] McGehee BE, Pollock JM, Maldjian JA. Brain perfusion imaging: how does it work and what should I use? J Magn Reson Imaging 2012;36(6):1257–72. Available from: https://doi.org/10.1002/jmri.23645.
[33] Roberts DA, Detre JA, Bolinger L, Insko EK, Leigh Jr. JS. Quantitative magnetic resonance imaging of human brain perfusion at 1.5 T using steady-state inversion of arterial water. Proc Natl Acad Sci U S A 1994;91 (1):33–7. Available from: https://doi.org/10.1073/pnas.91.1.33.
[34] Williams DS, Detre JA, Leigh JS, Koretsky AP. Magnetic resonance imaging of perfusion using spin inversion of arterial water. Proc Natl Acad Sci U S A 1992;89(1):212–16. Available from: https://doi.org/10.1073/pnas.89.1.212.
[35] Ye FQ, Berman KF, Ellmore T, Esposito G, van Horn JD, Yang Y, et al. H(2)(15)O PET validation of steady-state arterial spin tagging cerebral blood flow measurements in humans. Magn Reson Med 2000;44(3):450–6. Available from: https://doi.org/10.1002/1522-2594(200009)44:3 < 450::aid-mrm16 > 3.0.co;2-0.
[36] Detre JA, Rao H, Wang DJ, Chen YF, Wang Z. Applications of arterial spin labeled MRI in the brain. J Magn Reson Imaging 2012;35(5):1026–37. Available from: https://doi.org/10.1002/jmri.23581.

References

[37] Golay X, Hendrikse J, Lim TCC. Perfusion imaging using arterial spin labeling. Top Magn Reson Imaging 2004;15(1). Available from: https://journals.lww.com/topicsinmri/Fulltext/2004/02000/Perfusion_Imaging_Using_Arterial_Spin_Labeling.3.aspx.

[38] Alsop DC, Detre JA, Golay X, Gunther M, Hendrikse J, Hernandez-Garcia L, et al. Recommended implementation of arterial spin-labeled perfusion MRI for clinical applications: a consensus of the ISMRM perfusion study group and the European consortium for ASL in dementia. Magn Reson Med 2015;73(1):102–16. Available from: https://doi.org/10.1002/mrm.25197.

[39] Wang J, Zhang Y, Wolf RL, Roc AC, Alsop DC, Detre JA. Amplitude-modulated continuous arterial spin-labeling 3.0-T perfusion MR imaging with a single coil: feasibility study. Radiology 2005;235:218–28. Available from: https://doi.org/10.1148/radiol.2351031663.

[40] Ge Y, Law M, Johnson G, Herbert J, Babb JS, Mannon LJ, et al. Dynamic susceptibility contrast perfusion MR imaging of multiple sclerosis lesions: characterizing hemodynamic impairment and inflammatory activity. AJNR Am J Neuroradiol 2005;26(6):1539–47. Available from: https://www.ncbi.nlm.nih.gov/pubmed/15956527.

[41] Sheng H, Zhao B, Ge Y. Blood perfusion and cellular microstructural changes associated with iron deposition in multiple sclerosis lesions. Front Neurol 2019;10:747. Available from: https://doi.org/10.3389/fneur.2019.00747.

[42] Glover GH. Overview of functional magnetic resonance imaging. Neurosurg Clin N Am 2011;22(2):133–9. Available from: https://doi.org/10.1016/j.nec.2010.11.001 vii.

[43] Ogawa S, Menon RS, Tank DW, Kim SG, Merkle H, Ellermann JM, et al. Functional brain mapping by blood oxygenation level-dependent contrast magnetic resonance imaging. a comparison of signal characteristics with a biophysical model. Biophys J 1993;64(3):803–12. Available from: https://doi.org/10.1016/S0006-3495(93)81441-3.

[44] Lidén M, Adrian D, Widell J, Uggla B, Thunberg P. Quantitative T2* imaging of iron overload in a non-dedicated center – normal variation, repeatability and reader variation. Eur J Radiol Open 2021;8100357. Available from: https://doi.org/10.1016/j.ejro.2021.100357.

[45] Alústiza Echeverría JM, Castiella A, Emparanza JI. Quantification of iron concentration in the liver by MRI. Insights Imaging 2012;3(2):173–80. Available from: https://doi.org/10.1007/s13244-011-0132-1.

[46] Jackson JD, Fox RF. Classical electrodynamics, 3rd ed. Am J Phys 1999;67(9):841–2. Available from: https://doi.org/10.1119/1.19136.

[47] Deistung A, Schweser F, Reichenbach JR. Overview of quantitative susceptibility mapping. NMR Biomed 2017;30(4)e3569. Available from: https://doi.org/10.1002/nbm.3569.

[48] Haacke EM, Liu S, Buch S, Zheng W, Wu D, Ye Y. Quantitative susceptibility mapping: current status and future directions. Magn Reson Imaging 2015;33(1):1–25. Available from: https://doi.org/10.1016/j.mri.2014.09.004.

[49] Liu C, Wei H, Gong NJ, Cronin M, Dibb R, Decker K. Quantitative susceptibility mapping: contrast mechanisms and clinical applications. Tomography 2015;1(1):3–17. Available from: https://doi.org/10.18383/j.tom.2015.00136.

[50] Reichenbach JR, Schweser F, Serres B, Deistung A. Quantitative susceptibility mapping: concepts and applications. Clin Neuroradiol 2015;25(2):225–30. Available from: https://doi.org/10.1007/s00062-015-0432-9.

[51] Wang Y, Liu T. Quantitative susceptibility mapping (QSM): decoding MRI data for a tissue magnetic biomarker. Magn Reson Med 2015;73(1):82–101. Available from: https://doi.org/10.1002/mrm.25358.

[52] Eskreis-Winkler S, Zhang Y, Zhang J, Liu Z, Dimov A, Gupta A, et al. The clinical utility of QSM: disease diagnosis, medical management, and surgical planning. NMR Biomed 2017;30(4). Available from: https://doi.org/10.1002/nbm.3668.

[53] Wang Y, Spincemaille P, Liu Z, Dimov A, Deh K, Li J, et al. Clinical quantitative susceptibility mapping (QSM): biometal imaging and its emerging roles in patient care. J Magn Reson Imaging 2017;46(4):951–71. Available from: https://doi.org/10.1002/jmri.25693.

[54] Chen W, Gauthier SA, Gupta A, Comunale J, Liu T, Wang S, et al. Quantitative susceptibility mapping of multiple sclerosis lesions at various ages. Radiology 2014;271(1):183–92. Available from: https://doi.org/10.1148/radiol.13130353.

[55] Kaunzner UW, Kang Y, Zhang S, Morris E, Yao Y, Pandya S, et al. Quantitative susceptibility mapping identifies inflammation in a subset of chronic multiple sclerosis lesions. Brain 2019;142(1):133–45. Available from: https://doi.org/10.1093/brain/awy296.

[56] Langkammer C, Liu T, Khalil M, Enzinger C, Jehna M, Fuchs S, et al. Quantitative susceptibility mapping in multiple sclerosis. Radiology 2013;267(2):551–9. Available from: https://doi.org/10.1148/radiol.12120707.
[57] Schweser F, Raffaini Duarte Martins AL, Hagemeier J, Lin F, Hanspach J, Weinstock-Guttman B, et al. Mapping of thalamic magnetic susceptibility in multiple sclerosis indicates decreasing iron with disease duration: a proposed mechanistic relationship between inflammation and oligodendrocyte vitality. Neuroimage 2018;167:438–52. Available from: https://doi.org/10.1016/j.neuroimage.2017.10.063.
[58] Wisnieff C, Ramanan S, Olesik J, Gauthier S, Wang Y, Pitt D. Quantitative susceptibility mapping (QSM) of white matter multiple sclerosis lesions: interpreting positive susceptibility and the presence of iron. Magn Reson Med 2015;74(2):564–70. Available from: https://doi.org/10.1002/mrm.25420.
[59] Zivadinov R, Tavazzi E, Bergsland N, Hagemeier J, Lin F, Dwyer MG, et al. Brain iron at quantitative MRI is associated with disability in multiple sclerosis. Radiology 2018;289(2):487–96. Available from: https://doi.org/10.1148/radiol.2018180136.
[60] Deh K, Ponath GD, Molvi Z, Parel GCT, Gillen KM, Zhang S, et al. Magnetic susceptibility increases as diamagnetic molecules breakdown: myelin digestion during multiple sclerosis lesion formation contributes to increase on QSM. J Magn Reson Imaging 2018;48(5):1281–7. Available from: https://doi.org/10.1002/jmri.25997.
[61] Gillen KM, Mubarak M, Park C, Ponath G, Zhang S, Dimov A, et al. QSM is an imaging biomarker for chronic glial activation in multiple sclerosis lesions. Ann Clin Transl Neurol 2021;8(4):877–86. Available from: https://doi.org/10.1002/acn3.51338.
[62] Frischer JM, Weigand SD, Guo Y, Kale N, Parisi JE, Pirko I, et al. Clinical and pathological insights into the dynamic nature of the white matter multiple sclerosis plaque. Ann Neurol 2015;78(5):710–21. Available from: https://doi.org/10.1002/ana.24497.
[63] Hametner S, Wimmer I, Haider L, Pfeifenbring S, Brück W, Lassmann H. Iron and neurodegeneration in the multiple sclerosis brain. Ann Neurol 2013;74(6):848–61. Available from: https://doi.org/10.1002/ana.23974.
[64] Absinta M, Sati P, Schindler M, Leibovitch EC, Ohayon J, Wu T, et al. Persistent 7-tesla phase rim predicts poor outcome in new multiple sclerosis patient lesions. J Clin Invest 2016;126(7):2597–609. Available from: https://doi.org/10.1172/jci86198.
[65] Dal-Bianco A, Grabner G, Kronnerwetter C, Weber M, Höftberger R, Berger T, et al. Slow expansion of multiple sclerosis iron rim lesions: pathology and 7 T magnetic resonance imaging. Acta Neuropathol 2017;133(1):25–42. Available from: https://doi.org/10.1007/s00401-016-1636-z.

CHAPTER

9

Magnetic resonance imaging of the multiple sclerosis lesions

Massimiliano Calabrese and Agnese Tamanti
University of Verona, Verona, Italy

OUTLINE

Introduction	159	Chronic active lesions	165
White matter lesions	160	Cortical lesions	166
Magnetic resonance sequences for T2 white matter lesions imaging	162	*Magnetic resonance imaging sequences for cortical lesion detection*	167
Black holes	163	Conclusions	169
Contrast-enhancing lesions	164	References	169

Introduction

Multiple sclerosis (MS) is associated with various clinical and radiological features. Although the most evident radiological manifestation of MS is the presence of plaques in the white matter (WM), cortical demyelination, and extralesional abnormalities, characterized by axonal and dendritic injury, neuronal loss, and inflammatory infiltrates, have been evidenced [1,2]. Magnetic resonance imaging (MRI) is crucial for diagnosing and monitoring MS pathology. However, while MRI has been able to detect WM lesions since its earliest introduction into the clinical environment, the assessment of cortical pathology and the identification of subtypes of MS lesions characterized by different tissue damage and activity has required years of technological development and advancement. It has been demonstrated that field strength, pulse sequence, and acquisition parameters are all factors that can influence the ability to detect MS lesions, to assess the lesions' activity and the different stages of their evolution. Since one of the most valuable features of MRI is the variety of images contrasts that can be produced, this chapter will describe the radiological characteristics of T2 MS plaques, black holes, cortical lesions (CLs), active contrast-enhancing and chronic active lesions, as well as the advantages and disadvantages of the MR sequences most commonly used for the detection and characterization of lesion types.

White matter lesions

The presence of lesions in the central nervous system is the hallmark of MS and represents the basis of MS diagnostic criteria [3]. In fact, the unmatched in vivo visualization of lesion burden and activity achieved with the introduction of MRI can be considered the first revolution in the diagnosis and treatment of MS. As technology improved, MRI quickly became one of the most essential paraclinical diagnostic and monitoring tools.

The last neuroimaging criteria for the diagnosis of MS focuses on the number and topography of MS lesions and their characteristic dissemination in space and time [3–5]. However, MS lesions also exhibit distinctive features that help the differentiation from other WM lesions as the ovoid shape, the formation along the deep medullary veins localized perpendicular to the lateral ventricles or the involvement of juxtacortical U-fibers [6].

Since the 19th century, we know that MS lesions are centered around small veins where inflammation occurs [7,8]. It is now hypothesized that the degree of vascularization may be the biological reason for the prevalence of MS lesions in those regions. This hypothesis is supported by evidence that WM lesions occur more frequently in the areas characterized by higher perfusion, such as the sublobar region. In agreement with this hypothesis, fewer lesions were found in areas with lower perfusion, such as the external capsule, fornix-stria terminalis, and cingulum-hippocampus [9,10]. The perivenular distribution of MS lesions prompted the introduction of the central vein sign (CVS): a radiological feature detectable on susceptibility-based MRI sequences used to improve the differentiation of MS from other WM diseases of the central nervous system [8]. The use of CVS in previous studies and meta-analysis have demonstrated a remarkable diagnostic performance to distinguish MS patients from mimic diseases, with a pooled specificity and sensitivity of 92% and 95%, respectively [11,12]. However, further studies are necessary to standardize the acquisition protocols and to ascertain the optimal cut-off values for the proportion of lesions showing this imaging biomarker.

The continual technical MRI improvements have helped the understanding of neuroinflammatory disease mechanisms. However, the contribution of "conventional" MRI metrics to the prognosis of the disease, and especially its relationship with long-term disability, has been controversial for a long time. In fact, the mismatch between the radiological progression in terms of the number and volume of WM lesions and the variety of MS clinical outcomes has raised the so-called "clinical-radiological paradox."

In a cross-sectional study of 1312 patients from 11 trials, T2 lesion load was only weakly correlated with the expanded disability status scale (EDSS) score [13], and such a relationship even decreased when patients with higher EDSS (<4.5) were considered. This observation suggested that WM lesions influence the early disability accumulation, usually linked with relapses, rather than the progression independent of relapse, which characterizes the late disease phases [13].

A discussion on whether relapses and disability progression (derived from EDSS assessments) might represent a more adequate measure of disease evolution than MRI [14] was induced by the observed lack of predictive power of the 2-years change in total lesion load for the disability accumulated at the end of the study including relapsing–remitting multiple sclerosis (RRMS) patients [15]. Such observation was also confirmed in a recent study aimed at identifying the predictors of EDSS milestones: the T2 lesion volume (but not the number) showed relevance only in the prediction of the EDSS = 3; the EDSS 4 and 6 were associated with MRI variables more representative of neurodegenerative processes, such as the spinal cord area and the cerebellar or the thalamic volume [16].

In a recent study having almost 30 years of disease follow-up, the WM lesion number was excluded from the model predicting the variance of the EDSS, which included the CLs counts, the cervical spinal cord volume and the gray matter (GM) volume [17].

Nevertheless, recent observations suggest that lesion location, type, and severity, more than lesion number or volume, are good prognostic factors of disability progression [18–23].

Although WM lesions are disseminated throughout the brain, there is a long-known preference for specific brain regions such as the supratentorial brain, particularly the frontal lobe and the sublobar area [24,25], which has also been identified previously as a region of particular interest for worsening in MS [26].

In a recent study involving the 2355 RRMS patients included in the FREEDOMS and FREEDOMS II studies, the authors investigated the correlation between the WM lesion location and the disability in patients treated with Fingolimod versus placebo [26]. They identified sublobar lesions as critical contributors to physical and cognitive disability. Lesions in the sublobar region (including the corpus callosum, WM around the deep GM and the lateral ventricles, and near the insula) showed significant and highly reproducible associations with most of the tested physical and cognitive disability measures. Moreover, these associations were stronger than the correlations between the average whole brain lesion burden (without distinguishing lesion location) and the same disability measures. A possible explanation for the prominent role of the sublobar region could be the relay function of the thalamus.

Another critical region for the WM lesion includes the spinal cord, whose involvement is an important cause of disability in MS patients. Spinal cord lesions are frequently observed, and pathologic changes are reported in 80%–90% of patients using conventional MRI [27,28]. Most focal lesions are confined to the cervical spinal cord and are located in dorsal and lateral WM. These lesions are infrequently seen in anterior columns or central cord areas and usually do not respect anatomic boundaries between GM and WM.

Bonacchi et al. [29] retrospectively studied 120 patients with MS (58 with RRMS and 62 with progressive multiple sclerosis [PMS]) and 30 age- and sex-matched healthy control participants. A disease subtype analysis revealed that cervical spinal cord GM T2 lesion volume was the most crucial variable for predicting disability in patients with RRMS. In contrast, brain and cervical spinal cord atrophy were key in patients with PMS. Both association and multivariable analyses showed that cervical spinal cord GM lesion and volumetric measures explained more variance in disability outcomes than did global or tissue-specific assessments in the brain. In a logistic regression analysis, cervical spinal cord GM atrophy and cervical spinal cord GM T2 lesion volume were independent predictors of disease subtype (RRMS vs PMS).

Magnetic resonance sequences for T2 white matter lesions imaging

T2 lesions have become the hallmark of MS pathology in part due to their visibility since the advent of imaging with computer tomography (CT) in 1976 [30]. However, in the 1980s, MRI had already surpassed the ability of CT to detect WM abnormalities in MS. In 1984 it was first demonstrated on a fixed ex vivo brain sample, the correspondence between histologically identified demyelinating MS plaques and an increase of the transverse relaxation time (T2) with respect to the T2 of normal-appearing WM. Multiple MR sequences have been introduced and used to visualize and assess the WM lesion load in MS patients (see Fig. 9.1).

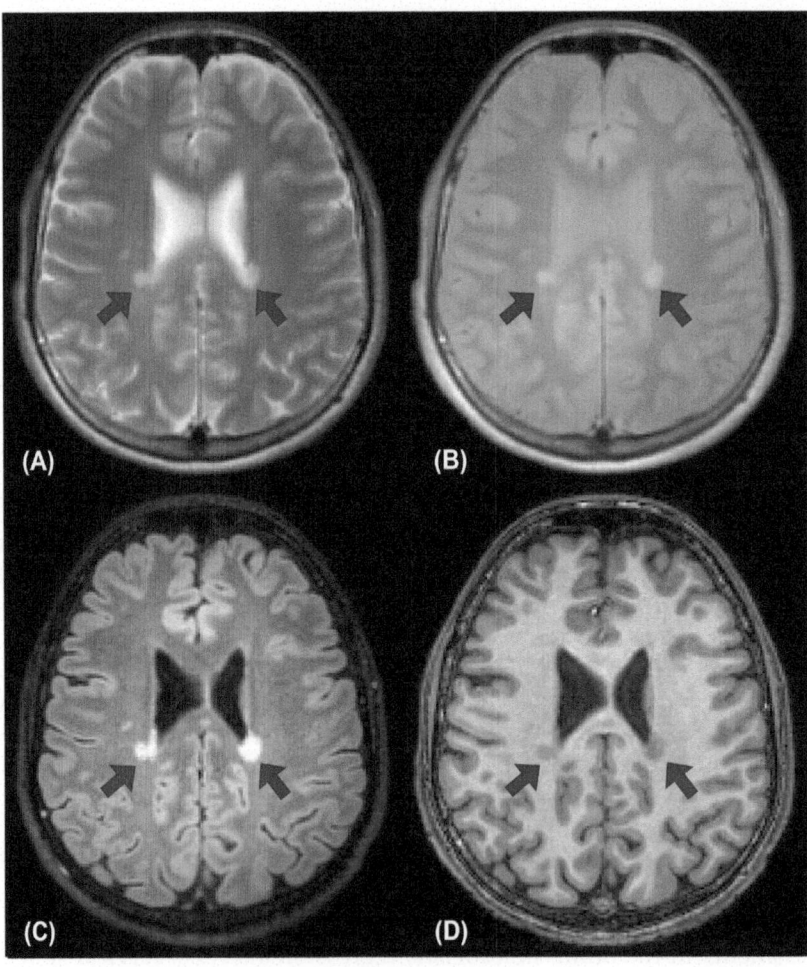

FIGURE 9.1 Periventricular lesions appearance in (A) T2-weighted, (B) PD-weighted, (C) FLAIR, and (D) T1-weighted MPRAGE images.

T2-weighted (T2w) images are traditionally obtained from spin echo sequences with a long repetition time (TR) and long echo time (TE). In these acquisition conditions, only tissues with long T2 relaxation time retain signal, thus, MS lesions tend to appear hyperintense [31]. For the same reason, cerebrospinal fluid (CSF) also shows a high signal in T2w images, making it difficult to distinguish periventricular lesions using only images with this type of weighting.

Proton density-weighted (PDw) images can be used, especially in combination with T2w images, for the delineation of periventricular lesions as they improve the differentiation between CSF and lesional tissue. Since PD weighting is obtained using a sequence design with long TR and short TE (thus minimizing the influence of both T1 and T2 relaxation times), PDw images can be acquired in a single excitation along with T2w images in a PDw/T2w dual-echo acquisition [31].

FLAIR has also been introduced to produce T2w images while nulling the otherwise hyperintense signal from the CSF and improve lesion detection and delineation, especially in the supratentorial area [32] and for lesion volume quantification [33]. FLAIR sequences are composed of an initial inversion recovery pulse followed, after an inversion time (TI), by a spin echo readout [34]. The TI is timed to have the readout echo centered on the null point of the recovering longitudinal magnetization of CSF. Due to the long T1 relaxation time characterizing CSF, long TIs are necessary for proper attenuation of its signal, thus, the traditional spin-echo readout is often substituted with either turbo or fast spin echo to shorten acquisitions times [34].

Black holes

T1-weighted (T1w) acquisitions are mainly influenced by the longitudinal magnetization and are the sequences most commonly used for structural imaging because of the excellent contrast between structures predominantly composed of fat (i.e., myelin, seen brighter due to its lower T1 relaxation time values) and water-predominant structures (i.e., cortex, seen darker because of the longer T1 time). Demyelination and axonal loss are pathological processes that, by destroying the myelin and axonal structures, increase the water content causing hypointense areas on T1w images [35].

During their evolution, newly formed contrast-enhancing lesions show different signal patterns on unenhanced T1w images. Through the time of enhancement, 20% of the lesions appear isointense, while 80% appear hypointense in comparison to ordinary WM (wet black holes). Once contrast enhancement ends, the progressive remyelination and reabsorption of edema induce 40% of the acute black holes to become isointense. Finally, over a 6-month period, less than 40% of these lesions evolve into persistent or chronic black holes due to permanent demyelination and severe axonal loss [36].

T1w MRI acquisitions are generally characterized by short TR/TE and include spin-echo as well as gradient-echo, both of which may be used to assess the presence of gadolinium-enhancing lesions as well as black holes in nonenhancing images. The T1w images first developed were based on the traditional spin-echo sequence. However, at the beginning of the '90s a three-dimensional (3D) magnetization-prepared rapid gradient echo (MPRAGE) sequence was demonstrated to produce T1w images of comparable, if not superior, quality with respect to its predecessor [37,38]. Thus, **MPRAGE** has become one of the most used sequences for anatomical imaging as the slightly lower signal-to-noise ratio, expected with a gradient-echo readout, is compensated by an increase in the contrast-to-noise ratio. However, MPRAGE images are sensitive to inhomogeneities of the B1 field, especially at high magnetic fields ($\geq$ 3 T).

Magnetization prepared two rapid acquisition gradient echoes (MP2RAGE) has been developed to inherently compensate for B1 inhomogeneities and allow submillimetric acquisition also at ultrahigh field ($\geq$7 T) [39]. MP2RAGE produces high-quality structural images, also sensitive to CLs while also allowing the quantification of T1 relaxation time that may provide information to stage and monitor the evolution of MS lesions [40].

Contrast-enhancing lesions

The relapsing forms of MS, especially in the early stages, exhibit acute inflammatory events, sometimes associated with clinical signs and symptoms, that are characterized by perivascular inflammation related to the breakdown of the blood−brain barrier (BBB).

The BBB leakage is associated with the extravasation of gadolinium-based contrast agents into the surrounding parenchyma producing hyperintensities on postcontrast T1w images due to the strong reduction in T1 relaxation time caused by contrast agents [41,42]. Local contrast enhancement tends to be associated with the formation of new MS plaques and lesions in acute activity [43].

According to their pattern of contrast uptake on static MRI, contrast-enhancing (gad +) lesions have been classified as nodular or ring-like (see Fig. 9.2). Although ring-like lesions tend to show more significant short-term tissue destruction and edema [44] and a slower resolution [45], both patterns tend to ultimately leave similar long-term footprints within 1 year [46] and do not show apparent histological differences [46,47]. In fact, the two patterns seem to be mostly influenced by the lesion size and timing of scanning after gadolinium administration [46,47].

New contrast-enhanced lesions are always associated with a hyperintense lesion in the same location on T2w images that usually shrink in size over time (some months) and decrease their intensity as edema resolves and some tissue repair occurs (extensive or partial remyelination) [48,49]. The resulting permanent "footprint" in T2w images is usually much smaller than the prior inflammatory event. These lesions may also evolve in black holes in T1w images [41].

Gadolinium-enhancing lesions are often clinically silent, in fact, they are five to ten times more common than clinical relapses, and show weak correlation with disability [50].

Although the presence of gadolinium-enhancing lesions is associated with ongoing disease activity [51] and short-term clinical relapses [52], their number did not improve the predictive accuracy of the disability prediction [15].

Gadolinium enhancement of parenchymal lesions becomes much less common in the later stages of RRMS and PMS phenotypes. The change in the prevalence of gad + lesions might represent a switch from adaptive to innate immunity as the main driver of disease worsening [53].

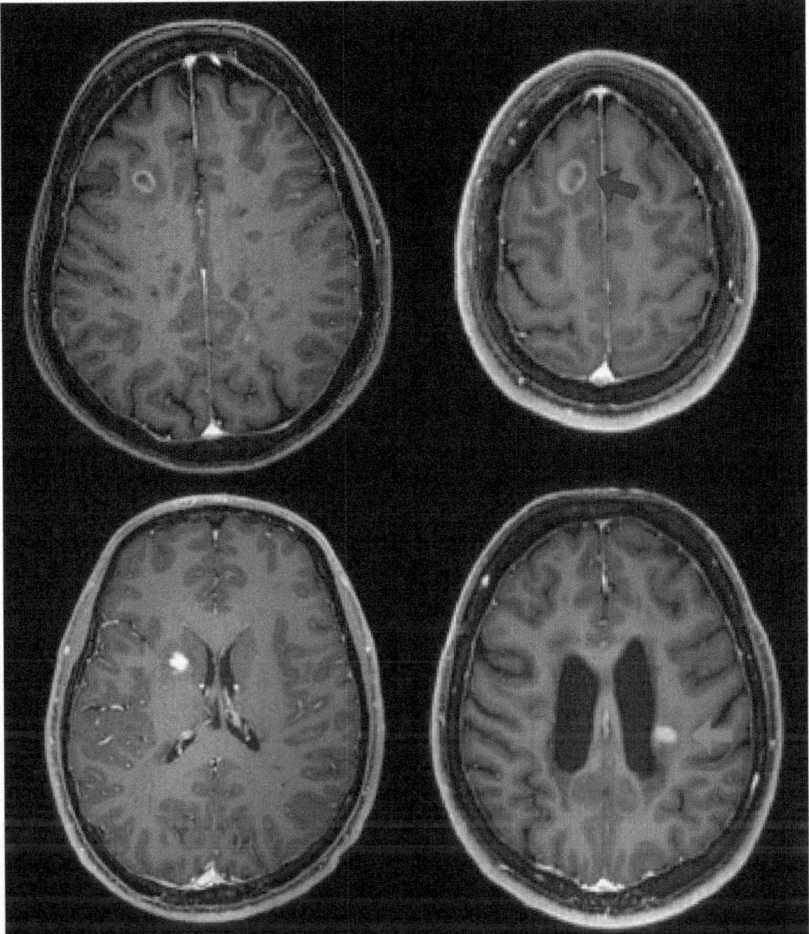

FIGURE 9.2 Examples of ring-shaped (red arrows) and nodular-shaped (green arrows) contrast-enhancing lesions.

Chronic active lesions

While the relapsing–remitting phase of MS is characterized by the presence of acute MS plaques with BBB breakdown, the progressive phenotypes demonstrate a higher proportion of chronic active lesions [54]. Chronic active lesions (also referred to as slowly expanding, smoldering, mixed active/inactive lesions) indicate lesions characterized by a hypocellular and inactive center and with activated microglia/macrophages distributed along the lesion border [54,55]. Furthermore, this chronic inflammation is associated with a partially closed BBB, hindering the possibility of identifying this type of lesion through contrast agent enhancement. Thus, alternative strategies have been developed to investigate these lesions. On the one hand, the characteristic slow expansion has inspired the use of the Jacobian determinant of the deformation between multiple follow-up scans to evaluate the local expansion and select slowly expanding lesions [56]. On the other hand, the iron-enriched proinflammatory microglia accumulating at the rim of these lesions show a paramagnetic behavior that can be exploited using susceptibility-based MR methods such as filtered phase, quantitative susceptibility mapping, T2* and susceptibility-weighted imaging [57–59]. Chronic active and slowly expanding lesions demonstrated an association with disability and poor clinical outcomes and thus may become an imaging marker for disease progression [57,59,60].

Cortical lesions

The cortical involvement in MS has been pathologically known for a long time as it was first noticed in a study dated back to 1916 [61], although a more detailed topography of CLs was described at the end of the 1990s. With the development of MRI sequences sensitive to cortical pathology, such as the first introduction of double inversion recovery sequences to study MS cases in 1998 [62], it became possible to also study cortical involvement in vivo. Nowadays, CLs are known to be present since the early phases of MS [63], although they accumulate over time. Thus, a greater number of lesions is associated with a long history of the disease and with the progressive phenotypes of MS [64]. CLs have been identified in 64% of the patients with relapsing–remitting, in 70% of secondary progressive multiple sclerosis (SPMS), in up to 80% of primary progressive patients and in more than 30% of patients with clinically isolated syndrome suggestive of MS [63,65].

More recently, CLs have been suggested as one of the primary neuropathological substrates of the disease progression, characterized by the progression of both cognitive impairment and irreversible physical disability [66–68]. CLs at disease onset predicted a greater risk of developing SPMS and a shorter latency to progression [69]. Patients with a high frequency of relapses had a larger volume of CLs at clinical onset, developed more focal cortical damage and experienced more severe cortical atrophy over time [70]. In a 5-year longitudinal study, the accumulation of CLs was associated with an increase in EDSS score and the risk of entering the progressive phase of the disease [66].

On the other hand, after 15 years of the disease, patients who still demonstrated a moderate disability without cognitive impairment (probably "Benign MS") show a remarkably lower CL number and volume increase compared to early RRMS [71].

Despite the benefits established in previous studies, imaging of CLs has always been considered technically challenging, especially because of their characteristics and location. In fact, CLs are usually small [63,72], and they are generally characterized by very small perivascular infiltrates of inflammatory cells and rarely show BBB breakdown, especially in the progressive forms of MS [1]. Furthermore, CLs show slight differences in their relaxation times as compared to the surrounding normal-appearing gray matter (NAGM) [73] that, in combination with the partial volume effect as well as the susceptibility and flow artifacts at the cortex/CSF interface, result in a poor contrast-to-noise ratio [74–76]. Also, the location across the cortex may influence the lesions' visibility: while leukocortical lesions (affecting both the juxtacortical WM and the inner layers of the cortex) are more easily detected and demonstrated sensitivities up to 100% at 7 T on postmortem studies

[77] and comparable performances at 3 T [78], the identification of intracortical and subpial lesions is much more challenging due to their small size and the more sparse myelination of the cortical layers extending from the pial surface causing lower contrast [77].

Due to these characteristics, the scoring of histologically validated CLs in ex vivo experiments still underestimates their number and extension, with prospective sensitivity varying from 10% to 48% depending on magnetic field strength and MRI acquisition [77,79—83].

In fact, multiple studies evidenced that most CLs are still not detected by any MRI technique [63,72,84,85], although on average 7 T MRI detects 52% more CLs with respect to the best performing 3 T MRIs [64], and imaging protocols including double inversion recovery (DIR), PSIR, MPRAGE, and MP2RAGE sequences may substantially improve in vivo detection [85—89].

Magnetic resonance imaging sequences for cortical lesion detection

DIR, PSIR, T2*-weighted (T2*w), T1w-MPRAGE, and MP2RAGE are the MRI sequences most commonly used for the detection of CLs that have been introduced and tested in the last two decades. Examples of some of the MRI modalities most used for the identification of CLs are shown in Fig. 9.3.

DIR is an MRI sequence where the signal arising from both the CSF and the WM is attenuated to enhance the delineation of the GM and to produce a higher lesion-to-background and GM-to-WM contrast [34,90]. The nullification of these signals is achieved by properly timing two inversion pulses depending on the differences between the T1 relaxation times of CSF, WM, and GM followed by either the conventional spin-echo readout or faster multispin-echo approaches (such as fast spin echo [FSE] or turbo spin echo [TSE]). This approach, while suppressing CSF and WM, highly also reduces the contribution from the GM-producing images with generally low SNR. Furthermore, the inflow of unsuppressed blood and CSF from adjacent slices can produce artifacts on DIR sequences. However, the use of 3D acquisitions with nonselective pulses rather than 2D acquisitions can greatly reduce the presence of this artifact [34] at the expense of longer acquisition times. Nonetheless, DIR has been demonstrated to improve cortical and intracortical lesions detection with respect to T2w and FLAIR images [72,73,91].

PSIR is an inversion recovery sequence where, rather than retrieving the magnitude of the image reconstruction, both the negative and positive components of the longitudinal magnetization are combined using a phase-sensitive reconstruction algorithm [92]. Phase-sensitive reconstructions also reduce the background noise, thus improving the signal-to-noise ratio [93] and allowing clinically feasible acquisition times [89,94]. By preserving the information regarding the sign of the magnetization, PSIR sequences double the dynamic range of signal intensity available and provide a better GM/WM matter contrast with respect to traditional T1w sequences [92]. PSIR images have been demonstrated to improve the detection of CLs in vivo [89,94] and in retrospective scoring on histologically validated CLs [79], while a performance similar to DIR sequences was evidenced in ex vivo prospective scoring [79]. However, because of the higher signal and contrast-to-noise ratio, PSIR sequences seem most useful in classifying CL types depending on their

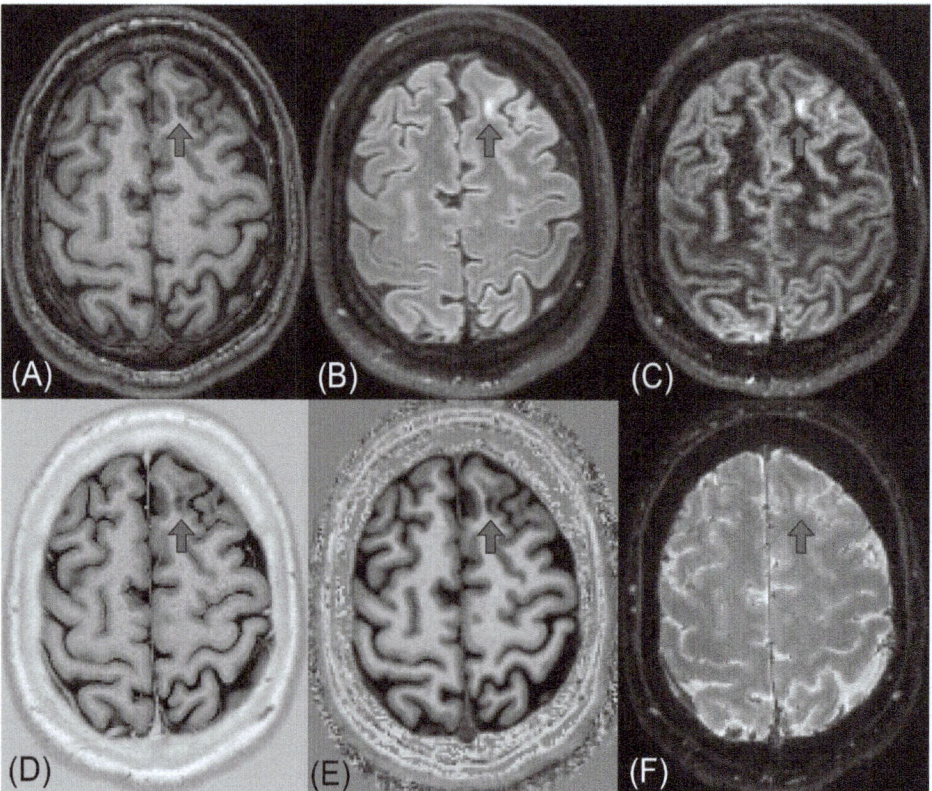

FIGURE 9.3 Example of cortical lesion appearance on (A) MPRAGE, (B) FLAIR, (C) DIR, (D) PSIR, (E) MP2RAGE, and (F) T2*-weighted sequences at 3 T.

location and in delineating lesions' boundaries, especially if used in combination with other sequences such as DIR or FLAIR [68].

T2*w sequence is another sequence used to detect CLs, especially at high and ultrahigh fields, as it provides high resolution (with submillimetric in-plane voxel size), good GM/WM contrast and lesion/NAGM contrast. For these reasons and because of the strong inter-rater agreement, T2*w sequences have been proposed as a gold standard at 7 T [85,95]. However, this sequence requires long acquisition times and did not significantly outperform the prospective detection rates of conventional sequences for histologically verified lesions [96] as it seems less sensitive to small lesions. Nonetheless, T2*w performance in CLs detection improves if combined with MP2RAGE [87].

T1w MPRAGE and **MP2RAGE** have been demonstrated to be sensitive to CLs, to improve the delineation of lesions boundaries as compared to the combination of PSIR and DIR sequences [88], to improve the classification in CLs subtypes [87] and to identify CLs with comparable or higher performances with respect to T2*w images at 7 T in clinically feasible acquisition times [86]. Thus, these sequences may give useful contributions to the study of CLs with the added value of providing high contrast-to-noise ratio structural imaging.

Despite the many sequences proposed and also investigated for high and ultrahigh field applications, a consensus for the best-performing sequence for CL detection and delineation is yet to be reached and included in the international guidelines [97].

Conclusions

The pathological effects of MS are many and vary according to the type and duration of the disease. The clinical outcomes of MS are influenced by WM lesion location, nature, extent, and intensity; however, WM abnormalities are definitely not the sole contributors to MS pathology. In fact, also GM lesions are a frequent and early phenomenon in the disease course but become more evident in people with PMS, where they may become more extensive than WM lesions. Lesions themselves are also heterogeneous in terms of degrees of inflammation, demyelination and remyelination, axonal loss, and spatial localization, thus may differently impact on disability. MRI can produce a variety of images sensitive to the different aspects of the MS pathology; however, each sequence is influenced by specific abnormalities in the tissue that makes the optimization of MRI protocols a priority as much as a challenge.

References

[1] Lassmann H. Pathology and disease mechanisms in different stages of multiple sclerosis. J Neurol Sci 2013;333(1):1—4. Available from: https://doi.org/10.1016/j.jns.2013.05.010.
[2] Magliozzi R, Howell OW, Reeves C, Roncaroli F, Nicholas R, Serafini B, et al. A Gradient of neuronal loss and meningeal inflammation in multiple sclerosis. Ann Neurol 2010;68(4):477—93. Available from: https://doi.org/10.1002/ana.22230.
[3] Thompson AJ, Banwell BL, Barkhof F, Carroll WM, Coetzee T, Comi G, et al. Diagnosis of multiple sclerosis: 2017 revisions of the McDonald criteria. Lancet Neurol 2018;17(2):162—73. Available from: https://doi.org/10.1016/S1474-4422(17)30470-2.
[4] Filippi M, Rocca MA, Ciccarelli O, De Stefano N, Evangelou N, Kappos L, et al. MRI criteria for the diagnosis of multiple sclerosis: MAGNIMS consensus guidelines. Lancet Neurol 2016;15(3):292—303. Available from: https://doi.org/10.1016/S1474-4422(15)00393-2.
[5] Barkhof F, Filippi M, Miller DH, Scheltens P, Campi A, Polman CH, et al. Comparison of MRI criteria at first presentation to predict conversion to clinically definite multiple sclerosis. Brain: A J Neurol 1997;120(Pt 11):2059—69. Available from: https://doi.org/10.1093/brain/120.11.2059.
[6] Geraldes R, Ciccarelli O, Barkhof F, De Stefano N, Enzinger C, Filippi M, et al.MAGNIMS Study Group The current role of MRI in differentiating multiple sclerosis from its imaging mimics. Nat Rev Neurol 2018;14(4):199—213. Available from: https://doi.org/10.1038/nrneurol.2018.14.
[7] Charcot J.-M. (1825-1893) A. du texte. Leçons sur les maladies du système nerveux: Faites à la Salpêtrière. Tome 2 / par J.-M. Charcot, ... recueillies et publ. par Bourneville,...; 1875. https://gallica.bnf.fr/ark:/12148/bpt6k98763k.
[8] Tallantyre EC, Brookes MJ, Dixon JE, Morgan PS, Evangelou N, Morris PG. Demonstrating the perivascular distribution of MS lesions in vivo with 7-Tesla MRI. Neurology 2008;70(22):2076—8. Available from: https://doi.org/10.1212/01.wnl.0000313377.49555.2e.
[9] Giezendanner S, Fisler MS, Soravia LM, Andreotti J, Walther S, Wiest R, et al. Microstructure and cerebral blood flow within white matter of the human brain: a TBSS analysis. PLoS ONE 2016;11(3):e0150657. Available from: https://doi.org/10.1371/journal.pone.0150657.
[10] Holland CM, Charil A, Csapo I, Liptak Z, Ichise M, Khoury SJ, et al. The relationship between normal cerebral perfusion patterns and white matter lesion distribution in 1,249 patients with multiple sclerosis. J Neuroimaging: J Am Soc Neuroimaging 2012;22(2):129—36. Available from: https://doi.org/10.1111/j.1552-6569.2011.00585.x.

[11] Castellaro M, Tamanti A, Pisani AI, Pizzini FB, Crescenzo F, Calabrese M. The use of the central vein sign in the diagnosis of multiple sclerosis: a systematic review and meta-analysis. Diagnostics 2020;10(12):1025. Available from: https://doi.org/10.3390/diagnostics10121025.

[12] Sati P, Oh J, Constable RT, Evangelou N, Guttmann CRG, Henry RG, et al. The central vein sign and its clinical evaluation for the diagnosis of multiple sclerosis: a consensus statement from the North American imaging in multiple sclerosis cooperative. Nat Rev Neurol 2016;12(12):714−22. Available from: https://doi.org/10.1038/nrneurol.2016.166.

[13] Li DKB, Held U, Petkau J, Daumer M, Barkhof F, Fazekas F, et al. MRI T2 lesion burden in multiple sclerosis: a plateauing relationship with clinical disability. Neurology 2006;66(9):1384−9. Available from: https://doi.org/10.1212/01.wnl.0000210506.00078.5c.

[14] Barkhof F, Filippi M. MRI−the perfect surrogate marker for multiple sclerosis? Nat Rev Neurol 2009;5(4):182−3. Available from: https://doi.org/10.1038/nrneurol.2009.31.

[15] Daumer M, Neuhaus A, Morrissey S, Hintzen R, Ebers GC. MRI as an outcome in multiple sclerosis clinical trials. Neurology 2009;72(8):705−11. Available from: https://doi.org/10.1212/01.wnl.0000336916.38629.43.

[16] Hidalgo de la Cruz M, Valsasina P, Meani A, Gallo A, Gobbi C, Bisecco A, et al. Differential association of cortical, subcortical and spinal cord damage with multiple sclerosis disability milestones: a multiparametric MRI study. Multiple Scler J 2022;28(3):406−17. Available from: https://doi.org/10.1177/13524585211020296.

[17] Haider L, Prados F, Chung K, Goodkin O, Kanber B, Sudre C, et al. Cortical involvement determines impairment 30 years after a clinically isolated syndrome. Brain 2021;144(5):1384−95. Available from: https://doi.org/10.1093/brain/awab033.

[18] Altermatt A, Gaetano L, Magon S, Häring DA, Tomic D, Wuerfel J, et al. Clinical correlations of brain lesion location in multiple sclerosis: voxel-based analysis of a large clinical trial dataset. Brain Topogr 2018;31(5):886−94. Available from: https://doi.org/10.1007/s10548-018-0652-9.

[19] Charil A, Zijdenbos AP, Taylor J, Boelman C, Worsley KJ, Evans AC, et al. Statistical mapping analysis of lesion location and neurological disability in multiple sclerosis: application to 452 patient data sets. NeuroImage 2003;19(3):532−44. Available from: https://doi.org/10.1016/s1053-8119(03)00117-4.

[20] Dalton CM, Bodini B, Samson RS, Battaglini M, Fisniku LK, Thompson AJ, et al. Brain lesion location and clinical status 20 years after a diagnosis of clinically isolated syndrome suggestive of multiple sclerosis. Mult Scler (Houndmills, Basingstoke, Engl) 2012;18(3):322−8. Available from: https://doi.org/10.1177/1352458511420269.

[21] Kincses ZT, Ropele S, Jenkinson M, Khalil M, Petrovic K, Loitfelder M, et al. Lesion probability mapping to explain clinical deficits and cognitive performance in multiple sclerosis. Mult Scler (Houndmills, Basingstoke, Engl) 2011;17(6):681−9. Available from: https://doi.org/10.1177/1352458510391342.

[22] Rossi F, Giorgio A, Battaglini M, Stromillo ML, Portaccio E, Goretti B, et al. Relevance of brain lesion location to cognition in relapsing multiple sclerosis. PLoS One 2012;7(11):e44826. Available from: https://doi.org/10.1371/journal.pone.0044826.

[23] Vellinga MM, Geurts JJG, Rostrup E, Uitdehaag BMJ, Polman CH, Barkhof F, et al. Clinical correlations of brain lesion distribution in multiple sclerosis. J Magnetic Reson Imaging: JMRI 2009;29(4):768−73. Available from: https://doi.org/10.1002/jmri.21679.

[24] Brownell B, Hughes JT. The distribution of plaques in the cerebrum in multiple sclerosis. J Neurol, Neurosurg, Psychiatry 1962;25:315−20. Available from: https://doi.org/10.1136/jnnp.25.4.315.

[25] Ikuta F, Zimmerman HM. Distribution of plaques in seventy autopsy cases of multiple sclerosis in the United States. Neurology 1976;26(6 PT 2):26−8. Available from: https://doi.org/10.1212/wnl.26.6_part_2.26.

[26] Gaetano L, Häring DA, Radue E-W, Mueller-Lenke N, Thakur A, Tomic D, et al. Fingolimod effect on gray matter, thalamus, and white matter in patients with multiple sclerosis. Neurology 2018;90(15):e1324−32. Available from: https://doi.org/10.1212/WNL.0000000000005292.

[27] Ciccarelli O, Cohen JA, Reingold SC, Weinshenker BGInternational Conference on Spinal Cord Involvement and Imaging in Multiple Sclerosis and Neuromyelitis Optica Spectrum Disorders. Spinal cord involvement in multiple sclerosis and neuromyelitis optica spectrum disorders. Lancet Neurol 2019;18(2):185−97. Available from: https://doi.org/10.1016/S1474-4422(18)30460-5.

[28] Gass A, Rocca MA, Agosta F, Ciccarelli O, Chard D, Valsasina P, et al. MRI monitoring of pathological changes in the spinal cord in patients with multiple sclerosis. Lancet Neurol 2015;14(4):443−54. Available from: https://doi.org/10.1016/S1474-4422(14)70294-7.

[29] Bonacchi R, Pagani E, Meani A, Cacciaguerra L, Preziosa P, De Meo E, et al. Clinical relevance of multiparametric MRI assessment of cervical cord damage in multiple sclerosis. Radiology 2020;296(3):605–15. Available from: https://doi.org/10.1148/radiol.2020200430.

[30] Gyldensted C. Computer tomography of the cerebrum in multiple sclerosis. Neuroradiology 1976;12 (1):33–42. Available from: https://doi.org/10.1007/BF00344224.

[31] Jung BA, Weigel M. Spin echo magnetic resonance imaging. J Magn Reson Imaging 2013;37(4):805–17. Available from: https://doi.org/10.1002/jmri.24068.

[32] Wattjes MP, Lutterbey GG, Harzheim M, Gieseke J, Träber F, Klotz L, et al. Imaging of inflammatory lesions at 3.0 Tesla in patients with clinically isolated syndromes suggestive of multiple sclerosis: a comparison of fluid-attenuated inversion recovery with T2 turbo spin-echo. Eur Radiol 2006;16(7):1494–500. Available from: https://doi.org/10.1007/s00330-005-0082-4.

[33] Filippi M, Rocca MA. Conventional MRI in multiple sclerosis. J Neuroimaging 2007;17:3S–9S. Available from: https://doi.org/10.1111/j.1552-6569.2007.00129.x.

[34] Saranathan M, Worters PW, Rettmann DW, Winegar B, Becker J. Physics for clinicians: fluid-attenuated inversion recovery (FLAIR) and double inversion recovery (DIR) Imaging: FLAIR and DIR imaging. J Magn Reson Imaging 2017;46(6):1590–600. Available from: https://doi.org/10.1002/jmri.25737.

[35] Hemond CC, Bakshi R. Magnetic resonance imaging in multiple sclerosis. Cold Spring Harb Perspect Med 2018;8(5):a028969. Available from: https://doi.org/10.1101/cshperspect.a028969.

[36] Sahraian MA, Radue E-W, Haller S, Kappos L. Black holes in multiple sclerosis: definition, evolution, and clinical correlations. Acta Neurol Scand 2010;122(1):1–8. Available from: https://doi.org/10.1111/j.1600-0404.2009.01221.x.

[37] Brant-Zawadzki M, Gillan GD, Nitz WR. MP RAGE: a three-dimensional, T1-weighted, gradient-echo sequence—initial experience in the brain. Radiology 1992;182(3):769–75. Available from: https://doi.org/10.1148/radiology.182.3.1535892.

[38] Mugler JP, Brookeman JR. Three-dimensional magnetization-prepared rapid gradient-echo imaging (3D MP RAGE). Magn Reson Med 1990;15(1):152–7. Available from: https://doi.org/10.1002/mrm.1910150117.

[39] Marques JP, Kober T, Krueger G, van der Zwaag W, Van de Moortele P-F, Gruetter R. MP2RAGE, a self bias-field corrected sequence for improved segmentation and T1-mapping at high field. NeuroImage 2010;49 (2):1271–81. Available from: https://doi.org/10.1016/j.neuroimage.2009.10.002.

[40] Kober T, Granziera C, Ribes D, Browaeys P, Schluep M, Meuli R, et al. MP2RAGE multiple sclerosis magnetic resonance imaging at 3 T. Invest Radiol 2012;47(6):346–52. Available from: https://doi.org/10.1097/RLI.0b013e31824600e9.

[41] Bou Fakhredin R, Saade C, Kerek R, El-Jamal L, Khoury SJ, El-Merhi F. Imaging in multiple sclerosis: a new spin on lesions. J Med Imaging Radiat Oncol 2016;60(5):577–86. Available from: https://doi.org/10.1111/1754-9485.12498.

[42] Warntjes JBM, Tisell A, Landtblom A-M, Lundberg P. Effects of gadolinium contrast agent administration on automatic brain tissue classification of patients with multiple sclerosis. AJNR Am J Neuroradiol 2014;35 (7):1330–6. Available from: https://doi.org/10.3174/ajnr.A3890.

[43] Lassmann H. The pathologic substrate of magnetic resonance alterations in multiple sclerosis. Neuroimaging Clin North Am 2008;18(4):563–76. Available from: https://doi.org/10.1016/j.nic.2008.06.005 ix.

[44] Rovira A, Alonso J, Cucurella G, Nos C, Tintoré M, Pedraza S, et al. Evolution of multiple sclerosis lesions on serial contrast-enhanced T1-weighted and magnetization-transfer MR images. AJNR: Am J Neuroradiol 1999;20(10):1939–45.

[45] Minneboo A, Uitdehaag BMJ, Ader HJ, Barkhof F, Polman CH, Castelijns JA. Patterns of enhancing lesion evolution in multiple sclerosis are uniform within patients. Neurology 2005;65(1):56–61. Available from: https://doi.org/10.1212/01.wnl.0000167538.24338.bb.

[46] Davis M, Auh S, Riva M, Richert ND, Frank JA, McFarland HF, et al. Ring and nodular multiple sclerosis lesions. Neurology 2010;74(10):851–6. Available from: https://doi.org/10.1212/WNL.0b013e3181d31df5.

[47] Gaitán MI, Shea CD, Dphil IEE, Stone RD, Fenton KM, Bielekova B, et al. Evolution of the blood-brain barrier in newly forming multiple sclerosis lesions. Ann Neurol 2011;70(1):22–9. Available from: https://doi.org/10.1002/ana.22472.

[48] Meier DS, Guttmann CRG. Time-series analysis of MRI intensity patterns in multiple sclerosis. NeuroImage 2003;20(2):1193–209. Available from: https://doi.org/10.1016/S1053-8119(03)00354-9.

[49] Meier DS, Weiner HL, Guttmann CRG. MR imaging intensity modeling of damage and repair in multiple sclerosis: relationship of short-term lesion recovery to progression and disability. AJNR Am J Neuroradiol 2007;28(10):1956–63. Available from: https://doi.org/10.3174/ajnr.A0701.

[50] McFarland HF. Examination of the role of magnetic resonance imaging in multiple sclerosis: a problem-orientated approach. Ann Indian Acad Neurol 2009;12(4):254–63. Available from: https://doi.org/10.4103/0972-2327.58284.

[51] Molyneux PD, Filippi M, Barkhof F, Gasperini C, Yousry TA, Truyen L, et al. Correlations between monthly enhanced MRI lesion rate and changes in T2 lesion volume in multiple sclerosis. Ann Neurol 1998;43(3):332–9. Available from: https://doi.org/10.1002/ana.410430311.

[52] Kappos L, Moeri D, Radue EW, Schoetzau A, Schweikert K, Barkhof F, et al. Predictive value of gadolinium-enhanced magnetic resonance imaging for relapse rate and changes in disability or impairment in multiple sclerosis: a meta-analysis. Lancet 1999;353(9157):964–9. Available from: https://doi.org/10.1016/S0140-6736(98)03053-0.

[53] Weiner HL. The challenge of multiple sclerosis: how do we cure a chronic heterogeneous disease? Ann Neurol 2009;65(3):239–48. Available from: https://doi.org/10.1002/ana.21640.

[54] Frischer JM, Weigand SD, Guo Y, Kale N, Parisi JE, Pirko I, et al. Clinical and pathological insights into the dynamic nature of the white matter multiple sclerosis plaque. Ann Neurol 2015;78(5):710–21. Available from: https://doi.org/10.1002/ana.24497.

[55] Kuhlmann T, Ludwin S, Prat A, Antel J, Brück W, Lassmann H. An updated histological classification system for multiple sclerosis lesions. Acta Neuropathol 2017;133(1):13–24. Available from: https://doi.org/10.1007/s00401-016-1653-y.

[56] Elliott C, Wolinsky JS, Hauser SL, Kappos L, Barkhof F, Bernasconi C, et al. Slowly expanding/evolving lesions as a magnetic resonance imaging marker of chronic active multiple sclerosis lesions. Mult Scler J 2019;25(14):1915–25. Available from: https://doi.org/10.1177/1352458518814117.

[57] Absinta M, Sati P, Masuzzo F, Nair G, Sethi V, Kolb H, et al. Association of chronic active multiple sclerosis lesions with disability in vivo. JAMA Neurol 2019;76(12):1474. Available from: https://doi.org/10.1001/jamaneurol.2019.2399.

[58] Dal-Bianco A, Grabner G, Kronnerwetter C, Weber M, Höftberger R, Berger T, et al. Slow expansion of multiple sclerosis iron rim lesions: pathology and 7 T magnetic resonance imaging. Acta Neuropathol 2017;133(1):25–42. Available from: https://doi.org/10.1007/s00401-016-1636-z.

[59] Kaunzner UW, Kang Y, Zhang S, Morris E, Yao Y, Pandya S, et al. Quantitative susceptibility mapping identifies inflammation in a subset of chronic multiple sclerosis lesions. Brain 2019;142(1):133–45. Available from: https://doi.org/10.1093/brain/awy296.

[60] Calvi A, Tur C, Chard D, Stutters J, Ciccarelli O, Cortese R, et al. Slowly expanding lesions relate to persisting black-holes and clinical outcomes in relapse-onset multiple sclerosis. NeuroImage: Clin 2022;35103048. Available from: https://doi.org/10.1016/j.nicl.2022.103048.

[61] Dawson JW. The histology of disseminated sclerosis. Edinb Med J 1916;17(4):229–41.

[62] Bedell BJ, Narayana PA. Implementation and evaluation of a new pulse sequence for rapid acquisition of double inversion recovery images for simultaneous suppression of white matter and CSF. J Magn Reson Imaging 1998;8(3):544–7. Available from: https://doi.org/10.1002/jmri.1880080305.

[63] Calabrese M, De Stefano N, Atzori M, Bernardi V, Mattisi I, Barachino L, et al. Detection of cortical inflammatory lesions by double inversion recovery magnetic resonance imaging in patients with multiple sclerosis. Arch Neurol 2007;64(10):1416–22. Available from: https://doi.org/10.1001/archneur.64.10.1416.

[64] Madsen MAJ, Wiggermann V, Bramow S, Christensen JR, Sellebjerg F, Siebner HR. Imaging cortical multiple sclerosis lesions with ultra-high field MRI. NeuroImage: Clin 2021;32102847. Available from: https://doi.org/10.1016/j.nicl.2021.102847.

[65] Magliozzi R, Reynolds R, Calabrese M. MRI of cortical lesions and its use in studying their role in MS pathogenesis and disease course. Brain Pathol 2018;28(5):735–42. Available from: https://doi.org/10.1111/bpa.12642.

[66] Calabrese M, Poretto V, Favaretto A, Alessio S, Bernardi V, Romualdi C, et al. Cortical lesion load associates with progression of disability in multiple sclerosis. Brain 2012;135(10):2952–61. Available from: https://doi.org/10.1093/brain/aws246.

[67] Harrison DM, Roy S, Oh J, Izbudak I, Pham D, Courtney S, et al. Association of cortical lesion burden on 7-T magnetic resonance imaging with cognition and disability in multiple sclerosis. JAMA Neurol 2015;72(9):1004–12. Available from: https://doi.org/10.1001/jamaneurol.2015.1241.

[68] Nelson F, Datta S, Garcia N, Rozario NL, Perez F, Cutter G, et al. Intracortical lesions by 3T magnetic resonance imaging and correlation with cognitive impairment in multiple sclerosis. Mult Scler J 2011;17(9):1122–9. Available from: https://doi.org/10.1177/1352458511405561.

[69] Pisani AI, Scalfari A, Crescenzo F, Romualdi C, Calabrese M. A novel prognostic score to assess the risk of progression in relapsing − remitting multiple sclerosis patients. Eur J Neurol 2021;28(8):2503–12. Available from: https://doi.org/10.1111/ene.14859.

[70] Scalfari A, Romualdi C, Nicholas RS, Mattoscio M, Magliozzi R, Morra A, et al. The cortical damage, early relapses, and onset of the progressive phase in multiple sclerosis. Neurology 2018;90(24):e2107–18. Available from: https://doi.org/10.1212/WNL.0000000000005685.

[71] Calabrese M, Filippi M, Rovaris M, Bernardi V, Atzori M, Mattisi I, et al. Evidence for relative cortical sparing in benign multiple sclerosis: a longitudinal magnetic resonance imaging study. Mult Scler (Houndmills, Basingstoke, Engl) 2009;15(1):36–41. Available from: https://doi.org/10.1177/1352458508096686.

[72] Seewann A, Kooi E-J, Roosendaal SD, Pouwels PJW, Wattjes MP, Valk P, et al. Postmortem verification of MS cortical lesion detection with 3D DIR. Neurology 2012;78(5):302–8. Available from: https://doi.org/10.1212/WNL.0b013e31824528a0.

[73] Geurts JJG, Pouwels PJW, Uitdehaag BMJ, Polman CH, Barkhof F, Castelijns JA. Intracortical lesions in multiple sclerosis: improved detection with 3D double inversion-recovery MR imaging. Radiology 2005;236(1):254–60. Available from: https://doi.org/10.1148/radiol.2361040450.

[74] Kidd D, Barkhof F, McConnell R, Algra PR, Allen IV, Revesz T. Cortical lesions in multiple sclerosis. Brain 1999;122(1):17–26. Available from: https://doi.org/10.1093/brain/122.1.17.

[75] Schmierer K, Parkes HG, So P-W, An SF, Brandner S, Ordidge RJ, et al. High field (9.4 Tesla) magnetic resonance imaging of cortical grey matter lesions in multiple sclerosis. Brain 2010;133(3):858–67. Available from: https://doi.org/10.1093/brain/awp335.

[76] Tardif CL, Bedell BJ, Eskildsen SF, Collins DL, Pike GB. Quantitative magnetic resonance imaging of cortical multiple sclerosis pathology. Mult Scler Int 2012;2012:1–13. Available from: https://doi.org/10.1155/2012/742018.

[77] Kilsdonk ID, Jonkman LE, Klaver R, van Veluw SJ, Zwanenburg JJM, Kuijer JPA, et al. Increased cortical grey matter lesion detection in multiple sclerosis with 7 T MRI: a post-mortem verification study. Brain 2016;139(5):1472–81. Available from: https://doi.org/10.1093/brain/aww037.

[78] Maranzano J, Dadar M, Rudko DA, De Nigris D, Elliott C, Gati JS, et al. Comparison of multiple sclerosis cortical lesion types detected by multicontrast 3T and 7T MRI. AJNR Am J Neuroradiol 2019;40(7):1162–9. Available from: https://doi.org/10.3174/ajnr.A6099.

[79] Bouman PM, Steenwijk MD, Pouwels PJW, Schoonheim MM, Barkhof F, Jonkman LE, et al. Histopathology-validated recommendations for cortical lesion imaging in multiple sclerosis. Brain 2020;. Available from: https://doi.org/10.1093/brain/awaa233.

[80] Bagnato F, Yao B, Cantor F, Merkle H, Condon E, Montequin M, et al. Multisequence-imaging protocols to detect cortical lesions of patients with multiple sclerosis: observations from a post-mortem 3 Tesla imaging study. J Neurol Sci 2009;282(1–2):80–5. Available from: https://doi.org/10.1016/j.jns.2009.03.021.

[81] Geurts JJG, Blezer ELA, Vrenken H, van der Toorn A, Castelijns JA, Polman CH, et al. Does high-field MR imaging improve cortical lesion detection in multiple sclerosis? J Neurol 2008;255(2):183–91. Available from: https://doi.org/10.1007/s00415-008-0620-5.

[82] Seewann A, Vrenken H, Kooi E-J, van der Valk P, Knol DL, Polman CH, et al. Imaging the tip of the iceberg: Visualization of cortical lesions in multiple sclerosis. Mult Scler J 2011;17(10):1202–10. Available from: https://doi.org/10.1177/1352458511406575.

[83] Yao B, Hametner S, Gelderen Pv, Merkle H, Chen C, Lassmann H, et al. 7 Tesla magnetic resonance imaging to detect cortical pathology in multiple sclerosis. PLoS ONE 2014;9(10):e108863. Available from: https://doi.org/10.1371/journal.pone.0108863.

[84] Geurts JJG, Roosendaal SD, Calabrese M, Ciccarelli O, Agosta F, Chard DT, et al.MAGNIMS Study Group Consensus recommendations for MS cortical lesion scoring using double inversion recovery MRI. Neurology 2011;76(5):418–24. Available from: https://doi.org/10.1212/WNL.0b013e31820a0cc4.

[85] Mainero C, Benner T, Radding A, Kouwe Av d, Jensen R, Rosen BR, et al. In vivo imaging of cortical pathology in multiple sclerosis using ultra-high field MRI. Neurology 2009;73(12):941–8. Available from: https://doi.org/10.1212/WNL.0b013e3181b64bf7.

[86] Cocozza S, Cosottini M, Signori A, Fleysher L, El Mendili MM, Lublin F, et al. A clinically feasible 7-Tesla protocol for the identification of cortical lesions in multiple sclerosis. Eur Radiol 2020;30(8):4586−94. Available from: https://doi.org/10.1007/s00330-020-06803-y.

[87] Beck ES, Sati P, Sethi V, Kober T, Dewey B, Bhargava P, et al. Improved visualization of cortical lesions in multiple sclerosis using 7T MP2RAGE. AJNR. Am J Neuroradiol 2018;39(3):459−66. Available from: https://doi.org/10.3174/ajnr.A5534.

[88] Nelson F, Poonawalla A, Hou P, Wolinsky JS, Narayana PA. 3D MPRAGE improves classification of cortical lesions in multiple sclerosis. Mult Scler (Houndmills, Basingstoke, Engl) 2008;14(9):1214−19. Available from: https://doi.org/10.1177/1352458508094644.

[89] Sethi V, Yousry TA, Muhlert N, Ron M, Golay X, Wheeler-Kingshott C, et al. Improved detection of cortical MS lesions with phase-sensitive inversion recovery MRI. J Neurol Neurosurg Psychiatry 2012;83(9):877−82. Available from: https://doi.org/10.1136/jnnp-2012-303023.

[90] Soares BP, Porter SG, Saindane AM, Dehkharghani S, Desai NK. Utility of double inversion recovery MRI in paediatric epilepsy. Br J Radiol 2016;89(1057):20150325. Available from: https://doi.org/10.1259/bjr.20150325.

[91] Elkholy SF, Sabet MA, Mohammad ME, Asaad REI. Comparative study between double inversion recovery (DIR) and fluid-attenuated inversion recovery (FLAIR) MRI sequences for detection of cerebral lesions in multiple sclerosis. Egypt J Radiol Nucl Med 2020;51(1):188. Available from: https://doi.org/10.1186/s43055-020-00298-9.

[92] Hou P, Hasan KM, Sitton CW, Wolinsky JS, Narayana PA. Phase-sensitive T1 inversion recovery imaging: a time-efficient interleaved technique for improved tissue contrast in neuroimaging. AJNR Am J Neuroradiol 2005;26(6):1432.

[93] Bernstein MA, Thomasson DM, Perman WH. Improved detectability in low signal-to-noise ratio magnetic resonance images by means of a phase-corrected real reconstruction. Med Phys 1989;16(5):813−17. Available from: https://doi.org/10.1118/1.596304.

[94] Harel A, Ceccarelli A, Farrell C, Fabian M, Howard J, Riley C, et al. Phase-sensitive inversion-recovery MRI improves longitudinal cortical lesion detection in progressive MS. PLoS ONE 2016;11(3):e0152180. Available from: https://doi.org/10.1371/journal.pone.0152180.

[95] Nielsen AS, Kinkel RP, Tinelli E, Benner T, Cohen-Adad J, Mainero C. Focal cortical lesion detection in multiple sclerosis: 3 Tesla DIR versus 7 Tesla FLASH-T2. J Magn Reson Imaging: JMRI 2012;35(3):537−42. Available from: https://doi.org/10.1002/jmri.22847.

[96] Jonkman LE, Klaver R, Fleysher L, Inglese M, Geurts JJG. Ultra-high-field MRI visualization of cortical multiple sclerosis lesions with T2 and T2*: a postmortem MRI and histopathology study. Am J Neuroradiol 2015;36(11):2062−7. Available from: https://doi.org/10.3174/ajnr.A4418.

[97] Wattjes MP, Ciccarelli O, Reich DS, Banwell B, de Stefano N, Enzinger C, et al. 2021 MAGNIMS−CMSC−NAIMS consensus recommendations on the use of MRI in patients with multiple sclerosis. Lancet Neurol 2021;20(8):653−70. Available from: https://doi.org/10.1016/S1474-4422(21)00095-8.

CHAPTER 10

Spinal cord imaging in multiple sclerosis

Lorena Lorefice[1] and Giuseppe Fenu[2]

[1]Department of Medical Sciences and Public Health, Multiple Sclerosis Center, Binaghi Hospital, ASL Cagliari, University of Cagliari, Cagliari, Italy [2]Department of Neurosciences, ARNAS Brotzu, Cagliari, Italy

OUTLINE

Introduction	175
Characteristics of spinal cord lesions in multiple sclerosis: multiple sclerosis diagnosis and differential diagnosis	176
Practical considerations in spinal cord clinical imaging	180
Spinal cord magnetic resonance imaging assessment for patient's prognosis and disease monitoring	182
Future directions	183
Conclusions	185
References	185

Introduction

Spinal cord (SC) involvement is common in multiple sclerosis (MS) and includes a variety of pathological processes, including demyelination, axonal loss, and gliosis [1], which can be studied and quantified in vivo using magnetic resonance imaging (MRI) [2]. Conventional MRI is routinely used in clinical settings to detect focal lesions, both in the diagnostic phase and during the disease course, and provides important information on prognosis, response to disease-modifying therapies (DMTs), and MS evolution [3]. However, SC pathology is difficult to detect, characterize, and quantify because of limitations in the sensitivity and specificity of SC MRI, and because the SC is a small and mobile structure. Thus, SC MRI images are often affected by motion artifacts [4]. Recent advances in the MRI acquisition and postprocessing protocols and the latest consensus guidelines from the magnetic resonance imaging in multiple sclerosis (MAGNIMS) and Consortium of Multiple Sclerosis Centers (CMSCs) have overcome some limitations associated with SC MRI and optimized its use [5,6].

This chapter describes the radiological manifestations of SC involvement in MS, the characteristics of the lesions, the protocols used to study the SC, the role of SC involvement in establishing the diagnosis of MS, and the aspects related to differential diagnosis. Beyond this, a part of the SC study on MRI disease monitoring and the prognostic value of SC involvement will be examined, considering future perspectives.

Characteristics of spinal cord lesions in multiple sclerosis: multiple sclerosis diagnosis and differential diagnosis

Two different patterns of signal abnormalities have been described in the SC of patients with MS: focal lesions and diffuse SC abnormalities [7]. Focal SC lesions, generally present throughout the SC, are most commonly located in the periphery of the SC (mainly posterior and lateral), whereas these lesions rarely involve central gray matter. Typically, these lesions are wedge-shaped on axial images, ovoid on sagittal images, small, extend to less than two vertebral segments, and occupy less than half of the cord area. The presence of multiple asymmetric SC lesions strongly suggests MS [8]. On the other hand, mainly described in the progressive forms of MS, diffuse SC abnormalities are typically seen as abnormal areas of subtle increases in signal intensity between those of focal lesions and the normal-appearing SC, lacking well-demarcated borders from the adjacent normal-appearing cord [9]. Beyond these, especially in patients with long MS duration, lesions with longitudinal extension can be found, mostly indicating the presence of several confluent lesions for which identification by high-resolution axial MRI is extremely useful [10]. Considering their clinical significance, well-established characteristics, and frequent occurrence in patients suspected of having MS, the evaluation of focal SC lesions has been incorporated into the diagnostic criteria for MS since 2001, and subsequent revisions to these criteria have continued to include this assessment [11]. A hyperintense lesion in the cervical, thoracic, or lumbar SC seen on T2 plus short tau inversion recovery (STIR), proton-density images, other appropriate sequences, or in two planes on T2 images, is considered to define the SC involvement in MS [12]. The distinction between symptomatic and asymptomatic MRI lesions is unnecessary for the diagnosis of MS according to the recently revised MS diagnostic criteria [11]. Conversely, diffuse SC abnormalities have limited detection specificity and reliability; therefore, they are not included in the diagnostic criteria for MS [11].

In this framework, the SC is one of the four anatomical locations incorporated into the 2017 McDonald diagnostic criteria for MS to document dissemination in space in patients with clinically isolated syndromes (CIS) suggestive of MS [11]. New or gadolinium-enhancing lesions can be used to prove dissemination over time, with gadolinium enhancement present in most acute SC lesions with variable patterns: homogeneous and ring-enhancing lesions in approximately 20% of the lesions [9]. Both symptomatic and asymptomatic SC lesions can be counted to define dissemination in space and time. However, to establish the diagnosis of MS when the first MRI does not fulfill the criteria, it should be noted that the current MAGNIMS–CMSC–NAIMS consensus guidelines do not indicate that SC MRI is routinely used as a followup imaging because of insufficient documentation [13], and thus should be considered on a case-by-case basis, raising the controversial issue of using SC MRI for disease monitoring [6]. In contrast, the diagnostic criteria for primary progressive multiple sclerosis (PPMS) in the 2017 McDonald's criteria indicated that the PPMS can be diagnosed in individuals with 1 year of disability progression in the presence of at least two of the following: two or more T2 hyperintense SC lesions, one or more T2 hyperintense lesions in the periventricular, cortical, juxtacortical, or infratentorial regions, and the presence of cerebrospinal fluid (CSF)—specific oligoclonal bands [11]. This highlights the importance of distinguishing longitudinally extensive lesions from confluent SC lesions, particularly in patients with progressive course [12,14]. A revision of the diagnostic criteria for MS is anticipated, with the involvement of the SC remaining as a key topographical site of damage. In a clinical setting, the SC MRI is highly indicated for patients who present with symptoms involving the SC at the disease onset. Its primary purpose is to rule out alternative diagnoses and provide further clarification in cases where there is progression of symptoms from the disease onset. Moreover, the SC MRI plays a crucial role when brain MRI findings are equivocal, such as in the differential diagnosis of cerebrovascular diseases, inflammatory disorders, age-related abnormalities, and lesions associated with migraines [15]. It is also recommended to improve the diagnostic definition when brain MRI results are inconclusive, such as when there is incomplete fulfilment of the diagnostic criteria for spatial dissemination. However, for all patients with MS at the disease onset, it is strongly advised as the presence of SC lesions serves as a negative prognostic factor for disability accumulation [16].

A standardized approach to MRI evaluation in clinical practice, including image acquisition protocols and scan timing, is crucial for MS diagnosis and disease monitoring [6,17,18]. Recently, the MAGNIMS—CMSC—NAIMS panel of experts in the diagnosis and management of patients with MS discussed and agreed on new or modified recommendations on the use of brain and SC MRI in clinical practice, resulting in some recommendations [6]. First, even though the 3 T scanners offer an increased identification rate for MS lesions and potentially shorter scanning times than lower magnetic field strengths, there is no substantiated proof that the 3 T MRI results in superior detection of SC focal lesions [19]. Second, according to the SC MRI protocol, at the time of diagnosis, at least two sagittal T2-weighted sequences, fast or turbo spin-echo (FSE/TSE), proton density-weighted sequences (TSE or FSE), or STIR sequences are recommended, whereas sagittal T1-weighted sequences (TSE or FSE) are recommended after the contrast [6]. The acquisition of a T2-weighted sequence alone is not considered sufficient because of its low sensitivity in identifying the SC alterations; another sequence is considered necessary to confirm the presence of alterations and exclude the presence of artifacts [20,21]. Although considered optional, other sequences have shown utility for better characterization of the lesions. Among these, T2-weighted axial echo (TSE or FSE) or gradient-recalled echo can help characterize and confirm the lesions detected on sagittal imaging, better document the location and extent of the lesions, and show their utility for detecting small lesions not evident in sagittal sequences [20,21]. The images obtained transverse to the SC can provide better image quality than sagittal or coronal slices because there is less field variation across the slice thickness. Thus, this is particularly effective when the slices are aligned with either the intervertebral disks or the centers of the vertebral bodies [22]; however,

1. Conventional MRI use in multiple sclerosis

this choice of slice orientation and positioning can limit the anatomical coverage of the images. Optional sequences include precontrast sagittal T1-weighted (TSE or FSE) and axial T1-weighted (TSE or FSE) sequences. The combination of the recommended and optional sequences improves the diagnostic definition, especially in cases that require an adequate differential diagnosis [9]. Neuromyelitis optica spectrum disorder (NMOSD) and acute disseminated encephalomyelitis produce SC lesions that extend longitudinally and exhibit a similar appearance to the confluent lesions of MS [23]. Characteristically, acute NMOSD lesions extend to more than three vertebral segments, but to a lesser extent in approximately 15% of aquaporin-4-Ab-positive (AQP4) NMOSD. The NMOSD lesions can occur in any SC segment but are most frequently found in the cervical region in AQP4-NMOSD and have a caudal predilection in myelin oligodendrocyte antibody disease (MOGAD), where they occur in approximately 75% of the cases compared to 20% of MS patients [24]. These lesions can be unilateral or peripheral, and even if more frequent they are located centrally and can also involve the gray matter with an "H pattern" in MOGAD cases. Acute NMOSD lesions appear hyperintense on T2-weighted images and hypointense on T1-weighted images. In approximately 90% of AQP4-NMOSD cases, extremely hyperintense lesions (bright spotty lesions) are described [25]; these lesions appear with variable patterns of gadolinium enhancement, predominantly exhibiting ring enhancements on contrast-enhanced T1-weighted sequences [26]. Gadolinium enhancement is usually present but at a lower frequency in cases of MOGAD [27]. Therefore, SC lesions in MS differ from those in NMOSD primarily based on the lesion length on sagittal images. Table 10.1 details the MRI characteristics of SC lesions in MS and NMOSD, also showed in Fig. 10.1. An appropriate differential diagnosis based on clinical evaluation, MRI

TABLE 10.1 MRI characteristics of spinal cord lesions in MS and NMOSD.

	MS	AQP4-NMOSD	MOGAD
Characteristics of lesions	Multiple asymmetric SC lesions extend <2 vertebral segments in cervical, thoracic region Generally located in the periphery of the spinal cord (mainly posteriorly and laterally) and occupying less than half of the cord area Hyperintense in T2 sequences, wedge-shaped on axial images, and ovoid on sagittal images	Acute lesions (single or multiple) extend >3 vertebral segments; most frequently in cervical region Most frequently located centrally, sometimes unilateral or even peripheral Extremely hyperintense in T2 sequences (bright spotty lesions)	Acute lesions (single or multiple) extend >3 vertebral segments; most frequently in caudal region Most frequently located centrally with gray matter involvement ("H pattern"), unilateral or even peripheral Hyperintense in T2 sequences
Gadolinium enhancement	Typically, Gd + in most acute lesions with variable pattern: homogeneous, ring-enhancing	Usually, Gd + in acute lesions with variable patterns	Sometimes, Gd + in acute lesions with variable patterns
Chronic evolution	Commonly evolve into T1 hypointense lesions; longitudinal extension of lesions can be found (presence of several confluent lesions)	Chronic lesions tend to be replaced by long segments of atrophy or myelomalacia (pseudosyrinx)	Lesions tend to resolve

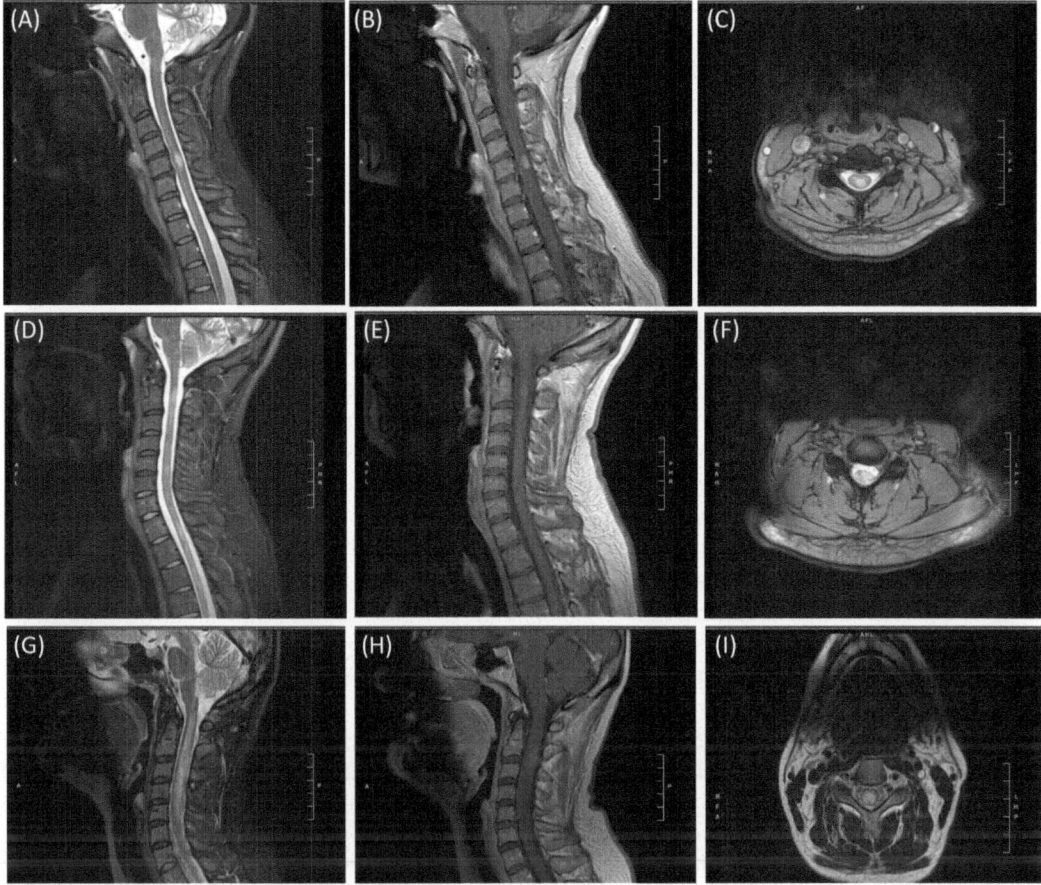

FIGURE 10.1 Short tau inversion recovery sagittal image, sagittal T1-weighted sequences after contrast, and T2-weighted axial images achieved using three-tesla MRI. Images A, B, C show focal SC lesions in MS; images D, E, F show confluent SC lesions in MS; images G, H, I show longitudinally extensive lesions in NMOSD.

studies, and laboratory tests to assess the presence of specific NMOSD antibodies (AQP4 and MOG) should be recorded in individuals presenting with idiopathic transverse myelitis. Although asymptomatic SC injuries can occur in patients with NMOSD, they are less common than those in patients with MS, and their ability to predict future disease activity and disability is unclear [28].

Recent findings have highlighted the importance of considering the length of the SC lesions when distinguishing between these three diseases, particularly during the acute phase [23]. In AQP4-NMOSD, approximately 85% of the acute cervical lesions span more than three vertebral segments, whereas they are typically rare in MOGAD, in which the lesions are more frequent in the caudal region and MS. Although longitudinally extensive hazy T2 hyperintensities can be observed beyond the focal lesions in MS, the chronic lesions in AQP4-NMOSD and MOGAD tend to be shorter, which complicates the differentiation between the three diseases. In cases of MOGAD, where the lesions tend to resolve instead of enlarging and the occurrence of new lesions is uncommon, conducting a single followup MRI of the brain and cervical cord can be beneficial for confirming the diagnosis and facilitating an accurate differentiation between MOGAD, AQP4-NMOSD, and relapsing—remitting multiple sclerosis (RRMS) in the nonacute phase [29].

Additionally, SC involvement has been detected in many other inflammatory and noninflammatory central nervous system diseases. The SC involvement is a common occurrence in systemic lupus erythematosus, and patients are often categorized based on MRI findings of the SC. The classification distinguishes between gray- and white-matter-predominant myelitis with SC swelling, more frequent longitudinally extensive transverse myelitis in gray matter subtypes, and more frequent gadolinium enhancement in white matter myelitis. The cervical to mid-thoracic regions are most commonly affected [30]. Longitudinally extensive transverse myelitis is well documented in systemic Sjögren's syndrome, although several patients show clinical and immunological characteristics of NMOSD, making the diagnosis even more challenging [31]. Another underrecognized cause of longitudinally extensive myelitis, which often mimics NMOSD, is SC neurosarcoidosis. It is characterized by linear subpial, leptomeningeal, and nerve root enhancements and the "trident sign" (subpial enhancement combined with the enhancement of the central spinal canal) on axial images [28]. However, even if the linear dorsal subpial enhancement over two vertebral segments is typical of sarcoidosis, it can occur in patients with insufficient vitamin B12 levels [32]. Neuro-Behçet's disease is typically characterized by one or multiple noncontiguous lesions that can extend over two or three segments and show gadolinium enhancement and surrounding swelling. Two distinctive imaging features of myelopathy associated with Behçet's disease were described according to T2-weighted axial images: "Bagel sign" pattern: a central lesion with hypointense core and hyperintense rim with or without contrast enhancement, and "Motor neuron" pattern: a symmetric involvement of the anterior horn cells. Notably, Bagel's sign has not been observed in other forms of longitudinal myelopathy apart from Behçet's disease [33]. Finally, in addition to inflammatory myelopathies, SC involvement with hyperintense lesions in T2 sequences has been observed in viral myelitis [34], infarction [35], and spondylotic compressive myelopathy, with the lesions sometimes spanning more than three vertebral segments. For the latter condition, recent research has highlighted specific and often-overlooked radiological characteristics. Specifically, the presence of transverse pancake-like gadolinium enhancement, both adjacent to and just below the point of maximum stenosis, along with spindle-shaped T2 hyperintensity at the midpoint of the lesion, indicates spondylosis as the underlying cause of myelopathy [36]. Thus, it is essential to explore the causes of SC involvement through a comprehensive assessment of the medical history, physical examination, MRI, and other diagnostic investigations.

Practical considerations in spinal cord clinical imaging

The SC MRI is not as habitually utilized as the brain MRI in clinical practice to assess MS activity and evolution; it extends the imaging acquisition time, consequently increasing costs. Furthermore, the SC imaging typically has more limitations than that of the brain and is therefore generally preferred. The SC is a slender anatomical structure, measuring approximately 12–15 mm in diameter, surrounded by CSF within the spinal canal. The dynamic movement of the CSF flow, arterial pulsations, and physiological processes, such as heartbeat and breathing, result in notable displacement of the SC within the spinal canal, complicating SC imaging acquisition [37]. Thus SC imaging should consider the small dimensions of its cross-sectional area, the spatial variation in the magnetic field strength in this region, and physiological motions. The methods to mitigate these artifacts have been effectively implemented, such as cardiac gating, presaturation slabs, fast imaging sequences, and customized phase array coils [38]. Moreover, revised recommendations have been suggested to address the technical factors that affect lesion detection, and to establish optimized diagnostic protocols, thereby standardizing the lesion assessment [38,39].

Since the lesions can occur at each level, the imaging of the entire SC (cervical and thoracic) is recommended; however, these segments should not be acquired in a single sequence. As the MRI most often detects lesions of the SC in patients with MS in the cervical region [40], a single sagittal acquisition from cervical level 1 to the upper thoracic SC might be sufficient to improve the image resolution because of the smaller field of view compared to the whole SC coverage. Thus, this improves the confidence and utility of SC imaging. The conventional dual-echo spin-echo is the gold standard for sagittal imaging. As recommended, the protocols to detect the focal lesions, including the combination of T2-weighted sequences with STIR or sensitive inversion recovery sequences, offer the greatest contrast between the lesions and surrounding tissue, and the greatest sensitivity and specificity. The abnormalities identified on sagittal images should preferably be validated through the acquisition of axial images. To detect the lesions in the axial plane, it is essential to use high-resolution sequences because of the relatively small cross-sectional area of the SC, which typically ranges from 50 to 100 mm^2, depending on the level [41]. The axial imaging covering the entire SC can be obtained within an acceptable time frame with parallel imaging acceleration, and it detects more abnormalities than the sagittal imaging, particularly the small abnormalities in the peripheral regions of the SC [42] mainly occur in the lateral (68%–75% of the lesions) or dorsal columns (26%–38% of the lesions) [40].

While SC imaging is essential for the diagnostic evaluation of MS at the onset for several reasons (such as increased sensitivity, specificity, accuracy of diagnostic criteria, identification of asymptomatic lesions, exclusion of alternative diagnoses, and determination of patient prognosis), its application is not always required for disease and treatment monitoring in individuals with stable MS [6]. Conversely, in patients experiencing SC relapse, it is highly recommended to better evaluate the severity of the disease activity, and in those with disease progression, to exclude alternative diagnoses [6].

According to the recommendations for the use of gadolinium-based contrast agents in the diagnosis and monitoring of MS, the use of sagittal gadolinium-enhanced T1-weighted spin-echo sequences for SC MRI acquisition is advised for diagnostic purposes. However, gadolinium is not recommended for the routine monitoring of otherwise stable patients with MS because of its utility, cost, and safety [6,17]. Enhanced MS lesions are less commonly observed in the SC than in the brain. However, they are frequently associated with new clinical symptoms and are often accompanied by enhanced brain lesions [43]. Thus, if the clinical activity or disease progression cannot be explained by brain MRI findings, SC MRI is required [6].

1. Conventional MRI use in multiple sclerosis

Spinal cord magnetic resonance imaging assessment for patient's prognosis and disease monitoring

Several studies have demonstrated the prognostic value of SC lesions in both CIS and early MS. The presence of SC lesions increased the risk of MS in all CIS cases, especially in nonspinal CIS patients who did not meet the brain MRI diagnostic criteria and was associated with a shorter conversion to clinically definite MS, indicating that SC MRI assessment is a powerful tool in the diagnosis and prognosis [44]. Thus, SC lesions are independent predictors of MS in all CIS cases and contribute to the accumulation of disability. This is demonstrated by the data indicating that the presence of SC lesions increases the risk of achieving an expanded disability status scale (EDSS) [45] score ≥ 3.0 at 2 years, especially in nonspinal CIS patients [46], and the data indicating a higher number of baseline SC lesions in CIS patients who develop a secondary progressive course at 15 years compared with those who remain relapsing and remitting [47]. Similar data were reported in studies that included patients with early RRMS, indicating a negative prognostic role of SC damage [48]. The role of asymptomatic SC lesions in predicting the degree of future disability has also been demonstrated [49], highlighting the importance of SC MRI assessment for disease monitoring. Nevertheless, SC imaging for this purpose is less routinely used for several reasons, most notably because incorporating SC MRI into routine disease monitoring significantly increases the acquisition time and costs. Furthermore, the frequency of new SC lesions is significantly lower than that of new brain lesions [50]. Additionally, there is a correlation between the development of new lesions in the brain and the appearance of new lesions in the SC, which could reduce the necessity of performing SC MRI [6,51]. Despite these considerations, SC MRI is frequently performed in clinical practice because of its recognized independent value in monitoring treatment responses and disease progression, as supported by more recent studies. A retrospective analysis of the presence of no evidence of disease activity (NEDA-3 status) [52], evaluated by the three assessment components (absence of clinical relapses; no EDSS progression; no radiological activity) in a cohort of people with MS under first-line DMTs, also considering SC damage, concluded that SC imaging led to a slight but significant change in the proportion of patients classified as clinically and radiologically stable according to the NEDA-3 definition [53]. A recent study evaluated the frequency of SC inflammatory MRI activity occurring independently of brain activity and investigated the association between MRI activity in the SC, alone or in combination with brain activity, relapse, and disability accrual. This study found that SC activity (new T2 lesions) occurs independently of the brain activity in 32.3% of patients with MS, with 16.1% of patients presenting with asymptomatic, isolated SC activity (gadolinium [Gd] + lesions), and SC lesion burden associated with a higher risk of concomitant relapses and disability progression [54]. This finding was in line with the previous results that reported asymptomatic cord lesions in one-fourth of clinically stable MS patients and SC lesion activity alone in 10% of clinically stable MS patients [13,55,56]. The presence of new T2 lesions at the SC level is associated with an increased risk of disability accrual compared to new T2 lesions in the brain alone [54,57,58]. Due to its prognostic significance regarding relapses and disability progression, the detection of new asymptomatic and isolated SC lesions substantially impacted treatment choices [59]. Conversely, SC Gd + lesions did not predict the worsening of future disability.

This highlights the role of chronic inflammatory processes (SC slowly expanding lesions) rather than acute inflammation as drivers of long-term disability progression [60].

In this context, the identification of the evolution of SC damage, irrespective of brain lesions, could significantly influence the management of MS and allow a more accurate risk stratification and treatment optimization [61].

Future directions

MS is a condition that has primarily reaped the benefits of sophisticated SC imaging methods, including measurements of SC atrophy, diffusion tensor imaging (DTI), magnetization transfer ratio (MTR), myelin visualization, proton magnetic resonance spectroscopy (Mrs), and functional magnetic resonance imaging (fMRI) [39]. The advances in quantitative imaging techniques to evaluate neuroaxonal integrity, myelin content, metabolic changes, and functional connectivity, have provided new insights into the mechanisms of damage in MS (Table 10.2). First, the quantification of SC atrophy, with the measurement of cross-sectional SC area over time, is the most attractive advanced technique for clinical applications because of its association with disability in MS. SC atrophy is a common occurrence among individuals with MS, and it is evident from the early stages of the disease, including CIS [62,63]. However, it is more pronounced in patients with progressive disease than in those with RRMS, representing a marker of disease progression [64,65]. The rate of SC atrophy is greater than that of brain atrophy (-1.78% vs -0.5% per year) [65], and reflects a neuroaxonal loss. Moreover, a recent study demonstrated the differential impact of brain and SC damage on the worsening of the disability in MS, revealing the leading role of SC atrophy in determining motor disability [66,67]. SC atrophy is generally measured as the cross-sectional area at the cervical level, which is least affected by movement artifacts, yields the most reproducible results, and provides the best clinical correlates [44]. Atrophy assessment can be performed on a variety of sequences acquired by

TABLE 10.2 Quantitative spinal cord imaging techniques.

Advanced MRI technique	Aims of the technique application
SC atrophy	To measure the cross-sectional area at the cervical level, as a surrogate measure of neuroaxonal loss
DTI	To study neuroaxonal integrity with characterization of specific tissue and cell types changes associated with MS pathology
MTR	To estimate myelin/water content and neuroaxonal integrity
MWI	To quantify myelin water fraction, a surrogate measure of myelin content
Proton Mrs	To estimate the levels of key metabolites (N-acetyl-aspartate, myoinositol, glutamate, and its precursors, sodium)
fMRI	To explore compensatory changes of demyelination/axonal loss, network reorganization, and tissue plasticity

different MRI scanners using different methods for SC image segmentation and atrophy calculation (manual, semiautomated, and automated methods) [68]. An increasing number of studies have focused on the importance of SC atrophy as a biomarker of disability progression [69,70]. Among the SC imaging metrics, atrophy is distinct, as it has been employed as a secondary endpoint in clinical trials focusing on progressive MS. SC atrophy may have clinical relevance and enhanced sensitivity for the surveillance of patients with progressive MS, thus making it a potential candidate for a primary endpoint in phase 2 trials [71]. Conversely, its use to improve disease management has not yet entered clinical practice because of its cost, acquisition time, and other technical aspects.

Advanced imaging techniques are now used in investigative studies to explore the microstructural abnormalities related to neurodegeneration, with the benefit of better understanding the mechanisms of disease evolution and developing new targets for therapeutics [68,72]. These techniques include those that study neuroaxonal integrity (DTI), myelin content (MTR), myelin water imaging (MWI), metabolic changes (Mrs), and functional connectivity; the clinical maturity of these methods remains limited [72]. The DTI allows for the characterization of specific tissue- and cell-type changes associated with SC pathology, providing evidence of the strong contribution of SC microstructural changes to irreversible disability [73,74]. Despite the simple acquisition of DTI images of the SC, this examination has a low pathological specificity as its main limitation. Further advances in acquisition, preprocessing, and biophysical modeling are needed to make SC diffusion MRI a useful biomedical tool for MS. Among the MRI techniques that reflect the myelin content, MTR is the most promising; it indirectly estimates myelin content, neuroaxonal integrity, and water content. A correlation between lower MTR and higher lesion load has been previously reported. Lower MTR values have been observed in patients with MS and a higher EDSS, independent of the lesion load [44], allowing the capture of aspects of MS pathology that are not detectable with conventional MRI. Similarly, reduced MTR values were observed in MS patients independent of SC atrophy [75]. Beyond this, MWI for the quantification of myelin water fraction, a surrogate measure of myelin content, was recently identified by MAGNIMS [72], along with MTR, as among the most "mature," and toward clinical applicability and utility among several advanced SC imaging techniques. The MTR and MWI appear to offer complementary insights. Although MTR lacks pathological specificity, it is widely accessible and quick to acquire. In contrast, MWI provides a more specific marker for myelin but requires more complex postprocessing procedures [68]. Finally, Mrs estimates the levels of metabolites by proton magnetic resonance spectroscopy, identifying the levels of the main markers of neuronal integrity (N-acetyl-aspartate, myoinositol, glutamate, and its precursors). One particularly promising application of Mrs is the quantification of sodium, which may contribute to neuroinflammation and neurodegeneration in MS [76]; research on this topic is currently ongoing.

Finally, given that the SC plays a pivotal role in transmitting and conducting neural impulses to and from the brain and serves as the principal juncture between the central and peripheral nervous systems, novel investigations into the functional linkage of the SC have emerged. This was accomplished by assessing inherent fluctuations in the SC blood oxygenation levels using fMRI. These studies suggest that, similar to the brain, cord activity at rest is organized into distinct, synchronized functional networks among gray matter regions, most likely related to the motor and sensory systems [39,72]. Recently, a study showed that SC lesions are associated with local connectivity alterations with differential effects depending on columnar localization and suggested that functional SC networks are generally intact in RRMS, and that the lesions are associated with focal abnormalities in intrinsic connectivity [77]. Given these promising results, fMRI is a promising method for exploring compensatory changes in demyelination/axonal loss, network reorganization, and SC plasticity.

In summary, novel approaches to quantitative MRI can offer insights that could enhance patient stratification, assessment of treatment response, and tracking of subclinical disease progression. However, current quantitative MRI techniques still require developmental strides to reach full clinical maturity [72].

Conclusions

Currently, SC imaging is attainable and serves as a relevant resource for the diagnostic evaluation of individuals with MS, offering insights into disease evolution, prediction, and ongoing condition monitoring [78]. Conventional SC MRI is a valuable tool for assessing individuals with MS because of its high sensitivity in visualizing demyelinating lesions within this structure. However, an accurate approach is crucial for preventing artifacts and ensuring the acquisition of high-quality images. Advanced quantitative techniques, particularly SC atrophy measurements [79], offer the ability to gauge the extent of tissue damage and provide a reliable assessment of the true disease burden. Given that SC atrophy mirrors neurodegenerative processes and advances more rapidly in patients manifesting disability accumulation and in progressive forms, the quantification of SC atrophy could serve as a meaningful measure for tracking the impact of new therapies that are not solely focused on preventing inflammation and demyelination. At present, the evaluation of SC atrophy and the utilization of other innovative quantitative MRI techniques have the potential to identify clinically significant markers for disease progression. However, this necessitates validation prior to its establishment as a study endpoint in clinical investigations, and it still needs to acquire clinical maturity [72].

References

[1] Filippi M, Bar-Or A, Piehl F, et al. Multiple sclerosis. Nat Rev Dis Primers 2018;4(1):43.
[2] Bonacchi R, Pagani E, Meani A, et al. Clinical relevance of multiparametric MRI assessment of cervical cord damage in multiple sclerosis. Radiology 2020;296(3):605−15.
[3] Rocca MA, Margoni M, Battaglini M, et al. Emerging perspectives on MRI application in multiple sclerosis: moving from pathophysiology to clinical practice. Radiology 2023;307(5):e221512.
[4] Stankiewicz JM, Neema M, Alsop DC, et al. Spinal cord lesions and clinical status in multiple sclerosis: a 1.5 T and 3 T MRI study. J Neurol Sci 2009;279:99−105.
[5] Chen Y, Haacke EM, Bernitsas E. Imaging of the spinal cord in multiple sclerosis: past, present, future. Brain Sci 2020;10(11):857.
[6] Wattjes MP, Ciccarelli O, Reich DS, et al. MAGNIMS-CMSC-NAIMS consensus recommendations on the use of MRI in patients with multiple sclerosis 2021. Lancet Neurol 2021;20(8):653−70.
[7] Combes B, Kerbrat A, Ferre JC, et al. Focal and diffuse cervical spinal cord damage in patients with early relapsing-remitting MS: a multicentre magnetisation transfer ratio study. Mult Scler 2019;25(8):1113−23.

[8] Eden D, Gros C, Badji A, et al. Spatial distribution of multiple sclerosis lesions in the cervical spinal cord. Brain 2019;142(3):633−46.
[9] Lycklama G, Thompson A, Filippi M, et al. Spinal-cord MRI in multiple sclerosis. Lancet Neurol 2003; 2(9):555−62.
[10] Elliott C, Arnold DL, Chen H, et al. Patterning chronic active demyelination in slowly expanding/evolving white matter MS lesions. AJNR Am J Neuroradiol 2020;41(9):1584−91.
[11] Thompson AJ, Banwell BL, Barkhof F, et al. Diagnosis of multiple sclerosis: 2017 revisions of the McDonald criteria. Lancet Neurol 2018;17(2):162−73.
[12] Bergers E, Bot JCJ, van der Valk P, et al. Diffuse signal abnormalities in the spinal cord in multiple sclerosis: direct postmortem in situ magnetic resonance imaging correlated with in vitro high-resolution magnetic resonance imaging and histopathology. Ann Neurol 2002;51:652−6.
[13] Zecca C, Disanto G, Sormani MP, Riccitelli GC, Cianfoni A, Del Grande F, et al. Relevance of asymptomatic spinal MRI lesions in patients with multiple sclerosis. Mult Scler 2016;22(6):782−91.
[14] Kitley J, Waters P, Woodhall M, et al. Neuromyelitis optica spectrum disorders with aquaporin4 and myelin-oligodendrocyte glycoprotein antibodies. A comparative study. JAMA Neurol 2014;71:276−83.
[15] Calabrese M, Marastoni D, Crescenzo F, et al. Early multiple sclerosis: diagnostic challenges in clinically and radiologically isolated syndrome patients. Curr Opin Neurol 2021;34(3):277−85.
[16] Sombekke MH, Wattjes MP, Balk LJ, et al. Spinal cord lesions in patients with clinically isolated syndrome: a powerful tool in diagnosis and prognosis. Neurology 2013;80(1):69−75.
[17] Traboulsee A, Li D. Addressing concerns regarding the use of gadolinium in a standardized MRI protocol for the diagnosis and follow-up of multiple sclerosis. AJNR Am J Neuroradiol 2016;37(12):E82−3.
[18] Rovira A, De Stefano N. MRI monitoring of spinal cord changes in patients with multiple sclerosis. Curr Opin Neurol 2016;29:445−52.
[19] Hagens MHJ, Burggraaff J, Kilsdonk ID, et al. Three-Tesla MRI does not improve the diagnosis of multiple sclerosis: a multicenter study. Neurology 2018;91:e249−57.
[20] Philpott C, Brotchie P. Comparison of MRI sequences for evaluation of multiple sclerosis of the cervical spinal cord at 3 T. Eur J Radiol 2011;80:780−5.
[21] Bot JC, Barkhof F, Lycklama à Nijeholt GJ, et al. Comparison of a conventional cardiac-triggered dual spin-echo and a fast STIR sequence in detection of spinal cord lesions in multiple sclerosis. Eur Radiol 2000;10:753−8.
[22] Stroman PW, Ryner LN. Functional MRI of motor and sensory activation in the human spinal cord. Magn Reson Imaging 2001;19(1):27−32.
[23] Ciccarelli O, Cohen JA, Reingold SC, et al. Spinal cord involvement in multiple sclerosis and neuromyelitis optica spectrum disorders. Lancet Neurol 2019;18:185−97.
[24] Kim HJ, Paul F, Lana-Peixoto MA, et al. MRI characteristics of neuromyelitis optica spectrum disorder: an international update. Neurology 2015;84(11):1165−73.
[25] Hyun JW, Kim SH, Jeong IH, et al. Bright spotty lesions on the spinal cord: an additional MRI indicator of neuromyelitis optica spectrum disorder? J Neurol Neurosurg Psychiatry 2015;86(11):1280−2.
[26] Zalewski NL, Morris PP, Weinshenker BG, et al. Ringenhancing spinal cord lesions in neuromyelitis optica spectrum disorders. J Neurol Neurosurg Psychiatry 2017;88:218−25.
[27] Kitley JL, Leite MI, George JS, et al. The differential diagnosis of longitudinally extensive transverse myelitis. Mult Scler 2012;18:271−85.
[28] Flanagan EP, Kaufmann TJ, Krecke KN, et al. Discriminating long myelitis of neuromyelitis optica from sarcoidosis. Ann Neurol 2016;79(3):437−47.
[29] Cortese R, Battaglini M, Prados F, et al. MAGNIMS Study Group. Clinical and MRI measures to identify non-acute MOG-antibody disease in adults. Brain 2023;146(6):2489−501. Available from: https://doi.org/10.1093/brain/awac480 PMID: 36515653.
[30] Jain S, Khormi A, Sangle SR, et al. Transverse myelitis associated with systemic lupus erythematosus (SLE-TM): a review article. Lupus 2023; 9612033231185612.
[31] Kim SM, Waters P, Vincent A, et al. Sjogren's syndrome myelopathy: spinal cord involvement in Sjogren's syndrome might be a manifestation of neuromyelitis optica. Mult Scler 2009;15(9):1062−8.
[32] Paliwal VK, Malhotra HS, Chaurasia RN, et al. "Anchor"-shaped bright posterior column in a patient with vitamin B12 deficiency myelopathy. Postgrad Med J 2009;85(1002):186.

[33] Uygunoglu U, Zeydan B, Ozguler Y, et al. Myelopathy in Behçet's disease: the bagel sign. Ann Neurol 2017;82(2):288−98.
[34] Goh C, Phal PM, Desmond PM. Neuroimaging in acute transverse myelitis. Neuroimaging Clin N Am 2011;21(4):951−73. Available from: https://doi.org/10.1016/j.nic.2011.07.010.
[35] Zalewski NL, Rabinstein AA, Wijdicks EFM, et al. Spontaneous posterior spinal artery infarction: an under-recognized cause of acute myelopathy. Neurology 2018;91(9):414−17.
[36] Flanagan EP, Krecke KN, Marsh RW, et al. Specific pattern of gadolinium enhancement in spondylotic myelopathy. Ann Neurol 2014;76(1):54−65.
[37] Mikulis DJ, Wood ML, Zerdoner OA, et al. Oscillatory motion of the normal cervical spinal cord. Radiology 1994;192(1):117−21.
[38] Stroman PW, Wheeler-Kingshott C, Bacon M, et al. The current state-of-the-art of spinal cord imaging: methods. Neuroimage 2014;84:1070−81.
[39] Wheeler-Kingshott CA, Stroman PW, Schwab JM, et al. The current state-of-the-art of spinal cord imaging: applications. Neuroimage 2014;84:1082−93.
[40] Weier K, Mazraeh J, Naegelin Y, et al. Biplanar MRI for the assessment of the spinal cord in multiple sclerosis. Mult Scler 2012;18(11):1560−9.
[41] Bronskill MJ, McVeigh ER, Kucharczyk W, et al. Syrinx-like artifacts on MR images of the spinal cord. Radiology 1988;166(2):485−8.
[42] Breckwoldt MO, Gradl J, Hähnel S, et al. Increasing the sensitivity of MRI for the detection of multiple sclerosis lesions by long axial coverage of the spinal cord: a prospective study in 119 patients. J Neurol 2017;264(2):341−9.
[43] Thorpe JW, Kidd D, Moseley IF, et al. Spinal MRI in patients with suspected multiple sclerosis and negative brain MRI. Brain 1996;119:709−14.
[44] Oh J, Saidha S, Chen M, et al. Spinal cord quantitative MRI discriminates between disability levels in multiple sclerosis. Neurology 2013;80:540−7.
[45] Kurtzke JF. Rating neurologic impairment in multiple sclerosis: an expanded disability status scale (EDSS). Neurology 1983;33(11):1444−52.
[46] Arrambide G, Rovira A, Sastre-Garriga J, et al. Spinal cord lesions: a modest contributor to diagnosis in clinically isolated syndromes but arelevant prognostic factor. Mult Scler 2018;24(3):301−12.
[47] Brownlee WJ, Altmann DR, Prados F, et al. Early imaging predictors of long-term outcomes in relapse-onset multiple sclerosis. Brain 2019;142(8):2276−87.
[48] D'Amico E, Patti F, Leone C, et al. Negative prognostic impact of MRI spinal lesions in the early stages of relapsing-remitting multiple sclerosis. Mult Scler J Exp Transl Clin 2016;2 2055217316631565.
[49] Nijeholt GJ, van Walderveen MA, Castelijns JA, et al. Brain and spinal cord abnormalities in multiple sclerosis. Correlation between MRI parameters, clinical subtypes and symptoms. Brain 1998;121(Pt 4):687−97.
[50] Bot JC, Barkhof F, Polman CH, et al. Spinal cord abnormalities in recently diagnosed MS patients: added value of spinal MRI examination. Neurology 2004;62(2):226−33. Available from: https://doi.org/10.1212/wnl.62.2.226.
[51] Tench CR, Morgan PS, Jaspan T, et al. Spinal cord imaging in multiple sclerosis. J Neuroimaging 2005;15(4 Suppl):94S−102S.
[52] Giovannoni G, Tomic D, Bright JR, et al. "No evident disease activity": the use of combined assessments in the management of patients with multiple sclerosis. Mult Scler 2017;23(9):1179−87.
[53] Di Sabatino E, Gaetani L, Sperandei S, et al. The no evidence of disease activity (NEDA) concept in MS: impact of spinal cord MRI. J Neurol 2022;269(6):3129−35.
[54] Ruggieri S, Prosperini L, Petracca M, et al. The added value of spinal cord lesions to disability accrual in multiple sclerosis. J Neurol 2023;270(10):4995−5003.
[55] Andelova M, Uher T, Krasensky J, et al. Additive effect of spinal cord volume, diffuse and focal cord pathology on disability in multiple sclerosis. Front Neurol. 2019;10:820.
[56] Rovaris M, Judica E, Sastre-Garriga J, et al. Large-scale, multicentre, quantitative MRI study of brain and cord damage in primary progressive multiple sclerosis. Mult Scler 2008;14(4):455−64.
[57] Kappos L, Moeri D, Radue EW, et al. Predictive value of gadolinium-enhanced magnetic resonance imaging for relapse rate and changes in disability or impairment in multiple sclerosis: a meta-analysis. Gadolinium MRI Meta-analysis Group. Lancet 1999;353(9157):964−9.

[58] Zhang Y, Cofield S, Cutter G, Krieger S, Wolinsky JS, Lublin F. Predictors of disease activity and worsening in relapsing-remitting multiple sclerosis. Neurol Clin Pract. 2022;12(4):e58e65.

[59] Lorefice L, Piras C, Sechi V, et al. Spinal cord MRI activity in multiple sclerosis: Predictive value for relapses and impact on treatment decisions. J Neurol Sci 2024;462:123057. Available from: https://doi.org/10.1016/j.jns.2024.123057.

[60] Preziosa P, Pagani E, Meani A, et al. Slowly expanding lesions predict 9-Year multiple sclerosis disease progression. Neurol Neuroimmunol Neuroinflamm 2022;9(2):e1139.

[61] Cortese R, Ciccarelli O. Clinical monitoring of multiple sclerosis should routinely include spinal cord imaging - yes. Mult Scler 2018;24(12):1536–7.

[62] Biberacher V, Boucard CC, Schmidt P, et al. Atrophy and structural variability of the upper cervical cord in early multiple sclerosis. Mult Scler 2015;21:875–84.

[63] Hagström IT, Schneider R, Bellenberg B, et al. Relevance of early cervical cord volume loss in the disease evolution of clinically isolated syndrome and early multiple sclerosis: a 2-year follow-up study. J Neurol 2017;264:1402–12.

[64] Valsasina P, Aboulwafa M, Preziosa P, et al. Cervical cord T1weighted hypointense lesions at MR imaging in multiple sclerosis: relationship to cord atrophy and disability. Radiology 2018;288:234–44.

[65] Bonati U, Fisniku LK, Altmann DR, et al. Cervical cord and brain grey matter atrophy independently associate with long-term MS disability. J Neurol Neurosurg Psychiatry 2011;82:471–2.

[66] Zivadinov R, Bergsland N. Cervical spinal cord lesions and atrophy versus brain measures in explaining physical disability in multiple sclerosis. Radiology 2020;296(3):616–18.

[67] Ruggieri S, Petracca M, De Giglio L, et al. A matter of atrophy: differential impact of brain and spine damage on disability worsening in multiple sclerosis. J Neurol 2021;268(12):4698–706.

[68] Moccia M, Ruggieri S, Ianniello A, et al. Advances in spinal cord imaging in multiple sclerosis. Ther Adv Neurol Disord 2019;12:1–19.

[69] Bischof A, Papinutto N, Keshavan A, et al. Spinal cord atrophy predicts progressive disease in relapsing multiple sclerosis. Ann Neurol 2022;91(2):268–81.

[70] Zeydan B, Rocca MA, Kantarci OH, Filippi M. Spinal cord atrophy is a preclinical marker of progressive MS. Ann Neurol 2022;91(5):734–5.

[71] Cawley N, Tur C, Prados F, et al. Spinal cord atrophy as a primary outcome measure in phase II trials of progressive multiple sclerosis. Mult Scler 2018;24(7):932–41.

[72] Granziera C, Wuerfel J, Barkhof F, et al. Quantitative magnetic resonance imaging towards clinical application in multiple sclerosis. Brain 2021;144(5):1296–311.

[73] Kearney H, Schneider T, Yiannakas MC, et al. Spinal cord grey matter abnormalities are associated with secondary progression and physical disability in multiple sclerosis. J Neurol Neurosurg Psychiatry 2021;86:608–14.

[74] Kearney H, Schneider T, Yiannakas MC, et al. Spinal cord grey matter abnormalities are associated with secondary progression and physical disability in multiple sclerosis. J Neurol Neurosurg Psychiatry 2015;86(6):608–14.

[75] Combës B, Kerbrat A, Ferré JC, et al. Focal and diffuse cervical spinal cord damage in patients with early relapsing–remitting MS: a multicentre magnetisation transfer ratio study. Mult Scler 2018; 1352458518781999.

[76] Huhn K, Engelhorn T, Linker RA, Nagel AM. Potential of sodium MRI as a biomarker for neurodegeneration and neuroinflammation in multiple sclerosis. Front Neurol 2019;10:84.

[77] Conrad BN, Barry RL, Rogers BP, et al. Multiple sclerosis lesions affect intrinsic functional connectivity of the spinal cord. Brain 2018;141(6):1650–64.

[78] Rocca MA, Preziosa P, Filippi M. What role should spinal cord MRI take in the future of multiple sclerosis surveillance? Expert Rev Neurother 2020;20(8):783–97.

[79] Rocca MA, Battaglini M, Benedict RH, et al. Brain MRI atrophy quantification in MS: from methods to clinical application. Neurology 2017;88(4):403–13.

PART 2

Non-conventional MRI use in multiple sclerosis

CHAPTER 11

Magnetization transfer imaging in multiple sclerosis

Matteo Mancini and Mara Cercignani

Cardiff University Brain Research Imaging Centre, Cardiff University, Cardiff, Wales, United Kingdom

OUTLINE

Introduction	192
History	192
Magnetization transfer experiment	192
Clinical use of magnetization transfer contrast	193
Quantifying the magnetization transfer effect	193
Magnetization transfer ratio	194
Quantitative magnetization transfer models	194
Magnetization transfer saturation	196
Inhomogeneous magnetization transfer	197
Validation of magnetization transfer–derived parameters as myelin markers	198
Magnetization transfer ratio in multiple sclerosis	198
Lesions	198
Normal-appearing white matter	198
Gray matter	199
Quantitative magnetization transfer in multiple sclerosis	200
Magnetization transfer saturation in multiple sclerosis	200
Inhomogeneous magnetization transfer in multiple sclerosis	200
Magnetization transfer in spinal cord and optic nerve	201
Spinal cord	201
Optic nerve	201
Conclusions	202
References	202

Introduction

Magnetization transfer (MT) refers to the exchange of magnetization occurring between spins sitting in different molecular environments. This exchange means that any external interaction with one of the two proton pools results in changes to the other pool. As protons attached to large molecules are typically "magnetic resonance imaging (MRI)–invisible" due to their short T_2, MT can be exploited to indirectly probe them.

History

The ability of nuclei in *different molecular compounds* to exchange magnetization was first demonstrated using nuclear magnetic resonance (NMR) spectroscopy in a system consisting of two weakly coupled sets of spins [1]. The concept was later extended to protons with the same resonant frequency and different transverse relaxation time T_2 [2], such as water protons and macromolecular protons in collagen or muscle. This phenomenon is typically explained as follows: the relatively "immobile" macromolecular protons produce a signal that decays too fast to be detected with standard MR sequences. This is due to their very short T_2, which causes their signal to decay faster than the typical echo times achievable on a clinical scanner. In biological tissue, macromolecular protons are mainly found in lipids and proteins. Water protons, by contrast, are the main contributors to the signal typically visible on magnetic resonance imaging (MRI). They are in constant motion, and when they come in contact with macromolecular protons, they can exchange magnetization with them. Through this magnetization exchange, the magnetization state of the water pool protons can affect that of the macromolecular protons, and vice versa. The myelin bilayer is made up of approximately 80% lipid and 20% protein [3], which makes MT of great relevance for the study of demyelinating disorders. In 1989, magnetization transfer *imaging* was introduced by Wolff and Balaban [4].

Magnetization transfer experiment

There are different ways of probing MT, but the simplest one relies on the differently shaped resonance lines of water (liquid pool) and macromolecular protons (Fig. 11.1). The width of such lines is inversely proportional to the protons T_2, and therefore macromolecular protons tend to have much broader line shape than water protons. This property can be exploited to selectively interact with the macromolecular proton pool by using radio frequency (RF) pulses tuned few kilohertz (kHz) away from the central resonance frequency (also known as *off-resonance* pulses). These pulses "saturate" the macromolecular pool, that is, cause the number of up-spins and down-spins to be the same, thus nulling the magnetization vector for the macromolecular protons. This saturation can be partially transferred to the liquid protons via MT, thus reducing the measured MRI signal (see Fig. 11.1). The resulting attenuation depends on the density of macromolecular protons and their ability to exchange magnetization. As a consequence, tissues with higher concentration of macromolecules will appear darker than tissues with primarily liquid content. This phenomenon creates a source of contrast in MRI which is independent from T_1 and T_2, and, importantly, is sensitive to the myelin content in the brain.

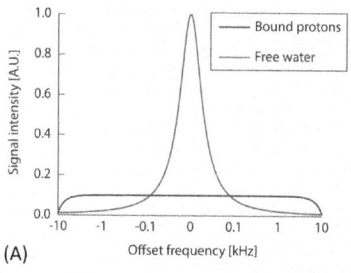 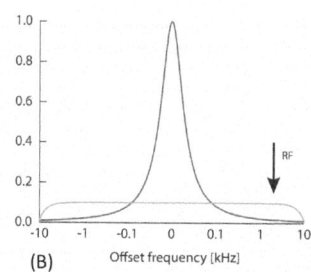 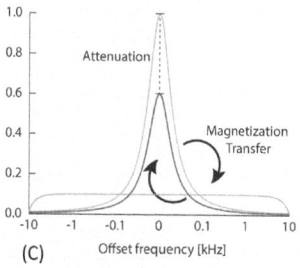

FIGURE 11.1 Qualitative representation of the spectra of free water and bound protons before and after RF irradiation. The spectra of the two pools are centered around the same frequency, but the free water protons present a narrow line shape (blue), while the macromolecular protons (red) present a broad line shape (A). If an RF pulse is applied with a frequency offset larger than 1–2 kHz, the free water pool is unaffected, while the macromolecular pool magnetization is saturated (B). As a result of the magnetization transfer between macromolecular protons and free water, the saturation is partially transferred to the latter pool, and attenuates the signal from the water itself (C).

In the early NMR experiments, off-resonance saturation was achieved using the so-called continuous wave irradiation, that is, applying constant RF energy for a long period (typically of the order of 1 second) with relatively low power, using a second transmitter system. Due to large power deposition into the patient, this method is not practical or safe for clinical use, and requires dedicated hardware, not available as standard on clinical MRI equipment. For this reason, on human MRI scanners, the so-called *pulsed MT* is used [5]. In this case, regular RF pulses (similar to those used for excitation and refocusing but typically larger) are used. They are tuned a few kHz (usually 1–2 kHz) from the resonant frequency and occur at regular intervals so that their effect can add up and create sufficient saturation. MT modules can be added to almost any MRI sequence, but they are most commonly used in spoiled gradient echo (fast low angle shot [FLASH], spoiled gradient-recalled [SPGR]) for efficiency. Alternative approaches to probe the MT effects rely on on-resonance saturation [6,7].

Clinical use of magnetization transfer contrast

MT increases the contrast between tissues rich in macromolecular content and fluids. This feature finds applications both in the brain and the body. Examples include musculoskeletal imaging (to provide increased contrast between cartilage and synovial fluid) and time-of-flight MRI angiography, where it helps to suppress background tissue [8]. In conjunction with gadolinium injections, it can improve the visibility of gadolinium-enhancing lesions [9].

Quantifying the magnetization transfer effect

The sensitivity of MT MRI to myelin density in the brain quickly attracted the interest of researchers in the field of multiple sclerosis (MS). The availability of a noninvasive technique potentially able to quantify the degree of demyelination clearly offers great

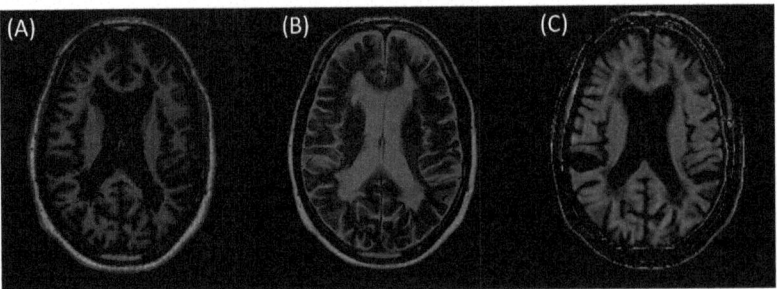

FIGURE 11.2 T1-weighted (A), T2-weighted (B), and magnetization transfer ratio (MTR) (C) images of a patient with secondary progressive multiple sclerosis (MS). Demyelinating lesions can be clearly seen on all scans, and they appear as hypointensities on the MTR map. In this example, the MTR of the normal appearing white matter ranges between 39 and 46 percentage units (p.u.), while the average values within lesions ranges between 17 and 28 p.u.

potential, and also the challenge of developing suitable methods to translate MT contrast into a quantitative and meaningful measurement. Increasingly complex approaches were proposed over the years, with variable degree of requirements in terms of specialized software and scan time.

Magnetization transfer ratio

The simplest approach proposed to quantify the MT effects in biological tissues is the magnetization transfer ratio (MTR). To compute MTR, only two proton-density acquisitions are required, one with an off-resonance saturation pulse and the other without. From the resulting images, MTR is then calculated pixel-wise using the following formula:

$$\text{MTR} = \frac{M_0 - M_S}{M_0} \cdot 100$$

where M_s and M_0 refer, respectively, to the images acquired with and without saturation. An example of an MTR map is shown in Fig. 11.2 (panel C). Each voxel in the resulting MTR map will contain a value from 0 to 100, providing a measure of the related macromolecular content. In brain imaging, the cerebrospinal fluid (CSF) will show values close to zero—because of the absence of macromolecules—while white and gray matter areas will show values depending on the acquisition parameters, with white matter showing higher MTR than gray matter. The drawbacks of this very simple approach are (1) the lack of a direct and clear biological or physical interpretation [10], and (2) the dependency on the acquisition parameters, in particular the magnetic field strength [11], the off-resonance pulse characteristics [12], and both the repetition time and flip angles [13].

Quantitative magnetization transfer models

To overcome the limitations of MTR, it is necessary to analytically describe the MT phenomena. With this goal in mind, the methods developed for quantitative magnetization transfer (qMT) leverage compartment modeling to formulate and fit the physical parameters underlying those phenomena. Historically, the first qMT method has been developed by Henkelman et al. [14] for continuous wave RF irradiation experiments using agar gels. In these experiments, the system under study is composed of two compartments or pools: the liquid pool (A) and the macromolecular pool (B). Each pool can be described by the density of spins (M_0^A, M_0^A) and the proportion of saturated spins, which depends on the past irradiation phenomena. The irradiation itself also causes multiple magnetization modifications and exchange phenomena—specifically, after the irradiation pulse:

- T1 relaxation causes the longitudinal magnetization of both pools to increase (with relaxation rates R_A and R_B, respectively);
- absorption of off-resonance irradiation counteracts T1 relaxation by saturation and therefore reducing the longitudinal magnetization (with rates R_{RFA} and R_{RFB});
- the two pools exchange magnetization with rate constant R, which results in the superposition of the exchanges A → B (with rate R_{M0B}) and B → A (with rate R_{M0A}).

Coupled Bloch equations can then be written for this system that combines all these different effects. In the continuous wave regime adopted by Henkelman et al., this system can be fully described by solving analytically in the steady state the coupled Bloch equations (for a detailed derivation see Cercignani et al. [15]), providing estimates of each pool's parameters. As mentioned, continuous wave experiments are not viable for clinical applications, and unfortunately pulsed MT experiments would require a numerical approach to solve the same equations, which is computationally expensive. For this reason, several approximations have been proposed for the two-pool model in pulsed regime [16–18].

Fitting the two-pool model pixel-wise results in five different parameter maps: the macromolecular pool fraction; the transverse relaxation times for each pool (T_{2A} and T_{2B}); the pseudo-first-order exchange rates between the pools (RM_{0A} and RM_{0B}, also known as k_r and k_f, respectively). Among these maps, the one that has received the most attention is the macromolecular pool fraction, as it has been shown to reflect myelin content [19]. Fig. 11.3 shows the main parametric maps derived from qMT.

The fundamental assumption of the two-pool model is that there is a single compartment of water. This is not the case in brain tissue, where water molecules can belong to the intra- and extra-cellular space, or be trapped in the myelin sheath wrapping around the axons. It is interesting to notice that a different quantitative MRI approach aiming at quantifying myelin, namely T_2 relaxometry, offers a complementary perspective: in T_2 relaxometry, one can estimate the size of the intra/extra-cellular and myelin water compartments, at the cost of neglecting the magnetization exchange between pools. A combination of these two approaches, resulting in a four-pool model, has been proposed [20,21], but the number of parameters to be estimated imposes extremely long scan times, which do not make it a viable approach for clinical applications.

Another important issue to keep in mind is related to partial voluming effects: as the spatial resolution for qMT acquisitions can go from 1 to 3 mm, the presence of CSF in a voxel can introduce a bias in the fitting, as the MT effect in CSF is negligible. To overcome this issue, Mossahebi et al. [22] have proposed a three-pool model, where the third pool represents a nonexchanging water compartment and takes into account CSF contributions.

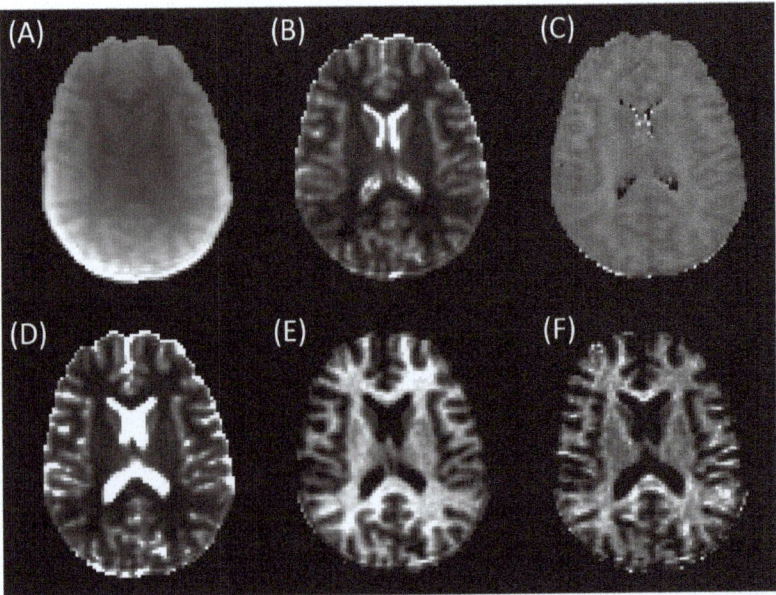

FIGURE 11.3 Quantitative magnetization transfer (MT) parametric maps derived from fitting an approximation of Henkelman's model to human brain data. Proton density (A), T_{1A} (B), T_{2B} (C), T_{2A} (D), F (E), and RM_{0A} (F).

Arguably, the main drawback limiting a wider use of qMT is the long acquisition time required. To overcome this drawback, Yarnykh and Yuan have proposed additional constraints to shorten the overall acquisition [23]. The same group also proposed a reduced protocol (consisting of one MT-weighted measurement and a reference scan) to estimate exclusively the macromolecular pool fraction [24,25].

Magnetization transfer saturation

In order to overcome the limits of MTR and keep at the same time a relatively shorter acquisition time compared to qMT, a different method focuses on the magnetization transfer saturation (MT_{sat})—a phenomenological parameter with no biophysical meaning, but overcoming some of the main shortfalls of MTR. As proposed by Helms et al. [13], this approach leverages a protocol that requires three commonly available acquisition sequences (T1-weighted, PD-weighted, MT-weighted) and allows to estimate not only MT_{sat}, but also the longitudinal relaxation time T_1 and the apparent proton density. Despite lacking a direct biological interpretation, MT_{sat} inherently corrects for T_1 relaxation and on-resonance excitation phenomena. It should be remarked that changes to the acquisition sequence will result in changes to the estimated MT_{sat}.

This approach has become increasingly popular in the last year, and a contributing factor for its popularity was the release of multiparameter mapping (MPM) protocols and processing tools from Weiskopf and colleagues [26,27]. The MPM protocols also include solutions based on vendor sequences already available in MRI scanners (https://hmri-group.github.io/hMRI-toolbox/), bypassing the need for tailored sequence development.

Inhomogeneous magnetization transfer

Inhomogeneous magnetization transfer (ihMT) was reported for the first time in 2005 [28] as an incidental finding during an arterial spin labeling scan. As explained in the introduction to this chapter, in a typical MT experiment off-resonance pulses are used to saturate the macromolecular pool: such pulses are tuned either at a negative (i.e., few kHz below the Larmor frequency) or a positive (i.e., few kHz above the Larmor frequency) frequency. In the experiment performed by Alsop et al. [28], instead, pulses with power evenly split between positive and negative frequency offsets (dual-frequency irradiation) were alternated (Fig. 11.4), and the resulting images showed a marked reduction of signal within the white matter, suggesting high specificity for myelin. The prevalent explanation for this augmented specificity is that MT relies on a combination of spin interactions. Some of these effects are negligible in isotropic liquids, but can leave a residual component for protons with a preferential molecular orientation, such as myelin bilayers. The dual-frequency irradiation used for ihMT effectively removes these effects while the single-frequency saturation (used for regular MT) does not. A way of highlighting such effects is therefore to compare images obtained with the two types of irradiation. An inhomogeneous magnetization transfer ratio (ihMTR) map is computed by obtaining images with positive (MT_+), negative (MT_-) and dual irradiation ($MT_\mp$, $MT_\pm$) and then combining them as follows:

$$ihMTR = \frac{(MT_+ + M_-) - (MT_\mp + M_\pm)}{MT_0}$$

where MT_0 is the reference image, without any saturation.

IhMT is still in its infancy and more work is needed to understand its underlying mechanisms, validate it, and develop reliable acquisition techniques that can be translated into clinical applications. Nevertheless, it is a promising approach to improve specificity to myelin.

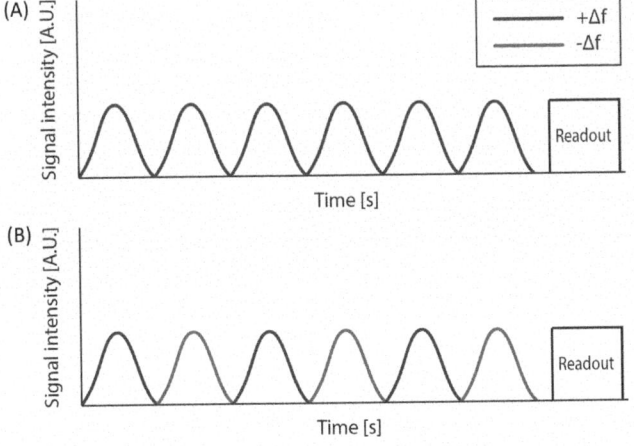

FIGURE 11.4 A conventional (A) vs inhomogeneous (B) MT experiment. In the former case, the MT pulses are all tuned at the same offset frequency, while in the latter case, positive and negative offsets are alternated.

Validation of magnetization transfer—derived parameters as myelin markers

Several studies have attempted to establish the relationship between histological measures of myelin and quantities derived from MT techniques. Both rodent and postmortem human studies have confirmed a moderate association between MTR and myelin content [29,30]. The macromolecular pool fraction derived from qMT also showed good correlation with myelin content in experimental models of both, demyelination and remyelination [31–33]. However, it is known that MT-derived parameters also correlate with edema and inflammation. In experimental allergic encephalomyelitis, MTR decreases were reversed with suppression of inflammation [34]. Correlations with the number of inactive macrophages within lesions, and with other indices of inflammation have also been reported [35]. These findings are not surprising, as the MTR measures a ratio between macromolecular and water protons, and therefore changes to its value are driven by changes to both pools. Despite these limitations, a recent meta-analysis [19] has confirmed that MT-based measures have the highest correlations with myelin content among all the MRI techniques that have been proposed as proxies of myelin.

Magnetization transfer ratio in multiple sclerosis

Lesions

Reductions in the MTR were observed from the early days within MS lesions [5,36,37], consistent with demyelination. A variable range of MTR values can be found in lesions (see Fig. 11.2 for an example), with lower values in lesions appearing as hypointense on T1-weighted scans [38]. The lower MTR typically observed in T1-hypointense lesions is consistent with the recognition that these so-called "balck holes" are the fraction of lesions with the most severe axonal loss and demyelination. Longitudinal MTR assessments have shown rapid changes over time in newly enhancing lesions. Dramatic reductions in MTR at lesion appearance have been consistently reported [39]. Within these lesions, the MTR tend to recover, at least partially, over a few weeks. Such rapid changes suggest that the initial drop might be driven by an increase in water content due to edema, rather than by a reduction in myelin. Conversely, persistent MTR reductions after 3–6 months suggest a substantial loss of myelin [40] and the presence of permanent tissue damage (see Fig. 11.5). MTR reductions have also been detected retrospectively in the normal-appearing white matter (NAWM) before lesion formation [41,42]. These changes in MTR are probably explained by a combination of edema and gliosis, followed by demyelination. In ring-enhancing lesions, the lowest MTR values are typically found in the center, where myelin loss is likely to have occurred, while inflammatory processes are taking place at the periphery [38,43].

Normal-appearing white matter

Although the MTR is typically larger in the NAWM compared to T_2-hyperintense lesions, abnormally low MTR values are found also in the NAWM compared to the white matter of

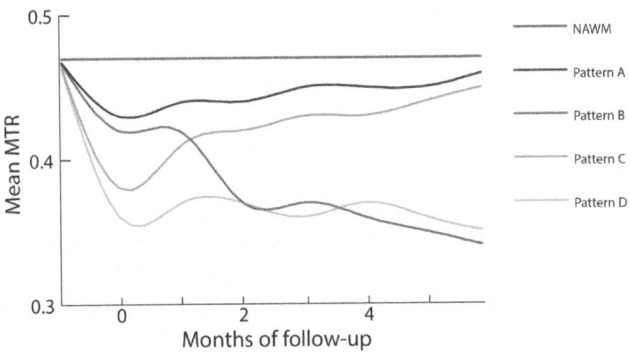

FIGURE 11.5 MTR value evolution in active MS lesions from appearance over 6 months of followup. The four lines correspond to lesions showing different patterns of appearance on T1-weighted images: pattern A (dark red), initially isointense lesions remain isointense; pattern B (light red), initially isointense lesions become hypointense; pattern C (orange), initially hypointense lesions become isointense; and pattern D (yellow), initially hypointense lesions remain hypointense. Note the "dip" in MTR occurring approximately at lesion appearance (month 0). Source: *Adapted with permission from van Waesberghe JH, van Walderveen MA, Castelijns JA, Scheltens P, Lycklama à Nijeholt GJ, Polman CH, et al. Patterns of lesion development in multiple sclerosis: longitudinal observations with T1-weighted spin-echo and magnetization transfer MR. AJNR Am J Neuroradiol 1998;19(4):675−83.*

healthy controls [5], with typically decreasing values approaching T2-visible lesions [44]. MTR reductions in the NAWM tend to be more widespread in secondary progressive multiple sclerosis (SPMS) than relapsing−remitting multiple sclerosis (RRMS) [45], and not uniform within the white matter. Specifically, a gradient of decreasing MTR toward the ventricles was demonstrated [46], with greater MTR reductions in SPMS compared with RRMS patients. This finding supports the hypothesis that MS pathology might be linked to CSF-mediated factors [47]. This MTR gradient toward the ventricles can be already observed soon after a clinically isolated syndrome, and is predictive of the development of MS, independently of WM lesions [48], and of further relapses after treatment with alemtuzumab [49]. The interpretation for these MTR changes in the absence of macroscopic lesions is multifaceted: we have already discussed how low MTR values can be retrospectively identified in the NAWM before the appearance of a lesion—some studies have reported changes appearing up to 1.5 years before the lesion appears [42]. However, low MTR values can also result from diffuse changes, including edema, gliosis, and myelin thinning. The recurring finding of lower MTR closer to visible lesions [44,50] suggests that it may be caused by a combination of secondary events, including retrograde degeneration.

Gray matter

MTR was also one of the first quantitative approaches used to assess tissue damage induced by MS to the gray matter, showing reduced values in the gray matter of people with MS, including those minimally disabled [51]. As gray matter lesions are frequent in MS [52] but almost MRI-invisible (particularly in the cortex), access to a quantitative technique able to assess tissue integrity is of great value. Reduced MTR can also be explained by subtle diffuse pathology or by a combination of both. Several studies of gray matter MTR were conducted in primary progressive multiple sclerosis (PPMS), as typically these patients have lower white matter lesion load but greater disability than people with RRMS. In PPMS, MTR reductions in the gray matter are more widespread than localized atrophy and correlate well with clinical scores [53]. In a longitudinal study in a PPMS cohort, baseline gray matter MTR best predicted progression over 3 years [54]. One potential confound when measuring cortical MTR is partial volume with CSF, which might produce artificially low value in patients with MS then in healthy controls, as they are more likely to show cortical atrophy. It is therefore recommended to adjust for this effect. With recent advances in image analysis more sophisticated approaches have become available, and it was shown that, just like in the NAWM, MTR is not uniform across the cortex: when segmenting the cortex into outer and inner bands, Samson et al. demonstrated a decrease in outer cortical MTR of SPMS and RRMS patients compared to healthy controls, which was not observed in PPMS [55]. This finding could reflect subpial pathology [52].

Quantitative magnetization transfer in multiple sclerosis

qMT studies in clinical populations have been limited so far, primarily due to the relatively long scan times, and the lack of off-the-shelf acquisition protocols and image analysis packages. Although the latter issue has been partially addressed (notable open-source projects include: qMRLab—https://qmrlab.org; and QUIT—https://quit.readthedocs.io), commercial sequences for qMT are not yet available. Nonetheless, initial, proof-of-concept studies have shown that the macromolecular pool size ratio, F, is decreased in lesions and NAWM, as expected in the presence of demyelination [56–58]. A recent systematic review provides an overview of the most consistently reported findings when using MTR and qMT in RRMS [59]. Overall, the macromolecular proton fraction (F) and the forward exchange rate (k_f) appear consistently reduced in lesions compared to the NAWM, and appear sensitive to lesion severity. In the NAWM changes to qMT parameters are more subtle and less reproducible. A study performed at 7 T investigated qMT in cortical gray matter and found a significant reduction in the exchange ratio of MS patients which was strongly correlated with cognitive performance [60].

Magnetization transfer saturation in multiple sclerosis

One study looking at MT_{sat} in MS showed reduced values in both NAWM and gray matter [61]. The MT_{sat} within lesions was tightly associated with cognitive scores.

Inhomogeneous magnetization transfer in multiple sclerosis

As ihMT is a relatively novel approach, most of the work so far has focused on method development. Initial applications to RRMS have demonstrated reduced ihMTR values within MS lesions compared to the NAWM, and lower values in the NAWM of patients compared to the white matter of healthy controls. The association with clinical disability was consistently stronger for ihMTR than MTR [62,63].

Magnetization transfer in spinal cord and optic nerve

Quantitative approaches to quantify myelin in the spinal cord and optic nerve are of great relevance for MS research: spinal cord and optic nerve lesions are typically more symptomatic than brain lesions, and correlate better with the degree of physical disability. Nevertheless, implementing reliable techniques for measuring MT effects in these anatomical locations is challenging, due to not only their specific anatomy, but also their tendency to move. The small section of both the cord and the optic nerve call for high spatial resolution, which, on turn, yields reduced signal-to-noise ratio (SNR) and increased motion sensitivity. Additional complications come from the surrounding tissue (bone, fluid, and fat) that may confound the experiment.

Spinal cord

In the cord, the majority of investigations have focused on the cervical portion. A simple approach for quantification is MT_{CSF} [64], which quantifies MT effects relative to the CSF signal. The voxelwise signal is normalized to that measured in a given region of interest (ROI) encompassing CSF only, where MT should be zero. MT_{CSF} showed tract-specific signal abnormalities in the dorsal and lateral columns of the spinal cord in MS patients [65]. One limitation of MT_{CSF} is its dependency on T_1 and T_2 contrast, which might be altered in the presence of inflammation. Attempts to set up protocols for qMT in the cord have been mainly hampered by the scan time that can quickly become prohibitively long. Attempts to compensate for this problem have initially focused on ROI-based (rather than voxelwise) fit, due to the relatively low SNR. Faster readouts, such as echo-planar imaging offer an attractive alternative and can be combined with reduced field-of-view acquisition [66]. Finally, single-point qMT approaches have been proposed [67]. Even ihMT has been applied to the spinal cord [68], in combination with diffusion tensor imaging but only in healthy volunteers. Two single slices at C2 and C5 were imaged, showing higher ihMTR at C2 compared to C5.

Optic nerve

Despite the optic nerve being even more challenging than the cord to image, early attempts to measure the MTR of the optic nerve suggested reduced values in patients with MS and optic neuritis compared to healthy controls [69], and moderate correlations with visual evoked potentials. More recently, an approach toward qMT of the nerve has been proposed using Dixon fat–water separation to remove fatty tissue from MT images [70].

Despite these isolated examples, qMT for these structures is limited by technical challenges and long scan times. Optimized acquisitions, as well as approaches based on reduced models allow the scan time to be reduced and/or the precision to be increased.

Conclusions

MT techniques offer a unique source of contrast and its sensitivity to myelin has been repeatedly demonstrated against histology. Simple approaches such as MTR or MT_{sat} make it a feasible approach in clinical population, and the MTR has been included as a secondary outcome in some pharmacological trials. One of the challenges that remain to be addressed is the standardization across scanner models and manufacturers, which limit adoption in the clinic. More complex modeling of the MT phenomenon has the advantage of providing estimates of true biophysical parameters, with a clear biological interpretability. However, they require long scan times, which are hardly implementable in the clinic. A further limitation of MT-derived parameters is its sensitivity to inflammation, which can bias myelin assessments. Despite these limitations, MT remains one of the most promising ways of measuring myelin noninvasively, and the recent introduction of ihMT promises to improve specificity.

References

[1] Forsén S, Hoffman RA. Study of moderately rapid chemical exchange reactions by means of nuclear magnetic double resonance. J Chem Phys 1963;39(11):2892–901. Available from: https://doi.org/10.1063/1.1734121.

[2] Edzes HT, Samulski ET. The measurement of cross-relaxation effects in the proton NMR spin-lattice relaxation of water in biological systems: hydrated collagen and muscle. J Magn Reson (1969) 1978;31(2):207–29. Available from: https://doi.org/10.1016/0022-2364(78)90185-3.

[3] Laule C, Vavasour IM, Kolind SH, Li DKB, Traboulsee TL, Moore GRW, et al. Magnetic resonance imaging of myelin. Neurotherapeutics 2007;4(3):460–84. Available from: https://doi.org/10.1016/j.nurt.2007.05.004.

[4] Wolff SD, Balaban RS. Magnetization transfer contrast (MTC) and tissue water proton relaxation in vivo. Magn Reson Med 1989;10(1):135–44. Available from: https://doi.org/10.1002/mrm.1910100113.

[5] Dousset V, Grossman RI, Ramer KN, Schnall MD, Young LH, Gonzalez-Scarano F, et al. Experimental allergic encephalomyelitis and multiple sclerosis: lesion characterization with magnetization transfer imaging. Radiology 1992;182(2):483–91. Available from: https://doi.org/10.1148/radiology.182.2.1732968.

[6] Gochberg DF, Gore JC. Quantitative magnetization transfer imaging via selective inversion recovery with short repetition times. Magn Reson Med 2007;57(2):437–41. Available from: https://doi.org/10.1002/mrm.21143.

[7] Bieri O, Scheffler K. Optimized balanced steady-state free precession magnetization transfer imaging. Magn Reson Med 2007;58(3):511–18. Available from: https://doi.org/10.1002/mrm.21326.

[8] Edelman RR, Ahn SS, Chien D, Li W, Goldmann A, Mantello M, et al. Improved time-of-flight MR angiography of the brain with magnetization transfer contrast. Radiology 1992;184(2):395–9. Available from: https://doi.org/10.1148/radiology.184.2.1620835.

[9] Finelli DA, Hurst GC, Gullapali RP, Bellon EM. Improved contrast of enhancing brain lesions on postgadolinium, T1-weighted spin-echo images with use of magnetization transfer. Radiology 1994;190(2):553–9. Available from: https://doi.org/10.1148/radiology.190.2.8284415.

[10] Henkelman RM, Stanisz GJ, Graham SJ. Magnetization transfer in MRI: a review. NMR Biomed 2001;14(2):57–64. Available from: https://doi.org/10.1002/nbm.683.

[11] Cercignani M, Symms MR, Ron M, Barker GJ. 3D MTR measurement: From 1.5T to 3.0T. NeuroImage 2006;31(1):181–6. Available from: https://doi.org/10.1016/j.neuroimage.2005.11.028.

[12] Martirosian P, Boss A, Deimling M, Kiefer B, Schraml C, Schwenzer NF, et al. Systematic variation of off-resonance prepulses for clinical magnetization transfer contrast imaging at 0.2, 1.5, and 3.0 Tesla. Invest Radiol 2008;43(1):16–26. Available from: https://doi.org/10.1097/RLI.0b013e3181559949.

[13] Helms G, Dathe H, Kallenberg K, Dechent P. High-resolution maps of magnetization transfer with inherent correction for RF inhomogeneity and T1 relaxation obtained from 3D FLASH MRI. Magn Reson Med 2008;60(6):1396–407. Available from: https://doi.org/10.1002/mrm.21732.

[14] Henkelman RM, Huang X, Xiang QS, Stanisz GJ, Swanson SD, Bronskill MJ. Quantitative interpretation of magnetization transfer. Magn Reson Med 1993;29(6):759–66. Available from: https://doi.org/10.1002/mrm.1910290607.

[15] Cercignani M, Dowell NG, Tofts PS. Quantitative MRI of the brain: principles of physical measurement: principles of physical measurement. 2nd ed. CRC Press; 2018.
[16] Ramani A, Dalton C, Miller DH, Tofts PS, Barker GJ. Precise estimate of fundamental in-vivo MT parameters in human brain in clinically feasible times. Magn Reson Imaging 2002;20(10):721–31. Available from: https://doi.org/10.1016/S0730-725X(02)00598-2.
[17] Sled JG, Pike GB. Quantitative interpretation of magnetization transfer in spoiled gradient echo MRI sequences. J Magn Reson 2000;145(1):24–36. Available from: https://doi.org/10.1006/jmre.2000.2059.
[18] Yarnykh VL. Pulsed Z-spectroscopic imaging of cross-relaxation parameters in tissues for human MRI: theory and clinical applications. Magn Reson Med 2002;47(5):929–39. Available from: https://doi.org/10.1002/mrm.10120.
[19] Mancini M, Karakuzu A, Cohen-Adad J, Cercignani M, Nichols TE, Stikov N. An interactive meta-analysis of MRI biomarkers of myelin. eLife 2020;9e61523. Available from: https://doi.org/10.7554/eLife.61523.
[20] Levesque IR, Pike GB. Characterizing healthy and diseased white matter using quantitative magnetization transfer and multicomponent T(2) relaxometry: a unified view via a four-pool model. Magn Reson Med 2009;62(6):1487–96. Available from: https://doi.org/10.1002/mrm.22131.
[21] Stanisz GJ, Kecojevic A, Bronskill MJ, Henkelman RM. Characterizing white matter with magnetization transfer and T(2). Magn Reson Med 1999;42(6):1128–36. Available from: https://doi.org/10.1002/(sici)1522-2594(199912)42:6<1128::aid-mrm18>3.0.co;2-9.
[22] Mossahebi P, Alexander AL, Field AS, Samsonov AA. Removal of cerebrospinal fluid partial volume effects in quantitative magnetization transfer imaging using a three-pool model with nonexchanging water component. Magn Reson Med 2015;74(5):1317–26. Available from: https://doi.org/10.1002/mrm.25516.
[23] Yarnykh VL, Yuan C. Cross-relaxation imaging reveals detailed anatomy of white matter fiber tracts in the human brain. NeuroImage 2004;23(1):409–24. Available from: https://doi.org/10.1016/j.neuroimage.2004.04.029.
[24] Yarnykh VL. Fast macromolecular proton fraction mapping from a single off-resonance magnetization transfer measurement. Magn Reson Med 2012;68(1):166–78. Available from: https://doi.org/10.1002/mrm.23224.
[25] Yarnykh VL. Time-efficient, high-resolution, whole brain three-dimensional macromolecular proton fraction mapping. Magn Reson Med 2016;75(5):2100–6. Available from: https://doi.org/10.1002/mrm.25811.
[26] Tabelow K, Balteau E, Ashburner J, Callaghan MF, Draganski B, Helms G, et al. hMRI – a toolbox for quantitative MRI in neuroscience and clinical research. NeuroImage 2019;194:191–210. Available from: https://doi.org/10.1016/j.neuroimage.2019.01.029.
[27] Weiskopf N, Suckling J, Williams G, Correia MM, Inkster B, Tait R, et al. Quantitative multi-parameter mapping of R1, PD(*), MT, and R2(*) at 3T: a multi-center validation. Front Neurosci 2013;7:95. Available from: https://doi.org/10.3389/fnins.2013.00095.
[28] Alsop DC, De Bazelaire C, Garcia M, Duhamel G. Inhomogenous magnetization transfer imaging: a potentially specific marker for myelin. Proceedings of the 13th annual meeting of ISMRM. 2005. Miami, FL. Abstract 2224.
[29] Deloire-Grassin MSA, Brochet B, Quesson B, Delalande C, Dousset V, Canioni P, et al. In vivo evaluation of remyelination in rat brain by magnetization transfer imaging. J Neurol Sci 2000;178(1):10–16. Available from: https://doi.org/10.1016/S0022-510X(00)00331-2.
[30] Schmierer K, Scaravilli F, Altmann DR, Barker GJ, Miller DH. Magnetization transfer ratio and myelin in postmortem multiple sclerosis brain. Ann Neurol 2004;56(3):407–15. Available from: https://doi.org/10.1002/ana.20202.
[31] Ou X, Sun SW, Liang HF, Song SK, Gochberg DF. The MT pool size ratio and the DTI radial diffusivity may reflect the myelination in shiverer and control mice. NMR Biomed 2009;22(5):480–7. Available from: https://doi.org/10.1002/nbm.1358.
[32] Ou X, Sun SW, Liang HF, Song SK, Gochberg DF. Quantitative magnetization transfer measured pool-size ratio reflects optic nerve myelin content in ex vivo mice. Magn Reson Med 2009;61(2):364–71. Available from: https://doi.org/10.1002/mrm.21850.
[33] Turati L, Moscatelli M, Mastropietro A, Dowell NG, Zucca I, Erbetta A, et al. In vivo quantitative magnetization transfer imaging correlates with histology during de- and remyelination in cuprizone-treated mice. NMR Biomed 2015;28(3):327–37. Available from: https://doi.org/10.1002/nbm.3253.
[34] Gareau PJ, Rutt BK, Karlik SJ, Mitchell JR. Magnetization transfer and multicomponent T2 relaxation measurements with histopathologic correlation in an experimental model of MS. J Magn Reson Imaging 2000;11(6):586–95. Available from: https://doi.org/10.1002/1522-2586(200006)11:6<586::aid-jmri3>3.0.co;2-v.

[35] Blezer EL, Bauer J, Brok HP, Nicolay K, t Hart BA. Quantitative MRI-pathology correlations of brain white matter lesions developing in a non-human primate model of multiple sclerosis. NMR Biomed 2007;20(2):90−103. Available from: https://doi.org/10.1002/nbm.1085.

[36] Gass A, Barker GJ, Kidd D, Thorpe JW, MacManus D, Brennan A, et al. Correlation of magnetization transfer ratio with clinical disability in multiple sclerosis. Ann Neurol 1994;36(1):62−7. Available from: https://doi.org/10.1002/ana.410360113.

[37] Campi A, Filippi M, Comi C, Scotti G, Gerevini S, Dousset V. Magnetisation transfer ratios of contrast-enhancing and nonenhancing lesions in multiple sclerosis. Neuroradiology 1996;38(2):115−19. Available from: https://doi.org/10.1007/BF00604792.

[38] Hiehle Jr. JF, Grossman RI, Ramer KN, Gonzalez-Scarano F, Cohen JA. Magnetization transfer effects in MR-detected multiple sclerosis lesions: comparison with gadolinium-enhanced spin-echo images and nonenhanced T1-weighted images. AJNR Am J Neuroradiol 1995;16(1):69−77.

[39] Filippi M, Rocca MA, Rizzo G, Horsfield MA, Rovaris M, Minicucci L, et al. Magnetization transfer ratios in multiple sclerosis lesions enhancing after different doses of gadolinium. Neurology 1998;50(5):1289−93. Available from: https://doi.org/10.1212/wnl.50.5.1289.

[40] van Waesberghe JH, van Walderveen MA, Castelijns JA, Scheltens P, Lycklama à Nijeholt GJ, Polman CH, et al. Patterns of lesion development in multiple sclerosis: longitudinal observations with T1-weighted spin-echo and magnetization transfer MR. AJNR Am J Neuroradiol 1998;19(4):675−83.

[41] Filippi M, Rocca MA, Martino G, Horsfield MA, Comi G. Magnetization transfer changes in the normal appearing white matter precede the appearance of enhancing lesions in patients with multiple sclerosis. Ann Neurol 1998;43(6):809−14. Available from: https://doi.org/10.1002/ana.410430616.

[42] Pike GB, Stefano ND, Narayanan S, Worsley KJ, Pelletier D, Francis GS, et al. Multiple sclerosis: magnetization transfer MR imaging of white matter before lesion appearance on T2-weighted images. Radiology 2000;215(3):824−30. Available from: https://doi.org/10.1148/radiology.215.3.r00jn02824.

[43] Petrella JR, Grossman RI, McGowan JC, Campbell G, Cohen JA. Multiple sclerosis lesions: relationship between MR enhancement pattern and magnetization transfer effect. AJNR Am J Neuroradiol 1996;17(6):1041−9.

[44] Filippi M, Campi A, Dousset V, Baratti C, Martinelli V, Canal N, et al. A magnetization transfer imaging study of normal-appearing white matter in multiple sclerosis. Neurology 1995;45(3 Pt 1):478−82. Available from: https://doi.org/10.1212/wnl.45.3.478.

[45] Traboulsee A, Dehmeshki J, Peters KR, Griffin CM, Brex PA, Silver N, et al. Disability in multiple sclerosis is related to normal appearing brain tissue MTR histogram abnormalities. Multiple Scler J 2003;9(6):566−73. Available from: https://doi.org/10.1191/1352458503ms958oa.

[46] Liu Z, Pardini M, Yaldizli Ö, Sethi V, Muhlert N, Wheeler-Kingshott CAM, et al. Magnetization transfer ratio measures in normal-appearing white matter show periventricular gradient abnormalities in multiple sclerosis. Brain 2015;138(5):1239−46. Available from: https://doi.org/10.1093/brain/awv065.

[47] Magliozzi R, Howell OW, Reeves C, Roncaroli F, Nicholas R, Serafini B, et al. A Gradient of neuronal loss and meningeal inflammation in multiple sclerosis. Ann Neurol 2010;68(4):477−93. Available from: https://doi.org/10.1002/ana.22230.

[48] Brown JWL, Pardini M, Brownlee WJ, Fernando K, Samson RS, Prados Carrasco F, et al. An abnormal periventricular magnetization transfer ratio gradient occurs early in multiple sclerosis. Brain 2016;140(2):387−98. Available from: https://doi.org/10.1093/brain/aww296.

[49] Brown JWL, Prados Carrasco F, Eshaghi A, Sudre CH, Button T, Pardini M, et al. Periventricular magnetisation transfer ratio abnormalities in multiple sclerosis improve after alemtuzumab. Multiple Scler J 2020;26(9):1093−101. Available from: https://doi.org/10.1177/1352458519852093.

[50] Vrenken H, Geurts JJ, Knol DL, Polman CH, Castelijns JA, Pouwels PJ, et al. Normal-appearing white matter changes vary with distance to lesions in multiple sclerosis. AJNR Am J Neuroradiol 2006;27(9):2005−11.

[51] Davies GR, Ramió-Torrentà L, Hadjiprocopis A, Chard DT, Griffin CMB, Rashid W, et al. Evidence for grey matter MTR abnormality in minimally disabled patients with early relapsing-remitting multiple sclerosis. J Neurol Neurosurg Psychiatry 2004;75(7):998−1002. Available from: https://doi.org/10.1136/jnnp.2003.021915.

[52] Bø L, Vedeler CA, Nyland HI, Trapp BD, Mørk SJ. Subpial demyelination in the cerebral cortex of multiple sclerosis patients. J Neuropathol Exp Neurol 2003;62(7):723−32. Available from: https://doi.org/10.1093/jnen/62.7.723.

[53] Khaleeli Z, Cercignani M, Audoin B, Ciccarelli O, Miller DH, Thompson AJ. Localized grey matter damage in early primary progressive multiple sclerosis contributes to disability. NeuroImage 2007;37(1):253−61. Available from: https://doi.org/10.1016/j.neuroimage.2007.04.056.

[54] Khaleeli Z, Altmann DR, Cercignani M, Ciccarelli O, Miller DH, Thompson AJ. Magnetization transfer ratio in gray matter: a potential surrogate marker for progression in early primary progressive multiple sclerosis. Arch Neurol 2008;65(11):1454–9. Available from: https://doi.org/10.1001/archneur.65.11.1454.

[55] Samson RS, Cardoso MJ, Muhlert N, Sethi V, Wheeler-Kingshott CA, Ron M, et al. Investigation of outer cortical magnetisation transfer ratio abnormalities in multiple sclerosis clinical subgroups. Multiple Scler J 2014;20(10):1322–30. Available from: https://doi.org/10.1177/1352458514522537.

[56] Davies GR, Ramani A, Dalton CM, Tozer DJ, Wheeler-Kingshott CA, Barker GJ, et al. Preliminary magnetic resonance study of the macromolecular proton fraction in white matter: a potential marker of myelin? Multiple Scler J 2003;9(3):246–9. Available from: https://doi.org/10.1191/1352458503ms911oa.

[57] Levesque IR, Giacomini PS, Narayanan S, Ribeiro LT, Sled JG, Arnold DL, et al. Quantitative magnetization transfer and myelin water imaging of the evolution of acute multiple sclerosis lesions. Magn Reson Med 2010;63(3):633–40. Available from: https://doi.org/10.1002/mrm.22244.

[58] Cercignani M, Basile B, Spanò B, Comanducci G, Fasano F, Caltagirone C, et al. Investigation of quantitative magnetisation transfer parameters of lesions and normal appearing white matter in multiple sclerosis. NMR Biomed 2009;22(6):646–53. Available from: https://doi.org/10.1002/nbm.1379.

[59] York EN, Thrippleton MJ, Meijboom R, Hunt DPJ, Waldman AD. Quantitative magnetization transfer imaging in relapsing-remitting multiple sclerosis: a systematic review and meta-analysis. Brain Commun 2022;4(2)fcac088. Available from: https://doi.org/10.1093/braincomms/fcac088.

[60] McKeithan LJ, Lyttle BD, Box BA, O'Grady KP, Dortch RD, Conrad BN, et al. 7T quantitative magnetization transfer (qMT) of cortical gray matter in multiple sclerosis correlates with cognitive impairment. NeuroImage 2019;203116190. Available from: https://doi.org/10.1016/j.neuroimage.2019.116190.

[61] Lommers E, Guillemin C, Reuter G, Fouarge E, Delrue G, Collette F, et al. Voxel-Based quantitative MRI reveals spatial patterns of grey matter alteration in multiple sclerosis. Hum Brain Mapp 2021;42(4):1003–12. Available from: https://doi.org/10.1002/hbm.25274.

[62] Van Obberghen E, Mchinda S, le Troter A, Prevost VH, Viout P, Guye M, et al. Evaluation of the sensitivity of inhomogeneous magnetization transfer (ihMT) MRI for multiple sclerosis. Am J Neuroradiol 2018;39(4):634–41. Available from: https://doi.org/10.3174/ajnr.A5563.

[63] Zhang L, Wen B, Chen T, Tian H, Xue H, Ren H, et al. A comparison study of inhomogeneous magnetization transfer (ihMT) and magnetization transfer (MT) in multiple sclerosis based on whole brain acquisition at 3.0 T. Magn Reson Imaging 2020;70:43–9. Available from: https://doi.org/10.1016/j.mri.2020.03.010.

[64] Smith SA, Golay X, Fatemi A, Jones CK, Raymond GV, Moser HW, et al. Magnetization transfer weighted imaging in the upper cervical spinal cord using cerebrospinal fluid as intersubject normalization reference (MTCSF imaging). Magn Reson Med 2005;54(1):201–6. Available from: https://doi.org/10.1002/mrm.20553.

[65] Zackowski KM, Smith SA, Reich DS, Gordon-Lipkin E, Chodkowski BA, Sambandan DR, et al. Sensorimotor dysfunction in multiple sclerosis and column-specific magnetization transfer-imaging abnormalities in the spinal cord. Brain 2009;132(5):1200–9. Available from: https://doi.org/10.1093/brain/awp032.

[66] Battiston M, Grussu F, Ianus A, Schneider T, Prados F, Fairney J, et al. An optimized framework for quantitative magnetization transfer imaging of the cervical spinal cord in vivo. Magn Reson Med 2018;79(5):2576–88. Available from: https://doi.org/10.1002/mrm.26909.

[67] Smith AK, By S, Lyttle BD, Dortch RD, Box BA, McKeithan LJ, et al. Evaluating single-point quantitative magnetization transfer in the cervical spinal cord: application to multiple sclerosis. Neuroimage Clin 2017;16:58–65. Available from: https://doi.org/10.1016/j.nicl.2017.07.010.

[68] Taso M, Girard OM, Duhamel G, Le Troter A, Feiweier T, Guye M, et al. Tract-specific and age-related variations of the spinal cord microstructure: a multi-parametric MRI study using diffusion tensor imaging (DTI) and inhomogeneous magnetization transfer (ihMT). NMR Biomed 2016;29(6):817–32. Available from: https://doi.org/10.1002/nbm.3530.

[69] Trip S, Schlottmann P, Jones S, Li W-Y, Garway-Heath D, Thompson A, et al. Optic nerve magnetization transfer imaging and measures of axonal loss and demyelination in optic neuritis. Multiple Scler J 2007;13(7):875–9. Available from: https://doi.org/10.1177/1352458507076952.

[70] Smith AK, Dortch RD, Dethrage LM, Lyttle BD, Kang H, Welch EB, et al. Incorporating dixon multi-echo fat water separation for novel quantitative magnetization transfer of the human optic nerve in vivo. Magn Reson Med 2017;77(2):707–16. Available from: https://doi.org/10.1002/mrm.26164.

CHAPTER 12

Susceptibility weighted imaging in multiple sclerosis

Sagar Buch and E. Mark Haacke

Department of Neurology, Wayne State University, Detroit, MI, United States

OUTLINE

Introduction	207	Water content as a new biomarker for multiple sclerosis lesions	211
Imaging biomarkers in multiple sclerosis	208		
SWI-FLAIR or FLAIR*	209	Microvascular in vivo contrast revealed origins	214
Quantitative susceptibility mapping	210	Conclusion	216
Introduction to STrategically Acquired Gradient Echo Imaging	211	References	216

Introduction

Multiple sclerosis (MS) is a chronic inflammatory and neurodegenerative disease of the central nervous system, affecting both the brain and the spinal cord, and it is characterized by inflammation, demyelination, oligodendrocyte loss, axonal and neuronal degeneration, gliosis, and remyelination [1,2]. Demyelinating plaques (MS lesions) are typical and result in episodes of neurologic deficit with recovery (relapses) and accumulation of sustained disability with the passage of time. Several magnetic resonance imaging (MRI) methods have been proposed to image MS lesions, which are characterized by myelin loss and later axonal loss [1,3]. MS diagnosis is based on typical clinical symptoms and radiologic findings, incorporating the dissemination in space and time of the MS lesions. Due to the scientific advances in recent years, the MS diagnostic criteria have undergone multiple revisions

[4−7]. Additionally, more sensitive imaging methods to distinguish the types of MS lesions are being explored to improve the diagnostic process.

One of these advanced techniques is susceptibility weighted imaging (SWI), which is a noninvasive MRI technique that utilizes both MR signal magnitude and phase information. SWI combines a T_2^*-weighted magnitude image with a filtered phase image acquired using a gradient echo (GRE) sequence to enhance image contrast. The magnitude image already provides some susceptibility contrast via the dephasing effect that occurs in the presence of locally varying fields across a voxel. SWI further enhances the contrast between tissues of differing susceptibility by applying a phase mask to the original magnitude image. For this reason, SWI has become established in the neuroimaging field, as it highlights small structures with high susceptibility such as veins and microbleeds that are otherwise difficult to detect by conventional MRI. Since the first description in 1997 [8], SWI has proven useful in a multitude of clinical applications including high-resolution MR venography, imaging of traumatic intracranial hemorrhage, visualizing blood products, and vascularization of tumors or assessing iron deposits in the brain [9−11].

Imaging biomarkers in multiple sclerosis

The role of SWI in MS has gained attention because it offers additional information about MS white matter (WM) lesions, which cannot be appreciated on conventional T1- and T2-weighted images currently used to diagnose and monitor patients [3]. The inclusion of the SWI sequence also enables the detection of two commonly evaluated imaging biomarkers associated with MS. First, the evidence from studies designed at multiple field strengths [12−16] shows that MS lesions form around small veins, a phenomenon termed "the central vein sign (CVS)" [17]. An example image is shown in Fig. 12.1. Studies of patients with established disease have proposed a 40% threshold of WM lesions with the CVS to differentiate MS from other disorders that can mimic MS on MR imaging [16,18]. The presence of abnormal veins is a possible explanation of the widely held hypothesis that the formation of an MS lesion depends on the entry of inflammatory cells from the systemic circulation into the brain parenchyma possibly from a disrupted endothelium of the veins, known as the inside-out theory [19]. However, not all MS lesions possess a lesion-centric major vein.

Second, lesions with hyperintense rims or rim lesions (RL) in QSM, reflecting either local demyelination or, if the lesion susceptibility is higher than 50 ppb, iron deposition within the microglia and macrophagic cells at the edge of MS lesions. These lesion types have been identified on SWI in all subtypes of MS [20−23]. However, this imaging feature seems to be absent in other diseases such as neuromyelitis optica spectrum disorder [24] and ischemic lesions [25]. This finding suggests that both the CVS and RLs might be specific features of MS lesions, which could be applied diagnostically to monitor lesions longitudinally.

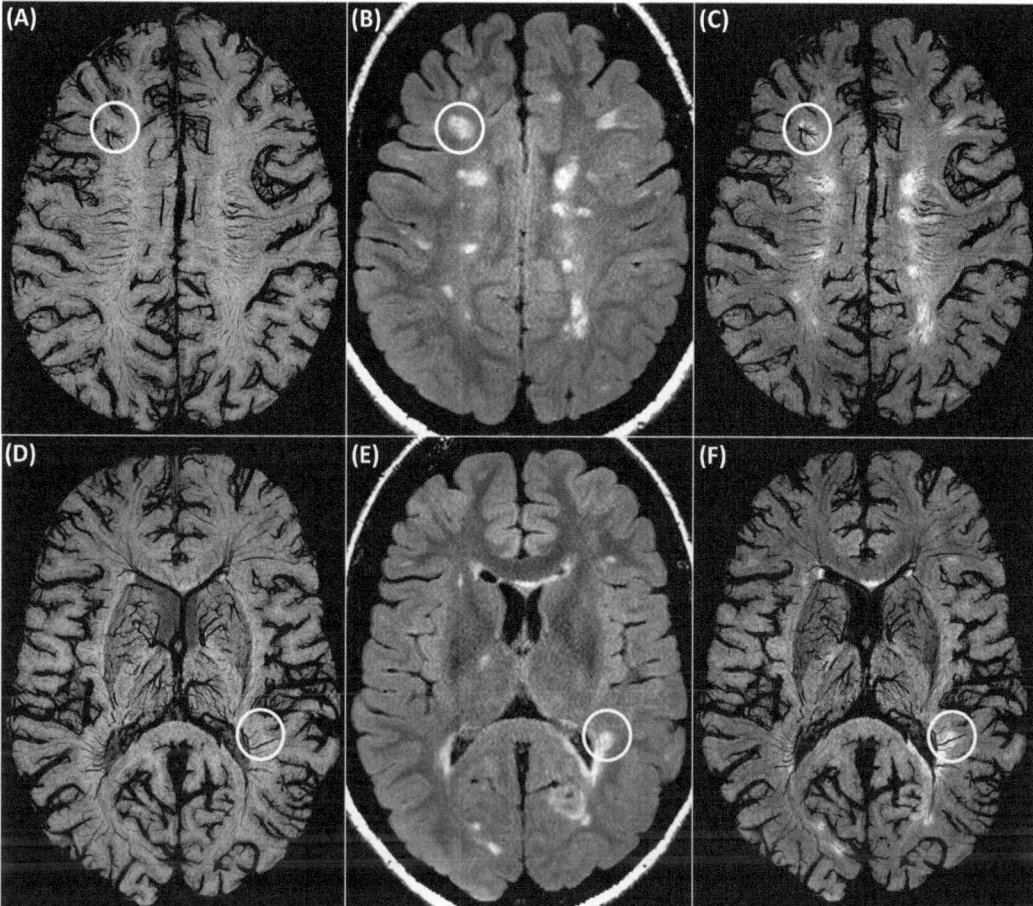

FIGURE 12.1 (A and D) SWI data, (B and E) original FLAIR data, and (C and F) the combined SWI-FLAIR images. SWI-FLAIR results demonstrate the identification of a developmental venous anomaly (top row) and a central vein sign (bottom row). The images acquired with a Ferumoxytol dose of 4 mg/kg and are maximum/minimum intensity projected over two slices (or with an effective slice thickness of 2 mm).

SWI-FLAIR or FLAIR*

Among conventional MRI techniques, T2-weighted fluid-attenuated inversion recovery (FLAIR) is considered one of the most useful contrast techniques for visualizing white matter hyperintensities (WMHs) that represent MS lesions [26]. Because of its high sensitivity to WM abnormalities and its excellent suppression of cerebrospinal fluid signal, brain imaging with T2-weighted FLAIR is used routinely to diagnose disease and to evaluate changes in lesion load [27]. However, FLAIR imaging cannot provide specific information about lesion pathology such as the vascularity of the lesions. On the other hand, SWI has been shown to provide specific information about parenchymal veins [8,9], hemorrhage and calcification [28,29], tissue iron deposition [30], and iron-laden macrophages [31]. Unlike FLAIR images, T2*-weighted and SWI images lack cerebrospinal fluid suppression and are, therefore, less capable of demonstrating contrast between lesions and surrounding tissues, making the detection of lesions more difficult. However, depending on the flip angle used, SWI can resemble FLAIR in terms of lesion depiction especially at longer echo times (TEs).

To overcome these issues, two research groups have recently proposed that FLAIR and T2* or SWI images can be combined into a single image. In the first method, high resolution, 3D, 1 mm isotropic or 0.55 mm isotropic 3D segmented echo planar imaging (SEPI, for vein detection) sequences—both acquired in less than 10 minutes—are used to assess MS lesions [32]. The second method utilizes the coregistered SWI and FLAIR data to generate SWI-FLAIR images to highlight the vascular information on the WM hyperintensity [33]. The drawback of the FLAIR* method is the fact that data for only a single TE is acquired, which limits the application of the sequence based on the susceptibility of the tissue-of-interest. On the other hand, the SWI sequence can be obtained with multiple TEs. Another drawback with the SEPI approach is that both arterial and venous signals will be hypointense, whereas the GRE-based SWI acquisition has bright arterial signal on the first TE due to the flow compensation gradients. This natural separation of arterial and venous signals makes identifying the CVS easier. Furthermore, studies have shown that SWI is more sensitive than T2*-weighted images in detecting hemorrhagic lesions [34–36], cavernous malformations [37], microbleeds [38,39], and small venous networks [11,40–42]. Examples of the SWI-FLAIR data, generated from the postcontrast SWI data combined with the noncontrast FLAIR data, are shown in Fig. 12.1. The postcontrast SWI data were acquired after the administration of the ultrasmall superparamagnetic iron oxide agent, Ferumoxytol (dose = 4 mg/kg).

Quantitative susceptibility mapping

Although SWI and high-pass filtered phase images help to depict susceptibility-related tissue changes directly (such as in the MS lesions to discern possible changes in iron content [3]), they do not provide quantitative measures for intrinsic tissue properties. Quantitative susceptibility mapping (QSM) is a method that uses the original phase data (after unwrapping and filtering it) to directly reconstruct the magnetic source images. This makes it possible to overcome nonlocal geometric restrictions that limit SWI and phase imaging. R2* (1/T2*) maps and QSM have both been used to provide measurements of susceptibility of different structures of the brain; and QSM has become a powerful competitor to T2* mapping for quantifying susceptibility [43]. Recently, it has been shown to have a number of implications in understanding MS lesion pathology. It has been shown in other studies that R2* is highly sensitive to iron changes in the brain, but QSM has proven to be more sensitive and has better reproducibility than R2*. Further, R2* cannot differentiate between paramagnetic iron and diamagnetic calcifications. QSM, on the other hand, can differentiate the two, showing iron as hyperintense signal, and calcium as hypointense signal. Clinical studies have also demonstrated that QSM had higher sensitivity and specificity to phase imaging in identifying calcifications and intracranial hemorrhages [44]. In

MS, QSM can be used to measure the loss of myelin. With the help of QSM, one can also distinguish seemingly active lesions (with high uniform susceptibility or rim enhancement, Fig. 12.2C) and chronic lesions (with negligible susceptibility difference as compared to the surrounding WM, Fig. 12.2G) [45].

Introduction to STrategically Acquired Gradient Echo Imaging

Recently, there has been a major effort in the community for rapid, multicontrast imaging. One such method referred to as STrategically Acquired Gradient Echo (STAGE) imaging has been shown to provide 15 pieces of quantitative and qualitative data [46–48]. STAGE also corrects for both B1 transmit and B1 receive field effects producing uniform images independent of field strength, manufacturer or coil type. As such, it can provide standardized imaging across manufactures that allows for collaborative work across sites and produces similar quantitative data [49]. Depending on the resolution used, STAGE can be run in 2–4 minutes per flip angle making it a very efficient means to collect multiple contrasts. The multiple flip angle, multiple echo data can then be processed to produce a number of imaging biomarkers critical to studying tissue abnormalities such as angiographic images, quantitative results such as water content, T1, T2*, and susceptibility; and tissue maps; as illustrated in Fig. 12.2 and Fig. 12.3.

Water content as a new biomarker for multiple sclerosis lesions

There is some evidence that the presence of "black holes" in T1-weighted imaging may hold some useful information in terms of radiologically correlating with expanded disability status scale (EDSS) [51,52]. T1 Black holes have been graded at different levels with the blackest of them being thought to be more representative of tissue damage. To understand this, one must turn to the tissue properties that define the signal for a T1W image. The higher the water content one might think the higher the signal, but this is counterbalanced by the fact that a T1W image suppresses signal from tissues with long T1 and, for fast 3D gradient echo methods, in the limit of a 90 degrees flip angle the expected signal depends on 1/T1. Proton spin density (PSD) maps are directly related to water content and show hyperintense signal in the WM lesion territories reflecting increased water content inside the lesion. PSD maps can also help differentiate cerebro-spinal fluid (CSF) and WM lesions. Using the tissue properties measured from STAGE imaging, one can correlate R1 (1/T1) and R2* (1/T2*) with water content or PSD maps via a parameter referred to here as beta (β), which is equal to 1/PSD. This multiparametric assessment of the MS lesions is demonstrated in Fig. 12.3.

In Fig. 12.4, the STAGE-derived R1 and R2* for lesions and the contralateral normal appearing WM regions show a linear correlation with β. Using these relationships, we can simulate the signal for WM, lesions, and CSF using the standard equations for inversion recovery and spin echo sequences to obtain the contrast between WM and lesions at different TI values and TEs. It is common to use a long TE to enhance the contrast between lesions and WM especially in FLAIR and long TE spin echo imaging.

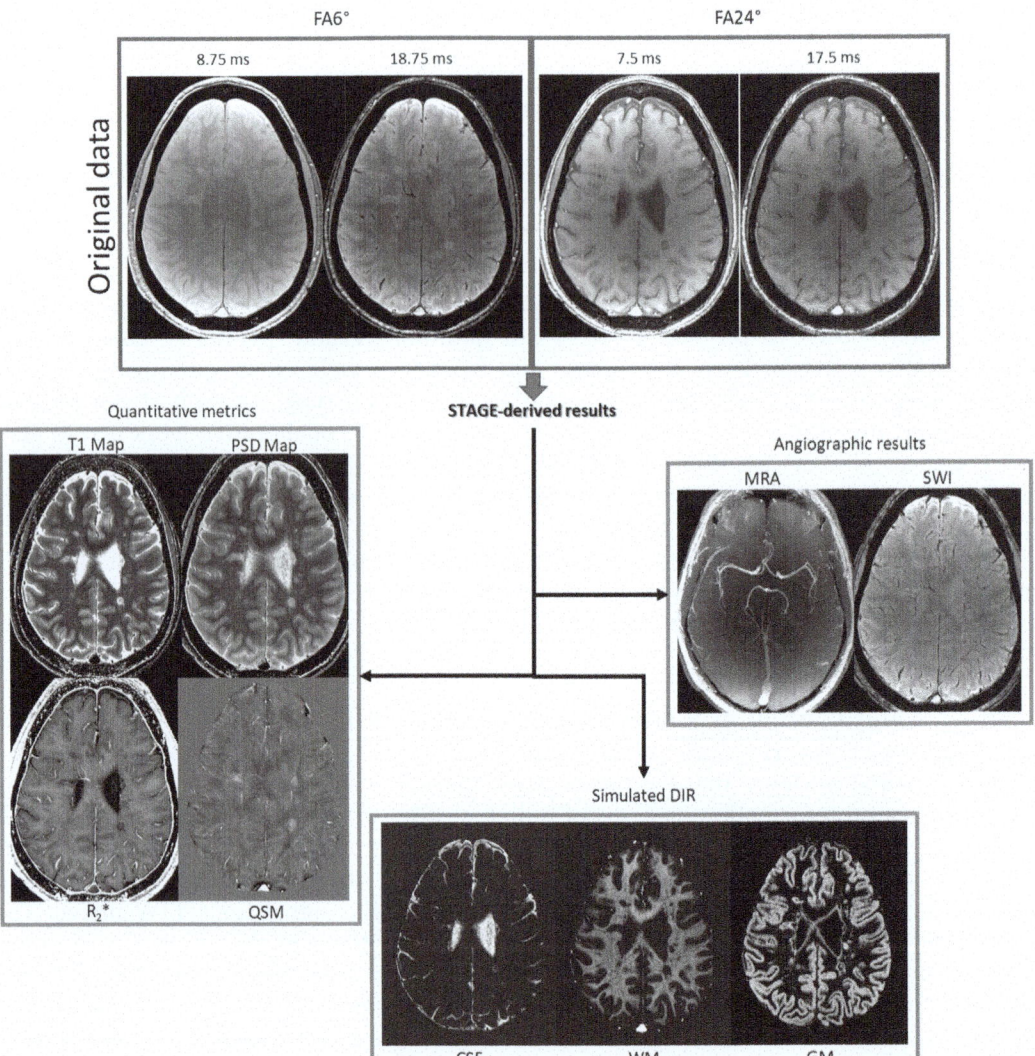

FIGURE 12.2 An overview of the STAGE-derived results. The STAGE imaging protocol involves running ideally two fully flow compensated multiecho sequences [50] with a low FA (spin density weighted) and a high FA (T1 weighted). These data are then processed to find the B1 transmit and B1 receive fields and creates new intensity corrected images. From these images, one can calculate water content, T1, T2*, and susceptibility. In this flow diagram, the original data are shown followed by the postprocessing pipeline results produces quantitative results (QSM, R2*, T1, and PSD maps), angiographic images (bright blood MRA for arteries and dark blood SWI for veins), and simulated double inversion recovery tissue maps (WM, CSF, and GM).

By utilizing the STAGE-derived spin density results for a given case, the traditional FLAIR data with the STAGE-based simulated FLAIR data (Fig. 12.5). It is clear that those lesions with the highest water content on the PSD maps are the most hypointense on the T1W images (not only because of the T1 weighting but also because a short TE is usually used for this sequence).

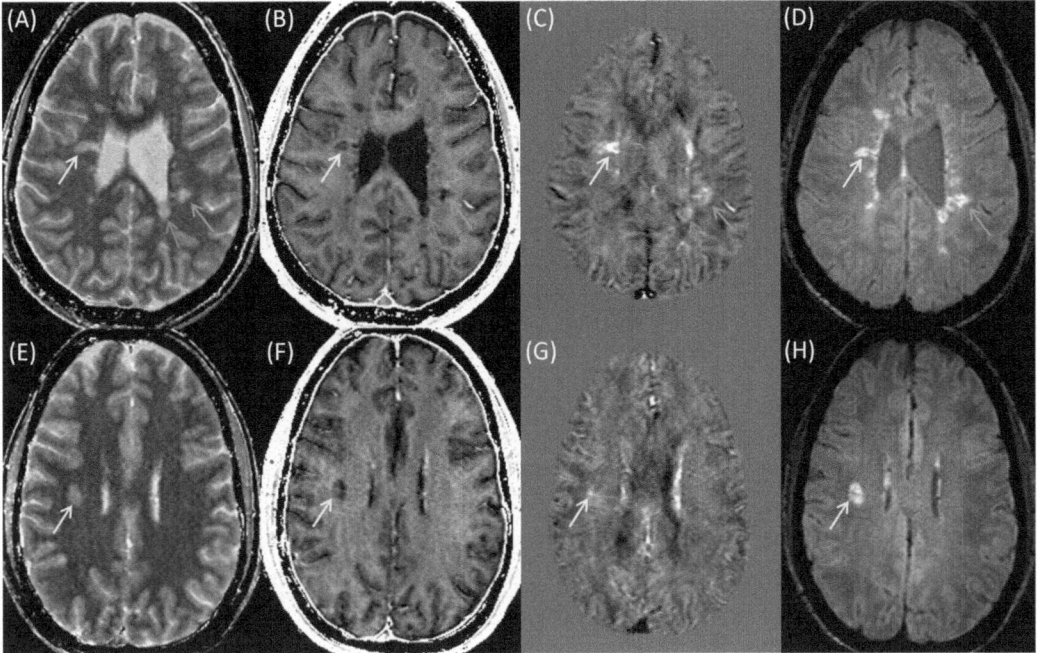

FIGURE 12.3 [50] Multiple WMHs are seen on PSD maps (A and E), R2* maps (B and F), QSM (C and G), and SWI-FLAIR data (D and H). The lesions shown by the yellow and orange arrows in the top row (A–D) appear bright in the QSM images (C) showing increased susceptibility in those lesions. The yellow arrow in the bottom row highlights a lesion that does not appear in the QSM data (it has no susceptibility difference as compared to the surrounding white matter). The green, yellow, and orange arrows show example lesions that are clearly seen on the PSD maps (A and E) and the SWI-FLAIR images (D and H). Note that the brighter lesions in the PSD maps indicate higher water content and these lesions appear the most hypointense in the R2* maps (B and F).

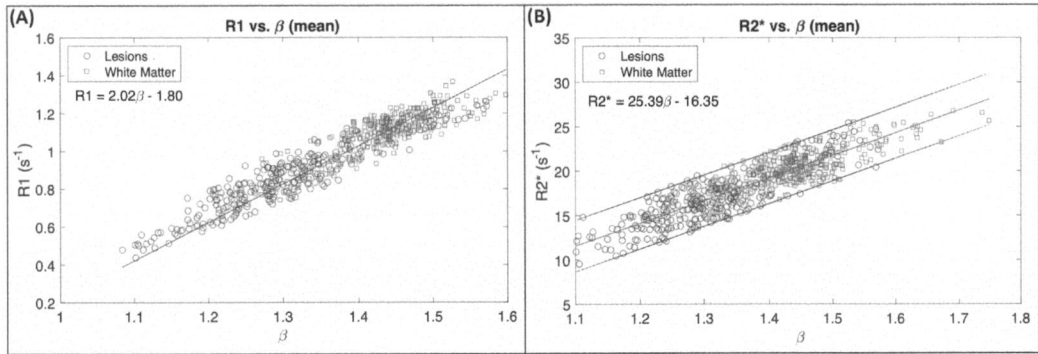

FIGURE 12.4 Plots of (A) R1 versus β (1/PSD) and (B) R2* versus β for normal appearing white matter (red squares) and MS lesions (blue circles). The correlation in (A) is fixed to go through the coordinates (1, 0.22), a known point representing CSF tissue properties. The plot in (B) was produced after iteratively using a 95% confidence interval to remove outliers. From these plots, it can be seen that the higher the water content (the lower β), the higher the T1 value (the lower R1), and the higher the T2* value (the lower R2*). Knowing these values, one can predict the effect of longer T1 and TEs on lesion conspicuity.

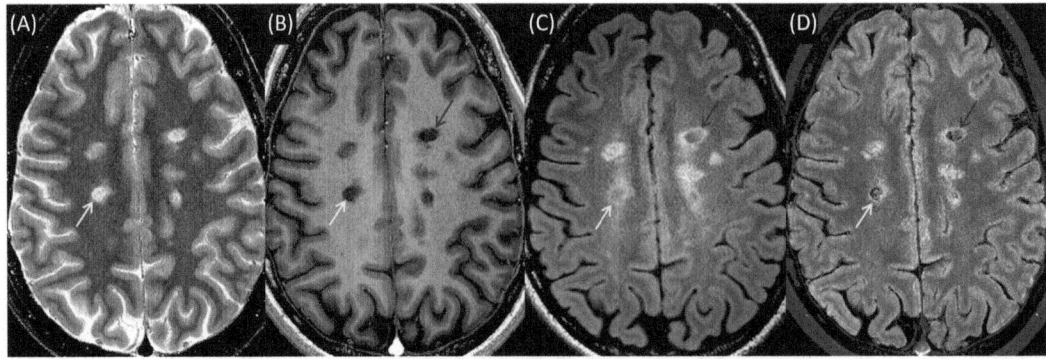

FIGURE 12.5 The STAGE water content (PSD) image (A), the STAGE T1W image (B), the FLAIR (C), and the STAGE simulated FLAIR image (D). The lesions marked by the yellow and red arrows highlight the lesions with the highest water content, which are the brightest on the PSD map (A) and darkest in the T1W image (B) depicting high water content. The black holes inside the lesions seen in FLAIR (C) and simulated FLAIR (D) images also correspond to high water content in that part of the lesion.

Microvascular in vivo contrast revealed origins

Visualizing small arteries in MRI has always been a challenge. Although SWI can visualize medullary veins easily at 3 T and venules easily at 7 T when high resolution is used, imaging small arteries even with a T1 reducing contrast agent is more difficult because of the limited change in contrast available from the usual Gadolinium based agents. The reason SWI does such a good job is due to the fact that the deoxyhemoglobin concentration is enough to create a significant susceptibility shift of roughly 450 ppb. Arteries on the other hand have basically zero susceptibility change relative to the surrounding tissue. To overcome this limitation, microscopic in vivo contrast revealed origins (MICRO) imaging uses Ferumoxytol, an ultrasmall particle iron oxide, to endow the arteries with a nonzero susceptibility. The very strong dephasing that results makes it possible to image vessels as small as 50–100 μm in size, by using an imaging resolution of 220 μm at 3 T [53–55]. This approach makes it possible to study microvascular disease in a way only matched previously in the literature using cadaver brain studies.

Although the presence of Ferumoxytol increases visibility for both the arterial and venous vascular network, there is a much higher probability of a vein being present inside the MS lesions as opposed to an artery. Most of the MS plaques are distributed in the WM [4,56] and the arterial blood in the WM is supplied through arterial branches arising from the cerebral arteries, which are roughly 100 μm in size making them very difficult to visualize, especially once they branch into much smaller arterioles and capillaries [57]. On the other hand, the medullary veins, subependymal veins, and other subcortical WM veins have their primary confluence in the periventricular region, where WM lesion occurrence is the highest [56,58] making them much larger, at roughly 300 μm in size, than the arterial network in the same region. Additionally, the venous blood volume is roughly four times that of the arterial vasculature [59–62], leaving little signal to arise from the arteries. Nevertheless, there is still a need to confirm whether the vascular anomalies have only venous origins. Hence, the precontrast SWI data should be used to confirm that these vascular anomalies are part of the venous

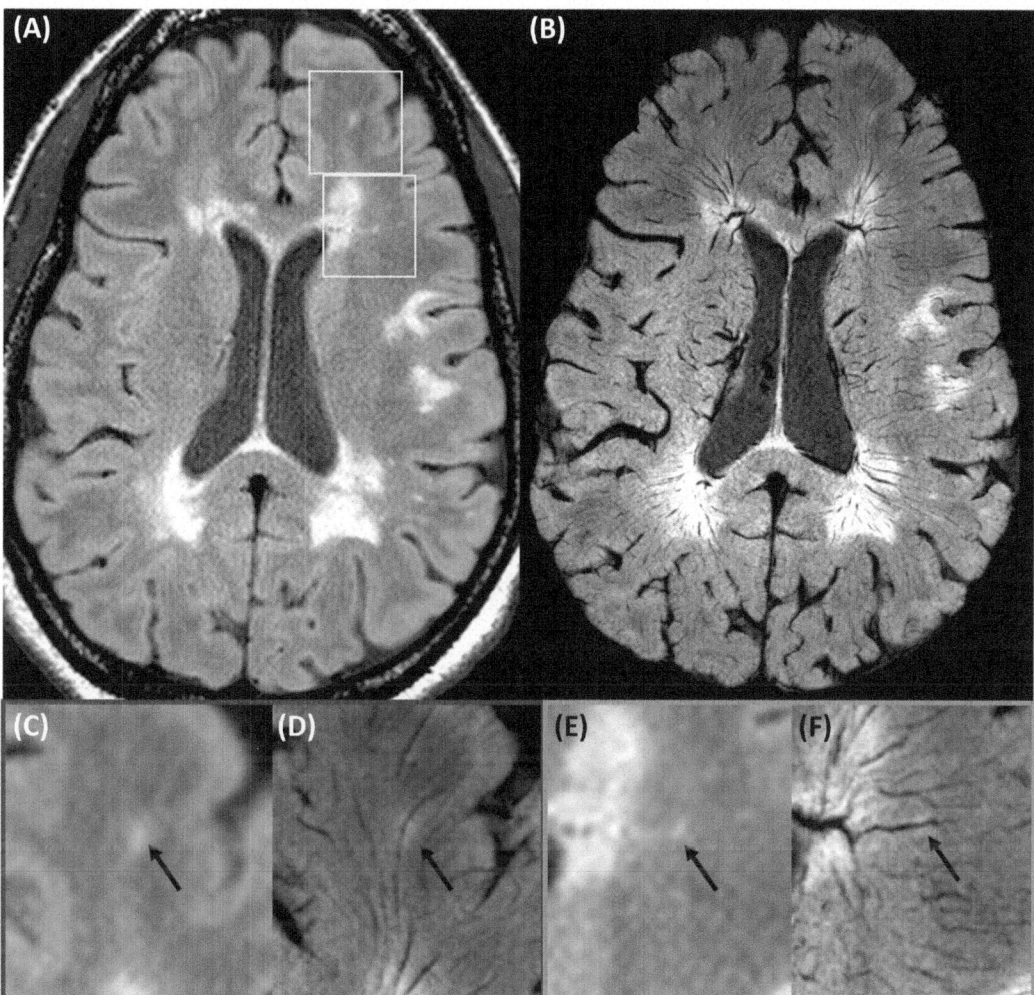

FIGURE 12.6 Inflammation along the vessel wall. (A) Original FLAIR data and (B) SWI-FLAIR data. The zoomed insets show two examples small WMH on the FLAIR (C and E) and SWI-FLAIR (D and F). The SWI-FLAIR data highlight the presence of these small WMHs along the vessel wall of a small vessel.

network [33]. We have seen a number of key venous vascular abnormalities in our MICRO imaging data for the MS patients including angiomas, microangiomas, engorged/dilated vessels, small ovoid lesions and perpendicular venous connections [33]. Additionally, the small ovoid enhancement (<3 mm in length) can be detected along the vessel wall of several small vessels on the combined SWI-FLAIR data (Fig. 12.6). These local enhancements are generally not observed or discarded in MS lesion-based analyses due to their small size or their inability to appear with sufficiently high contrast on the FLAIR data. The location of these lesions suggest collagenous thickening of the vessel wall and these lesions may enlarge and coalesce into a larger plaque over time [63].

Conclusion

MR continues to develop in its capabilities with more and more imaging biomarkers being developed. SWI, STAGE, and MICRO help complement other technologies such as FLAIR, T1W imaging, diffusion, and myelin water fraction imaging. They provide information about the vascular and inflammatory demyelinating nature of the lesions. Mapping the tissue properties of the lesions makes it possible to better understand the role of black holes versus high signal in T2W imaging and signal enhancements in FLAIR, that is, that they are all driven by the water content in the tissue. The water content itself may provide a biomarker for tissue damage and tissue atrophy. Taken together the new imaging approaches introduced in this chapter may lead to better correlations with clinical outcomes.

References

[1] Keegan BM, Noseworthy JH. Multiple sclerosis. Annu Rev Med 2002;53:285–302. Available from: https://doi.org/10.1146/annurev.med.53.082901.103909.

[2] Prineas JW, Kwon EE, Cho ES, Sharer LR, Barnett MH, Oleszak EL, et al. Immunopathology of secondary-progressive multiple sclerosis. Ann Neurol 2001;50(5):646–57. Available from: https://doi.org/10.1002/ana.1255.

[3] Haacke EM, Makki M, Ge Y, Maheshwari M, Sehgal V, Hu J, et al. Characterizing iron deposition in multiple sclerosis lesions using susceptibility weighted imaging. J Magnetic Reson Imaging 2009;29(3):537–44. Available from: https://doi.org/10.1002/jmri.21676.

[4] McDonald WI, Compston A, Edan G, Goodkin D, Hartung HP, Lublin FD, et al. Recommended diagnostic criteria for multiple sclerosis: guidelines from the International Panel on the diagnosis of multiple sclerosis. Ann Neurol 2001;50(1):121–7. Available from: https://doi.org/10.1002/ana.1032.

[5] Polman CH, Reingold SC, Banwell B, Clanet M, Cohen JA, Filippi M, et al. Diagnostic criteria for multiple sclerosis: 2010 revisions to the McDonald criteria. Ann Neurol 2011;69(2):292–302. Available from: https://doi.org/10.1002/ana.22366.

[6] Polman CH, Reingold SC, Edan G, Filippi M, Hartung H-P, Kappos L, et al. Diagnostic criteria for multiple sclerosis: 2005 revisions to the "McDonald Criteria. Ann Neurol 2005;58(6):840–6. Available from: https://doi.org/10.1002/ana.20703.

[7] Thompson AJ, Banwell BL, Barkhof F, Carroll WM, Coetzee T, Comi G, et al. Diagnosis of multiple sclerosis: 2017 revisions of the McDonald criteria. Lancet Neurol 2018;17(2):162–73. Available from: https://doi.org/10.1016/S1474-4422(17)30470-2.

[8] Reichenbach JR, Venkatesan R, Schillinger DJ, Kido DK, Haacke EM. Small vessels in the human brain: MR venography with deoxyhemoglobin as an intrinsic contrast agent. Radiology 1997;204(1):272–7. Available from: https://doi.org/10.1148/radiology.204.1.9205259.

[9] Haacke E, Mittal S, Wu Z, Neelavalli J, Cheng Y-CN. Susceptibility-weighted imaging: technical aspects and clinical applications, part 1. Am J Neuroradiol 2009;30(1):19–30. Available from: https://doi.org/10.3174/ajnr.A1400.

[10] Liu S, Buch S, Chen Y, Choi H-S, Dai Y, Habib C, et al. Susceptibility-weighted imaging: current status and future directions. NMR Biomed 2017;30(4). Available from: https://doi.org/10.1002/nbm.3552.

[11] Mittal S, Wu Z, Neelavalli J, Haacke EM. Susceptibility-weighted imaging: technical aspects and clinical applications, part 2. AJNR: Am J Neuroradiol 2009;30(2):232–52. Available from: https://doi.org/10.3174/ajnr.A1461.

[12] Absinta M, Sati P, Gaitán MI, Maggi P, Cortese ICM, Filippi M, et al. Seven-tesla phase imaging of acute multiple sclerosis lesions: a new window into the inflammatory process. Ann Neurol 2013;74(5):669–78. Available from: https://doi.org/10.1002/ana.23959.

[13] Lane JI, Bolster B, Campeau NG, Welker KM, Gilbertson JR. Characterization of multiple sclerosis plaques using susceptibility-weighted imaging at 1.5 T: can perivenular localization improve specificity of imaging criteria? J Computer Assist Tomogr 2015;39(3):317–20. Available from: https://doi.org/10.1097/RCT.0000000000000233.

[14] Samaraweera APR, Clarke MA, Whitehead A, Falah Y, Driver ID, Dineen RA, et al. The central vein sign in multiple sclerosis lesions is present irrespective of the T2* sequence at 3 T. J Neuroimaging: Off J Am Soc Neuroimaging 2017;27(1):114–21. Available from: https://doi.org/10.1111/jon.12367.

[15] Tallantyre EC, Brookes MJ, Dixon JE, Morgan PS, Evangelou N, Morris PG. Demonstrating the perivascular distribution of MS lesions in vivo with 7-Tesla MRI. Neurology 2008;70(22):2076–8. Available from: https://doi.org/10.1212/01.wnl.0000313377.49555.2e.

[16] Tallantyre EC, Dixon JE, Donaldson I, Owens T, Morgan PS, Morris PG, et al. Ultra-high-field imaging distinguishes MS lesions from asymptomatic white matter lesions. Neurology 2011;76(6):534–9. Available from: https://doi.org/10.1212/WNL.0b013e31820b7630.

[17] Sati P, Oh J, Constable RT, Evangelou N, Guttmann CRG, Henry RG, et al. The central vein sign and its clinical evaluation for the diagnosis of multiple sclerosis: A consensus statement from the North American Imaging in Multiple Sclerosis Cooperative. Nat Rev Neurol 2016;12(12):714–22. Available from: https://doi.org/10.1038/nrneurol.2016.166.

[18] Mistry N, Dixon J, Tallantyre E, Tench C, Abdel-Fahim R, Jaspan T, et al. Central veins in brain lesions visualized with high-field magnetic resonance imaging: a pathologically specific diagnostic biomarker for inflammatory demyelination in the brain. JAMA Neurol 2013;70(5):623–8. Available from: https://doi.org/10.1001/jamaneurol.2013.1405.

[19] Haacke EM, Bernitsas E, Subramanian K, Utriainen D, Palutla VK, Yerramsetty K, et al. A comparison of magnetic resonance imaging methods to assess multiple sclerosis lesions: implications for patient characterization and clinical trial design. Diagnostics (Basel, Switz) 2021;12(1):77. Available from: https://doi.org/10.3390/diagnostics12010077.

[20] Coffman CH, White R, Subramanian K, Buch S, Bernitsas E, Haacke EM. Quantitative susceptibility mapping of both ring and non-ring white matter lesions in relapsing remitting multiple sclerosis. Magnetic Reson Imaging 2022;91:45–51. Available from: https://doi.org/10.1016/j.mri.2022.05.009.

[21] Dal-Bianco A, Grabner G, Kronnerwetter C, Weber M, Höftberger R, Berger T, et al. Slow expansion of multiple sclerosis iron rim lesions: pathology and 7 T magnetic resonance imaging. Acta Neuropathol 2017;133 (1):25–42. Available from: https://doi.org/10.1007/s00401-016-1636-z.

[22] Harrison DM, Li X, Liu H, Jones CK, Caffo B, Calabresi PA, et al. Lesion heterogeneity on high-field susceptibility MRI is associated with multiple sclerosis severity. AJNR Am J Neuroradiol 2016;37(8):1447–53. Available from: https://doi.org/10.3174/ajnr.A4726.

[23] Zhang S, Nguyen TD, Hurtado Rúa SM, Kaunzner UW, Pandya S, Kovanlikaya I, et al. Quantitative susceptibility mapping of time-dependent susceptibility changes in multiple sclerosis lesions. AJNR: Am J Neuroradiol 2019;40(6):987–93. Available from: https://doi.org/10.3174/ajnr.A6071.

[24] Chawla S, Kister I, Sinnecker T, Wuerfel J, Brisset J-C, Paul F, et al. Longitudinal study of multiple sclerosis lesions using ultra-high field (7T) multiparametric MR imaging. PLoS One 2018;13(9):e0202918. Available from: https://doi.org/10.1371/journal.pone.0202918.

[25] Hosseini Z, Matusinec J, Rudko DA, Liu J, Kwan BYM, Salehi F, et al. Morphology-specific discrimination between MS white matter lesions and benign white matter hyperintensities using ultra-high-field MRI. AJNR Am J Neuroradiol 2018;39(8):1473–9. Available from: https://doi.org/10.3174/ajnr.A5705.

[26] Barkhof F, Scheltens P. Imaging of white matter lesions. Cerebrovasc Dis (Basel, Switz) 2002;13(Suppl 2):21–30. Available from: https://doi.org/10.1159/000049146.

[27] Simon JH, Li D, Traboulsee A, Coyle PK, Arnold DL, Barkhof F, et al. Standardized MR imaging protocol for multiple sclerosis: consortium of MS Centers consensus guidelines. AJNR Am J Neuroradiol 2006;27 (2):455–61.

[28] Scharf J, Bräuherr E, Forsting M, Sartor K. Significance of haemorrhagic lacunes on MRI in patients with hypertensive cerebrovascular disease and intracerebral haemorrhage. Neuroradiology 1994;36(7):504–8. Available from: https://doi.org/10.1007/BF00593508.

[29] Wu Z, Mittal S, Kish K, Yu Y, Hu J, Haacke EM. Identification of calcification with MRI using susceptibility-weighted imaging: a case study. J Magnetic Reson Imaging: JMRI 2009;29(1):177–82. Available from: https://doi.org/10.1002/jmri.21617.

[30] Haacke EM, Cheng NYC, House MJ, Liu Q, Neelavalli J, Ogg RJ, et al. Imaging iron stores in the brain using magnetic resonance imaging. Magnetic Reson Imaging 2005;23(1):1–25. Available from: https://doi.org/10.1016/j.mri.2004.10.001.

[31] Hammond KE, Metcalf M, Carvajal L, Okuda DT, Srinivasan R, Vigneron D, et al. Quantitative in vivo magnetic resonance imaging of multiple sclerosis at 7 Tesla with sensitivity to iron. Ann Neurol 2008;64(6):707–13. Available from: https://doi.org/10.1002/ana.21582.

[32] Sati P, George IC, Shea CD, Gaitán MI, Reich DS. FLAIR*: a combined MR contrast technique for visualizing white matter lesions and parenchymal veins. Radiology 2012;265(3):926–32. Available from: https://doi.org/10.1148/radiol.12120208.

[33] Buch S, Subramanian K, Jella PK, Chen Y, Wu Z, Shah K, et al. Revealing vascular abnormalities and measuring small vessel density in multiple sclerosis lesions using USPIO. NeuroImage: Clin 2020;102525. Available from: https://doi.org/10.1016/j.nicl.2020.102525.

[34] Tong KA, Ashwal S, Holshouser BA, Nickerson JP, Wall CJ, Shutter LA, et al. Diffuse axonal injury in children: clinical correlation with hemorrhagic lesions. Ann Neurol 2004;56(1):36–50. Available from: https://doi.org/10.1002/ana.20123.

[35] Tong KA, Ashwal S, Holshouser BA, Shutter LA, Herigault G, Haacke EM, et al. Hemorrhagic shearing lesions in children and adolescents with posttraumatic diffuse axonal injury: improved detection and initial results. Radiology 2003;227(2):332–9. Available from: https://doi.org/10.1148/radiol.2272020176.

[36] Wycliffe ND, Choe J, Holshouser B, Oyoyo UE, Haacke EM, Kido DK. Reliability in detection of hemorrhage in acute stroke by a new three-dimensional gradient recalled echo susceptibility-weighted imaging technique compared to computed tomography: a retrospective study. J Magnetic Reson Imaging: JMRI 2004;20(3):372–7. Available from: https://doi.org/10.1002/jmri.20130.

[37] Sparacia G, Speciale C, Banco A, Bencivinni F, Midiri M. Accuracy of SWI sequences compared to T2*-weighted gradient echo sequences in the detection of cerebral cavernous malformations in the familial form. Neuroradiol J 2016;29(5):326–35. Available from: https://doi.org/10.1177/1971400916665376.

[38] Cheng A-L, Batool S, McCreary CR, Lauzon ML, Frayne R, Goyal M, et al. Susceptibility-weighted imaging is more reliable than T2*-weighted gradient-recalled echo MRI for detecting microbleeds. Stroke 2013;44(10):2782–6. Available from: https://doi.org/10.1161/STROKEAHA.113.002267.

[39] Haacke EM, DelProposto ZS, Chaturvedi S, Sehgal V, Tenzer M, Neelavalli J, et al. Imaging cerebral amyloid angiopathy with susceptibility-weighted imaging. AJNR Am J Neuroradiol 2007;28(2):316–17.

[40] Lee BC, Vo KD, Kido DK, Mukherjee P, Reichenbach J, Lin W, et al. MR high-resolution blood oxygenation level-dependent venography of occult (low-flow) vascular lesions. AJNR Am J Neuroradiol 1999;20(7):1239–42.

[41] Reichenbach JR, Jonetz-Mentzel L, Fitzek C, Haacke EM, Kido DK, Lee BC, et al. High-resolution blood oxygen-level dependent MR venography (HRBV): a new technique. Neuroradiology 2001;43(5):364–9.

[42] Chavhan GB, Babyn PS, Thomas B, Shroff MM, Haacke EM. Principles, techniques, and applications of T2*-based MR imaging and its special applications. Radiographics 2009;29(5):1433–49. Available from: https://doi.org/10.1148/rg.295095034.

[43] Haacke EM, Liu S, Buch S, Zheng W, Wu D, Ye Y. Quantitative susceptibility mapping: current status and future directions. Magnetic Reson Imaging 2015;33(1):1–25. Available from: https://doi.org/10.1016/j.mri.2014.09.004.

[44] Wang Y, Spincemaille P, Liu Z, Dimov A, Deh K, Li J, et al. Clinical quantitative susceptibility mapping (QSM): biometal imaging and its emerging roles in patient care. J Magnetic Reson Imaging: JMRI 2017;46(4):951–71. Available from: https://doi.org/10.1002/jmri.25693.

[45] Liu C, Li W, Johnson GA, Wu B. High-field (9.4 T) MRI of brain dysmyelination by quantitative mapping of magnetic susceptibility. NeuroImage 2011;56(3):930–8. Available from: https://doi.org/10.1016/j.neuroimage.2011.02.024.

[46] Chen Y, Liu S, Wang Y, Kang Y, Haacke EM. STrategically Acquired Gradient Echo (STAGE) imaging, part I: creating enhanced T1 contrast and standardized susceptibility weighted imaging and quantitative susceptibility mapping. Magnetic Reson Imaging 2018;46:130–9. Available from: https://doi.org/10.1016/j.mri.2017.10.005.

[47] Haacke EM, Chen Y, Utriainen D, Wu B, Wang Y, Xia S, et al. STrategically Acquired Gradient Echo (STAGE) imaging, part III: technical advances and clinical applications of a rapid multi-contrast multi-parametric brain imaging method. Magnetic Reson Imaging 2020;65:15–26. Available from: https://doi.org/10.1016/j.mri.2019.09.006.

[48] Wang Y, Chen Y, Wu D, Wang Y, Sethi SK, Yang G, et al. STrategically Acquired Gradient Echo (STAGE) imaging, part II: Correcting for RF inhomogeneities in estimating T1 and proton density. Magnetic Reson Imaging 2018;46:140–50. Available from: https://doi.org/10.1016/j.mri.2017.10.006.

[49] He N, Wu B, Liu Y, Zhang C, Cheng J, Gao B, et al. STAGE as a multicenter, multivendor protocol for imaging Parkinson's disease: a validation study on healthy controls. Chin J Acad Radiol 2022;5(1):47–60. Available from: https://doi.org/10.1007/s42058-022-00089-3.

[50] Wu D, Liu S, Buch S, Ye Y, Dai Y, Haacke EM. A fully flow-compensated multiecho susceptibility-weighted imaging sequence: the effects of acceleration and background field on flow compensation. Magnetic Reson Med 2016;76(2):478–89. Available from: https://doi.org/10.1002/mrm.25878.

[51] van Walderveen MA, Kamphorst W, Scheltens P, van Waesberghe JH, Ravid R, Valk J, et al. Histopathologic correlate of hypointense lesions on T1-weighted spin-echo MRI in multiple sclerosis. Neurology 1998;50 (5):1282–8. Available from: https://doi.org/10.1212/wnl.50.5.1282.

[52] Zivadinov R, Stosic M, Cox JL, Ramasamy DP, Dwyer MG. The place of conventional MRI and newly emerging MRI techniques in monitoring different aspects of treatment outcome. J Neurol 2008;255(Suppl 1):61–74. Available from: https://doi.org/10.1007/s00415-008-1009-1.

[53] Liu S, Brisset J-C, Hu J, Haacke EM, Ge Y. Susceptibility weighted imaging and quantitative susceptibility mapping of the cerebral vasculature using ferumoxytol. J Magnetic Reson Imaging 2018;47(3):621–33. Available from: https://doi.org/10.1002/jmri.25809.

[54] Shen Y, Hu J, Eteer K, Chen Y, Buch S, Alhourani H, et al. Detecting sub-voxel microvasculature with USPIO-enhanced susceptibility-weighted MRI at 7 T. Magnetic Reson Imaging 2020;67:90–100. Available from: https://doi.org/10.1016/j.mri.2019.12.010.

[55] Buch S, Wang Y, Park M-G, Jella PK, Hu J, Chen Y, et al. Subvoxel vascular imaging of the midbrain using USPIO-Enhanced MRI. NeuroImage 2020;220117106. Available from: https://doi.org/10.1016/j.neuroimage.2020.117106.

[56] Vellinga MM, Geurts JJG, Rostrup E, Uitdehaag BMJ, Polman CH, Barkhof F, et al. Clinical correlations of brain lesion distribution in multiple sclerosis. J Magnetic Reson Imaging: JMRI 2009;29(4):768–73. Available from: https://doi.org/10.1002/jmri.21679.

[57] Nonaka H, Akima M, Hatori T, Nagayama T, Zhang Z, Ihara F. Microvasculature of the human cerebral white matter: arteries of the deep white matter. Neuropathology 2003;23(2):111–18. Available from: https://doi.org/10.1046/j.1440-1789.2003.00486.x.

[58] Charil A, Zijdenbos AP, Taylor J, Boelman C, Worsley KJ, Evans AC, et al. Statistical mapping analysis of lesion location and neurological disability in multiple sclerosis: application to 452 patient data sets. NeuroImage 2003;19(3):532–44. Available from: https://doi.org/10.1016/s1053-8119(03)00117-4.

[59] An H, Lin W. Cerebral venous and arterial blood volumes can be estimated separately in humans using magnetic resonance imaging. Magnetic Reson Med 2002;48(4):583–8. Available from: https://doi.org/10.1002/mrm.10257.

[60] Hua J, Liu P, Kim T, Donahue M, Rane S, Chen JJ, et al. MRI techniques to measure arterial and venous cerebral blood volume. NeuroImage 2019;187:17–31. Available from: https://doi.org/10.1016/j.neuroimage.2018.02.027.

[61] McCormick PW, Stewart M, Goetting MG, Balakrishnan G. Regional cerebrovascular oxygen saturation measured by optical spectroscopy in humans. Stroke 1991;22(5):596–602. Available from: https://doi.org/10.1161/01.str.22.5.596.

[62] Pollard V, Prough DS, DeMelo AE, Deyo DJ, Uchida T, Stoddart HF. Validation in volunteers of a near-infrared spectroscope for monitoring brain oxygenation in vivo. Anesthesia Analgesia 1996;82(2):269–77. Available from: https://doi.org/10.1097/00000539-199602000-00010.

[63] Bo L, Evangelou N, Tallantyre E. The neuropathology of progressive multiple sclerosis. In: Wilkins A, editor. Progressive multiple sclerosis. Springer; 2013. p. 51–70. Available from: https://doi.org/10.1007/978-1-4471-2395-8_4.

CHAPTER 13

Quantitative susceptibility mapping in multiple sclerosis

Ferdinand Schweser[1,2] and Alexander Rauscher[3,4,5]

[1]Department of Neurology, Buffalo Neuroimaging Analysis Center, Jacobs School of Medicine and Biomedical Sciences at the University at Buffalo, Buffalo, NY, United States [2]Center for Biomedical Imaging, Clinical and Translational Science Institute at the University at Buffalo, Buffalo, NY, United States [3]UBC MRI Research Centre, The University of British Columbia, Vancouver, BC, Canada [4]Department of Physics and Astronomy, The University of British Columbia, Vancouver, BC, Canada [5]Department of Pediatrics, The University of British Columbia, Vancouver, BC, Canada

OUTLINE

Introduction	222	Subvoxel distribution of iron in the deep gray matter	230
Fundamentals of quantitative susceptibility mapping	223	Gadolinium retention in the deep gray matter	230
Magnetic susceptibility of tissue	223		
Imaging	223	Confounding effects of myelin in the deep gray matter	231
Processing	225		
Interpretation of changes in tissue susceptibility	226	Normal appearing white matter	231
Nonheme iron concentrations in the deep gray matter	227	Focal white matter damage	232
		Identification of gadolinium-enhancing lesions	233
Findings in multiple sclerosis	227		
Comparison with other methods	228	Paramagnetic rim lesions	234
Regional content of nonheme iron in the deep gray matter	229		

Other applications of quantitative susceptibility mapping in multiple sclerosis	235	*Deep gray matter segmentation*	236
		Summary and outlook	236
Oxygen extraction fraction measurements	235	Acknowledgments	237
Microbleeds	235	References	237

Introduction

When tissue is exposed to the static magnetic field of magnetic resonance imaging (MRI), it becomes magnetized. This process of self-magnetization can be quantified using a unitless physics metric called magnetic susceptibility, typically symbolized by the Greek letter χ. Quantitative susceptibility mapping (QSM) is a relatively new technique that quantifies this magnetic susceptibility using MRI, generating susceptibility maps, wherein the image intensity of each voxel directly corresponds to the tissue's magnetic susceptibility value.

Variations in the local myelin and iron abundance are the primary drivers of differences in magnetic susceptibility among central nervous system (CNS) tissues [1]. QSM, with its heightened sensitivity to these variations, compared to alternative MRI techniques, has proven useful in investigating both the myelin-rich white matter (WM) and the iron-abundant deep gray matter (DGM) in MS. By providing a means to compare study outcomes across brain regions, groups, and time points, QSM has gained recognition as a powerful research tool, holding promise for applications in future personalized medicine.

The most commonly utilized acquisition pulse sequence for QSM is the gradient-recalled echo (GRE) sequence, identical to the one used in susceptibility-weighted imaging (SWI) (see Chapter X). The primary distinction between SWI and QSM lies in the postacquisition processing of raw MRI data. SWI processing enhances the visualization of small veins and hemorrhages by heuristically amplifying interfaces between tissues of different magnetic susceptibility [2], such as those between deoxygenated blood in veins and surrounding parenchyma. As a result, iron-containing structures in SWI appear larger than their actual dimensions, thereby aiding the identification of small structures like microhemorrhages and veins at or below the imaging resolution [3–5].

In contrast to SWI's heuristic, qualitative processing, QSM carries out an analytical, quantitative examination of the raw data [6–8]. This involves the numerical solving of a series of physics-based mathematical problems. Initially, it was anticipated that QSM would supersede SWI; however, it has become evident that the two methods serve complementary roles within the neuroimaging toolkit. While SWI is a practical clinical tool to visualize microhemorrhages and vascular abnormalities, the role of QSM is currently, apart from some special applications discussed below, primarily centered around quantitative investigations of tissue alterations in the research setting.

This chapter delineates the technical underpinnings of QSM before reviewing its applications in the investigation of white and gray matter (GM) in people with MS. Readers not interested in the technicalities of QSM are advised to proceed directly with subsection "Interpretation of changes in tissue susceptibility."

Fundamentals of quantitative susceptibility mapping

Magnetic susceptibility of tissue

Every substance magnetizes when exposed to an external magnetic field, such as the strong static main magnetic field in an MRI scanner. Magnetic susceptibility, which describes the magnitude and direction of this magnetization, is primarily determined by a substance's electron configuration. It can be expressed in its simplest form as the ratio of the induced magnetization M to the applied magnetic field B_0:

$$M = \chi \cdot B_0/\mu_0,$$

where μ_0 represents the vacuum permeability.

Exposure to a magnetic field induces a minimal magnetization in all matter, counteracting the applied field, a phenomenon known as Langevin diamagnetism ($\chi < 0$). Substances with unpaired valence electrons exhibit an additional paramagnetic magnetization, which typically exceeds Langevin diamagnetism by several orders of magnitude ($\chi > 0$).

The average susceptibility of biological tissues is dictated by the magnetic susceptibility and volume fraction of their components. Water, being diamagnetic ($\chi_{water} = -9.04$ ppm; all values reported in SI units in this chapter) and a primary constituent of biological tissues, strongly influences their average susceptibility. Observed susceptibility variations between tissue regions in the brain and spinal cord are predominantly attributed to differences in myelin, iron, and calcium content [1,9,10]. Myelin, though only slightly more diamagnetic than water ($\chi_{my} = -9.074$ to -9.053 ppm [11–14]), imparts a discernible diamagnetic shift in the WM due to its high abundance [1,10]. Paramagnetism of iron in deoxygenated heme [15,16] results in a considerable positive susceptibility shift in larger veins compared to the surrounding parenchyma (fully deoxygenated blood: $\chi_{deox} = -7.9$ ppm [17,18]), but contributions from the venous side of the microvasculature are generally insignificant [19–21]. Most nonheme iron, often labeled as tissue iron, is securely sequestered in the ferritin globular protein complex's crystalline core [22], where it exhibits strong paramagnetism ($\chi_{Fe} \leq 520$ ppm) [9,23,24]. Despite tissue iron concentrations being substantially lower than myelin concentrations, they induce noticeable positive shifts in tissue susceptibility [23]. Calcium in hydroxyapatite [25,26] is diamagnetic ($\chi_{ha} = -14.83$ ppm [27]), but its physiological concentrations in tissues are typically too low to significantly impact tissue susceptibility, except in calcified brain lesions and the choroid plexus [28–30].

Imaging

The spoiled GRE sequence, available on clinical MRI scanners and the backbone of data acquisition for QSM, is also employed for SWI and R_2^* mapping (cf. Chapters X and Y), facilitating the integration of all three imaging techniques within a single sequence acquisition. As with SWI and R_2^* mapping, QSM necessitates the configuration of the sequence with extended echo times (TEs) to sensitize the signal to magnetic field perturbations. While the parameters originally suggested for SWI may be employed for QSM [31], isotropic voxel sizes can enhance QSM processing [32]. Multiecho sequences, which concurrently capture several images at distinct TEs, offer improved noise characteristics in the final reconstructed maps

[33,34], permit more accurate QSM reconstruction techniques [35], and are a prerequisite for R_2^* calculation [36,37]. The GRE sequence particularly benefits from high magnetic field strength [38]. Increased signal strength can be exchanged for enhanced spatial resolution and faster acquisition. Typical TEs used are up to 40–50 ms at 1.5 T [39] and 20–50 ms at 3 and 7 T [40]. Most QSM studies have been performed at 3 and 7 T scanners.

The magnitude and phase components of the complex-valued GRE signal offer complementary information (see Fig. 13.1). While the magnitude primarily captures field perturbations at the subvoxel level (via the R_2^* effect), the phase component is dominated by long-ranging, voxel-scale spatial variations in the magnetic field. The Larmor relation [41] provides a straightforward mathematical connection that links local phase signal shifts ($\Delta\varphi$) at TE to local magnetic field perturbations (ΔB):

$$\Delta\varphi = \gamma \cdot TE \cdot \Delta B, \tag{13.1}$$

where γ represents the gyromagnetic ratio. This equation enables the estimation of the field, ΔB, from GRE phase images, $\Delta\varphi$, at TE.

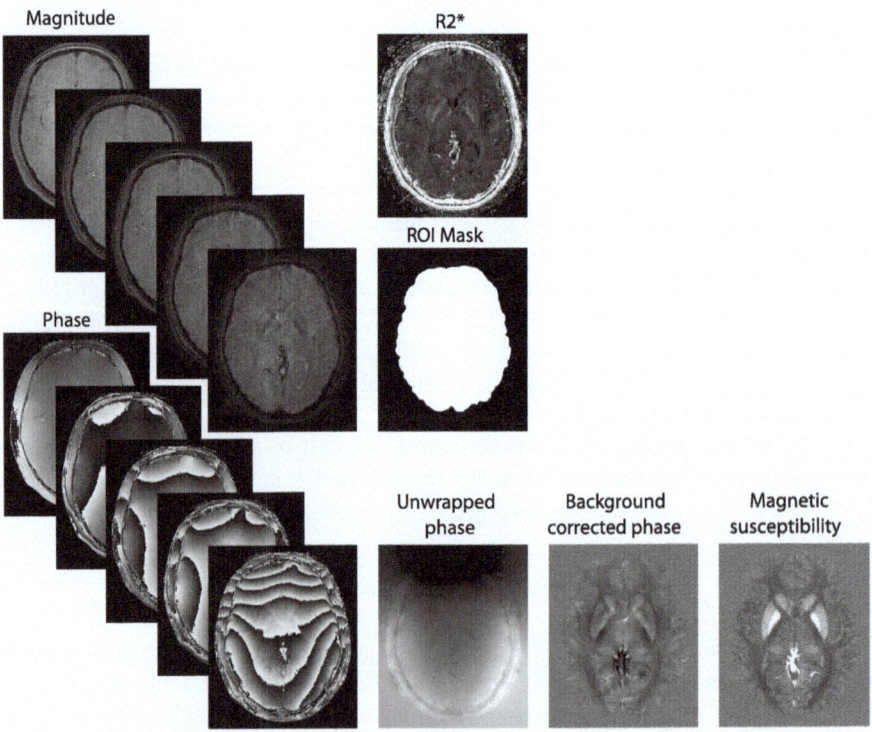

FIGURE 13.1 Schematic of the components of the quantitative susceptibility mapping pipeline. The multiecho GRE sequence yields a set of magnitude (top left) and phase (bottom left) images. Phase images are combined into a field map, which is unwrapped, background corrected (using a brain mask), and finally converted into a map of magnetic susceptibility. The magnitude echo images can be used to generate an R_2^* map. Source: *Courtesy of Christian Kames, University of British Columbia.*

By default, most MRI scanners only store magnitude images and discard the phase component. However, some scanners offer phase image reconstruction by default, and on others, this can be activated through research-level configurations. The presence of artifacts and other technical limitations related to scanner-based phase image reconstruction on many clinical MRI systems, particularly older models, currently pose a hurdle to the clinical translation of QSM. When scanner-based phase reconstruction is inadequate, data must be exported as raw k-space data or single-channel images, with phase images reconstructed offline [35]. A comprehensive review of the current state of GRE phase imaging and QSM for clinical research in the brain can be found in a consensus paper by the ISMRM Electro-Magnetic Tissue Properties (EMTP) study group that is under review at the time of this writing [42].

Several MRI scanners automatically generate SWI high-pass filtered phase images (Chapter X). Conventional QSM algorithms cannot utilize these filtered images, necessitating dedicated approaches that account for the filtering [43,44]. Alternatively, the unfiltered phase can be restored from the filtered phase and then used as an input to conventional QSM techniques [45]. Although these algorithms have demonstrated exceptional susceptibility mapping performance in methodological studies, their robustness and accuracy require validation in more extensive studies.

Processing

QSM seeks to determine a distribution of magnetic susceptibility, χ, that accounts for the detected magnetic field perturbation within the tissue, ΔB, as mirrored in the GRE phase images, $\Delta \varphi$: $\Delta \varphi \to \Delta B \to \chi$. The solution of this relation belongs to a class of mathematical problems known as inverse problems, which often pose substantial challenges due to their complexity. Magnetostatic theory provides a precise account of the field perturbation resulting from a given distribution of tissue magnetic susceptibility, which is the "forward problem." However, the inverse problem—discovering the susceptibility distribution responsible for an observed field perturbation—is highly sensitive to measurement noise and other errors in the observed field map (i.e., it is ill-posed) [46].

QSM involves the solving of multiple mathematical problems to transform the measured phase images into a susceptibility map. Herein, we offer a conceptual overview of the various steps involved, directing interested readers to referenced review articles for a more in-depth exploration. QSM addresses the following processing problems either discretely, in a step-by-step manner, or by integrating multiple steps into one mathematical problem. Several software toolboxes that consolidate all processing steps are publicly available at the website of the EMTP MR community, EMTP Hub (https://www.emtphub.org/magnetic-software-packages/).

The initial processing step of QSM is the conversion of the GRE phase images into a map of the magnetic field, ΔB, within the tissue. Based on Eq. (13.1), field mapping is accomplished by examining how swiftly GRE phase accumulates with increasing echo time. In multiecho data, fitting Eq. (13.1) to the measured phase shifts, $\Delta \varphi$, at different TEs directly yields ΔB. When only one echo is available, field mapping is based on the assumption that $\Delta \varphi$ is constant zero at TE = 0. Various methodologies improving noise characteristics and robustness have been proposed for field mapping [47–50]. Most methods involve resolving 2π-jumps or "wraps" in phase measurements—a process known as phase unwrapping [35]. Phase wraps occur when the dynamic range of phase exceeds an interval width of 2π, at which point the phase aliases back into the 2π-interval (Fig. 13.1). In brain imaging with long TEs, phase wrapping is almost inevitable.

The second processing step involves removing field perturbations from the field maps that originate from magnetic susceptibility variations outside the tissue region of interest (ROI)—for instance, the brain in a brain scan. These background fields are typically induced by tissue–air or tissue–bone interfaces and can be significantly stronger than the subtle field perturbations caused by the susceptibility variations within the ROI. Numerous background correction techniques have been proposed, differing mostly in robustness and accuracy near the ROI boundary region [51].

The final processing step consists of inverting the background-corrected field map into a magnetic susceptibility map. Algorithms manage the ill-posed nature of this problem using a strategy termed numerical regularization. Regularization applies mathematical constraints to the calculated solution, such as spatial smoothness, to limit noise and artifact amplification in the resultant susceptibility map. A considerable number of inversion algorithms have been proposed in the last decade. Benchmarking studies have revealed that the appearance and quantitative accuracy of the computed susceptibility maps can vary considerably between algorithms and based on algorithm-specific parameter choices [52,53]. The impact of differences between processing pipelines on study outcomes remains unclear [54,55]. Comparisons between studies and the pooling of results from different studies may necessitate special consideration of algorithmic differences.

Interpretation of changes in tissue susceptibility

Postmortem experiments have validated a linear association between tissue iron concentrations and susceptibility values derived from QSM within the DGM [56–58]. However, akin to numerous other quantitative MRI techniques [59–63], discerning the biological implications of contrast changes on susceptibility maps presents a challenge due to the interdependency of susceptibility on myelin and calcium.

The majority of QSM studies, particularly earlier ones, operated under the assumption that susceptibility alterations in the WM are primarily driven by myelin changes, while those in the GM are mainly influenced by iron. This assumption is based on two premises: firstly, iron concentrations in the WM are relatively low (40 ppm of tissue mass in the frontal WM vs 200 ppm in the globus pallidus [64]), and secondly, the GM tends to have lower levels of myelination. Consequently, an increase in WM susceptibility has often been interpreted as indicative of demyelination, while increasing GM susceptibility is often attributed to iron accumulation. However, this presumption is contested by the fact that certain WM regions have iron concentrations comparable to those in the DGM, such as the superficial WM U-fibers, and some DGM areas like the globus pallidus show relatively high myelination. The interpretive complexity is heightened further by the fact that both iron and myelin contents undergo alterations in MS [65–67].

Addressing the codependency of susceptibility on iron and myelin may be possible by integrating QSM with other quantitative MR metrics, which could yield biologically more specific outcome measures. This chapter will subsequently review several methods with relevance to MS that have potential in this context.

Nonheme iron concentrations in the deep gray matter

The DGM regions in the adult human brain normally contain the highest concentrations of nonheme iron [64,68]. The globus pallidus, with an iron concentration of 20 mg per 100 g of tissue, is the region with the highest concentration. This is followed in decreasing order by the red nucleus, substantia nigra, putamen, and the dentate nucleus, which has 10 mg of iron per 100 g of tissue [64,68]. It has been revealed through MRI studies that many neurological diseases are associated with changes in nonheme iron concentrations [69,70].

Drayer et al. were pioneers in identifying abnormal brain iron clinically using T_2-weighted MRI in people with MS [71]. Their studies reported elevated iron concentrations in the thalamus and putamen of people with MS [72,73]. Subsequently, Bakshi et al. quantified these alterations and established significant correlations with disease duration and neurological disability. Bakshi and colleagues also expanded the list of regions with abnormal iron concentrations in people with MS to include the globus pallidus, caudate, substantia nigra, and red nucleus [74].

QSM is the most recent MRI technique to not only confirm these initial findings but also provide more comprehensive insights into tissue iron dynamics in MS [23,75]. Techniques such as QSM, R_2^* mapping [76,77], and the quantitative evaluation of SWI filtered phase images [78–81] offer greater accuracy and reproducibility in estimating iron concentration compared to T2-weighted MRI and facilitate tracking of iron levels over time. QSM represents iron concentrations more directly than relaxation-based methods, such as R_2^*, as it is less sensitive to water content (e.g., in edema), the subvoxel distribution of iron [82–84], and the oxidation state of iron [85]—although the latter may be negligible in biological tissues [61]. Moreover, compared to filtered phase imaging, QSM is less susceptible to morphological changes, for instance, those related to atrophy [86].

Findings in multiple sclerosis

Magnetic susceptibility exhibits alterations in people with MS relative to controls across all disease phenotypes and stages. Clinically isolated syndrome studies demonstrate an increase in susceptibility relative to controls in the globus pallidus, caudate, putamen, and substantia nigra [87–89]. Conversely, a decrease in susceptibility in the thalamus was reported by some authors [90,91].

In parallel, studies of people with definite MS largely observe increased susceptibility in the globus pallidus, caudate, and putamen [82,87–89,92–98], which concurs with findings from other iron-sensitive MRI techniques. Findings regarding the thalamus are more inconsistent across studies. Some authors report increased susceptibility [89,92], while others report a decrease [90,91,94,95,97,99,100]. Research into intrathalamic susceptibility changes suggests the disease differentially affects thalamic nuclei [91,94], with the pulvinar subregions showing a significant linear decrease in susceptibility over the disease duration [91].

Clinical disability has been associated with lower susceptibility in the thalamus [94] and higher susceptibility in other DGM regions [90,92,94,101]. One study linked impaired inhibitory control with increased susceptibility in the caudate and the anterior part of the putamen [101], while another demonstrated that globus pallidus susceptibility could explain 10% of variance in total cognitive performance based on the Brief Repeatable Battery of Neuropsychological Tests, even after adjusting for regional atrophy [96].

MS phenotypes were found to exhibit differences in DGM susceptibility [95,99], and studies reported distinctive iron distribution patterns in people with neuromyelitis optica compared to typical MS patterns [90,102,103]. Overall, findings indicate that QSM may have utility in MS clinical diagnosis and in monitoring or predicting transitions from relapsing to progressive stages. However, the current evidence does not robustly support clinical translation, especially in translating population-level findings to individual subjects. Provided that population-level findings are systematically confirmed and the inconsistency between studies is elucidated, QSM could function as an element of an enhanced imaging biomarker for tracking disease progression in clinical trials.

The cellular substrates of the imaging findings of altered iron concentrations and the cause and role of these alterations in MS remain enigmatic [23,75,104]. The prevailing hypothesis postulates a causal relationship between increased iron observed on MRI and oxidative stress, which is implicated in cell injury [66,105]. Given the brain tissue's high sensitivity to oxidative stress due to a high concentration of polyunsaturated fatty acids— prone to lipid peroxidation [106]—free iron may amplify the generation of reactive oxygen species, specifically hydroxyl radicals, through Fenton and Haber-Weiss chemistry [107]. A recent study involving newly diagnosed people with MS found increased caudatal susceptibility and 8-hydroxy-2-deoxyguanosine [93], a marker of oxidative DNA damage in cerebrospinal fluid. However, while suggesting a relationship between brain iron on MRI and oxidative stress, interestingly, these two quantities were negatively correlated, leaving the interpretation of the findings unclear. Furthermore, the same study did not identify a correlation between iron levels and lipid peroxidation markers.

Another study [97] discovered that people with MS carrying the C282Y and H63D allele, polymorphisms of the HFE gene, exhibited higher susceptibility in the caudate and putamen, and that the C282Y allele modulated sex differences in the caudate. The study revealed that among people with progressive MS (but not relapsing MS or controls), those carrying the C282Y mutation had lower thalamic susceptibility than noncarriers.

As mentioned earlier, when interpreting QSM study findings, it's vital to consider that increased susceptibility could be linked to increased iron concentration, demyelination, or a combination of both. In the absence of concurrent use of other myelin-sensitive imaging techniques, it remains plausible that increased susceptibility in people with MS is affected, at least partially, by DGM demyelination [66], particularly in regions with relatively high myelin content, such as the globus pallidus and thalamus.

Comparison with other methods

Comparative studies have implied that QSM exhibits greater sensitivity than R_2 and R_2^* mapping in identifying pathological alterations in the DGM [89,92,96]. However, extrapolation of these findings must be done with caution as relaxation methods substantially depend on pulse sequence parameters, such as the number of echoes, and the specific fitting strategy utilized. Different acquisition and analysis strategies could yield varying conclusions regarding the sensitivity of the method [92].

The inverse effects of demyelination on susceptibility and relaxation rates provide a theoretical framework supporting QSM's higher sensitivity in situations featuring concurrent demyelination and iron accumulation, such as in the DGM of people with MS [66]. While both demyelination and increasing iron concentration additively lead to higher tissue susceptibility, their respective influences can (at least partially) counterbalance each other on relaxation rate maps and, thus, result in overall less substantial signal changes with these methods.

Technical methodologies designed to capitalize on these effects, enabling the differentiation of signal alterations associated with iron and myelin, are further discussed in subsequent sections.

Regional content of nonheme iron in the deep gray matter

Regional *concentration* of nonheme iron in the DGM is significantly affected by the atrophy observed in MS [108], with the rate and extent of atrophy differing across regions [109]. Fujiwara et al. [96] emphasized the necessity of adjusting iron-related imaging metrics to account for volume reduction associated with atrophy. Subsequent studies by [110] and [111] demonstrated that regional atrophy can yield elevated iron concentrations without any influx of iron into the region. These researchers posited that the increased iron concentration in the DGM of people with MS may be an artifact of a decreasing volume of iron-free parenchyma, leading to a higher density of iron within each imaging voxel. This hypothesis challenged the prevailing assumption that increased iron concentrations indicated on MRI are intrinsically linked to the disease process. Conversely, if iron concentrations were found to remain steady despite ongoing atrophy, it would imply a decrease in total iron content or mass, suggesting an efflux of iron from the region. Hernández-Torres et al., [110] utilized R_2^* relaxation rates to explore this hypothesis, calculating regional iron content rather than concentration in a cross-sectional study. Even though R_2^* rates suggested increased iron *concentrations* in the DGM of people with MS, the study found a decrease in iron *content* in both the caudate and thalamus. This concept was extended to QSM by Schweser et al. and Pontillo et al. [111,112], who confirmed a lower iron content in the thalamus (but not in the caudate) of people with MS and noted that increased susceptibility in the globus pallidus was driven by atrophy. On a whole-brain level, Hamdy et al. [113] found no significant differences in the total iron content between people with MS and controls. However, Schweser et al. [111] reported a longitudinal decrease in iron content throughout the DGM in people with MS over a period of 2 years.

These recent investigations of iron content have challenged the view that MRI-visible alterations in local concentrations of iron are a marker of iron-related toxicity, such as the promotion of oxidative stress in MS. Instead, they lend support to the hypothesis that MS is linked to a diminished availability of iron in iron-dependent pathways [67,91]. Histological studies have reported a decrease in the number of iron-rich oligodendrocytes in both WM and GM [66,67,114–116], as well as reduced nonheme iron within remaining cells [67]. However, as with the investigations into iron concentration, further research is needed to confirm these initial findings and to discern the confounding effects of myelin on the iron content metric.

Subvoxel distribution of iron in the deep gray matter

Observed discrepancies in DGM contrast between QSM and R_2^* [101], as well as the complementary sensitivity of QSM and R_2^* regarding the subvoxel distribution of iron (see above), present opportunities to isolate data on alterations in the subcellular distribution of iron by combining these methods. An innovative approach, based on the spatial correlation between QSM and R_2^*, was proposed by Taege et al. [82] and subsequently applied in a pilot study that included both people with MS and control subjects. The study's findings suggested that microdistribution is age-dependent, varies among anatomical regions, and undergoes changes in people with MS compared to controls. However, the method's specificity toward iron distribution requires further verification through ex vivo validation. Moreover, the impact of focal DGM demyelination [66] and the potential influences of heme-iron and cellular contributions to R_2^* [117] on the proposed spatial correlation approach remain to be elucidated.

Gadolinium retention in the deep gray matter

Repeated administrations of Gadolinium (Gd)-based contrast agents (GBCAs) have been associated with Gd deposition in the brain, particularly in the dentate nucleus and globus pallidus [118,119], revealing increased signal intensity on T_1-weighted images. This effect has primarily been reported with linear GBCAs, with macrocyclic GBCAs associated with fewer incidences.

Most studies on Gd retention have been reliant on the analysis of T_1 relaxation time-shortening impact, yet the intrinsic paramagnetic properties of Gd offer an opportunity for a more quantitative analysis using QSM [120]. Hinoda et al. [121] evaluated susceptibility changes due to Gd deposition in the dentate nucleus using QSM in people with MS following serial administration of GBCAs. Results demonstrated significantly higher susceptibility values on QSM in the dentate nucleus of the linear (but not macrocyclic) GBCA group compared to the non-GBCA group. Choi et al. [122] reported distinct changes in magnetic susceptibility on QSM after serial injections of gadobutrol, a macrocyclic GBCA. Interestingly, while no significant signal intensity changes were observed on T_1-weighted imaging, the study revealed an increase of 1.4 parts per billion in the magnetic susceptibility of the globus pallidus per gadobutrol administration. The study also found a trend of increased susceptibility at the dentate nucleus, albeit at about half the accumulation rate of the globus pallidus. This pattern, different from the linear GBCA findings, might suggest a different mechanism of deposition. Choi et al.'s results were later confirmed by Zhang et al. [123].

In essence, QSM has emerged as a sensitive and potentially superior method for the detection of Gd deposition in the brain following GBCA administration. The method offers a more quantitative and specific measure of Gd presence compared to T_1-weighted imaging, opening new doors for the investigation of GBCA-related brain changes in people with MS. However, more comprehensive studies are warranted to fully elucidate the potential and limitations of QSM in this context, including the differences between linear and macrocyclic GBCAs.

Confounding effects of myelin in the deep gray matter

The assessment of iron in the DGM via susceptibility-based techniques is based on the assumption that the myelination of GM is relatively low, and thus, does not significantly contribute to voxel susceptibility. While overall myelination is lower in GM compared to WM, certain DGM regions, such as the globus pallidus and parts of the thalamus, possess high myelination (Schaltenbrand et al., 1977). The influence of myelin on the bulk susceptibility of these areas remains an open question.

Studies have estimated the effect of myelin on DGM susceptibility and concluded that the diamagnetic property of myelin could lead to an underestimation of brain iron concentrations in a healthy brain [124]. Conversely, demyelination might result in overestimation of iron concentrations [1]. It's important to note that demyelinated lesions are prevalent in the DGM of people with MS, most frequently found in the hypothalamus and caudate, followed by thalamus, putamen, and pallidum [66].

Several correction techniques have been proposed to decrease QSM's myelin dependency and enhance its specificity to iron. These methods include the incorporation of other myelin-sensitive techniques such as magnetization transfer (MT) imaging [124] (though it should be noted that MT is also dependent on iron content [1,60,61]), R_2^* [124–126], and R1 [112]. Elkady et al. employed both R_2^* and QSM to monitor voxel-wise over-time changes in iron and myelin using a method referred to as discriminative analysis of regional evolution (DARE) [88,100,127]. DARE accounts for the temporal changes in QSM and R_2^* in people with MS based on alterations in both iron and myelin and demonstrated superior sensitivity toward over-time changes and associations with disability compared to QSM alone [88,100,127]. Notably, DARE disclosed a decrease in iron concentrations across all examined DGM regions in people with MS over a 5-year period [100], further reinforcing the notion of iron reduction in MS. Pontillo et al. [112] validated this decrease in iron concentrations in the thalamic nuclei by integrating QSM with R1 using a calibrated model. The study also indicated that atrophy is a prominent factor in increasing the iron concentration in the globus pallidus.

Since QSM and R_2^* can be obtained simultaneously from the same multiecho GRE sequence (see above), methods that separate the effects of myelin and iron on QSM using R_2^* can be conveniently implemented in clinical studies without extending the overall measurement duration. While protocols for the concurrent acquisition of QSM and R1 have been suggested [128–130], the associated pulse sequences have yet to achieve widespread availability.

Normal appearing white matter

Histological investigations by Hametner et al. [67] have not observed any variations in iron concentrations within the normal appearing white matter (NAWM) of people diagnosed with acute MS, as compared to controls, but indicated a decline in iron levels correlating with disease duration in progressive MS. These findings have been independently confirmed using quantitative X-ray fluorescence imaging [131], and in line with findings for the DGM, similar observations have been noted in oligodendrocytes [67,132].

In a comprehensive voxel-wise analysis of magnetic susceptibility in the NAWM, Rudko et al. [92] identified significantly reduced values in people with MS. The extent of these reduced regions exhibited a correlation with disability. In comparison with R_2^*, QSM demonstrated higher sensitivity in detecting group differences. Kor et al. [37] used a whole-brain analysis methods based on R_2^* and confirmed significantly reduced nonlesional WM iron concentration in people with MS compared to controls. Conversely, another study [133] reported increased susceptibility values in people with MS and found that values were predictive of disability worsening over time. Chen et al. [134] reported a dynamic susceptibility behavior in the NAWM during the relapsing phase of the disease course. NAWM susceptibility was significantly reduced relative to controls in people with MS without enhancing lesions, while it was similar in those with enhancing lesions. The authors attributed their observations to a global influx of iron into the NAWM during acute inflammation.

As observed in the DGM, the codependence of QSM on both iron and myelin makes the interpretation of NAWM susceptibility alterations challenging. The anisotropy of myelin susceptibility further complicates QSM interpretation within the WM, as susceptibility anisotropy causes WM susceptibility to depend on the major fiber direction relative to the main magnetic field [135]. Similarly, the R_2^* constant shows a strong orientation dependence [37,136,137] and exhibits a multiexponential decay behavior within the WM [138]. These factors make the separation of myelin and iron effects more difficult in the WM than in the DGM.

Most QSM studies of the WM have integrated susceptibility with techniques that assess myelin-related tissue compartments more robustly in the WM. These methods include myelin water fraction imaging [139], which is also influenced by tissue iron [59] and diffusion MRI (dMRI) [140]. In line with the study by Chen et al. [140] did not find susceptibility differences in NAWM of people with relapsing MS (80% without enhancing lesions), but comparison with dMRI metrics suggested decreased iron concentrations in the cingulate and forceps major.

Bergsland et al. [141] employed QSM with tract-based spatial statistics to study thalamic WM damage in people with MS. After controlling for volumes, the study reported decreased susceptibility and fractional anisotropy, with susceptibility being associated with information processing speed as assessed by the symbol digit modalities test. The authors highlighted the utility of a combined analysis of diffusion metrics and QSM for visualizing more extensive damage.

Focal white matter damage

A defining feature of MS is the focal myelin loss in WM [142]. Such WM lesions are easily observable in T_2-weighted and FLAIR MRI scans. However, these qualitative scans do not capture the high variability in the extent of demyelination within the lesions. Furthermore, lesions exhibit temporal variance, with some inactive lesions showing partial remyelination over time (forming a "shadow plaque"), while others remain demyelinated.

Owing to myelin's diamagnetic properties, QSM can serve as a sensitive indicator of both demyelination and remyelination suggesting that these histological lesion phenotypes can be differentiated on QSM. Multiple imaging studies have identified common lesion patterns using QSM either standalone or combined with other modalities [143–145]. On QSM, lesions have been described as isointense, hyperintense, or even featuring a hyperintense rim surrounding the lesion core relative to the surrounding WM [146–149].

Initially, increased magnetic susceptibility in MS lesions was hypothesized to stem from elevated iron levels ([150], fig. 4). However, the prevailing consensus is now that in WM MS lesions myelin breakdown and removal alone result in elevated QSM values, although increased iron can be present in some lesions [151,152]. In phantom experiments, Deh et al. demonstrated that the degradation of myelin basic protein and human myelin result in considerable increases in magnetic susceptibility [152]. Studies using animal models have supported this notion, demonstrating that myelin is a primary source of susceptibility contrast in WM [153,154]. Notably, a strong correlation ($R^2 = 0.93$) was found between magnetic susceptibility and Luxol fast blue staining in various WM regions in mouse brains, with no correlation found with iron staining [155]. This may be due to lower iron content in rodents compared to humans [114,156,157].

Monitoring MS lesions via monthly MRI scans from 6 months before and after enhancement with Gd, Wiggermann et al. [158] found that the formation of new MS lesions coincides with a localized increase in MR phase starting 3 months before Gd enhancement, persisting for at least 6 months postenhancement. Although phase is sensitive to the tissue changes during MS lesion formation, the nonlocal field effects present in phase can confound the use of phase images and it is therefore recommended that QSM is used rather than (filtered) phase images [159,160]. More recent studies using QSM have corroborated the findings by Wiggermann et al., reporting that the magnetic susceptibility of almost all nodular enhancing lesions is isointense (96.2% in [161]), then peaks/plateaus within 1–2 years, and then gradually reverts to normal appearing WM levels over several years [161,162]. However, QSM alone cannot distinguish between remyelination and the formation of a glial scar, requiring an additional myelin-sensitive scan [163]. QSM cannot discriminate between a decrease in myelin and an increase in iron either. However, by incorporating the signal relaxation, the problem of separating the susceptibility map into its positive (paramagnetic) and negative (diamagnetic) sources can be solved [164]. Yet, over 90% of lesions are characterized by iron loss, consistent with the loss of iron-bearing oligodendrocytes in MS [65,67,132,144,165].

Identification of gadolinium-enhancing lesions

Due to the Gd-retention described above, there is growing interest in identifying Gd enhancing MS lesions, that is, new active lesions, without the use of contrast agents [166]. When an MS lesion first enhances, it is isointense or slightly hyperintense on QSM and then becomes increasingly hyperintense, as the lesion transitions from enhancing to nonenhancing [161]. Using a combination of T_2 and QSM, Zhang et al. determined a sensitivity of 90.7% and a specificity 85.6% for predicting new enhancing lesions from these scans [167]. With a total of 54 enhancing lesions, this is a relatively small study and further research in large cohorts is needed to evaluate QSM for the identification of enhancing new lesions.

Paramagnetic rim lesions

A subset of focal lesions presents with paramagnetic rims on QSM (Fig. 13.2) [144,173]. These rims, associated with chronic inflammation, are marked by microglia and macrophages loaded with iron [67,132,142,144,145,173–177]. Chronic active lesions, termed paramagnetic rim lesions (PRLs), are seen in 50%–66% of people with MS and are predominantly located in the deep WM [146,175]. In an international trial in 438 patients, 52% of people with MS and only 7% of non-MS cases had PRLs, resulting in a high specificity of 93% [178]. PRLs exhibit a slow, transient expansion, no signs of remyelination, and they do not enhance with Gd, signifying their chronic nature. PRLs start with iron diffusely throughout the lesion, which is later confined to the lesion edge [175,179]. A study by Zhang et al. [161] noted a gradual reduction in PRL magnetic susceptibility over 6 years, post an initial peak between 1.5 and 4 years. In their study, out of 32 newly identified Gd-positive lesions, 16 were rim + . These rim + lesions were significantly larger than rim − lesions, a disparity that remained over time. While all lesions exhibited an increase in QSM until they peaked between 1 and 2 years, rim − lesions declined faster after peaking. Rim + lesions, on the other hand, sustained elevated QSM between 0.5 and 4 years, after which they declined.

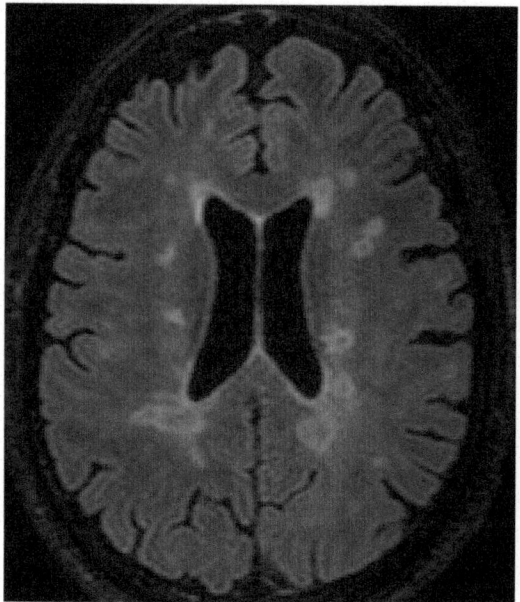 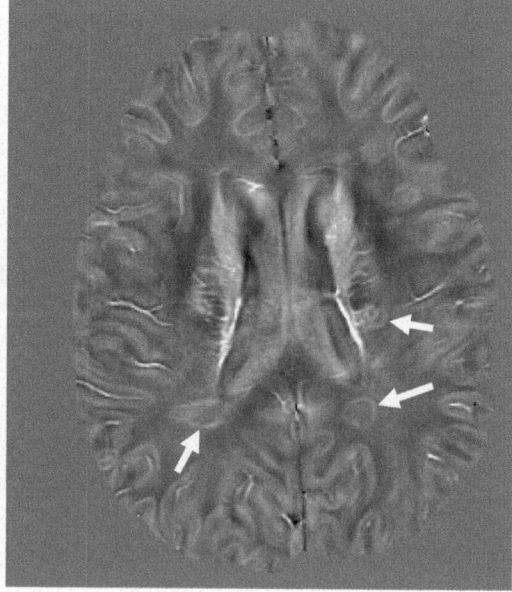

FIGURE 13.2 3D FLAIR (left) and QSM (right) of a patient with several paramagnetic rim lesions (arrows). Binary masks of the brain were computed using FSL's bet [168]. The wrapped phase was unwrapped using a hybrid Laplacian unwrapping [169] and background field correction method [170]. Nonharmonic background field contributions were removed using reSHARP [171], and the susceptibility maps were computed from the unwrapped, background field corrected fields using RTS [172]. The Julia toolbox QSM.jl (available at https://github.com/kamesy/QSM.jl) was used to process the data. *Source: Courtesy of Christian Kames, University of British Columbia.*

Given the chronic active nature of PRLs, the magnetic susceptibility of the paramagnetic rims may serve as potential markers for disease progression and treatment trials. For instance, a study comparing the magnetic susceptibility of rims in people with MS treated with glatiramer acetate (GA) versus dimethyl fumarate (DMF) found that the DMF group demonstrated a 2.27-fold greater rate of QSM reduction than the GA group [180]. Moreover, the relapse rate is positively correlated with the proportion of periventricular PRLs [181] and patients with at least one PRL perform worse on physical disability and cognitive assessments [182].

In summary, tracking new lesions and chronic active lesions with high remyelination potential appears promising for treatment trials of drugs aimed at promoting remyelination. QSM, with its high spatial resolution and sensitivity to changes in myelin content, is an essential tool for evaluating paramagnetic rims. At the time of writing, the North American Imaging in Multiple Sclerosis (NAIMS) Cooperative is working on a consensus statement for the imaging of chronic active lesions in MS, which will contain guidelines for the imaging and analysis of PRLs [183].

Other applications of quantitative susceptibility mapping in multiple sclerosis

Oxygen extraction fraction measurements

The gold-standard for measuring tissue oxygen extraction fraction (OEF)—a vital tissue viability marker—is ^{15}O-positron emission tomography (PET) [184]. However, this technique has limitations due to its relatively low spatial resolution and the necessity for a local cyclotron, restricting its availability. QSM, conversely, has been utilized in estimating OEF via widely available MRI with superior spatial resolution [167,185]. Recently, a combined approach of QSM with an MRI-OEF technique called quantitative blood oxygenation level dependency (qBOLD) [186] has been proposed for more realistic biophysics modeling of the OEF effect on MRI signal [187]. With the improvement of its robustness against measurement noise by machine learning algorithms [188–190], this integrative approach can detect disease-related OEF abnormalities in neurologic disorders [191–196]. Notably, Cho et al. [188] reported both global and localized reductions in OEF in people with MS, with an increase in OEF observed in the rim of chronic active lesions relative to the core. These findings indicate the potential of OEF mapping for monitoring tissue oxygenation in clinical settings. A fusion of the QSM-based OEF mapping with techniques for assessing cerebral blood flow, such as ASL, could enable local estimations of the cerebral metabolic rate of oxygen consumption in the future.

Microbleeds

Schweser et al. [28] illustrated that QSM can distinguish between calcifications and hemorrhages, thereby enhancing the specificity of microbleed detection compared to SWI. Zivadinov et al. [197] have utilized QSM in this regard, highlighting the potential, yet relatively unexplored, role of this methodology in MS research. The significance of microbleeds within the context of MS remains under debate [198,199].

Deep gray matter segmentation

FSL FIRST is the predominant automated segmentation algorithm for DGM [200], employing T1-weighted images to delineate anatomical boundaries. Nevertheless, it is becoming clear that DGM segmentation algorithms based on T_1-weighted images exhibit limitations potentially biasing clinical studies of diseases with significant brain atrophy and iron changes in the DGM, such as MS.

1. Depending on sequence parameters, the boundaries of some DGM regions are often inadequately defined on T_1-weighted images (e.g., globus pallidus and thalamus [201]), leading to segmentation inaccuracies. Given the high contrast between tissue regions on QSM (e.g., between the globus pallidus and internal capsule), even minor segmentation inaccuracies can significantly alter the mean susceptibility values estimated using these segments, leading to potential bias, increased variation, and decreased statistical power [202].
2. The T_1-weighted contrast is influenced by tissue iron concentration [63], which could amplify bias due to segmentation inaccuracies contingent on the regional iron concentration.
3. It has been demonstrated that FSL FIRST may underperform when brain anatomy significantly deviates from the dataset used to train the algorithm [202], such as in advanced MS. Atrophy-based segmentation inaccuracies could introduce bias into clinical studies.

The clear visual representation of DGM regions on susceptibility maps [203] justifies the use of this contrast in automated segmentation algorithms. The considerably stronger delineation of DGM anatomical boundaries on QSM reduces the likelihood of segmentation bias linked to changes in tissue iron compared to T_1-weighted images. Feng et al. [201,202] demonstrated that the performance of FSL FIRST could be enhanced by using a heuristic hybrid contrast created by merging T_1-weighted images and susceptibility maps and by improving the anatomical normalization step in FIRST.

Lim et al. [204] generated an atlas using T_1-weighted images and QSM of a single subject and then employed a multicontrast deformation approach, which used both T_1-weighted images and QSM, to apply the atlas to other subjects. Cobzas et al. [205] employed a multiatlas approach [206] that utilized both T1 and QSM-based atlases derived from multiple subjects. Their method outperformed both FSL FIRST and T_1-based multiatlas segmentation. Hanspach et al. [201] evaluated various approaches for generating population-based brain templates for voxel-wise analyses with QSM, demonstrating that QSM-based templates and templates using both T1w and QSM surpassed T1w templates. Zhang et al. subsequently employed this strategy to create longitudinal brain atlases of the normal brain [167].

Summary and outlook

The emerging role of QSM in the study of multiple sclerosis opens new avenues for enhanced understanding and diagnosis. QSM's ability to provide high-resolution information from a clinical 3D GRE scan facilitates a deeper comprehension of the complex manifestations of MS, including the characterization of PRLs, a feature seen as distinctively associated with MS.

An important advantage of QSM over other quantitative MRI techniques is the robustness of its signal-to-noise ratio, which remains reliable even under variable pathological conditions, including the complete loss of a specific tissue compartment like myelin. Since QSM relies on all spins within the tissue to probe the local magnetic susceptibility, reduction of a particular tissue compartment does not result in a reduction in the signal to noise ratio of the measurement. This feature contributes to QSM's data consistency, an attribute that has been systematically compared with other myelin-sensitive MRI measures, and also PET, in studies like the one conducted by van der Weijden et al. [163]. Such comparative explorations offer insightful assessments of the relative strengths of these investigative techniques. Another advantageous aspect of the QSM scan is that it can be used for R_2^* mapping and the visualization of minute venous vessels at the center of many MS lesions. This central vein sign is a potential radiological marker for MS [207].

Despite the promising characteristics of QSM, challenges persist in the standardization of the associated acquisition and processing methods, limiting its widespread application. However, ongoing community efforts [42,52,53], advancements in scanner technology, and efforts of all major vendors to include QSM in their product scanner are expected to gradually alleviate these issues. The application of QSM in the context of MS presents a potential for considerable influence on diagnostic strategies and a deeper understanding of the disease. Although the extent of QSM's future impact largely depends on the progress in development and standardization, its unique sensitivity, high spatial resolution, and versatility certainly render it a promising prospect for ongoing research in the field of MS.

Acknowledgments

We are grateful to Dr. Junghun Cho (University at Buffalo) for valuable feedback on the use of QSM for brain oxygenation measurement.

This publication was supported by the National Institute of Neurological Disorders and Stroke of the National Institutes of Health under Award Number R01NS114227 and the National Center for Advancing Translational Sciences of the National Institutes of Health under Award Number UL1TR001412. AR is supported by Canada Research Chairs under Award Number 950-230363. The content is solely the responsibility of the authors and does not necessarily represent the official views of the National Institutes of Health.

References

[1] Langkammer C, Krebs N, Goessler W, Scheurer E, Yen K, Fazekas F, et al. Susceptibility induced gray-white matter MRI contrast in the human brain. NeuroImage 2012;59:1413−19. Available from: https://doi.org/10.1016/j.neuroimage.2011.08.045.

[2] Ropele S. Applications of susceptibility-weighted imaging and mapping. Advances in magnetic resonance technology and applications. Elsevier; 2021. p. 359−69. Available from: https://doi.org/10.1016/B978-0-12-822479-3.00037-3.

[3] Reichenbach JR, Barth M, Haacke EM, Klarhöfer M, Kaiser WA, Moser E. High-resolution MR venography at 3.0 Tesla. J Comp Assist Tomogr 2000;24:949−57.

[4] Reichenbach JR, Essig M, Haacke EM, Lee BCP, Przetak C, Kaiser WA, et al. High-resolution venography of the brain using magnetic resonance imaging. Magn Reson Mater Phys 1998;6:62−9.

[5] Reichenbach JR, Venkatesan R, Schillinger D, Kido DK, Haacke EM. Small vessels in the human brain: MR-venography with deoxyhemoglobin as an intrinsic contrast agent. Radiology 1997;204:272−7.

[6] Ruetten PPR, Gillard JH, Graves MJ. Introduction to quantitative susceptibility mapping and susceptibility weighted imaging. Br J Radiol 2019;92:20181016. Available from: https://doi.org/10.1259/bjr.20181016.

[7] Schweser F, Deistung A, Reichenbach JR. Foundations of MRI phase imaging and processing for quantitative susceptibility mapping (QSM). Z Med Phys 2016;26:6—34. Available from: https://doi.org/10.1016/j.zemedi.2015.10.002.

[8] Shmueli K. Quantitative susceptibility mapping. In: Seiberlich N, Gulani V, Calamante F, Campbell-Washburn A, Doneva M, Hu HH, et al., editors. Advances in magnetic resonance technology and applications, quantitative magnetic resonance imaging. Academic Press; 2020. p. 819—38. Available from: https://doi.org/10.1016/B978-0-12-817057-1.00033-0.

[9] Duyn JH, Schenck JF. Contributions to magnetic susceptibility of brain tissue. NMR Biomed 2017;30:e3546. Available from: https://doi.org/10.1002/nbm.3546.

[10] Hametner S, Endmayr V, Deistung A, Palmrich P, Prihoda M, Haimburger E, et al. The influence of brain iron and myelin on magnetic susceptibility and effective transverse relaxation — a biochemical and histological validation study. NeuroImage 2018;179:117—33. Available from: https://doi.org/10.1016/j.neuroimage.2018.06.007.

[11] Schweser F, Sommer K, Deistung A, Reichenbach JR. Quantitative susceptibility mapping for investigating subtle susceptibility variations in the human brain. NeuroImage 2012;62:2083—100. Available from: https://doi.org/10.1016/j.neuroimage.2012.05.067.

[12] Stüber C, Morawski M, Schäfer A, Labadie C, Wähnert M, Leuze C, et al. Myelin and iron concentration in the human brain: a quantitative study of MRI contrast. NeuroImage 2014;93:95—106. Available from: https://doi.org/10.1016/j.neuroimage.2014.02.026.

[13] Wharton S, Bowtell R. Fiber orientation-dependent white matter contrast in gradient echo MRI. Proc Natl Acad Sci U S A 2012;109:18559—64. Available from: https://doi.org/10.1073/pnas.1211075109.

[14] Lee J, Shmueli K, Kang B-T, Yao B, Fukunaga M, van Gelderen P, et al. The contribution of myelin to magnetic susceptibility-weighted contrasts in high-field MRI of the brain. NeuroImage 2012;59:3967—75. Available from: https://doi.org/10.1016/j.neuroimage.2011.10.076.

[15] Pauling I, Coryell CDD, Pauling L. The magnetic properties and structure of hemoglobin, oxyhemoglobin and carbonmonoxyhemoglobin. Proc Natl Acad Sci U S A 1936;22:159.

[16] Pauling L. Magnetic properties and structure of oxyhemoglobin. Proc Natl Acad Sci U S A 1977;74:2612—13.

[17] Spees WM, Yablonskiy DA, Oswood MC, Ackerman JJH. Water proton MR properties of human blood at 1.5 Tesla: magnetic susceptibility, T1, T2, T2*, and non-Lorentzian signal behavior. Magn Reson Med 2001;45:533—42.

[18] Jain V, Abdulmalik O, Propert KJ, Wehrli FW. Investigating the magnetic susceptibility properties of fresh human blood for noninvasive oxygen saturation quantification. Magn Reson Med 2012;68:863—7. Available from: https://doi.org/10.1002/mrm.23282.

[19] Marques JP, Maddage R, Mlynarik V, Gruetter R. On the origin of the MR image phase contrast: an in vivo MR microscopy study of the rat brain at 14.1 T. NeuroImage 2009;46(2):345—52. Available from: https://doi.org/10.1016/j.neuroimage.2009.02.023.

[20] Petridou N, Wharton SJ, Lotfipour AK, Gowland P, Bowtell RW. Investigating the effect of blood susceptibility on phase contrast in the human brain. NeuroImage 2010;50(2):491—8. Available from: https://doi.org/10.1016/j.neuroimage.2009.12.052.

[21] Sedlacik J, Kutschbach C, Rauscher A, Deistung A, Reichenbach JR. Investigation of the influence of carbon dioxide concentrations on cerebral physiology by susceptibility-weighted magnetic resonance imaging (SWI). NeuroImage 2008;43(1):36—43. Available from: https://doi.org/10.1016/j.neuroimage.2008.07.008.

[22] Quintana C, Bellefqih S, Laval JY, Guerquin-Kern JL, Wu TD, Avila J, et al. Study of the localization of iron, ferritin, and hemosiderin in Alzheimer's disease hippocampus by analytical microscopy at the subcellular level. J Struct Biol 2006;153:42—54. Available from: https://doi.org/10.1016/j.jsb.2005.11.001.

[23] Ropele S, Langkammer C. Iron quantification with susceptibility. NMR Biomed 2017;30:e3534. Available from: https://doi.org/10.1002/nbm.3534.

[24] Schenck JF. Health and physiological effects of human exposure to whole-body four-tesla magnetic fields during MRI. Ann N Y Acad Sci 1992;649:285—301.

[25] Yamada N, Imakita S, Sakuma T, Takamiya M. Intracranial calcification on gradient-echo phase image: depiction of diamagnetic susceptibility. Radiology 1996;198:171—8.

[26] Makinen S, van Groen T, Clarke J, Thornell A, Corbett D, Hiltunen M, et al. Coaccumulation of calcium and b-amyloid in the thalamus after transient middle cerebral artery occlusion in rats. J Cerebr Blood Flow Metab 2008;28:263—8. Available from: https://doi.org/10.1038/sj.jcbfm.9600529.

[27] Hopkins JA, Wehrli FW. Magnetic susceptibility measurement of insoluble solids by NMR: magnetic susceptibility of bone. Magn Reson Med 1997;37:494–500.
[28] Schweser F, Deistung A, Lehr BW, Reichenbach JR. Differentiation between diamagnetic and paramagnetic cerebral lesions based on magnetic susceptibility mapping. Med Phys 2010;37:5165–78. Available from: https://doi.org/10.1118/1.3481505.
[29] Chen W, Zhu W, Kovanlikaya I, Kovanlikaya A, Liu T, Wang S, et al. Intracranial calcifications and hemorrhages: characterization with quantitative susceptibility mapping. Radiology 2014;270:496–505. Available from: https://doi.org/10.1148/radiol.13122640.
[30] Ciraci S, Gumus K, Doganay S, Dundar MS, Kaya Ozcora GD, Gorkem SB, et al. Diagnosis of intracranial calcification and hemorrhage in pediatric patients: Comparison of quantitative susceptibility mapping and phase images of susceptibility-weighted imaging. Diagn Interv Imaging 2017;4–11. Available from: https://doi.org/10.1016/j.diii.2017.05.004.
[31] Deistung A, Rauscher A, Sedlacik J, Stadler J, Witoszynskyj S, Reichenbach JR. Susceptibility weighted imaging at ultra high magnetic field strengths: theoretical considerations and experimental results. Magn Reson Med 2008;60:1155–68. Available from: https://doi.org/10.1002/mrm.21754.
[32] Karsa A, Punwani S, Shmueli K. The effect of low resolution and coverage on the accuracy of susceptibility mapping. Magn Reson Med 2019;81:1833–48. Available from: https://doi.org/10.1002/mrm.27542.
[33] Biondetti E, Karsa A, Thomas DL, Shmueli K. Investigating the accuracy and precision of TE-dependent versus multi-echo QSM using Laplacian-based methods at 3 T. Magn Reson Med 2020;84:3040–53. Available from: https://doi.org/10.1002/mrm.28331.
[34] Denk C, Rauscher A. Susceptibility weighted imaging with multiple echoes. J Magn Reson Imaging 2010;31:185–91. Available from: https://doi.org/10.1002/jmri.21995.
[35] Robinson SD, Bredies K, Khabipova D, Dymerska B, Marques JP, Schweser F. An illustrated comparison of processing methods for MR phase imaging and QSM: combining array coil signals and phase unwrapping. NMR Biomed 2017;30:e3601. Available from: https://doi.org/10.1002/nbm.3601.
[36] Khalil M, Langkammer C, Pichler A, Pinter D, Gattringer T, Bachmaier G, et al. Dynamics of brain iron levels in multiple sclerosis: a longitudinal 3T MRI study. Neurology 2015;84:2396–402. Available from: https://doi.org/10.1212/wnl.0000000000001679.
[37] Kor D, Birkl C, Ropele S, Doucette J, Xu T, Wiggermann V, et al. The role of iron and myelin in orientation dependent R_2^* of white matter. NMR Biomed 2019;:e4092. Available from: https://doi.org/10.1002/nbm.4092.
[38] Marques JP, Simonis FFJ, Webb AG. Low-field MRI: an MR physics perspective. J Magn Reson Imaging 2019;49:1528–42. Available from: https://doi.org/10.1002/jmri.26637.
[39] Ippoliti M, Adams LC, Winfried B, Hamm B, Spincemaille P, Wang Y, et al. Quantitative susceptibility mapping across two clinical field strengths: contrast-to-noise ratio enhancement at 1.5T. J Magn Reson Imaging 2018;. Available from: https://doi.org/10.1002/jmri.26045.
[40] Spincemaille P, Anderson J, Wu G, Yang B, Fung M, Li K, et al. Quantitative susceptibility mapping: MRI at 7T versus 3T. J Neuroimaging 2020;30:65–75. Available from: https://doi.org/10.1111/jon.12669.
[41] Hinshaw WS, Lent AH. An introduction to NMR imaging: from the bloch equation to the imaging equation. Proc IEEE 1983;71:338–50.
[42] QSM Consensus Organization Committee, Bilgic B, Costagli M, Chan K-S, Duyn J, Langkammer C, et al. Recommended implementation of quantitative susceptibility mapping for clinical research in the brain: a consensus of the ISMRM electro-magnetic tissue properties study group. Magn Reson Med 2024;91:1834–62.
[43] Beliveau V, Birkl C, Stefani A, Gizewski ER, Scherfler C. HFP-QSMGAN: QSM from homodyne-filtered phase images. Magn Reson Med 2022;88:1255–62. Available from: https://doi.org/10.1002/mrm.29260.
[44] Lu Z, Li J, Wang C, Ge R, Chen L, He H, et al. S2Q-net: mining the high-pass filtered phase data in susceptibility weighted imaging for quantitative susceptibility mapping. IEEE J Biomed Health Inf 2022;. Available from: https://doi.org/10.1109/JBHI.2022.3156548.
[45] Kames C, Doucette J, Birkl C, Rauscher A. Recovering SWI-filtered phase data using deep learning. Magn Reson Med 2022;87:948–59. Available from: https://doi.org/10.1002/mrm.29013.
[46] Chung J, Ruthotto L. Computational methods for image reconstruction. NMR Biomed 2017;30. Available from: https://doi.org/10.1002/nbm.3545.
[47] Wu B, Li W, Avram AV, Gho S-M, Liu C. Fast and tissue-optimized mapping of magnetic susceptibility and $T2^*$ with multi-echo and multi-shot spirals. NeuroImage 2012;59:297–305. Available from: https://doi.org/10.1016/j.neuroimage.2011.07.019.

[48] Chen L, Cai S, van Zijl PCM, Li X. Single-step calculation of susceptibility through multiple orientation sampling. NMR Biomed 2021;34:e4517. Available from: https://doi.org/10.1002/nbm.4517.

[49] Kressler B, de Rochefort L, Liu T, Spincemaille P, Jiang Q, Wang Y. Nonlinear regularization for per voxel estimation of magnetic susceptibility distributions from MRI field maps. IEEE Trans Med Imaging 2010;29:273−81. Available from: https://doi.org/10.1109/TMI.2009.2023787.

[50] Liu T, Wisnieff C, Lou M, Chen W, Spincemaille P, Wang Y. Nonlinear formulation of the magnetic field to source relationship for robust quantitative susceptibility mapping. Magn Reson Med 2013;69:467−76. Available from: https://doi.org/10.1002/mrm.24272.

[51] Schweser F, Robinson SD, de Rochefort L, Li W, Bredies K. An illustrated comparison of processing methods for phase MRI and QSM: removal of background field contributions from sources outside the region of interest. NMR Biomed 2017;30:e3604. Available from: https://doi.org/10.1002/nbm.3604.

[52] Langkammer C, Schweser F, Shmueli K, Kames C, Li X, Guo L, et al. Quantitative susceptibility mapping: report from the 2016 reconstruction challenge. Magn Reson Med 2018;79:1661−73. Available from: https://doi.org/10.1002/mrm.26830.

[53] QSM Challenge 2.0 Organization CommitteeBilgic B, Langkammer C, Marques JP, Meineke J, Milovic C, Schweser F. QSM reconstruction challenge 2.0: design and report of results. Magn Reson Med 2021;86:1241−55. Available from: https://doi.org/10.1002/mrm.28754.

[54] Salman F, Ramesh A, Prayer M, Jochmann T, Bergsland N, Dwyer MG, et al. Systematic assessment of published QSM inversion algorithms for detecting longitudinal changes in brain susceptibility. In: Proc Intl Soc Mag Reson Med. Toronto, ON, Canada, vol. 31; 2023.

[55] Choudhary P, Bergsland N, Dhamankar A, Dwyer M, Weinstock-Guttman B, Zivadinov R, et al. Are all susceptibility maps created equal? In: Proc Intl Soc Mag Reson Med. Paris, France, vol. 26; 2018. p. 2218.

[56] Sun H, Walsh AJ, Lebel RM, Blevins G, Catz I, Lu J-Q, et al. Validation of quantitative susceptibility mapping with Perls' iron staining for subcortical gray matter. NeuroImage 2015;105:486−92. Available from: https://doi.org/10.1016/j.neuroimage.2014.11.010.

[57] Zheng W, Nichol H, Liu S, Cheng Y-CN, Haacke EM. Measuring iron in the brain using quantitative susceptibility mapping and X-ray fluorescence imaging. NeuroImage 2013;78:68−74. Available from: https://doi.org/10.1016/j.neuroimage.2013.04.022.

[58] Langkammer C, Schweser F, Krebs N, Deistung A, Goessler W, Scheurer E, et al. Quantitative susceptibility mapping (QSM) as a means to measure brain iron? A post mortem validation study. NeuroImage 2012;62:1593−9. Available from: https://doi.org/10.1016/j.neuroimage.2012.05.049.

[59] Birkl C, Birkl-Toeglhofer AM, Endmayr V, Hoftberger R, Kasprian G, Krebs C, et al. The influence of brain iron on myelin water imaging. NeuroImage 2019;199:645. Available from: https://doi.org/10.1016/j.neuroimage.2019.05.042 −552.

[60] Smith SA, Bulte JWM, van Zijl PCM. Direct saturation MRI: theory and application to imaging brain iron. Magn Reson Med 2009;62:384−93. Available from: https://doi.org/10.1002/mrm.21980.

[61] Birkl C, Birkl-Toeglhofer AM, Kames C, Goessler W, Haybaeck J, Fazekas F, et al. The influence of iron oxidation state on quantitative MRI parameters in post mortem human brain. NeuroImage 2020;220:117080. Available from: https://doi.org/10.1016/j.neuroimage.2020.117080.

[62] Zhang J, Tao R, Liu C, Wu W, Zhang Y, Cui J, et al. Possible effects of iron deposition on the measurement of DTI metrics in deep gray matter nuclei: an in vitro and in vivo study. Neurosci Lett 2013;551:47−52. Available from: https://doi.org/10.1016/j.neulet.2013.07.003.

[63] Gelman N, Ewing JR, Gorell JM, Spickler EM, Solomon EG. Interregional variation of longitudinal relaxation rates in human brain at 3.0 T: relation to estimated iron and water contents. Magn Reson Med 2001;45:71−9.

[64] Hallgren B, Sourander P. The effect of age on the non-haemin iron in the human brain. J Neurochem 1958;3:41−51.

[65] Wiggermann V, Hametner S, Hernández-Torres E, Kames C, Endmayr V, Kasprian G, et al. Susceptibility-sensitive MRI of multiple sclerosis lesions and the impact of normal-appearing white matter changes. NMR Biomed 2017;30:1−12. Available from: https://doi.org/10.1002/nbm.3727.

[66] Haider L, Simeonidou C, Steinberger G, Hametner S, Grigoriadis N, Deretzi G, et al. Multiple sclerosis deep grey matter: the relation between demyelination, neurodegeneration, inflammation and iron. J Neurol Neurosurg Psychiatry 2014;85:1386−95. Available from: https://doi.org/10.1136/jnnp-2014-307712.

[67] Hametner S, Wimmer I, Haider L, Pfeifenbring S, Brück W, Lassmann H. Iron and neurodegeneration in the multiple sclerosis brain. Ann Neurol 2013;74:848−61. Available from: https://doi.org/10.1002/ana.23974.

[68] Langkammer C, Krebs N, Goessler W, Scheurer E, Ebner F, Yen K, et al. Quantitative MR imaging of brain iron: a postmortem validation study. Radiology 2010;257:455−62. Available from: https://doi.org/10.1148/radiol.10100495.
[69] Ravanfar P, Loi SM, Syeda WT, Van Rheenen TE, Bush AI, Desmond P, et al. Systematic review: quantitative susceptibility mapping (QSM) of brain iron profile in neurodegenerative diseases. Front Neurosci 2021;15:618435. Available from: https://doi.org/10.3389/fnins.2021.618435.
[70] Ghassaban K, Liu S, Jiang C, Haacke EM. Quantifying iron content in magnetic resonance imaging. NeuroImage 2019;187:77−92. Available from: https://doi.org/10.1016/j.neuroimage.2018.04.047.
[71] Drayer BP, Burger P, Darwin R, Riederer S, Herfkens R, Johnson GA. MRI of brain iron. Am J Roentgenol 1986;147:103−10.
[72] Drayer BP, Burger P, Hurwitz B, Dawson D, Cain J. Reduced signal intensity on MR images of thalamus and putamen in multiple sclerosis: increased iron content? AJR Am J Roentgenol 1987;149:357−63. Available from: https://doi.org/10.2214/ajr.149.2.357.
[73] Drayer BP, Burger P, Hurwitz B, Dawson D, Cain J, Leong J, et al. Magnetic resonance imaging in multiple sclerosis: decreased signal in thalamus and putamen. Ann Neurol 1987;22:546−50. Available from: https://doi.org/10.1002/ana.410220418.
[74] Bakshi R, Benedict RHB, Bermel Ra, Caruthers SD, Puli SR, Tjoa CW, et al. T2 hypointensity in the deep gray matter of patients with multiple sclerosis: a quantitative magnetic resonance imaging study. Arch Neurol 2002;59:62−8.
[75] Stüber C, Pitt D, Wang Y. Iron in multiple sclerosis and its noninvasive imaging with quantitative susceptibility mapping. Int J Mol Sci 2016;17:100. Available from: https://doi.org/10.3390/ijms17010100.
[76] Yao B, Li T, Gelderen PV, Shmueli K, de Zwart JA, Duyn JH. Susceptibility contrast in high field MRI of human brain as a function of tissue iron content. NeuroImage 2009;44:1259−66. Available from: https://doi.org/10.1016/j.neuroimage.2008.10.029.
[77] Gelman N, Gorell JM, Barker PB, Savage RM, Spickler EM, Windham JP, et al. MR imaging of human brain at 3.0 T: preliminary report on transverse relaxation rates and relation to estimated iron content. Radiology 1999;210:759−67.
[78] Ogg RJ, Langston JW, Haacke EM, Steen RG, Taylor JS. The correlation between phase shifts in gradient-echo MR images and regional brain iron concentration. Magn Reson Imaging 1999;17:1141−8.
[79] Walsh AJ, Wilman AH. Susceptibility phase imaging with comparison to R2 mapping of iron-rich deep grey matter. NeuroImage 2011;57:452−61. Available from: https://doi.org/10.1016/j.neuroimage.2011.04.017.
[80] Xu X, Wang Q, Zhang M. Age, gender, and hemispheric differences in iron deposition in the human brain: an in vivo MRI study. NeuroImage 2008;40:35−42. Available from: https://doi.org/10.1016/j.neuroimage.2007.11.017.
[81] Zivadinov R, Heininen-Brown M, Schirda CV, Poloni GU, Bergsland NP, Magnano CR, et al. Abnormal subcortical deep-gray matter susceptibility-weighted imaging filtered phase measurements in patients with multiple sclerosis: a case-control study. NeuroImage 2012;59:331−9. Available from: https://doi.org/10.1016/j.neuroimage.2011.07.045.
[82] Taege Y, Hagemeier J, Bergsland N, Dwyer MG, Weinstock-Guttman B, Zivadinov R, et al. Assessment of mesoscopic properties of deep gray matter iron through a model-based simultaneous analysis of magnetic susceptibility and R2* − a pilot study in patients with multiple sclerosis and normal controls. NeuroImage 2019;186:308−20. Available from: https://doi.org/10.1016/j.neuroimage.2018.11.011.
[83] Colgan TJ, Knobloch G, Reeder SB, Hernando D. Sensitivity of quantitative relaxometry and susceptibility mapping to microscopic iron distribution. Magn Reson Med 2020;83:673−80. Available from: https://doi.org/10.1002/mrm.27946.
[84] Balasubramanian M, Polimeni JR, Mulkern RV. *In vivo* measurements of irreversible and reversible transverse relaxation rates in human basal ganglia at 7 T: making inferences about the microscopic and mesoscopic structure of iron and calcification deposits. NMR Biomed 2019;32. Available from: https://doi.org/10.1002/nbm.4140.
[85] Dietrich O, Levin J, Ahmadi S-A, Plate A, Reiser MF, Bötzel K, et al. MR imaging differentiation of Fe2 + and Fe3 + based on relaxation and magnetic susceptibility properties. Neuroradiology 2017;59:403−9. Available from: https://doi.org/10.1007/s00234-017-1813-3.
[86] Schweser F, Dwyer MG, Deistung A, Reichenbach JR, Zivadinov R. Impact of tissue atrophy on high-pass filtered MRI signal phase-based assessment in large-scale group-comparison studies: a simulation study. Front Phys 2013;1:1−9. Available from: https://doi.org/10.3389/fphy.2013.00014.

[87] Al-Radaideh AM, Wharton SJ, Lim S-Y, Tench CR, Morgan PS, Bowtell RW, et al. Increased iron accumulation occurs in the earliest stages of demyelinating disease: an ultra-high field susceptibility mapping study in clinically isolated syndrome. Mult Scler 2013;19:896–903. Available from: https://doi.org/10.1177/1352458512465135.

[88] Elkady AM, Cobzas D, Sun H, Blevins G, Wilman AH. Progressive iron accumulation across multiple sclerosis phenotypes revealed by sparse classification of deep gray matter. J Magn Reson Imaging 2017;46:1464–73. Available from: https://doi.org/10.1002/jmri.25682.

[89] Langkammer C, Liu T, Khalil M, Enzinger C, Jehna M, Fuchs S, et al. Quantitative susceptibility mapping in multiple sclerosis. Radiology 2013;267:551–9. Available from: https://doi.org/10.1148/radiol.12120707.

[90] Pudlac A, Burgetova A, Dusek P, Nytrova P, Vaneckova M, Horakova D, et al. Deep gray matter iron content in neuromyelitis optica and multiple sclerosis. BioMed Res Int 2020;2020:1–6. Available from: https://doi.org/10.1155/2020/6492786.

[91] Schweser F, Raffaini Duarte Martins AL, Hagemeier J, Lin F, Hanspach J, Weinstock-Guttman B, et al. Mapping of thalamic magnetic susceptibility in multiple sclerosis indicates decreasing iron with disease duration: a proposed mechanistic relationship between inflammation and oligodendrocyte vitality. NeuroImage 2018;167:438–52. Available from: https://doi.org/10.1016/j.neuroimage.2017.10.063.

[92] Rudko DA, Solovey I, Gati JS, Kremenchutzky M, Menon RS. Multiple sclerosis: improved identification of disease-relevant changes in gray and white matter by using susceptibility-based MR imaging. Radiology 2014;272:851–64. Available from: https://doi.org/10.1148/radiol.14132475.

[93] Burgetova A, Dusek P, Uher T, Vaneckova M, Vejrazka M, Burgetova R, et al. Oxidative stress markers in cerebrospinal fluid of newly diagnosed multiple sclerosis patients and their link to iron deposition and atrophy. Diagnostics 2022;12:1365. Available from: https://doi.org/10.3390/diagnostics12061365.

[94] Zivadinov R, Tavazzi E, Bergsland N, Hagemeier J, Lin F, Dwyer MG, et al. Brain iron at quantitative MRI is associated with disability in multiple sclerosis. Radiology 2018;289:487–96. Available from: https://doi.org/10.1148/radiol.2018180136.

[95] Burgetova A, Dusek P, Vaneckova M, Horakova D, Langkammer C, Krasensky J, et al. Thalamic iron differentiates primary-progressive and relapsing-remitting multiple sclerosis. Am J Neuroradiol 2017;38:1079–86. Available from: https://doi.org/10.3174/ajnr.A5166.

[96] Fujiwara E, Kmech JA, Cobzas D, Sun H, Seres P, Blevins G, et al. Cognitive implications of deep gray matter iron in multiple sclerosis. Am J Neuroradiol 2017;38:942–8. Available from: https://doi.org/10.3174/ajnr.A5109.

[97] Hagemeier J, Ramanathan M, Schweser F, Dwyer MG, Lin F, Bergsland NP, et al. Iron-related gene variants and brain iron in multiple sclerosis and healthy individuals. NeuroImage Clin 2018;17:530–40. Available from: https://doi.org/10.1016/j.nicl.2017.11.003.

[98] Hagemeier J, Zivadinov R, Dwyer MG, Polak P, Bergsland N, Weinstock-Guttman B, et al. Changes of deep gray matter magnetic susceptibility over 2 years in multiple sclerosis and healthy control brain. NeuroImage Clin 2018;18:1007–16. Available from: https://doi.org/10.1016/j.nicl.2017.04.008.

[99] Pontillo G, Cocozza S, Lanzillo R, Russo C, Stasi MD, Paolella C, et al. Determinants of deep gray matter atrophy in multiple sclerosis: a multimodal MRI study. Am J Neuroradiol 2019;40:99–106. Available from: https://doi.org/10.3174/ajnr.A5915.

[100] Elkady AM, Cobzas D, Sun H, Seres P, Blevins G, Wilman AH. Five year iron changes in relapsing-remitting multiple sclerosis deep gray matter compared to healthy controls. Mult Scler Relat Disord 2019;33:107–15. Available from: https://doi.org/10.1016/j.msard.2019.05.028.

[101] Schmalbrock P, Prakash RS, Schirda B, Janssen A, Yang GK, Russell M, et al. Basal ganglia iron in patients with multiple sclerosis measured with 7T quantitative susceptibility mapping correlates with inhibitory control. Am J Neuroradiol 2016;37:439–46. Available from: https://doi.org/10.3174/ajnr.A4599.

[102] Yan Z, Liu H, Chen X, Zheng Q, Zeng C, Zheng Y, et al. Quantitative susceptibility mapping-derived radiomic features in discriminating multiple sclerosis from neuromyelitis optica spectrum disorder. Front Neurosci 2021;15:765634. Available from: https://doi.org/10.3389/fnins.2021.765634.

[103] Doring TM, Granado V, Rueda F, Deistung A, Reichenbach JR, Tukamoto G, et al. Quantitative susceptibility mapping indicates a disturbed brain iron homeostasis in neuromyelitis optica – a pilot study. PLoS ONE 2016;11:e0155027. Available from: https://doi.org/10.1371/journal.pone.0155027.

[104] Dales J-P, Desplat-Jégo S. Metal imbalance in neurodegenerative diseases with a specific concern to the brain of multiple sclerosis patients. Int J Mol Sci 2020;21:9105. Available from: https://doi.org/10.3390/ijms21239105.

[105] Siotto M, Filippi MM, Simonelli I, Landi D, Ghazaryan A, Vollaro S, et al. Oxidative stress related to iron metabolism in relapsing remitting multiple sclerosis patients with low disability. Front Neurosci 2019;13. Available from: https://doi.org/10.3389/fnins.2019.00086.

[106] Ayala A, Muñoz MF, Argüelles S. Lipid peroxidation: production, metabolism, and signaling mechanisms of malondialdehyde and 4-hydroxy-2-nonenal. Oxid Med Cell Longev 2014;2014:e360438. Available from: https://doi.org/10.1155/2014/360438.

[107] Halliwell B, Gutteridge JM. Oxygen toxicity, oxygen radicals, transition metals and disease. Biochem J 1984;219:1–14.

[108] Bermel RA, Bakshi R. The measurement and clinical relevance of brain atrophy in multiple sclerosis. Lancet Neurol 2006;5:158–70. Available from: https://doi.org/10.1016/S1474-4422(06)70349-0.

[109] Eshaghi A, Marinescu RV, Young AL, Firth NC, Prados F, Cardoso MJ, et al. Progression of regional grey matter atrophy in multiple sclerosis. Brain 2019;141(6):1665–77.

[110] Hernández-Torres E, Wiggermann V, Machan L, Sadovnick AD, Li DKB, Traboulsee A, et al. Increased mean R2* in the deep gray matter of multiple sclerosis patients: have we been measuring atrophy? J Magn Reson Imaging 2019;50:201–8. Available from: https://doi.org/10.1002/jmri.26561.

[111] Schweser F, Hagemeier J, Dwyer MG, Bergsland N, Hametner S, Weinstock-Guttman B, et al. Decreasing brain iron in multiple sclerosis: the difference between concentration and content in iron MRI. Hum Brain Mapp 2021;42:1463–74. Available from: https://doi.org/10.1002/hbm.25306.

[112] Pontillo G, Petracca M, Monti S, Quarantelli M, Criscuolo C, Lanzillo R, et al. Unraveling deep gray matter atrophy and iron and myelin changes in multiple sclerosis. Am J Neuroradiol 2021;. Available from: https://doi.org/10.3174/ajnr.A7093.

[113] Hamdy E, Galeel AA, Ramadan I, Gaber D, Mustafa H, Mekky J. Iron deposition in multiple sclerosis: overall load or distribution alteration? Eur Radiol Exp 2022;6:49. Available from: https://doi.org/10.1186/s41747-022-00279-9.

[114] Benkovic SA, Connor JR. Ferritin, transferrin, and iron in selected regions of the adult and aged rat brain. J Comp Neurol 1993;338:97–113. Available from: https://doi.org/10.1002/cne.903380108.

[115] Reinert A, Morawski M, Seeger J, Arendt T, Reinert T. Iron concentrations in neurons and glial cells with estimates on ferritin concentrations. BMC Neurosci 2019;20:25. Available from: https://doi.org/10.1186/s12868-019-0507-7.

[116] Connor JR, Menzies SL. Relationship of iron to oligondendrocytes and myelination. Glia 1996;17:83–93. Available from: https://doi.org/10.1002/(SICI)1098-1136.

[117] Yablonskiy DA, Wen J, Kothapalli SVVN, Sukstanskii AL. In vivo evaluation of heme and non-heme iron content and neuronal density in human basal ganglia. NeuroImage 2021;235:118012. Available from: https://doi.org/10.1016/j.neuroimage.2021.118012.

[118] Radbruch A, Weberling LD, Kieslich PJ, Eidel O, Burth S, Kickingereder P, et al. Gadolinium retention in the dentate nucleus and globus pallidus is dependent on the class of contrast agent. Radiology 2015;275:150337. Available from: https://doi.org/10.1148/radiol.2015150337.

[119] Robert P, Lehericy S, Grand S, Violas X, Fretellier N, Idée J-M, et al. T1-Weighted hypersignal in the deep cerebellar nuclei after repeated administrations of gadolinium-based contrast agents in healthy rats: difference between linear and macrocyclic agents. Invest Radiol 2015;50:473–80. Available from: https://doi.org/10.1097/RLI.0000000000000181.

[120] de Rochefort L, Nguyen TD, Brown R, Spincemaille P, Choi G, Weinsaft J, et al. In vivo quantification of contrast agent concentration using the induced magnetic field for time-resolved arterial input function measurement with MRI. Med Phys 2008;35:5328–39.

[121] Hinoda T, Fushimi Y, Okada Tomohisa, Arakawa Y, Liu C, Yamamoto A, et al. Quantitative assessment of gadolinium deposition in dentate nucleus using quantitative susceptibility mapping. J Magn Reson Imaging 2017;45:1352–8. Available from: https://doi.org/10.1002/jmri.25490.

[122] Choi Y, Jang J, Kim J, Nam Y, Shin N-Y, Ahn K-J, et al. MRI and quantitative magnetic susceptibility maps of the brain after serial administration of gadobutrol: a longitudinal follow-up study. Radiology 2020;297:143–50. Available from: https://doi.org/10.1148/radiol.2020192579.

[123] Zhang J, Xie L, Yang X, Xu L, Chen K, Luo Y, et al. Higher magnetic susceptibility of globus pallidus in patients after macrocyclic GBCAs: assessment using quantitative susceptibility mapping. Acta Radiol 2022;. Available from: https://doi.org/10.1177/02841851221147618.

[124] Schweser F, Deistung A, Lehr BW, Reichenbach JR. Quantitative imaging of intrinsic magnetic tissue properties using MRI signal phase: an approach to in vivo brain iron metabolism? NeuroImage 2011;54:2789–807. Available from: https://doi.org/10.1016/j.neuroimage.2010.10.070.

[125] Schweser F., Deistung A., Lehr B.W., Sommer K., Reichenbach J.R. SEMI-TWInS: simultaneous extraction of myelin and iron using a T2*-weighted imaging sequence. In: Proc Intl Soc Mag Reson Med. Montreal, CA, vol. 19; 2011. p. 120.

[126] Schweser F., Deistung A., Sommer K., Reichenbach J.R. Disentangling contributions from iron and myelin architecture to brain tissue magnetic susceptibility by using quantitative susceptibility mapping (QSM). In: Proc Intl Soc Mag Reson Med. Melbourne, Australia, vol. 20; 2012. p. 409.

[127] Elkady AM, Cobzas D, Sun H, Blevins G, Wilman AH. Discriminative analysis of regional evolution of iron and myelin/calcium in deep gray matter of multiple sclerosis and healthy subjects. J Magn Reson Imaging 2018;48:652–68. Available from: https://doi.org/10.1002/jmri.26004.

[128] Metere R, Kober T, Möller HE, Schäfer A. Simultaneous quantitative MRI mapping of T1, T2* and magnetic susceptibility with multi-echo MP2RAGE. PLoS One 2017;12:e0169265. Available from: https://doi.org/10.1371/journal.pone.0169265.

[129] Sun H, Cleary JO, Glarin R, Kolbe SC, Ordidge RJ, Moffat BA, et al. Extracting more for less: multi-echo MP2RAGE for simultaneous T1-weighted imaging, T1 mapping, mapping, SWI, and QSM from a single acquisition. Magn Reson Med 2020;83:1178–91. Available from: https://doi.org/10.1002/mrm.27975.

[130] Caan MWA, Bazin P, Marques JP, Hollander G, Dumoulin S, Zwaag W. MP2RAGEME: T1, T2*, and QSM mapping in one sequence at 7 tesla. Hum Brain Mapp 2018;. Available from: https://doi.org/10.1002/hbm.24490.

[131] Popescu BFGh, Frischer JM, Webb SM, Tham M, Adiele RC, Robinson CA, et al. Pathogenic implications of distinct patterns of iron and zinc in chronic MS lesions. Acta Neuropathol (Berl) 2017;134:45–64. Available from: https://doi.org/10.1007/s00401-017-1696-8.

[132] Bagnato F, Hametner S, Yao B, van Gelderen P, Merkle H, Cantor FK, et al. Tracking iron in multiple sclerosis: a combined imaging and histopathological study at 7 Tesla. Brain 2011;134:3599–612. Available from: https://doi.org/10.1093/brain/awr278.

[133] Pietroboni AM, Colombi A, Contarino VE, Russo FML, Conte G, Morabito A, et al. Quantitative susceptibility mapping of the normal-appearing white matter as a potential new marker of disability progression in multiple sclerosis. Eur Radiol 2022;. Available from: https://doi.org/10.1007/s00330-022-09338-6.

[134] Chen W, Zhang Y, Mu K, Pan C, Gauthier SA, Zhu W, et al. Quantifying the susceptibility variation of normal-appearing white matter in multiple sclerosis by quantitative susceptibility mapping. Am J Roentgenol 2017;1–6. Available from: https://doi.org/10.2214/AJR.16.16851.

[135] Lancione M, Tosetti M, Donatelli G, Cosottini M, Costagli M. The impact of white matter fiber orientation in single-acquisition quantitative susceptibility mapping. NMR Biomed 2017;1–8. Available from: https://doi.org/10.1002/nbm.3798.

[136] Denk C, Hernández-Torres E, MacKay A, Rauscher A. The influence of white matter fibre orientation on MR signal phase and decay. NMR Biomed 2011;24:246–52. Available from: https://doi.org/10.1002/nbm.1581.

[137] Bender B, Klose U. The in vivo influence of white matter fiber orientation towards B(0) on T2* in the human brain. NMR Biomed 2010;23:1071–6. Available from: https://doi.org/10.1002/nbm.1534.

[138] van Gelderen P, de Zwart JA, Lee J, Sati P, Reich DS, Duyn JH. Nonexponential T(2)* decay in white matter. Magn Reson Med 2011;000:1–8. Available from: https://doi.org/10.1002/mrm.22990.

[139] Yao Y, Nguyen TD, Pandya S, Zhang Y, Hurtado Rúa S, Kovanlikaya I, et al. Combining quantitative susceptibility mapping with automatic zero reference (QSM0) and myelin water fraction imaging to quantify iron-related myelin damage in chronic active MS lesions. Am J Neuroradiol 2018;39:303–10. Available from: https://doi.org/10.3174/ajnr.A5482.

[140] Yu FF, Chiang FL, Stephens N, Huang SY, Bilgic B, Tantiwongkosi B, et al. Characterization of normal-appearing white matter in multiple sclerosis using quantitative susceptibility mapping in conjunction with diffusion tensor imaging. Neuroradiology 2019;61:71–9. Available from: https://doi.org/10.1007/s00234-018-2137-7.

[141] Bergsland N, Schweser F, Dwyer MG, Weinstock-Guttman B, Benedict RHB, Zivadinov R. Thalamic white matter in multiple sclerosis: a combined diffusion-tensor imaging and quantitative susceptibility mapping study. Hum Brain Mapp 2018;39:4007–17. Available from: https://doi.org/10.1002/hbm.24227.

[142] Lassmann H. The pathologic substrate of magnetic resonance alterations in multiple sclerosis. Neuroimaging Clin N Am 2008;18:563—76. Available from: https://doi.org/10.1016/j.nic.2008.06.005.
[143] Wisnieff C, Ramanan S, Olesik J, Gauthier S, Wang Y, Pitt D. Quantitative susceptibility mapping (QSM) of white matter multiple sclerosis lesions: interpreting positive susceptibility and the presence of iron. Magn Reson Med 2015;74:564—70. Available from: https://doi.org/10.1002/mrm.25420.
[144] Dal-Bianco A, Grabner G, Kronnerwetter C, Weber M, Höftberger R, Berger T, et al. Slow expansion of multiple sclerosis iron rim lesions: pathology and 7T magnetic resonance imaging. Acta Neuropathol (Berl) 2016;. Available from: https://doi.org/10.1007/s00401-016-1636-z.
[145] Kaunzner UW, Kang Y, Zhang S, Morris E, Yao Y, Pandya S, et al. Quantitative susceptibility mapping identifies inflammation in a subset of chronic multiple sclerosis lesions. Brain 2019;142:133—45. Available from: https://doi.org/10.1093/brain/awy296.
[146] Chawla S, Kister I, Sinnecker T, Wuerfel J, Brisset J-C, Paul F, et al. Longitudinal study of multiple sclerosis lesions using ultra-high field (7T) multiparametric MR imaging. PLoS ONE 2018;13:e0202918. Available from: https://doi.org/10.1371/journal.pone.0202918.
[147] Chawla S, Kister I, Wuerfel J, Brisset J-C, Liu S, Sinnecker T, et al. Iron and non-iron-related characteristics of multiple sclerosis and neuromyelitis optica lesions at 7T MRI. Am J Neuroradiol 2016;37:1223—30. Available from: https://doi.org/10.3174/ajnr.A4729.
[148] Harrison DM, Li X, Liu H, Jones CK, Caffo B, Calabresi PA, et al. Lesion heterogeneity on high-field susceptibility MRI is associated with multiple sclerosis severity. Am J Neuroradiol 2016;37:1447—53. Available from: https://doi.org/10.3174/ajnr.A4726.
[149] Li X, Harrison DM, Liu H, Jones CK, Oh J, Calabresi PA, et al. Magnetic susceptibility contrast variations in multiple sclerosis lesions. J Magn Reson Imaging 2016;43:463—73. Available from: https://doi.org/10.1002/jmri.24976.
[150] Haacke EM, Makki M, Ge Y, et al. Characterizing iron deposition in multiple sclerosis lesions using susceptibility weighted imaging. J Magn Reson Imaging 2009;29(3):537—44. Available from: https://doi.org/10.1002/jmri.21676.
[151] Shmueli K, de Zwart JA, van Gelderen P, Li T-Q, Dodd SJ, Duyn JH. Magnetic susceptibility mapping of brain tissue in vivo using MRI phase data. Magn Reson Med 2009;62:1510—22. Available from: https://doi.org/10.1002/mrm.22135.
[152] Deh K, Ponath GD, Molvi Z, Parel G-CT, Gillen KM, Zhang S, et al. Magnetic susceptibility increases as diamagnetic molecules breakdown: myelin digestion during multiple sclerosis lesion formation contributes to increase on QSM. myelin breakdown in MS lesion formation. J Magn Reson Imaging 2018;48:1281—7. Available from: https://doi.org/10.1002/jmri.25997.
[153] Wang N, Zhuang J, Wei H, Dibb R, Qi Y, Liu C. Probing demyelination and remyelination of the cuprizone mouse model using multimodality MRI. J Magn Reson Imaging 2019;:jmri.26758. Available from: https://doi.org/10.1002/jmri.26758.
[154] Liu C, Li W, Johnson GA, Wu B. High-field (9.4T) MRI of brain dysmyelination by quantitative mapping of magnetic susceptibility. NeuroImage 2011;56:930—8. Available from: https://doi.org/10.1016/j.neuroimage.2011.02.024.
[155] Argyridis I, Li W, Johnson GA, Liu C. Quantitative magnetic susceptibility of the developing mouse brain reveals microstructural changes in the white matter. NeuroImage 2014;. Available from: https://doi.org/10.1016/j.neuroimage.2013.11.026.
[156] Pal A, Prasad R. Regional distribution of copper, zinc and iron in brain of wistar rat model for non-wilsonian brain copper toxicosis | SpringerLink. Indian J Clin Biochem 2015;93—8.
[157] Desmond KL, Al-Ebraheem A, Janik R, Oakden W, Kwiecien JM, Dabrowski W, et al. Differences in iron and manganese concentration may confound the measurement of myelin from R1 and R2 relaxation rates in studies of dysmyelination. NMR Biomed 2016;29:985—98. Available from: https://doi.org/10.1002/nbm.3549.
[158] Wiggermann V, Hernández-Torres E, Vavasour IM, Moore GRW, Laule C, MacKay AL, et al. Magnetic resonance frequency shifts during acute MS lesion formation. Neurology 2013;81:211—18. Available from: https://doi.org/10.1212/WNL.0b013e31829bfd63.
[159] Cronin MJ, Wharton SJ, Al-Radaideh AM, et al. A comparison of phase imaging and quantitative susceptibility mapping in the imaging of multiple sclerosis lesions at ultrahigh field. Magn Reson Mater Phy 2016;29:543—57. Available from: https://doi.org/10.1007/s10334-016-0560-5.

[160] Eskreis-Winkler S, Deh K, Gupta A, et al. Multiple sclerosis lesion geometry in quantitative susceptibility mapping (QSM) and phase imaging. J Magn Reson Imaging 2015;42(1):224−9. Available from: https://doi.org/10.1002/jmri.24745.

[161] Zhang Y, Gauthier SA, Gupta A, Chen W, Comunale J, Chiang GC-Y, et al. Quantitative susceptibility mapping and R2* measured changes during white matter lesion development in multiple sclerosis: myelin breakdown, myelin debris degradation and removal, and iron accumulation. Am J Neuroradiol 2016;37:1629−35. Available from: https://doi.org/10.3174/ajnr.A4825.

[162] Chen W, Gauthier SA, Gupta A, Comunale J, Liu T, Wang S, et al. Quantitative susceptibility mapping of multiple sclerosis lesions at various ages. Radiology 2014;271:183−92. Available from: https://doi.org/10.1148/radiol.13130353.

[163] van der Weijden CWJ, Biondetti E, Gutmann IW, Dijkstra H, McKerchar R, de Paula Faria D, et al. Quantitative myelin imaging with MRI and PET: an overview of techniques and their validation status. Brain 2022;. Available from: https://doi.org/10.1093/brain/awac436.

[164] Emmerich J, Bachert P, Ladd ME, Straub S. On the separation of susceptibility sources in quantitative susceptibility mapping: Theory and phantom validation with an in vivo application to multiple sclerosis lesions of different age. J Magn Reson 2021;330:107033. Available from: https://doi.org/10.1016/j.jmr.2021.107033.

[165] Walton JC, Kaufmann JC. Iron deposits and multiple sclerosis. Arch Pathol Lab Med 1984;108:755−6.

[166] Gupta A, Al-Dasuqi K, Xia F, et al. The use of noncontrast quantitative MRI to detect gadolinium-enhancing multiple sclerosis brain lesions: a systematic review and meta-analysis. Am J Neuroradiol 2017;38(7):1317−22. Available from: https://doi.org/10.3174/ajnr.A5209.

[167] Zhang Y, Wei H, Cronin MJ, He N, Yan F, Liu C. Longitudinal atlas for normative human brain development and aging over the lifespan using quantitative susceptibility mapping. NeuroImage 2018;171:176−89. Available from: https://doi.org/10.1016/j.neuroimage.2018.01.008.

[168] Smith SM, Zhang Y, Jenkinson M, et al. Accurate, robust, and automated longitudinal and cross-sectional brain change analysis. NeuroImage 2002;17(1):479−89. Available from: https://doi.org/10.1006/nimg.2002.1040.

[169] Song SMH, Napel S, Pelc NJ, Glover GH. Phase unwrapping of MR phase images using poisson equation. IEEE Trans Image Process 1995;4(5):667−76.

[170] Zhou D, Liu T, Spincemaille P, Wang Y. Background field removal by solving the Laplacian boundary value problem. NMR Biomed 2014;27(3):312−19. Available from: https://doi.org/10.1002/nbm.3064.

[171] Sun H, Wilman AH. Background field removal using spherical mean value filtering and Tikhonov regularization. Magn Reson Med 2014;71(3):1151−7. Available from: https://doi.org/10.1002/mrm.24765.

[172] Kames C, Wiggermann V, Rauscher A. Rapid two-step dipole inversion for susceptibility mapping with sparsity priors. NeuroImage 2018;167(June 2017):276−83. Available from: https://doi.org/10.1016/j.neuroimage.2017.11.018.

[173] Absinta M, Sati P, Schindler M, et al. Persistent 7-tesla phase rim predicts poor outcome in new multiple sclerosis patient lesions. J Clin Invest 2016;126(7):2597−609. Available from: https://doi.org/10.1172/JCI86198.

[174] Craelius W, Migdal MW, Luessenhop CP, Sugar A, Mihalakis I. Iron deposits surrounding multiple sclerosis plaques. Arch Pathol Lab Med 1982;106:397−9.

[175] Dal-Bianco A, Grabner G, Kronnerwetter C, Weber M, Kornek B, Kasprian G, et al. Long-term evolution of multiple sclerosis iron rim lesions in 7 T MRI. Brain 2021. Available from: https://doi.org/10.1093/brain/awaa436.

[176] Mehta V, Pei W, Yang G, Li S, Swamy E, Boster A, et al. Iron is a sensitive biomarker for inflammation in multiple sclerosis lesions. PLoS One 2013;8:1−10. Available from: https://doi.org/10.1371/journal.pone.0057573.

[177] Gillen KM, Mubarak M, Park C, et al. QSM is an imaging biomarker for chronic glial activation in multiple sclerosis lesions. Ann Clin Transl Neurol 2021;acn3.51338. Available from: https://doi.org/10.1002/acn3.51338.

[178] Maggi P, Sati P, Nair G, et al. Paramagnetic rim lesions are specific to multiple sclerosis: an international multicenter 3T MRI study. Ann Neurol 2020;ana.25877. Available from: https://doi.org/10.1002/ana.25877.

[179] Blindenbacher N, Brunner E, Asseyer S, Scheel M, Siebert N, Rasche L, et al. Evaluation of the 'ring sign' and the 'core sign' as a magnetic resonance imaging marker of disease activity and progression in clinically isolated syndrome and early multiple sclerosis. Mult Scler J - Exp Transl Clin 2020;6. Available from: https://doi.org/10.1177/2055217320915480.

[180] Zinger N, Ponath G, Sweeney E, Nguyen TD, Lo CH, Diaz I, et al. Dimethyl fumarate reduces inflammation in chronic active multiple sclerosis lesions. Neurol - Neuroimmunol Neuroinflammat 2022;9:e1138. Available from: https://doi.org/10.1212/NXI.0000000000001138.

[181] Guo Z, Long L, Qiu W, et al. The distributional characteristics of multiple sclerosis lesions on quantitative susceptibility mapping and their correlation with clinical severity. Front Neurol 2021;12:647519. Available from: https://doi.org/10.3389/fneur.2021.647519.

[182] Marcille M, Hurtado Rúa S, Tyshkov C, et al. Disease correlates of rim lesions on quantitative susceptibility mapping in multiple sclerosis. Sci Rep 2022;12(1):4411. Available from: https://doi.org/10.1038/s41598-022-08477-6.

[183] Bagnato F, Sati P, Hemond CC, et al. Imaging chronic active lesions in multiple sclerosis: a consensus statement. Brain 2024;awae013. Available from: https://doi.org/10.1093/brain/awae013.

[184] Baron J-C, Jones T. Oxygen metabolism, oxygen extraction and positron emission tomography: historical perspective and impact on basic and clinical neuroscience. NeuroImage, Neuroimaging: Then, Now Future 2012;61:492–504. Available from: https://doi.org/10.1016/j.neuroimage.2011.12.036.

[185] Zhang J, Liu T, Gupta A, Spincemaille P, Nguyen TD, Wang Y. Quantitative mapping of cerebral metabolic rate of oxygen ($CMRO_2$) using quantitative susceptibility mapping (QSM). Magn Reson Med 2015;74:945–52. Available from: https://doi.org/10.1002/mrm.25463.

[186] He X, Yablonskiy DA. Quantitative BOLD: mapping of human cerebral deoxygenated blood volume and oxygen extraction fraction: default state. Magn Reson Med 2007;57:115–26. Available from: https://doi.org/10.1002/mrm.21108.

[187] Cho J, Kee Y, Spincemaille P, Nguyen TD, Zhang J, Gupta A, et al. Cerebral metabolic rate of oxygen ($CMRO_2$) mapping by combining quantitative susceptibility mapping (QSM) and quantitative blood oxygenation level-dependent imaging (qBOLD). Magn Reson Med 2018;80:1595–604. Available from: https://doi.org/10.1002/mrm.27135.

[188] Cho J, Nguyen TD, Huang W, Sweeney EM, Luo X, Kovanlikaya I, et al. Brain oxygen extraction fraction mapping in patients with multiple sclerosis. J Cereb Blood Flow Metab 2021;. Available from: https://doi.org/10.1177/0271678X211048031.

[189] Cho J, Zhang J, Spincemaille P, et al. QQ-NET – using deep learning to solve quantitative susceptibility mapping and quantitative blood oxygen level dependent magnitude (QSM + qBOLD or QQ) based oxygen extraction fraction (OEF) mapping. Magn Reson Med 2022;87(3):1583–94. Available from: https://doi.org/10.1002/mrm.29057.

[190] Cho J, Zhang S, Kee Y, et al. Cluster analysis of time evolution (CAT) for quantitative susceptibility mapping (QSM) and quantitative blood oxygen level-dependent magnitude (qBOLD)-based oxygen extraction fraction (OEF) and cerebral metabolic rate of oxygen (CMRO 2) mapping. Magn Reson Med 2020;83(3):844–57. Available from: https://doi.org/10.1002/mrm.27967.

[191] Zhang J, Cho J, Zhou D, Nguyen TD, Spincemaille P, Gupta A, et al. Quantitative susceptibility mapping-based cerebral metabolic rate of oxygen mapping with minimum local variance. Magn Reson Med 2017;179:172–9. Available from: https://doi.org/10.1002/mrm.26657.

[192] Chiang GC, Cho J, Dyke J, et al. Brain oxygen extraction and neural tissue susceptibility are associated with cognitive impairment in older individuals. J Neuroimaging 2022;32(4):697–709. Available from: https://doi.org/10.1111/jon.12990.

[193] Shen N, Zhang S, Cho J, et al. Application of cluster analysis of time evolution for magnetic resonance imaging -derived oxygen extraction fraction mapping: a promising strategy for the genetic profile prediction and grading of glioma. Front Neurosci 2021;15:736891. Available from: https://doi.org/10.3389/fnins.2021.736891.

[194] Wu D, Zhou Y, Cho J, et al. The spatiotemporal evolution of MRI-derived oxygen extraction fraction and perfusion in ischemic stroke. Front Neurosci 2021;15:716031. Available from: https://doi.org/10.3389/fnins.2021.716031.

[195] Zhang Q, Sui C, Cho J, et al. Assessing cerebral oxygen metabolism changes in patients with preeclampsia using voxel-based morphometry of oxygen extraction fraction maps in magnetic resonance imaging. Korean J Radiol 2023;24(4):324–37. Available from: https://doi.org/10.3348/kjr.2022.0652.

[196] Zhuang H, Cho J, Chiang GCY, et al. Cerebral oxygen extraction fraction declines with ventricular enlargement in patients with normal pressure hydrocephalus. Clin Imaging 2023;97:22–7. Available from: https://doi.org/10.1016/j.clinimag.2023.02.001.

[197] Zivadinov R, Ramasamy DP, Benedict RHB, Polak P, Hagemeier J, Magnano CR, et al. Cerebral microbleeds in multiple sclerosis evaluated on susceptibility-weighted images and quantitative susceptibility maps: a case-control study. Radiology 2016;281:884–95. Available from: https://doi.org/10.1148/radiol.2016160060.

[198] Eisele P, Alonso A, Griebe M, Szabo K, Hennerici MG, Gass A. Investigation of cerebral microbleeds in multiple sclerosis as a potential marker of blood-brain barrier dysfunction. Mult Scler Relat Disord 2016;7:61–4. Available from: https://doi.org/10.1016/j.msard.2016.03.010.

[199] Ziliotto N, Bernardi F, Jakimovski D, et al. Hemostasis biomarkers in multiple sclerosis. Eur J Neurol 2018;. Available from: https://doi.org/10.1111/ene.13681.

[200] Patenaude B, Smith SM, Kennedy DN, Jenkinson M. A Bayesian model of shape and appearance for subcortical brain segmentation. NeuroImage 2011;56:907–22. Available from: https://doi.org/10.1016/j.neuroimage.2011.02.046.

[201] Hanspach J, Dwyer MG, Bergsland NP, Feng X, Hagemeier J, Bertolino N, et al. Methods for the computation of templates from quantitative magnetic susceptibility maps (QSM): toward improved atlas- and voxel-based analyses (VBA). J Magn Reson Imaging 2017;46:1474–84. Available from: https://doi.org/10.1002/jmri.25671.

[202] Feng X, Deistung A, Dwyer MG, Hagemeier J, Polak P, Lebenberg J, et al. An improved FSL-FIRST pipeline for subcortical gray matter segmentation to study abnormal brain anatomy using quantitative susceptibility mapping (QSM). Magn Reson Imaging 2017;39:110–22. Available from: https://doi.org/10.1016/j.mri.2017.02.002.

[203] Deistung A, Schäfer A, Schweser F, Biedermann U, Turner R, Reichenbach JR. Toward in vivo histology: a comparison of quantitative susceptibility mapping (QSM) with magnitude-, phase-, and R2*-imaging at ultra-high magnetic field strength. NeuroImage 2013;65:299–314. Available from: https://doi.org/10.1016/j.neuroimage.2012.09.055.

[204] Lim IAL, Faria AV, Li X, Hsu JTC, Airan RD, Mori S, et al. Human brain atlas for automated region of interest selection in quantitative susceptibility mapping: application to determine iron content in deep gray matter structures. NeuroImage 2013;82:449–69. Available from: https://doi.org/10.1016/j.neuroimage.2013.05.127.

[205] Cobzas D, Sun H, Walsh AJ, Lebel RM, Blevins G, Wilman AH. Subcortical gray matter segmentation and voxel-based analysis using transverse relaxation and quantitative susceptibility mapping with application to multiple sclerosis. J Magn Reson Imaging 2015;42:1601–10. Available from: https://doi.org/10.1002/jmri.24951.

[206] Heckemann RA, Hajnal JV, Aljabar P, Rueckert D, Hammers A. Automatic anatomical brain MRI segmentation combining label propagation and decision fusion. NeuroImage 2006;33:115–26. Available from: https://doi.org/10.1016/j.neuroimage.2006.05.061.

[207] Sati P, Oh J, Constable RT, Evangelou N, Guttmann CRG, Henry RG, et al. The central vein sign and its clinical evaluation for the diagnosis of multiple sclerosis: a consensus statement from the North American Imaging in Multiple Sclerosis Cooperative. Nat Rev Neurol 2016;12:714–22. Available from: https://doi.org/10.1038/nrneurol.2016.166.

CHAPTER 14

Functional magnetic resonance imaging in multiple sclerosis

Eva A. Krijnen[1,2] and Menno M. Schoonheim[1]

[1]MS Center Amsterdam, Anatomy and Neurosciences, Amsterdam Neuroscience, Amsterdam UMC Location VUmc, Amsterdam, The Netherlands [2]Department of Neurology, Massachusetts General Hospital, Harvard Medical School, Boston, MA, United States

OUTLINE

Introduction	249	Longitudinal resting-state functional magnetic resonance imaging	256
Functional reorganization: task-based functional magnetic resonance imaging	250	Dynamic functional connectivity	257
Changes to sensorimotor systems	250	The network collapse	257
Changes to cognitive systems	251	Advanced network analyses: network efficiency	257
Functional connectivity: resting-state functional magnetic resonance imaging	253	Conclusion	258
Changes to sensorimotor systems	253	References	259
Changes to cognitive systems	254		
Functional brain changes over time	255		
Longitudinal task-based functional magnetic resonance imaging	255		

Introduction

Multiple sclerosis (MS) commonly features cognitive and motor impairments, both of which remain poorly understood. To monitor patterns of central nervous system (CNS) damage thought to relate to these impairments in Ms patients, structural magnetic resonance imaging (MRI) has been largely applied in clinical practice. Despite great improvements in structural MRI techniques to visualize the extent of damage, such quantifications of macro- and microstructural damage seen on MRI cannot fully explain the extensive heterogeneity in clinical disability [1]. A potential explanation for this so-called clinico-radiological paradox is a functional reorganization of functional activity, that is, the ability of the brain to respond to CNS damage. As such, the brain is hypothesized to be able to modify the properties of its neural circuits, which is considered to provide some form of buffer capacity in different stages of the disease. This "brain plasticity" involves a balance between adaptive (limiting the clinical consequences of widespread CNS damage) and maladaptive mechanisms (actually partly responsible for worsening neurological deficits) [2,3]. This intriguing conundrum has driven the field to technological evolution in the last decade, mostly based on functional MRI (fMRI).

fMRI is a technique enabling noninvasive visualization of brain activity based on the blood-oxygenation level dependent (BOLD) response. This BOLD response allows for the mapping of fluctuations in oxy- and deoxyhemoglobin concentrations following neuronal activity, quantifying activation, and deactivation. This is only possible by taking advantage of the excessive vascular response, driving a large overshoot of oxygenated blood into an active area which increases signal and thus allows for the quantification of such differences in magnetic resonance signal within an area of interest. This technique can be applied during a given task [4,5] or, alternatively, during rest [6,7]. The former application, "task-based" fMRI, has mainly been used to map the level of activation of brain regions, during the performance of a specific task [5,8]. The latter application, "resting-state" (RS) fMRI, is primarily used to reconstruct patterns of functionally interacting brain regions, that is, functional connectivity (FC) and functional networks [7,9].

Despite being an indirect measure of neural activity, fMRI represents a powerful and suitable tool to evaluate brain plasticity in vivo, greatly improving the understanding of clinical deficits in people with MS. Based on our recent overviews of the current state of the literature [10–12], this chapter outlines some of the latest findings in Ms on functional reorganization in the brain using fMRI and their clinical relevance. Brain plasticity will be discussed in the context of task-based fMRI and RS fMRI, in the light of both motor and cognitive deficits. It should be noted that while this chapter focuses on motor function and cognition only, further insights can be gained by studying other symptoms like fatigue and visual problems in MS [13]. This chapter then provides a summary of findings from studies applying longitudinal task-based and RS fMRI, which allow for a more solid exploration on different chains of events leading to clinical progression. In the last section, we highlight novel advanced networks analyses designed to quantify how the network as a whole is altered in terms of efficiency and how this relates to clinical function [13].

Functional reorganization: task-based functional magnetic resonance imaging

Changes to sensorimotor systems

Task-based fMRI has been widely applied to map brain activity in MS both in the acute phase of the disease [14–16] and in clinically stable disease [17–20]. When studying the brain using motor tasks, research has mainly focused on upper limb tasks, including flexion-extension of fingers, and also tasks as motor imagery [21], hand writing [22], grip force [23], and joystick movements [24]. For obvious reasons, lower limb function remains understudied, given difficulties in task design despite clear clinical relevance. In general, regions subserving upper motor function show significantly higher regional fMRI activations in people with MS compared to healthy volunteers (HCs) in all MS phenotypes [20], in particular at early disease stages.

Organized from early to late disease stages, people with clinically isolated syndrome (CIS) typically show increased activity of the primary sensorimotor cortex, which is also seen in people with "benign" relapsing-remitting MS (RRMS) experiencing no to mild clinical symptoms [25,26]. In people with somewhat more advanced RRMS, additional regions are recruited, including the inferior frontal gyrus, supplementary motor area, and inferior parietal lobule, even during simple motor tasks [27–29]. In HC, these regions are generally only active during complex tasks [30,31]. It is hypothesized this increased reaching out of the brain toward additional areas might be a compensatory mechanism in people with RRMS to maintain optimal task performance despite widespread CNS damage [17,28,32].

Interestingly, people with progressive forms of MS frequently show increased fMRI signals in even more areas while performing a motor task, mostly including high-order regions, such as the precuneus, cerebellum, and middle temporal gyrus [17,19,33]. As these areas are not typically recruited at all in controls, this process seems rather unique to populations with more severe clinical impairments. What this particular process actually means remains controversial. On the one hand, this could be interpreted as beneficial plasticity only required in such a severely damaged brain. The depletion of the functional competence of the classical motor circuitries would thus require the involvement of these higher order brain regions to preserve a certain degree of clinical functioning. Alternatively, these changes could also be a sign of maladaptive change: a loss of inhibition of these brain regions that normally occurs in controls when performing a motor task only occurring in progressive disease. Alternatively, it could also be that these additional changes are not necessarily directly related to task functioning at all, neither beneficial nor maladaptive. Although generally the consensus is that these changes represent some form of beneficial change, additional research is required including longitudinal and possibly treatment-based designs. Interestingly, in addition to such hyperactivation, decreased activity has also been observed in classical motor areas, mostly in relation to poorer function. As these seem somewhat easier to interpret, hypoactivation might indicate that either structural disconnection, local depletion of functional circuits and/or neurodegeneration leads to such maladaptive mechanisms in the Ms brain [33].

Changes to cognitive systems

Task-based fMRI has also been applied using tasks based on various cognitive tasks. Tasks most commonly performed during task-based fMRI are the paced auditory serial addition test (PASAT), N-back, Go-NoGo, and Stroop task. Therefore attention, information processing speed, and working memory are the cognitive domains most often investigated in these studies. This selection of cognitive domains also encompasses those most frequently and severely impacted by Ms [34].

The PASAT involves a sequence of heard numbers that have to be added together, but in a specific order. As such, the PASAT assesses aspects of processing speed, attention, and working memory; hence, regions subserving these cognitive domains usually show alterations in fMRI activity, including the inferior frontal gyri [35], medial and dorsolateral prefrontal cortices [35–39], angular gyri and other parietal regions [35,36,39,40], and the cerebellum [37,40]. Generally, in people with MS who are cognitively preserved, increases in fMRI signaling were more evident compared to people with MS with cognitive impairment [38,39,41]. In fact, this hyperactivation in populations with preserved or mildly affected function is seen across cognitive tasks, in addition to motor paradigms.

The N-back task is similar in that it also involves a series of stimuli that are presented in a specific order, but now participants are required to respond to either the stimulus directly before (0-back) or further back (up to n-back), with varying levels of difficulty. While performing the N-back task, mainly measuring working memory, significant increases in fMRI activity are found in frontoparietal cortices of working memory circuitry in people with MS compared to HC [41–45]. With increasing task complexity, in particular cognitively impaired people with MS showed less activations of these cortices, which could be interpreted as a failure of compensatory hyperactivation. In addition, such more difficult levels feature a reduced ability to suppress activity of the default-mode network (DMN) [42,45], a key cognitive network that is supposed to be deactivated when performing such tasks. In fact, the DMN is often referred to with the controversial term "task-negative" network, in which the brain would be when doing nothing [46]. Interestingly, similar to aforementioned paradigms, fMRI signals were more evidently hyperactivated in cognitively preserved than in cognitively impaired people with MS, in comparison with HC [45]. This has shed new light on the role of the DMN (see Fig. 14.1), further stimulating cognitive network neuroscience.

The Go-NoGo task is designed such that participants should only respond to certain stimuli ("Go") and not during others ("No Go"). In the Stroop participants are shown a matrix of words, describing colors, they have to read out as fast as possible. Next, a matrix is shown with blocks of colors, without words, also to be spoken as fast as possible. Then, a matrix is shown with words, again describing colors, but now with incongruent font

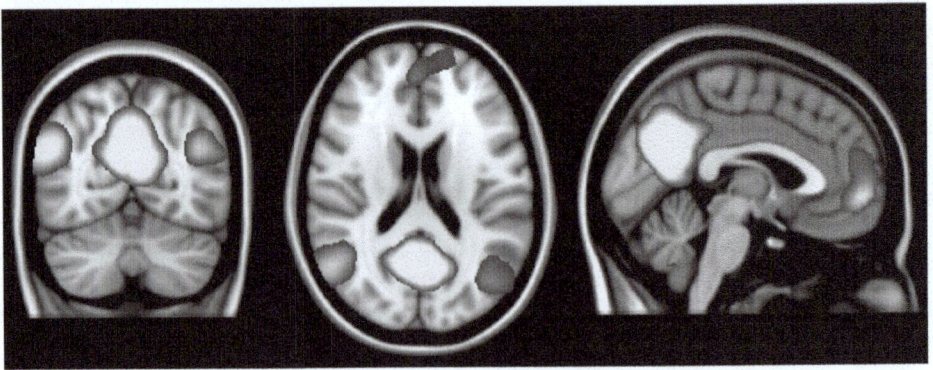

FIGURE 14.1   The default-mode network.

colors (e.g., the word blue in a yellow font). Here, participants have to name the color, not the word. As such, both tasks are examples of designs that quantify inhibition/impulsivity, attention and processing speed. In MS, we see increases in frontal, parietal and thalamic fMRI activity in people with MS with normal cognition compared to HC. As such, again similar to previous paradigms, while task performance was similar between MS phenotypes, activation patterns deviated with disease progression, as people with progressive MS demonstrated the most abnormal brain activity [47].

An episodic memory fMRI task has also been used, designed to first allow participants to remember certain images, which is followed by a recall period to identify successful encoding of information. During successful encoding in MS, increases in brain activity were seen in parahippocampal and anterior cingulate regions in cognitively preserved people with MS compared to HC. In contrast, cognitively impaired people with MS express parahippocampal and prefrontal activity decreases, along with posterior cingulate and precuneus activity increases [48]. The latter again indicates that hypoactivation is related to poorer functioning, while hyperactivation could indicate beneficial functional reorganization. However, the latter remains difficult to prove, given that patients with poor functioning also display hyperactivation, although not necessarily in the same areas.

Functional connectivity: resting-state functional magnetic resonance imaging

While task-based fMRI studies can shed light on which regions are active during specific tasks, how active regions communicate with the rest of the brain requires additional methodologies to unravel. The latter is typically referred to as FC, that is, the strength to which the activity between a pair of brain regions covaries or correlates over time. This approach has usually been applied with the use of RS fMRI, during which a participant does not perform any explicit task.

While task-based paradigms are typically highly similar in their findings, the field of connectivity in MS remains especially difficult to grasp, as varying results of both increased and decreased FC cooccur in the same patient group across different regions [49] and longitudinal work remains rare. Early RS fMRI studies showed an increased FC in people with CIS [50–53] and decreased FC in people with progressive MS [54]. This combination was thus interpreted as representing similar reorganization concepts as task-based paradigms. However, subsequent work indicated that most alterations, increased or decreased, actually relate to poorer functioning, with preserved patients showing a normal network [55]. Below we will discuss RS fMRI findings, once more in the light of both motor and cognitive impairment.

Changes to sensorimotor systems

Since RS fMRI does not involve the performance of a specific task that can be related to functional signals, typically researchers quantify metrics of connectivity from the scan and relate this to clinical profiles outside of the scanner, such as the severity of disability. Disability in MS is mostly quantified by the expanded disability status scale score which is mainly driven by physical (motor) impairment. Within connectivity literature, many studies have related changes in RS FC in MS within the sensorimotor system to clinical disability [56]. RS FC of the cerebellum, motor cortices, and other brain regions can show increases as well as decreases, which are associated with physical and motor impairment [57−62], that is, all changes are related to impairment. Interestingly, connectivity changes in disabled patients have also been reported to expand beyond primary and secondary motor areas, such as occipito-temporal areas related to visual processing [63]. Additionally, functional networks known to comprise cognitive processes also show such changes, such as the DMN and frontoparietal network (FPN). This could indicate that dysfunctional networks typically related to higher order processing could impact mobility, hypothesized to revolve around impaired motor planning, attention, and/or processing of information during movement. This finding is in contrast with task-based activation findings, where hyperactivation of such higher order areas might play some beneficial role. Finally, RS FC patterns of specific brain structures are found to relate differently to cognition and disability based on the specific affected area, such as the thalamus [64−66] and cerebellum [67,68]. For example, RS FC between the thalamus and temporal areas is increased in people with MS with cognitive impairment, whereas decreases in thalamic RS FC with caudate and cingulate cortex relate to worse motor function. Together, these insights have led to a reappraisal of the complexity of altered brain function in people with MS, indicating that functional activation grasps and connectivity grasp vastly different neuronal processes.

Changes to cognitive systems

Clinically, the evaluation of cognitive functioning in MS has drastically evolved in recent years. Where initially this was limited to the PASAT as a subtest of the MS functional composite, nowadays much more expansive test batteries are used [34], encompassing aforementioned cognitive domains such as information processing speed, memory, and attention. These typically include the brief repeatable battery of neuropsychological tests and minimal assessment of cognitive function in MS, both encompassing multiple cognitive domains. For clinical practice, these two batteries are usually too time-intensive, which has led to the development of the more compact brief international cognitive assessment for MS, focusing on information processing speed, verbal, and visuospatial memory. With this clinical development, connectivity studies typically investigate comparisons of cognitively impaired (usually defined as failing two domains or tests) to preserved patients.

Most RS FC research studying cognitive impairment in MS has shown a dysfunctional DMN, a network that was also discussed earlier in the context of cognitive tasks. Interestingly, studies focusing on cognition in MS using connectivity approaches show conflicting results in the DMN, being either increased [69−74] or decreased [54,75−79] in cognitively impaired patients. Similar to aforementioned disability results, regardless of directionality, essentially all alterations of RS FC of the DMN are related to the severity of cognitive disability. This implies that both hyper- and hypoconnectivity of the DMN seems to indicate that the brain network has lost some of its efficiency, which is crucial for normal cognitive functioning. As mentioned, the DMN was formerly seen as a mere "intrinsic" network, while clinical findings such as those in MS actually show that the

DMN features processes relevant for cognition instead. In fact, recent insights into normal brain functioning also highlight that the DMN does have a central function in cognitive processes: the temporal embedding of higher order extrinsic and intrinsic information [80]. It, therefore, seems logical if this process becomes inefficient in MS, cognitive impairment can ensue.

Similar to findings in disability, relevant RS FC alterations for cognition are not limited to only one functional network. The FPN has been found to play a key role in cognitive impairment as well [70,74,81–84], which, like the DMN, relates to cognitive impairment in both a hyper- and hypoconnected way [75]. In contrast to the DMN, the FPN has commonly been referred to as the "task-positive" network, together with the dorsal attention network (DAN), as it has long been known to be actively involved in cognitive processing while performing cognitive tasks [46,85]. The DMN and FPN comprise most of so-called brain "hubs" (strongly connected regions) [86] and have opposing roles [87]. Despite these different roles, they both present similar activity patterns in the MS brain, showing particularly FC disruptions with nonhub regions [74]. This potentially drives an altered balance between the two networks, leaving the global brain network imbalanced and losing normal flexible alterations of network states driven by intrinsic (DMN) and extrinsic (FPN) processes, both of which are crucial for cognition.

Taken together, task-based and RS studies have resulted in the main hypothesis that while hyperactivation could be beneficial and should perhaps be stimulated, connectivity patterns should stay as normal as possible, which is more difficult to achieve. The ability of the brain to preserve its FC, and thus its cognitive functioning, is commonly related to the concept of "cognitive reserve" (typically quantified as having higher levels of education) [88,89]. This buffer capacity of some people seems to allow the network to withstand the effects of structural damage for a longer period of time. Whether this buffer is also related to more extensive hyperactivation remains unclear at this time. As such, this relation might have implications for how cognitive reserve can modulate the susceptibility to cognitive impairment in people with MS [89], which warrants future study in the context of developing possible treatment strategies to stimulate such reserve capacity further, if possible.

Functional brain changes over time

Longitudinal task-based functional magnetic resonance imaging

Longitudinal task studies remain quite rare and often feature very small sample sizes, so their findings should be treated with some caution. Nonetheless, some interesting findings have been shown. For instance, at the time of an acute clinical relapse, MS patients seem to experience abnormally high fMRI activity in homologous brain areas of the unaffected hemisphere, which returns to normal during follow-up, in particular in patients who recover successfully [15,29,90]. Another application of such paradigms is in the context of cognitive rehabilitation. Here, task-related fMRI activation was increased after treatment compared to baseline, in one study involving parietal, temporal, dorsolateral prefrontal cortices, precuneus, and hippocampus, which was consistent with the trained cognitive domains, that is, attention, information processing speed, working memory, and executive functions [91–97]. As such, this could highlight the adaptive nature of hyperactivation in MS. Research investigating motor rehabilitation shows decreased activity of high-order, integrative regions [98–100] and restoration of lateralized sensorimotor network activity [101]. Observational studies confirmed these findings indicating that beneficial changes were less visible in patients who showed clinical progression. Interestingly, one study assessing longitudinal Go–NoGo performance showed widespread beneficial increases at baseline, as well as maladaptive increases in the DMN over time, which also warrants future study in larger samples.

Longitudinal resting-state functional magnetic resonance imaging

Patterns of RS FC changes have also been rarely longitudinally studied and again mostly in small samples. One study showed that patients displaying increasing FC over time mostly included patients with minimal disability at baseline, while those showing decreases over time already had worse disability at baseline, which was interpreted as a loss of plasticity over time. Other studies have shown mostly stable and reproducibly reduced frontal RS FC, related to worse PASAT scores in patients who remained clinically stable.

Interestingly, there seems to be a difference in longitudinal RS FC patterns between cognitive subgroups, which seems to indicate that specific FC findings are disease stage specific, which could explain aforementioned findings. A transition from preserved cognition toward mild cognitive impairment was initially related to disturbed functioning of the ventral attention network (VAN), after which a shift toward DMN dysfunction takes place when transitioning toward more severe cognitively impaired [102]. As the VAN plays a role in the balance between intrinsic and extrinsic stimuli and is thought to balance network dominance from FPN to the DMN (and vice versa) [103], processes relevant for maintaining overall network functioning might be progressively disrupted as patients cognitively progress. After this initial stage of network destabilization, this phenomenon is lost in more severe cognitive impairment, with DMN dysfunction, hyperactivation, and hyperconnectivity becoming more apparent.

Apart from observational studies, a few studies have also investigated rehabilitation effects. During cognitive rehabilitation, within DMN, RS FC appears to be increased in people with RRMS and related to improved PASAT performance in those patients [104]. This interesting finding seems to contrast observational work, although within DMN, FC is not as frequently related to cognitive impairment compared to FC of the DMN with the rest of the brain. Other work has shown that alterations in DMN RS FC help explain the maintenance of the effects of cognitive rehabilitation after several months and even their improvement on depression and quality of life, perhaps indicating the receptiveness of the brain to implement beneficial rehabilitation effects depending on the level of DMN dysfunction at baseline [105]. Increases in RS FC within the memory system following memory rehabilitation further highlight the need for more study to disentangle seemingly beneficial treatment-induced changes in FC from maladaptive FC changes as indicated by the large body of observational work.

Dynamic functional connectivity

Aforementioned insights indicate that the functional network becomes less efficient as patients progress, as increases and decreases in connectivity continue to accumulate. This could indicate that the brain network becomes less dynamic, as normally such high and low connectivity states are known to alternate depending on the current demand within the system. For instance, the activation of specific networks shifts during cognitive tasks [106], even during resting conditions. Recently, this "time-varying" or "dynamic" shaping of functional networks became technically possible to quantify [86], looking at the variability of FC and reoccurring FC patterns (i.e., states).

Using this technique, people with CIS showed specific alterations in network dynamics within states, which progressed over time despite clinical stability [107]. Subsequent work indicated that such state-specific changes are different across the different MS phenotypes [108]. Cognitively impaired patients also had a reduced dynamic FC of deep gray matter and DMN, spending less time in a highly connected state [109]. Apart from state-specific analyses, assessments of dynamic FC also indicated that people with progressive MS showed increased dynamic cerebellar connectivity with the FPN, which was also related to cognitive impairment [110]. Additional work in cognitive impairment showed reduced network variations not seen in preserved patients [111]. For instance, the DMN showed reduced dynamics specific to cognitive impairment, together with reductions in FPN and visual areas [112]. The DMN also showed a loss of normal interplay with other networks using such a dynamic approach, possibly indicating that other networks are not able to (re)activate or suppress the DMN as also indicated by aforementioned task-based work [112].

The network collapse

At early phases of MS when structural brain damage is still limited, compensatory local hyperactivation in brain regions contribute to ensure brain network efficiency is maintained, as FC remains largely normal (see Fig. 14.2). After some time, when structural damage progresses, these compensatory mechanisms begin to fail, leading to a loss of network efficiency and dynamics over time. This concept is especially relevant for cognition, which fully relies on an efficient processing of information, which dynamically alternates between the different cognitive networks. As network efficiency loss exceeds a certain threshold, this eventually will lead to a so-called "network collapse" featuring an accelerated clinical progression as the network can no longer maintain its crucial efficient processing of information, involving a rigid, hyperactive and hyperconnected DMN.

Advanced network analyses: network efficiency

The hypothesis of a network collapse seems to fit with current literature, apart from the quantification of "network efficiency," which requires an assessment of the entire brain network and how connectivity changes impact the network as a whole. This requires the

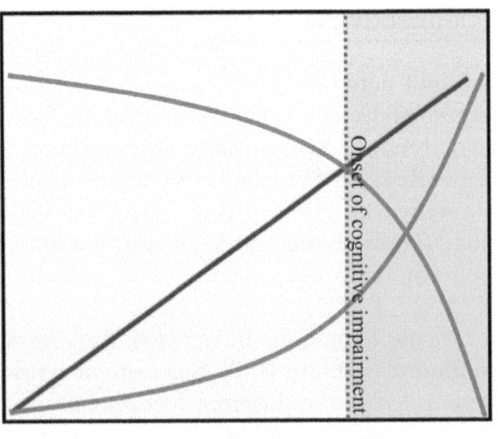

FIGURE 14.2 The hypothesis of a "network collapse" underlying clinical progression in multiple sclerosis. Source: *Adapted with permission from Schoonheim MM, Meijer KA, Geurts JJ. Network collapse and cognitive impairment in multiple sclerosis. Front Neurol 2015;6:82. https://doi.org/10.3389/fneur.2015.00082.*

researcher to incorporate RS FC data into a more complex system, and analyze whole-network features rather than individual connections [55]. In graph theoretical analysis, the brain is represented as a network (graph) of interconnected brain regions (nodes). This method allows to evaluate measures such as "path length" (the level of integration), "clustering coefficient" (the level of segregation), "modularity" (the level of integration of subnetworks into the entire network), or "centrality" (the level of hub-ness) [86].

Using graph theory, a more random path length in people with early MS has been found [113], resulting in a less efficient network which was related to cognitive impairment. In addition, more segregated subnetworks (such as the DMN and FPN) also related to worse cognitive functioning. Hub-analyses showed altered hub patterns related to cognitive impairment in terms of spatial location and strength [114,115]. For instance, centrality analyses indicated DMN and ventral stream network centrality increases were related to cognitive impairment, whereas sensorimotor decreases were related to disability [116–118]. Increased network clustering within the motor system was also related to more severe disability.

These studies provide a few examples of how the field is trying to grasp how local changes impact global network efficiency and how these changes relate to clinical symptoms. Developing an optimal measure of efficiency remains challenging, however. Thankfully, the field of network neuroscience continues to grow with novel measures developed continuously to fill this gap.

Conclusion

Taken together, functional activation, connectivity, and network analyses have greatly enhanced our understanding of mechanisms underlying clinical preservation and impairments in MS. The current body of work suggests that MS involves a complex combination of beneficial functional reorganization—mostly in the form of local hyperactivation during tasks—and maladaptive hyperconnectivity centered around the DMN. These changes are seen for a wide range of clinical symptoms, such as disability and cognitive impairment. As the field continues to grow, especially work involving longitudinal evaluations of larger cohorts as well as studies focusing on understanding treatment effects remains crucial. With sufficient time and methodological innovation, brain function might find a place in monitoring individual patients, aimed at stimulating beneficial reorganization and preventing the hypothesized network collapse from developing.

References

[1] Filippi M, Brück W, Chard D, Fazekas F, Geurts JJG, Enzinger C, et al. Association between pathological and MRI findings in multiple sclerosis. Lancet Neurol 2019;18(2):198–210. Available from: https://doi.org/10.1016/S1474-4422(18)30451-4.

[2] Rocca MA, De Meo E, Filippi M. Functional MRI in investigating cognitive impairment in multiple sclerosis. Acta Neurol Scand 2016;134(Suppl 200):39–46. Available from: https://doi.org/10.1111/ane.12654.

[3] Schoonheim MM, Geurts JJ, Barkhof F. The limits of functional reorganization in multiple sclerosis. Neurology 2010;74(16):1246–7. Available from: https://doi.org/10.1212/WNL.0b013e3181db9957.

[4] Friston KJ, Holmes AP, Poline JB, Grasby PJ, Williams SC, Frackowiak RS, et al. Analysis of fMRI time-series revisited. Neuroimage 1995;2(1):45–53. Available from: https://doi.org/10.1006/nimg.1995.1007.

[5] Ogawa S, Menon RS, Tank DW, Kim SG, Merkle H, Ellermann JM, et al. Functional brain mapping by blood oxygenation level-dependent contrast magnetic resonance imaging. A comparison of signal characteristics with a biophysical model. Biophys J 1993;64(3):803–12. Available from: https://doi.org/10.1016/S0006-3495(93)81441-3.

[6] Biswal B, Yetkin FZ, Haughton VM, Hyde JS. Functional connectivity in the motor cortex of resting human brain using echo-planar MRI. Magn Reson Med 1995;34(4):537–41. Available from: https://doi.org/10.1002/mrm.1910340409.

[7] Biswal BB, Mennes M, Zuo XN, Gohel S, Kelly C, Smith SM, et al. Toward discovery science of human brain function. Proc Natl Acad Sci U S A 2010;107(10):4734–9. Available from: https://doi.org/10.1073/pnas.0911855107.

[8] Cordes D, Haughton VM, Arfanakis K, Wendt GJ, Turski PA, Moritz CH, et al. Mapping functionally related regions of brain with functional connectivity MR imaging. AJNR Am J Neuroradiol 2000;21(9):1636–44.

[9] van den Heuvel MP, Hulshoff Pol HE. Exploring the brain network: a review on resting-state fMRI functional connectivity. Eur Neuropsychopharmacol 2010;20(8):519–34. Available from: https://doi.org/10.1016/j.euroneuro.2010.03.008.

[10] Chard DT, Alahmadi AAS, Audoin B, Charalambous T, Enzinger C, Hulst HE, et al. Mind the gap: from neurons to networks to outcomes in multiple sclerosis. Nat Rev Neurol 2021;17(3):173–84. Available from: https://doi.org/10.1038/s41582-020-00439-8.

[11] Jandric D, Doshi A, Scott R, Paling D, Rog D, Chataway J, et al. A systematic review of resting-state functional mri connectivity changes and cognitive impairment in multiple sclerosis. Brain Connect 2022;12(2):112–33. Available from: https://doi.org/10.1089/brain.2021.0104.

[12] Rocca MA, Schoonheim MM, Valsasina P, Geurts JJG, Filippi M. Task- and resting-state fMRI studies in multiple sclerosis: From regions to systems and time-varying analysis. Current status and future perspective. Neuroimage Clin 2022;35103076. Available from: https://doi.org/10.1016/j.nicl.2022.103076.

[13] Manjaly ZM, Harrison NA, Critchley HD, Do CT, Stefanics G, Wenderoth N, et al. Pathophysiological and cognitive mechanisms of fatigue in multiple sclerosis. J Neurol Neurosurg Psychiatry 2019;90(6):642–51. Available from: https://doi.org/10.1136/jnnp-2018-320050.

[14] Werring DJ, Bullmore ET, Toosy AT, Miller DH, Barker GJ, MacManus DG, et al. Recovery from optic neuritis is associated with a change in the distribution of cerebral response to visual stimulation: a functional magnetic resonance imaging study. J Neurol Neurosurg Psychiatry 2000;68(4):441–9. Available from: https://doi.org/10.1136/jnnp.68.4.441.

[15] Mezzapesa DM, Rocca MA, Rodegher M, Comi G, Filippi M. Functional cortical changes of the sensorimotor network are associated with clinical recovery in multiple sclerosis. Hum Brain Mapp 2008;29(5):562–73. Available from: https://doi.org/10.1002/hbm.20418.

[16] Rombouts SA, Lazeron RH, Scheltens P, Uitdehaag BM, Sprenger M, Valk J, et al. Visual activation patterns in patients with optic neuritis: an fMRI pilot study. Neurology 1998;50(6):1896–9. Available from: https://doi.org/10.1212/wnl.50.6.1896.

[17] Rocca MA, Falini A, Colombo B, Scotti G, Comi G, Filippi M. Adaptive functional changes in the cerebral cortex of patients with nondisabling multiple sclerosis correlate with the extent of brain structural damage. Ann Neurol 2002;51(3):330–9. Available from: https://doi.org/10.1002/ana.10120.

[18] Rocca MA, Matthews PM, Caputo D, Ghezzi A, Falini A, Scotti G, et al. Evidence for widespread movement-associated functional MRI changes in patients with PPMS. Neurology 2002;58(6):866–72. Available from: https://doi.org/10.1212/wnl.58.6.866.

[19] Rocca MA, Gavazzi C, Mezzapesa DM, Falini A, Colombo B, Mascalchi M, et al. A functional magnetic resonance imaging study of patients with secondary progressive multiple sclerosis. Neuroimage 2003;19(4):1770–7. Available from: https://doi.org/10.1016/s1053-8119(03)00242-8.

[20] Rocca MA, Colombo B, Falini A, Ghezzi A, Martinelli V, Scotti G, et al. Cortical adaptation in patients with MS: a cross-sectional functional MRI study of disease phenotypes. Lancet Neurol 2005;4(10):618–26. Available from: https://doi.org/10.1016/S1474-4422(05)70171-X.

[21] Tacchino A, Saiote C, Brichetto G, Bommarito G, Roccatagliata L, Cordano C, et al. Motor imagery as a function of disease severity in multiple sclerosis: an fMRI study. Front Hum Neurosci 2017;11:628. Available from: https://doi.org/10.3389/fnhum.2017.00628.

[22] Bonzano L, Bisio A, Pedullà L, Brichetto G, Bove M. Right inferior parietal lobule activity is associated with handwriting spontaneous tempo. Front Neurosci 2021;15656856. Available from: https://doi.org/10.3389/fnins.2021.656856.

[23] Strik M, Shanahan CJ, van der Walt A, Boonstra FMC, Glarin R, Galea MP, et al. Functional correlates of motor control impairments in multiple sclerosis: a 7 Tesla task functional MRI study. Hum Brain Mapp 2021;42(8):2569–82. Available from: https://doi.org/10.1002/hbm.25389.

[24] Boonstra FM, Noffs G, Perera T, Jokubaitis VG, Vogel AP, Moffat BA, et al. Functional neuroplasticity in response to cerebello-thalamic injury underpins the clinical presentation of tremor in multiple sclerosis. Mult Scler 2020;26(6):696–705. Available from: https://doi.org/10.1177/1352458519837706.

[25] Rico A, Zaaraoui W, Franques J, Attarian S, Reuter F, Malikova I, et al. Motor cortical reorganization is present after a single attack of multiple sclerosis devoid of cortico-spinal dysfunction. MAGMA 2011;24(2):77–84. Available from: https://doi.org/10.1007/s10334-010-0232-9.

[26] Rocca MA, Mezzapesa DM, Falini A, Ghezzi A, Martinelli V, Scotti G, et al. Evidence for axonal pathology and adaptive cortical reorganization in patients at presentation with clinically isolated syndromes suggestive of multiple sclerosis. Neuroimage 2003;18(4):847–55. Available from: https://doi.org/10.1016/s1053-8119(03)00043-0.

[27] Colorado RA, Shukla K, Zhou Y, Wolinsky JS, Narayana PA. Multi-task functional MRI in multiple sclerosis patients without clinical disability. Neuroimage 2012;59(1):573–81. Available from: https://doi.org/10.1016/j.neuroimage.2011.07.065.

[28] Lenzi D, Conte A, Mainero C, Frasca V, Fubelli F, Totaro P, et al. Effect of corpus callosum damage on ipsilateral motor activation in patients with multiple sclerosis: a functional and anatomical study. Hum Brain Mapp 2007;28(7):636–44. Available from: https://doi.org/10.1002/hbm.20305.

[29] Reddy H, Narayanan S, Woolrich M, Mitsumori T, Lapierre Y, Arnold DL, et al. Functional brain reorganization for hand movement in patients with multiple sclerosis: defining distinct effects of injury and disability. Brain 2002;125(Pt 12):2646–57. Available from: https://doi.org/10.1093/brain/awf283.

[30] Filippi M, Rocca MA, Mezzapesa DM, Falini A, Colombo B, Scotti G, et al. A functional MRI study of cortical activations associated with object manipulation in patients with MS. Neuroimage 2004;21(3):1147–54. Available from: https://doi.org/10.1016/j.neuroimage.2003.10.023.

[31] Rocca MA, Tortorella P, Ceccarelli A, Falini A, Tango D, Scotti G, et al. The "mirror-neuron system" in MS: A 3 tesla fMRI study. Neurology 2008;70(4):255–62. Available from: https://doi.org/10.1212/01.wnl.0000284667.29375.7e.

[32] Rocca MA, Gallo A, Colombo B, Falini A, Scotti G, Comi G, et al. Pyramidal tract lesions and movement-associated cortical recruitment in patients with MS. Neuroimage 2004;23(1):141–7. Available from: https://doi.org/10.1016/j.neuroimage.2004.05.005.

[33] Ciccarelli O, Toosy AT, Marsden JF, Wheeler-Kingshott CM, Miller DH, Matthews PM, et al. Functional response to active and passive ankle movements with clinical correlations in patients with primary progressive multiple sclerosis. J Neurol 2006;253(7):882–91. Available from: https://doi.org/10.1007/s00415-006-0125-z.

[34] Benedict RHB, Amato MP, DeLuca J, Geurts JJG. Cognitive impairment in multiple sclerosis: clinical management, MRI, and therapeutic avenues. Lancet Neurol 2020;19(10):860–71. Available from: https://doi.org/10.1016/S1474-4422(20)30277-5.

[35] Chiaravalloti N, Hillary F, Ricker J, Christodoulou C, Kalnin A, Liu WC, et al. Cerebral activation patterns during working memory performance in multiple sclerosis using FMRI. J Clin Exp Neuropsychol 2005;27(1):33–54. Available from: https://doi.org/10.1080/138033990513609.

[36] Staffen W, Mair A, Zauner H, Unterrainer J, Niederhofer H, Kutzelnigg A, et al. Cognitive function and fMRI in patients with multiple sclerosis: evidence for compensatory cortical activation during an attention task. Brain 2002;125(Pt 6):1275–82. Available from: https://doi.org/10.1093/brain/awf125.

[37] Audoin B, Ibarrola D, Ranjeva JP, Confort-Gouny S, Malikova I, Ali-Chérif A, et al. Compensatory cortical activation observed by fMRI during a cognitive task at the earliest stage of MS. Hum Brain Mapp 2003;20(2):51–8. Available from: https://doi.org/10.1002/hbm.10128.

[38] Forn C, Barros-Loscertales A, Escudero J, Belloch V, Campos S, Parcet MA, et al. Cortical reorganization during PASAT task in MS patients with preserved working memory functions. Neuroimage 2006;31(2):686–91. Available from: https://doi.org/10.1016/j.neuroimage.2005.12.030.

[39] Mainero C, Caramia F, Pozzilli C, Pisani A, Pestalozza I, Borriello G, et al. fMRI evidence of brain reorganization during attention and memory tasks in multiple sclerosis. Neuroimage 2004;21(3):858–67. Available from: https://doi.org/10.1016/j.neuroimage.2003.10.004.

[40] Forn C, Rocca MA, Valsasina P, Boscá I, Casanova B, Sanjuan A, et al. Functional magnetic resonance imaging correlates of cognitive performance in patients with a clinically isolated syndrome suggestive of multiple sclerosis at presentation: an activation and connectivity study. Mult Scler 2012;18(2):153–63. Available from: https://doi.org/10.1177/1352458511417744.

[41] Penner IK, Rausch M, Kappos L, Opwis K, Radü EW. Analysis of impairment related functional architecture in MS patients during performance of different attention tasks. J Neurol 2003;250(4):461–72. Available from: https://doi.org/10.1007/s00415-003-1025-0.

[42] Amann M, Dössegger LS, Penner IK, Hirsch JG, Raselli C, Calabrese P, et al. Altered functional adaptation to attention and working memory tasks with increasing complexity in relapsing-remitting multiple sclerosis patients. Hum Brain Mapp 2011;32(10):1704–19. Available from: https://doi.org/10.1002/hbm.21142.

[43] Sweet LH, Rao SM, Primeau M, Durgerian S, Cohen RA. Functional magnetic resonance imaging response to increased verbal working memory demands among patients with multiple sclerosis. Hum Brain Mapp 2006;27(1):28–36. Available from: https://doi.org/10.1002/hbm.20163.

[44] Cader S, Cifelli A, Abu-Omar Y, Palace J, Matthews PM. Reduced brain functional reserve and altered functional connectivity in patients with multiple sclerosis. Brain 2006;129(Pt 2):527–37. Available from: https://doi.org/10.1093/brain/awh670.

[45] Rocca MA, Valsasina P, Hulst HE, Abdel-Aziz K, Enzinger C, Gallo A, et al. Functional correlates of cognitive dysfunction in multiple sclerosis: a multicenter fMRI Study. Hum Brain Mapp 2014;35(12):5799–814. Available from: https://doi.org/10.1002/hbm.22586.

[46] Spreng RN. The fallacy of a "task-negative" network. Front Psychol 2012;3:145. Available from: https://doi.org/10.3389/fpsyg.2012.00145.

[47] Loitfelder M, Fazekas F, Petrovic K, Fuchs S, Ropele S, Wallner-Blazek M, et al. Reorganization in cognitive networks with progression of multiple sclerosis: insights from fMRI. Neurology 2011;76(6):526–33. Available from: https://doi.org/10.1212/WNL.0b013e31820b75cf.

[48] Hulst HE, Schoonheim MM, Roosendaal SD, Popescu V, Schweren LJ, van der Werf YD, et al. Functional adaptive changes within the hippocampal memory system of patients with multiple sclerosis. Hum Brain Mapp 2012;33(10):2268–80. Available from: https://doi.org/10.1002/hbm.21359.

[49] Rocca MA, Valsasina P, Leavitt VM, Rodegher M, Radaelli M, Riccitelli GC, et al. Functional network connectivity abnormalities in multiple sclerosis: correlations with disability and cognitive impairment. Mult Scler 2018;24(4):459–71. Available from: https://doi.org/10.1177/1352458517699875.

[50] Shu N, Duan Y, Xia M, Schoonheim MM, Huang J, Ren Z, et al. Disrupted topological organization of structural and functional brain connectomes in clinically isolated syndrome and multiple sclerosis. Sci Rep 2016;629383. Available from: https://doi.org/10.1038/srep29383.

[51] Liu Y, Dai Z, Duan Y, Huang J, Ren Z, Liu Z, et al. Whole brain functional connectivity in clinically isolated syndrome without conventional brain MRI lesions. Eur Radiol 2016;26(9):2982−91. Available from: https://doi.org/10.1007/s00330-015-4147-8.
[52] Liu Y, Wang H, Duan Y, Huang J, Ren Z, Ye J, et al. Functional brain network alterations in clinically isolated syndrome and multiple sclerosis: a graph-based connectome study. Radiology 2017;282(2):534−41. Available from: https://doi.org/10.1148/radiol.2016152843.
[53] Roosendaal SD, Schoonheim MM, Hulst HE, Sanz-Arigita EJ, Smith SM, Geurts JJ, et al. Resting state networks change in clinically isolated syndrome. Brain 2010;133(Pt 6):1612−21. Available from: https://doi.org/10.1093/brain/awq058.
[54] Rocca MA, Valsasina P, Absinta M, Riccitelli G, Rodegher ME, Misci P, et al. Default-mode network dysfunction and cognitive impairment in progressive MS. Neurology 2010;74(16):1252−9. Available from: https://doi.org/10.1212/WNL.0b013e3181d9ed91.
[55] Schoonheim MM, Meijer KA, Geurts JJ. Network collapse and cognitive impairment in multiple sclerosis. Front Neurol 2015;6:82. Available from: https://doi.org/10.3389/fneur.2015.00082.
[56] Pinter D, Beckmann CF, Fazekas F, Khalil M, Pichler A, Gattringer T, et al. Morphological MRI phenotypes of multiple sclerosis differ in resting-state brain function. Sci Rep 2019;9(1)16221. Available from: https://doi.org/10.1038/s41598-019-52757-7.
[57] Tommasin S, De Giglio L, Ruggieri S, Petsas N, Giannì C, Pozzilli C, et al. Relation between functional connectivity and disability in multiple sclerosis: a non-linear model. J Neurol 2018;265(12):2881−92. Available from: https://doi.org/10.1007/s00415-018-9075-5.
[58] Tona F, De Giglio L, Petsas N, Sbardella E, Prosperini L, Upadhyay N, et al. Role of cerebellar dentate functional connectivity in balance deficits in patients with multiple sclerosis. Radiology 2018;287(1):267−75. Available from: https://doi.org/10.1148/radiol.2017170311.
[59] Zhong J, Nantes JC, Holmes SA, Gallant S, Narayanan S, Koski L. Abnormal functional connectivity and cortical integrity influence dominant hand motor disability in multiple sclerosis: a multimodal analysis. Hum Brain Mapp 2016;37(12):4262−75. Available from: https://doi.org/10.1002/hbm.23307.
[60] Cordani C, Hidalgo de la Cruz M, Meani A, Valsasina P, Esposito F, Pagani E, et al. MRI correlates of clinical disability and hand-motor performance in multiple sclerosis phenotypes. Mult Scler 2021;27(8):1205−21. Available from: https://doi.org/10.1177/1352458520958356.
[61] Dogonowski AM, Siebner HR, Soelberg Sørensen P, Paulson OB, Dyrby TB, Blinkenberg M, et al. Resting-state connectivity of pre-motor cortex reflects disability in multiple sclerosis. Acta Neurol Scand 2013;128(5):328−35. Available from: https://doi.org/10.1111/ane.12121.
[62] Fu J, Chen X, Gu Y, Xie M, Zheng Q, Wang J, et al. Functional connectivity impairment of postcentral gyrus in relapsing-remitting multiple sclerosis with somatosensory disorder. Eur J Radiol 2019;118:200−6. Available from: https://doi.org/10.1016/j.ejrad.2019.07.029.
[63] Bollaert RE, Poe K, Hubbard EA, Motl RW, Pilutti LA, Johnson CL, et al. Associations of functional connectivity and walking performance in multiple sclerosis. Neuropsychologia 2018;117:8−12. Available from: https://doi.org/10.1016/j.neuropsychologia.2018.05.007.
[64] d'Ambrosio A, Hidalgo de la Cruz M, Valsasina P, Pagani E, Colombo B, Rodegher M, et al. Structural connectivity-defined thalamic subregions have different functional connectivity abnormalities in multiple sclerosis patients: Implications for clinical correlations. Hum Brain Mapp 2017;38(12):6005−18. Available from: https://doi.org/10.1002/hbm.23805.
[65] Giannì C, Belvisi D, Conte A, Tommasin S, Cortese A, Petsas N, et al. Altered sensorimotor integration in multiple sclerosis: a combined neurophysiological and functional MRI study. Clin Neurophysiol 2021;132(9):2191−8. Available from: https://doi.org/10.1016/j.clinph.2021.05.028.
[66] Schoonheim MM, Pinter D, Prouskas SE, Broeders TA, Pirpamer L, Khalil M, et al. Disability in multiple sclerosis is related to thalamic connectivity and cortical network atrophy. Mult Scler 2022;28(1):61−70. Available from: https://doi.org/10.1177/13524585211008743.
[67] Pasqua G, Tommasin S, Bharti K, Ruggieri S, Petsas N, Piervincenzi C, et al. Resting-state functional connectivity of anterior and posterior cerebellar lobes is altered in multiple sclerosis. Mult Scler 2021;27(4):539−48. Available from: https://doi.org/10.1177/1352458520922770.
[68] Sbardella E, Upadhyay N, Tona F, Prosperini L, De Giglio L, Petsas N, et al. Dentate nucleus connectivity in adult patients with multiple sclerosis: functional changes at rest and correlation with clinical features. Mult Scler 2017;23(4):546−55. Available from: https://doi.org/10.1177/1352458516657438.

[69] Soares JM, Conde R, Magalhães R, Marques P, Gomes L, Gonçalves Ó, et al. Alterations in functional connectivity are associated with white matter lesions and information processing efficiency in multiple sclerosis. Brain Imaging Behav 2021;15(1):375–88. Available from: https://doi.org/10.1007/s11682-020-00264-z.

[70] Tommasin S, De Giglio L, Ruggieri S, Petsas N, Giannì C, Pozzilli C, et al. Multi-scale resting state functional reorganization in response to multiple sclerosis damage. Neuroradiology 2020;62(6):693–704. Available from: https://doi.org/10.1007/s00234-020-02393-0.

[71] Veréb D, Kovács MA, Kocsis K, Tóth E, Bozsik B, Király A, et al. Functional connectivity lateralisation shift of resting state networks is linked to visuospatial memory and white matter microstructure in relapsing-remitting multiple sclerosis. Brain Topogr 2022;35(2):268–75. Available from: https://doi.org/10.1007/s10548-021-00881-x.

[72] Has Silemek AC, Fischer L, Pöttgen J, Penner IK, Engel AK, Heesen C, et al. Functional and structural connectivity substrates of cognitive performance in relapsing remitting multiple sclerosis with mild disability. Neuroimage Clin 2020;25102177. Available from: https://doi.org/10.1016/j.nicl.2020.102177.

[73] Hawellek DJ, Hipp JF, Lewis CM, Corbetta M, Engel AK. Increased functional connectivity indicates the severity of cognitive impairment in multiple sclerosis. Proc Natl Acad Sci U S A 2011;108(47):19066–71. Available from: https://doi.org/10.1073/pnas.1110024108.

[74] Meijer KA, Eijlers AJC, Douw L, Uitdehaag BMJ, Barkhof F, Geurts JJG, et al. Increased connectivity of hub networks and cognitive impairment in multiple sclerosis. Neurology 2017;88(22):2107–14. Available from: https://doi.org/10.1212/WNL.0000000000003982.

[75] Jandric D, Lipp I, Paling D, Rog D, Castellazzi G, Haroon H, et al. Mechanisms of network changes in cognitive impairment in multiple sclerosis. Neurology 2021;97(19):e1886–97. Available from: https://doi.org/10.1212/WNL.0000000000012834.

[76] Janssen AL, Boster A, Patterson BA, Abduljalil A, Prakash RS. Resting-state functional connectivity in multiple sclerosis: an examination of group differences and individual differences. Neuropsychologia 2013;51(13):2918–29. Available from: https://doi.org/10.1016/j.neuropsychologia.2013.08.010.

[77] Leavitt VM, Paxton J, Sumowski JF. Default network connectivity is linked to memory status in multiple sclerosis. J Int Neuropsychol Soc 2014;20(9):937–44. Available from: https://doi.org/10.1017/S1355617714000800.

[78] Louapre C, Perlbarg V, García-Lorenzo D, Urbanski M, Benali H, Assouad R, et al. Brain networks disconnection in early multiple sclerosis cognitive deficits: an anatomofunctional study. Hum Brain Mapp 2014;35(9):4706–17. Available from: https://doi.org/10.1002/hbm.22505.

[79] Bonavita S, Gallo A, Sacco R, Corte MD, Bisecco A, Docimo R, et al. Distributed changes in default-mode resting-state connectivity in multiple sclerosis. Mult Scler 2011;17(4):411–22. Available from: https://doi.org/10.1177/1352458510394609.

[80] Yeshurun Y, Nguyen M, Hasson U. The default mode network: where the idiosyncratic self meets the shared social world. Nat Rev Neurosci 2021;22(3):181–92. Available from: https://doi.org/10.1038/s41583-020-00420-w.

[81] Wojtowicz M, Mazerolle EL, Bhan V, Fisk JD. Altered functional connectivity and performance variability in relapsing-remitting multiple sclerosis. Mult Scler 2014;20(11):1453–63. Available from: https://doi.org/10.1177/1352458514524997.

[82] Marchesi O, Bonacchi R, Valsasina P, Preziosa P, Pagani E, Cacciaguerra L, et al. Functional and structural MRI correlates of executive functions in multiple sclerosis. Mult Scler 2022;28(5):742–56. Available from: https://doi.org/10.1177/13524585211033184.

[83] Petracca M, Saiote C, Bender HA, Arias F, Farrell C, Magioncalda P, et al. Synchronization and variability imbalance underlie cognitive impairment in primary-progressive multiple sclerosis. Sci Rep 2017;746411. Available from: https://doi.org/10.1038/srep46411.

[84] Riccitelli GC, Pagani E, Meani A, Valsasina P, Preziosa P, Filippi M, et al. Cognitive impairment in benign multiple sclerosis: a multiparametric structural and functional MRI study. J Neurol 2020;267(12):3508–17. Available from: https://doi.org/10.1007/s00415-020-10025-z.

[85] Marek S, Dosenbach NUF. The frontoparietal network: function, electrophysiology, and importance of individual precision mapping. Dialogues Clin Neurosci 2018;20(2):133–40.

[86] Bassett DS, Sporns O. Network neuroscience. Nat Neurosci 2017;20(3):353–64. Available from: https://doi.org/10.1038/nn.4502.

[87] Douw L, Wakeman DG, Tanaka N, Liu H, Stufflebeam SM. State-dependent variability of dynamic functional connectivity between frontoparietal and default networks relates to cognitive flexibility. Neuroscience 2016;339:12–21. Available from: https://doi.org/10.1016/j.neuroscience.2016.09.034.

[88] Fuchs TA, Benedict RHB, Bartnik A, Choudhery S, Li X, Mallory M, et al. Preserved network functional connectivity underlies cognitive reserve in multiple sclerosis. Hum Brain Mapp 2019;40(18):5231–41. Available from: https://doi.org/10.1002/hbm.24768.

[89] Bizzo BC, Arruda-Sanchez T, Tobyne SM, Bireley JD, Lev MH, Gasparetto EL, et al. Anterior insular resting-state functional connectivity is related to cognitive reserve in multiple sclerosis. J Neuroimaging 2021;31(1):98–102. Available from: https://doi.org/10.1111/jon.12779.

[90] Pantano P, Bernardi S, Tinelli E, Pontecorvo S, Lenzi D, Raz E, et al. Impaired cortical deactivation during hand movement in the relapsing phase of multiple sclerosis: a cross-sectional and longitudinal fMRI study. Mult Scler 2011;17(10):1177–84. Available from: https://doi.org/10.1177/1352458511411757.

[91] Cerasa A, Gioia MC, Valentino P, Nisticò R, Chiriaco C, Pirritano D, et al. Computer-assisted cognitive rehabilitation of attention deficits for multiple sclerosis: a randomized trial with fMRI correlates. Neurorehabil Neural Repair 2013;27(4):284–95. Available from: https://doi.org/10.1177/1545968312465194.

[92] Chiaravalloti ND, Wylie G, Leavitt V, Deluca J. Increased cerebral activation after behavioral treatment for memory deficits in MS. J Neurol 2012;259(7):1337–46. Available from: https://doi.org/10.1007/s00415-011-6353-x.

[93] Dobryakova E, Wylie GR, DeLuca J, Chiaravalloti ND. A pilot study examining functional brain activity 6 months after memory retraining in MS: the MEMREHAB trial. Brain Imaging Behav 2014;8(3):403–6. Available from: https://doi.org/10.1007/s11682-014-9309-9.

[94] Ernst A, Botzung A, Gounot D, Sellal F, Blanc F, de Seze J, et al. Induced brain plasticity after a facilitation programme for autobiographical memory in multiple sclerosis: a preliminary study. Mult Scler Int 2012;2012820240. Available from: https://doi.org/10.1155/2012/820240.

[95] Filippi M, Riccitelli G, Mattioli F, Capra R, Stampatori C, Pagani E, et al. Multiple sclerosis: effects of cognitive rehabilitation on structural and functional MR imaging measures—an explorative study. Radiology 2012;262(3):932–40. Available from: https://doi.org/10.1148/radiol.11111299.

[96] Huiskamp M, Dobryakova E, Wylie GD, DeLuca J, Chiaravalloti ND. A pilot study of changes in functional brain activity during a working memory task after mSMT treatment: The MEMREHAB trial. Mult Scler Relat Disord 2016;7:76–82. Available from: https://doi.org/10.1016/j.msard.2016.03.012.

[97] Sastre-Garriga J, Alonso J, Renom M, Arévalo MJ, González I, Galán I, et al. A functional magnetic resonance proof of concept pilot trial of cognitive rehabilitation in multiple sclerosis. Mult Scler 2011;17(4):457–67. Available from: https://doi.org/10.1177/1352458510389219.

[98] Tomassini V, Johansen-Berg H, Jbabdi S, Wise RG, Pozzilli C, Palace J, et al. Relating brain damage to brain plasticity in patients with multiple sclerosis. Neurorehabil Neural Repair 2012;26(6):581–93. Available from: https://doi.org/10.1177/1545968311433208.

[99] Zuber P, Tsagkas C, Papadopoulou A, Gaetano L, Huerbin M, Geiter E, et al. Efficacy of inpatient personalized multidisciplinary rehabilitation in multiple sclerosis: behavioural and functional imaging results. J Neurol 2020;267(6):1744–53. Available from: https://doi.org/10.1007/s00415-020-09768-6.

[100] Péran P, Nemmi F, Dutilleul C, Finamore L, Falletta Caravasso C, Troisi E, et al. Neuroplasticity and brain reorganization associated with positive outcomes of multidisciplinary rehabilitation in progressive multiple sclerosis: A fMRI study. Mult Scler Relat Disord 2020;42102127. Available from: https://doi.org/10.1016/j.msard.2020.102127.

[101] Bonzano L, Pedullà L, Tacchino A, Brichetto G, Battaglia MA, Mancardi GL, et al. Upper limb motor training based on task-oriented exercises induces functional brain reorganization in patients with multiple sclerosis. Neuroscience 2019;410:150–9. Available from: https://doi.org/10.1016/j.neuroscience.2019.05.004.

[102] Huiskamp M, Eijlers AJC, Broeders TAA, Pasteuning J, Dekker I, Uitdehaag BMJ, et al. Longitudinal network changes and conversion to cognitive impairment in multiple sclerosis. Neurology 2021;97(8):e794–802. Available from: https://doi.org/10.1212/WNL.0000000000012341.

[103] Uddin LQ. Salience processing and insular cortical function and dysfunction. Nat Rev Neurosci 2015;16(1):55–61. Available from: https://doi.org/10.1038/nrn3857.

[104] Parisi L, Rocca MA, Valsasina P, Panicari L, Mattioli F, Filippi M. Cognitive rehabilitation correlates with the functional connectivity of the anterior cingulate cortex in patients with multiple sclerosis. Brain Imaging Behav 2014;8(3):387–93. Available from: https://doi.org/10.1007/s11682-012-9160-9.

[105] Parisi L, Rocca MA, Mattioli F, Copetti M, Capra R, Valsasina P, et al. Changes of brain resting state functional connectivity predict the persistence of cognitive rehabilitation effects in patients with multiple sclerosis. Mult Scler 2014;20(6):686–94. Available from: https://doi.org/10.1177/1352458513505692.

[106] Braun U, Schäfer A, Walter H, Erk S, Romanczuk-Seiferth N, Haddad L, et al. Dynamic reconfiguration of frontal brain networks during executive cognition in humans. Proc Natl Acad Sci U S A 2015;112(37):11678–83. Available from: https://doi.org/10.1073/pnas.1422487112.

[107] Rocca MA, Hidalgo de La Cruz M, Valsasina P, Mesaros S, Martinovic V, Ivanovic J, et al. Two-year dynamic functional network connectivity in clinically isolated syndrome. Mult Scler 2020;26(6):645–58. Available from: https://doi.org/10.1177/1352458519837704.

[108] Hidalgo de la Cruz M, Valsasina P, Sangalli F, Esposito F, Rocca MA, Filippi M. Dynamic functional connectivity in the main clinical phenotypes of multiple sclerosis. Brain Connect 2021;11(8):678–90. Available from: https://doi.org/10.1089/brain.2020.0920.

[109] d'Ambrosio A, Valsasina P, Gallo A, De Stefano N, Pareto D, Barkhof F, et al. Reduced dynamics of functional connectivity and cognitive impairment in multiple sclerosis. Mult Scler 2020;26(4):476–88. Available from: https://doi.org/10.1177/1352458519837707.

[110] Schoonheim MM, Douw L, Broeders TA, Eijlers AJ, Meijer KA, Geurts JJ. The cerebellum and its network: disrupted static and dynamic functional connectivity patterns and cognitive impairment in multiple sclerosis. Mult Scler 2021;27(13):2031–9. Available from: https://doi.org/10.1177/1352458521999274.

[111] Lin SJ, Vavasour I, Kosaka B, Li DKB, Traboulsee A, MacKay A, et al. Education, and the balance between dynamic and stationary functional connectivity jointly support executive functions in relapsing-remitting multiple sclerosis. Hum Brain Mapp 2018;39(12):5039–49. Available from: https://doi.org/10.1002/hbm.24343.

[112] Eijlers AJC, Wink AM, Meijer KA, Douw L, Geurts JJG, Schoonheim MM. Reduced network dynamics on functional MRI signals cognitive impairment in multiple sclerosis. Radiology 2019;292(2):449–57. Available from: https://doi.org/10.1148/radiol.2019182623.

[113] Hardmeier M, Schoonheim MM, Geurts JJ, Hillebrand A, Polman CH, Barkhof F, et al. Cognitive dysfunction in early multiple sclerosis: altered centrality derived from resting-state functional connectivity using magneto-encephalography. PLoS One 2012;7(7):e42087. Available from: https://doi.org/10.1371/journal.pone.0042087.

[114] Buyukturkoglu K, Zeng D, Bharadwaj S, Tozlu C, Mormina E, Igwe KC, et al. Classifying multiple sclerosis patients on the basis of SDMT performance using machine learning. Mult Scler 2021;27(1):107–16. Available from: https://doi.org/10.1177/1352458520958362.

[115] Rocca MA, Valsasina P, Meani A, Falini A, Comi G, Filippi M. Impaired functional integration in multiple sclerosis: a graph theory study. Brain Struct Funct 2016;221(1):115–31. Available from: https://doi.org/10.1007/s00429-014-0896-4.

[116] Dekker I, Schoonheim MM, Venkatraghavan V, Eijlers AJC, Brouwer I, Bron EE, et al. The sequence of structural, functional and cognitive changes in multiple sclerosis. Neuroimage Clin 2021;29102550. Available from: https://doi.org/10.1016/j.nicl.2020.102550.

[117] Eijlers AJ, Meijer KA, Wassenaar TM, Steenwijk MD, Uitdehaag BM, Barkhof F, et al. Increased default-mode network centrality in cognitively impaired multiple sclerosis patients. Neurology 2017;88(10):952–60. Available from: https://doi.org/10.1212/WNL.0000000000003689.

[118] Schoonheim MM, Geurts J, Wiebenga OT, De Munck JC, Polman CH, Stam CJ, et al. Changes in functional network centrality underlie cognitive dysfunction and physical disability in multiple sclerosis. Mult Scler 2014;20(8):1058–65. Available from: https://doi.org/10.1177/1352458513516892.

CHAPTER 15

Perfusion-weighted imaging in multiple sclerosis

Maria Marcella Laganà* and Laura Pelizzari

IRCCS Fondazione Don Carlo Gnocchi ONLUS, Milan, Italy

OUTLINE

Magnetic resonance imaging techniques for estimating cerebral perfusion — 268	Perfusion differences across multiple sclerosis phenotypes — 276
Dynamic susceptibility contrast magnetic resonance imaging — 268	Physical disability — 276
Dynamic contrast enhancement magnetic resonance imaging — 270	Cognitive performance — 277
Arterial spin labeling — 272	Lesion evolution — 279
	Hypoperfusion and brain atrophy — 279
Magnetic resonance imaging perfusion studies in multiple sclerosis — 274	Cerebrovascular reactivity — 280
	Conclusions — 281
	References — 281

Brain perfusion has been assessed in multiple sclerosis (Ms) with positron emission tomography and single-photon emission tomography since the 1990s [1,2]. Since magnetic resonance imaging (MRI) is a nonionizing imaging technique widely used in Ms, in the last 15 years, various MRI sequences have also been adopted, revealing brain perfusion alterations in Ms [3]. The different MRI techniques that can estimate cerebral perfusion are explained in the first part of this chapter.

The numerous studies investigating perfusion in Ms mainly aimed to assess the potential link between brain perfusion, atrophy, Ms lesion formation and activity, and Ms phenotypes and disability. They are reported and commented on the second part of this chapter.

* Current affiliation: Canon Medical Systems srl, Rome, Italy

Magnetic resonance imaging techniques for estimating cerebral perfusion

Any cell in gray matter (GM) and white matter (WM) cannot survive without perfusion. For this reason, perfusion is a fundamental physiological parameter to assess the brain in healthy and pathological conditions [4].

Various MRI techniques can estimate brain perfusion, showing alterations in several diseases [3,5–12].

The main MRI techniques for evaluating brain perfusion are: dynamic susceptibility contrast (DSC) MRI, dynamic contrast enhancement (DCE) MRI, and arterial spin labeling (ASL) [13]. The former two require exogenous contrast agents, while the latter takes advantage of blood as an endogenous contrast agent.

Gadolinium (Gd)-based chelates are the most commonly used exogenous contrast agents for DSC and DCE MRI [14]. These substances produce a change in the local relaxation time of water (proportional to the contrast agent concentration) that in turn results in changing the detected Mr signal [13]. Both T2/T2* and T1 relaxation times are shortened by Gd-based chelates. Changes in local relaxation times depend on dipole–dipole interactions and on the susceptibility effects induced by Gd [14]. The former is the predominant cause of Gd-induced T1 shortening, which produces T1 relaxation enhancement measured with DCE MRI; the latter is the predominant cause of Gd-induced T2/T2* shortening, which induces signal loss on T2 and T2*-weighted images measured with DSC MRI [14].

Dynamic susceptibility contrast magnetic resonance imaging

Most of the Ms studies were performed using the DSC sequence [15–28].

The DSC technique allows assessing brain perfusion by taking advantage of the modification of T2 and T2* relaxation times induced by the contrast agent [13]. As flowing in the vessels, the Gd-based chelates modify the relaxation times of the surrounding tissue [29]. Dipole–dipole interactions occur over a microscopic distance so, when the contrast agent is confined in the vasculature, this effect is limited to the blood and almost not affecting the surrounding tissue [14]. Conversely, the susceptibility effect affects several micrometers of the tissue surrounding the vessel [14]. Therefore, in the case of intact blood–brain barrier (BBB) (i.e., when the blood is confined in the vasculature), T2 and T2* relaxation time shortening induced by the contrast agent is much greater than T1 shortening. This results in a T2/T2*-weighted signal drop concurrent with the passage of Gd-based chelates [29]. This signal drop is proportional to the change of the relaxation rate ($\Delta R2$ or $\Delta R2^*$), thus proportional to the concentration of the bolus of Gd-based chelates [13].

Therefore the concentration–time curve through the capillary bed can be derived from the DSC signal-time curve (Fig. 15.1). Then, the following local parameters can be estimated [13]:

- time to peak (TTP), defined as the time elapsed between the start of intravenous bolus administration and the peak of the concentration–time curve (which corresponds to the peak of the signal drop induced by the first passage of Gd bolus);
- arrival time (AT), defined as the time elapsed between the start of the intravenous bolus administration and the moment when the local concentration begins to increase;

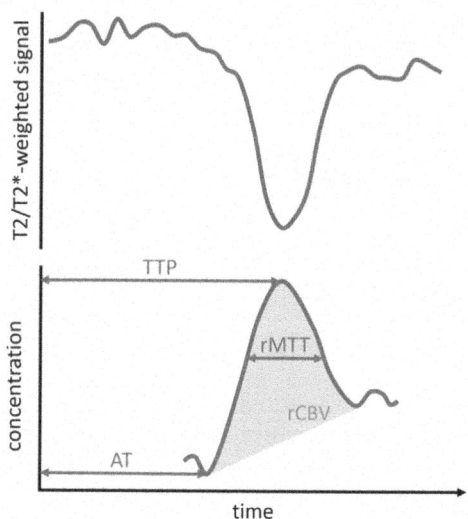

FIGURE 15.1 DSC signal-time curve and derived concentration–time curve. Time to peak (TTP), arrival time (AT), relative cerebral blood volume (rCBV), and relative mean transit time (rMTT) are represented on the concentration–time curve.

- relative cerebral blood volume (rCBV), calculated as the area under the concentration–time curve.
- relative mean transit time (rMTT), defined as the full width at half maximum of the concentration–time curve.

Additional quantitative indices can be derived with DSC if the arterial input (i.e., how the bolus arrives in the region of interest) is measured and mathematically described with an arterial input function (AIF) [13,30]:

- cerebral blood flow (CBF), defined as the volume of blood flowing in a certain mass of tissue over a certain time (mL blood/100 g/min) and computed with deconvolution techniques;
- cerebral blood flow (CBV), defined as the volume of blood per 100 g of tissue [mL blood/100 g], computed as the ratio between the integral of the concentration–time curve over time and the integral of the AIF over time.
- mean transit time (MTT), defined as the characteristic time the blood remains in the capillary bed and computed as the ratio between CBV and CBF.

As accurately measuring the AIF is challenging, DSC-derived relative measures are commonly used [13]. In group studies, rCBV and rCBF are often normalized by dividing them for rCBF and rCBV computed in a reference tissue (e.g., normal-appearing WM [NAWM]), to allow comparisons between subjects [28].

An accurate estimate of all DSC-derived parameters is based on a major assumption: Gd-based bolus distribution is limited to the vascular tree. However, in the case of BBB leakage, Gd-based chelates flood in the extravascular space, whose protons then experience relaxation enhancement due to dipole–dipole interactions [14]. Therefore T1 shortening, which was negligible in the case of intact BBB, becomes relevant when BBB integrity is disrupted, hampering the accuracy of derived hemodynamic parameters. One strategy that is used to overcome this limitation is setting the DSC sequence with a low flip angle, which reduces the sensitivity to T1 effects [30]. Another common way to minimize the T1-shortening effect is by injecting a predose bolus. Indeed, the relationship between Gd concentration and signal change due to the T1-shortening is not linear, but follows a biexponential decay [30]. Consequently, at a certain Gd-based chelates concentration, a further increase of Gd-based chelates does not produce a great signal change due to T1-shortening effect. This means that if the "main bolus" injection is performed after preenhancement, the derived signal change has to be ascribed to T2/T2* relaxation time shortening.

Another factor that can impact the accuracy of the estimate of perfusion parameters in DSC is recirculation, which can make it difficult to discriminate the end of the first pass of the bolus from the next peak given by the recirculation [29]. However, the use of curve fitting techniques allows to solve this issue [30].

Both T2*-weighted gradient echo and T2-weighted spin echo sequences can be used for DSC MRI. As the T2* change induced by the contrast agent is greater than the associated T2 change, gradient echo DSC sequences provide a greater contrast-to-noise ratio, thus requiring a lower concentration of Gd-based contrast agents [31]. However, spin echo DSC MRI is less sensitive to the signal generated by major vessels and allows for better imaging of the tissue surrounding the microvasculature [29]. Combined spin- and gradient-echo sequences merge the advantages of the two types of sequences, providing even additional information about vascular architecture (e.g., vessel size) [14]. As the maximum signal drop is produced when the Gd bolus first passes through the regional circulation, all DSC sequences are characterized by high temporal resolution (sampling rate < 2 seconds) and are generally very fast (duration = about 1 minute) [31]. Recent technological advances such as multi-echo and multiband sequences allowed an increase in temporal resolution and more accurate perfusion estimation with DSC MRI [14].

Dynamic contrast enhancement magnetic resonance imaging

As previously mentioned, on the one hand, the DSC technique exploits the modification of T2* tissue relaxation time to evaluate perfusion; on the other hand, DCE MRI exploits the modification of the T1 relaxation time of the tissue [14]. Although DCE is the most used MRI technique to assess perfusion in the majority of organs in the clinical setting, DCE is less commonly used than DSC for the evaluation of cerebral perfusion [32]. However, DCE is recognized as the best technique for assessing brain capillary permeability, as in DCE the enhancement is produced by the uptake of the contrast agent in the interstitium [32]. Therefore, unlike DSC, the presence of leakage increases the DCE signal [13]. As DCE takes advantage of the vascular-interstitial exchange, DCE techniques require longer acquisition times than DSC (about some minutes) [13].

DCE images can be primarily evaluated qualitatively, by observing the signal-time curve Fig. 15.2 and comparing the wash-in and wash-out slopes of different specific regions of interest [32]. For instance, this is commonly done by radiologists to distinguish a tumor (rapid wash-in) from necrotic tissue (slow wash-in) [13]. In addition to a purely qualitative analysis, it is possible to do a semiquantitative analysis. This approach involves the measurement of indices that can be directly derived from the signal-time curve, such as the TTP, AT, wash-in rate (i.e., the slope of the curve in the wash-in phase), and wash-out rate (i.e., the slope of the curve in the wash-out phase) [13]. All these measurements are relative and have no absolute meaning, but they can be used to compare two different regions of interest, more accurately than through a simple visual inspection of the curve.

Finally, quantitative evaluations of permeability and perfusion can also be performed, but three additional requirements are needed: (1) information about the T1 pre-contrast values, (2) description of the arterial input, and (3) the definition of a tracer kinetics model [32]. As for the T1 precontrast values, this assessment is needed, because it is not possible to directly derive the concentration–time curve from the signal-time curve. T1 mapping can be performed before the actual DCE scan to estimate T1 relaxation time locally. Inversion recovery sequences are the gold standard for T1 mapping but other faster sequences (e.g., Lock-Locker, Variable Flip Angle) are commonly preferred in clinical settings [14]. In order to further reduce the global scan duration, T1 mapping might not be estimated for all brain voxels and a single T1 may be used for concentration calibration in DCE. However, this approach can introduce bias, so it is generally discouraged [32]. The AIF can be determined in several different ways, but the most common is defining a region of interest in a large vessel and measuring the signal in that area [14]. As for the tracer kinetics model, the simplest model that can be used for DCE MRI is the two-compartment Tofts model, which assumes passive diffusion (i.e., driven by concentration gradient) of the Gd-based chelates from the vasculature to the extravascular extracellular space (Fig. 15.3) [32]. In this model, K_{trans} represents the transfer coefficient between plasma and extracellular extravascular space and it is proportional to blood flow, vascular surface, and vascular permeability. Furthermore, reflux from the extravascular extracellular space to the vasculature is assumed to occur and it is described with k_{EP}. As the Gd-based chelates are assumed to be distributed in the plasma and extravascular extracellular space only (i.e., it does not enter cells), the concentration depends on the fractional volumes of plasma (v_p) and extravascular extracellular space (v_e).

More complex multicompartmental models or model-free approaches have been introduced recently. Second-generation models allow us to derive additional parameters, such as CBF, CBV, and BBB permeability–surface area product [32]. However, defining a

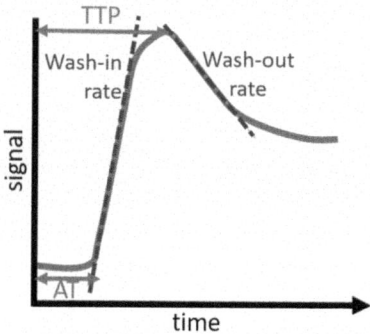

FIGURE 15.2 DCE signal-time curve. Time to peak (TTP), arrival time (AT), wash-in rate, and wash-out rate are represented.

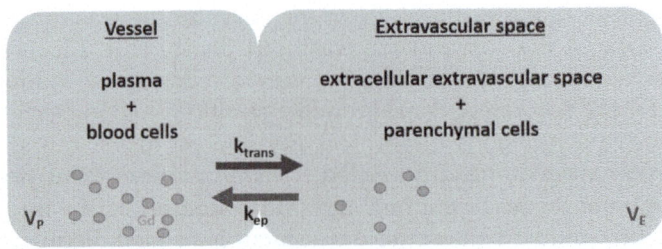

FIGURE 15.3 Tofts model for DCE MRI. Vasculature (represented in pink) and extravascular space (represented in gray) are the only compartments assumed. The Gd-based chelate (in green) is diluted in the plasma and it diffuses in the extracellular extravascular space. Diffusion from plasma to extracellular extravascular space occurs according to k_{trans}; diffusion from the extacellular extravascular space to plasma occurs according to k_{ep}. The plasma volume v_p and the interstitial volume v_e represent the measure the volumes of the two regions as a fraction of the total tissue volume.

suitable tracer kinetics model is still a challenge, which makes it difficult to compare DCE-derived permeability and perfusion indices reported by studies [13].

DCE has mainly been used to characterize perfusion and BBB permeability in enhancing versus non-enhancing lesions [33–35], as detailed in the paragraph "lesion evolution." In particular, Gaitan et al. [35] studied the development and expansion of new Ms lesions, focusing on the dynamics of BBB permeability and describing the spatiotemporal evolution of different lesion kinds.

Arterial spin labeling

ASL is a perfusion MRI modality alternative to DSC and DCE, it does not require the use of an exogenous contrast agent [13], so it was widely used in Ms studies [36–45].

In ASL, radiofrequency pulses are used to selectively "label" the blood flowing in the neck, thus allowing noninvasive brain perfusion measures.

Before labeling, all spins are aligned to the magnetic static field (B0, inferior–superior direction). Then radiofrequency pulses are used to invert the magnetization of the spins of the blood flowing in the neck (i.e., "labeling/tagging") [46]. If sufficient time passes after labeling, the labeled blood is spread out in the brain through the vascular tree and produces a signal drop proportional to its local concentration [46,47]. The time between labeling and imaging recording is called post-labeling delay (PLD, or inversion time) and has to be carefully set: the optimal PLD is not only long enough to let all the labeled blood reach the brain (i.e., longer than the arterial transit time) but also short enough to exploit a relatively large amount of nondecayed signal [13]. PLD of 1800 Ms was recommended for healthy adults under 70 years old, while PLD of 2000 Ms was recommended for subjects over 70 years and for patients. Multidelay sequences (including both short and long PLDs) were also developed and encouraged to provide protocols less sensitive to differences in arterial transit time among subjects [48].

By subtracting a tag image (i.e., obtained after labeling) from a control image (i.e., same signal, but with no labeling), a perfusion-weighted image can be obtained, as it mirrors the amount of tracer that has locally accumulated. Multiple control-tag pairs are generally acquired to increase the signal-to-noise ratio, and result in sequences that generally last 5−10 minutes [31].

Depending on how the labeling is performed, ASL can be classified into continuous ASL (CASL), pulsed ASL (pASL), and pseudo-continuous ASL (pCASL) (Fig. 15.4) [46]. In CASL labeling is performed with a continuous radiofrequency field, which results in a labeling plane, while in pASL sharp short radiofrequency pulses (i.e., a few milliseconds) are applied to a thick slab positioned at the neck level. In pCASL labeling is performed with a series of radiofrequency pulses, in a thin labeling plane. CASL provides efficient labeling but requires special hardware and may exceed the specific absorption rate. pASL is characterized by a lower signal-to-noise ratio than CASL, but it can be set even on standard clinical scanners and it produces relatively low power deposition. pCASL merges the advances of CASL and pASL and it was recommended as a labeling approach in the ISMRM ASL consensus paper [48].

Both 2D (EPI or spiral) and 3D (GRASE or RARE) readout schemes can be used to implement ASL sequences, but 3D readout without flow suppression gradients, which provides minimal signal dropout, was recommended by the ISMRM perfusion study group and the European Consortium for ASL in Dementia [48].

Once perfusion-weighted images have been produced (in arbitrary units), absolute CBF can be locally quantified with a kinetic model, as long as a calibration image was acquired [46]. The calibration image is acquired with the same resolution and setting of ASL sequence, but without background suppression and with longer repetition time, namely a proton-density-weighted image. Calibration is performed by assuming that the equilibrium magnetization of the blood (which conveys the information about the concentration of the tracer in the blood) is associated with the magnetization of the tissue (which can be

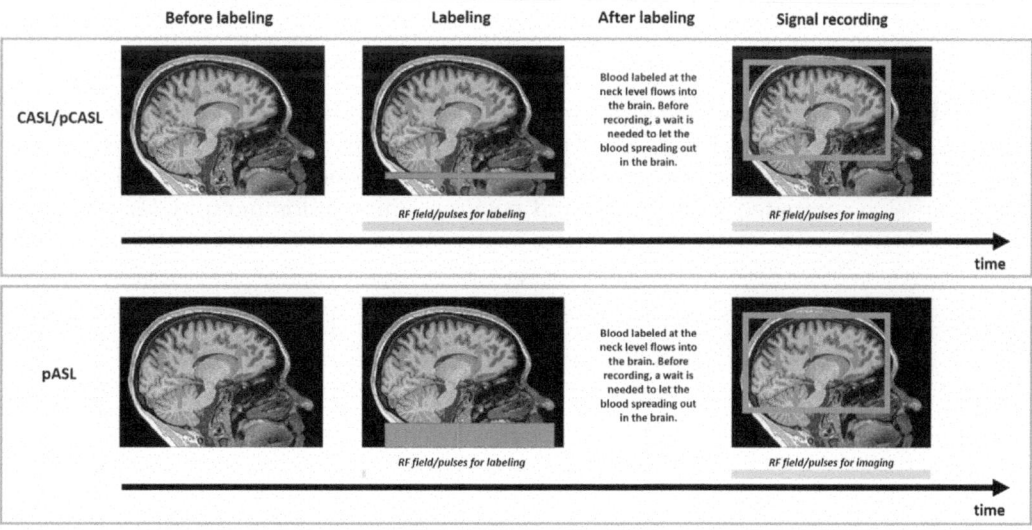

FIGURE 15.4 Representation of CASL/pCASL and pASL labeling schemes.

estimated from proton-density-weighted data) via a coefficient relating the density of water in the tissue to the one in the blood (partition coefficient) [46].

As for contrast-enhanced MRI perfusion techniques, the kinetic model has to take into account that the signal depends on (1) the delivery of the labeled blood in the brain (i.e., AIF) and (2) on the decay of the magnetization of the labeled blood occurring from the time of labeling and the time of signal recording (i.e., residue function) [46]. The previous depends on the acquisition parameters (e.g., labeling scheme, sequence efficiency) and arterial transit time, which produces a shift between the beginning of labeling and the arrival of the tagged blood in the brain. Assuming that most labeled molecules are in the tissue when the signal is recorded (well-mixed model assumption), the residue function can be described as a simple T1 decay [48]. The profile of the derived magnetization signal (i.e., difference of magnetization between tag and control images) is represented in Fig. 15.5.

Once the model has been defined, CBF can be computed using the deconvolution process. Although arterial transit time (ATT) is neglected in the ASL simplest kinetic models, this parameter can also be locally quantified with more complex models, especially when multi-PLD sequences are acquired [46].

ASL is generally acquired to assess CBF in resting conditions. However, it can also be used to assess the CBF change in response to a stimulus (e.g., task, drug). Indeed, ASL was recently proposed as an alternative sequence for functional MRI. Compared to the blood oxygen-dependent signal, ASL allows flexible design and, given its quantitative nature, it can be used to detect even subtle changes [47,46]. Besides task-functional MRI, ASL can be used to assess the physiological response to vasoactive substances, namely cerebrovascular reactivity (CVR). By acquiring the same sequence in the presence and in the absence of a vasoactive stimulus (e.g., acetazolamide, breathing gas mixture with increased CO_2 content) and computing the difference of CBF in the two conditions, the capacity of the vasculature to increase perfusion can be quantified (Fig. 15.6) [46].

Magnetic resonance imaging perfusion studies in multiple sclerosis

Cerebral hypoperfusion in Ms patients has been described since 1980 using single-photon emission computed tomography and positron emission tomography [1,2,49–51]. A

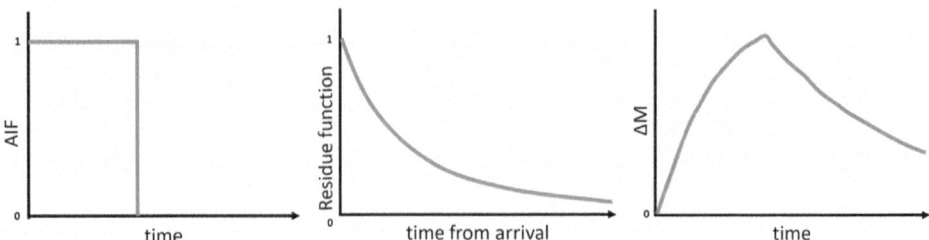

FIGURE 15.5 Arterial input fuction (AIF), residue function and Δ magnetization (control and tag difference) profiles over time in a pCASL sequence.

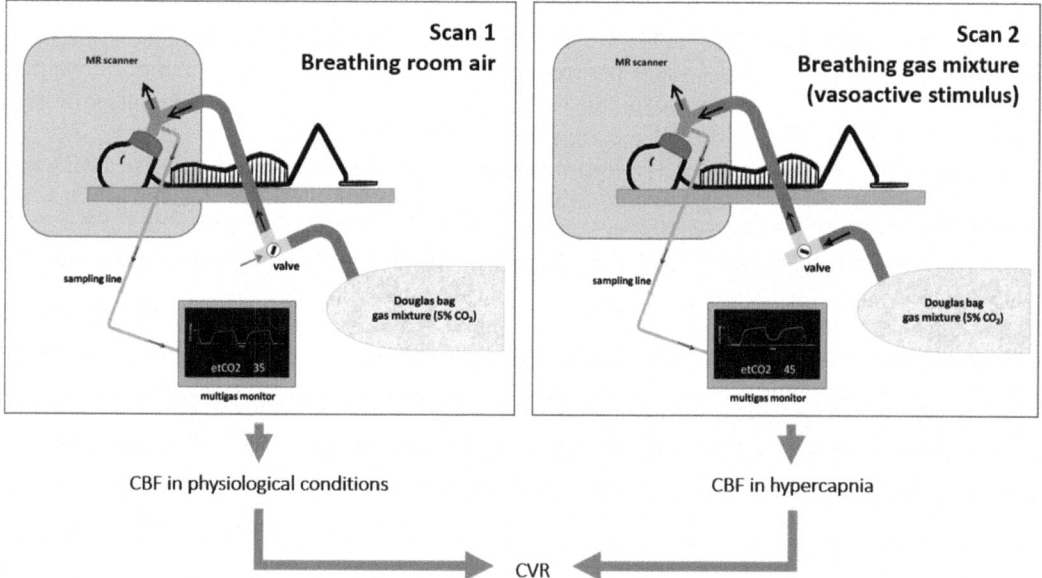

FIGURE 15.6 Example of design for cerebrovascular reactivity (CVR) assessment with ASL.

recent study [52] using the former technique showed GM atrophy and hypoperfusion in different brain regions, suggesting they may occur independently.

These findings brought researchers investigating brain perfusion with noninvasive techniques such as transcranial Doppler and MRI. With the former, it was recently shown that lower cerebral arterial blood flow was associated with higher serum neurofilament light chain levels, a biomarker of neuronal damage [53]. Another recent study [54] using the same technique revealed lower cerebral arterial blood flow in cognitively impaired Ms patients.

In the last 15 years, numerous studies used also MRI to study brain perfusion in Ms [3,55,56]. Most of the MRI findings suggested reduced cerebral perfusion in Ms, especially in the progressive phenotypes and in cognitively impaired patients [3,44,56]. Hypoperfusion has been described even in the absence of structural alterations [44,57], as suggested by the previously reported single-photon emission computed tomography study [52]. Moreover, perfusion changes precede lesion formation [58]. As with ultrasound [53], also using DSC MRI it was shown [28] that higher serum neurofilament light chain levels were associated with lower perfusion. MRI allowed finding the brain regions where the two factors were related to each other: thalamus longer MTT and a lower nCBV and nCBF were related to higher neurofilament levels.

A recent study revealed a mismatch also between functional and perfusion alterations in Ms patients [59].

Multimodal studies [60,61], using diffusion and perfusion indices or evaluating N-acetylaspartate levels with Mr spectroscopy and perfusion, showed that alterations in the NAWM are compatible with ischemia, rather than with hypoperfusion secondary to axonal damage. The higher level of endothelin-1 vasoconstricting peptide has been hypothesized to be responsible for hypoperfusion [38,62]. A study showed higher endothelin-1 levels in 10 patients with Ms compared to 10 healthy controls, which was restored to normal values after bosentan administration, an endothelin receptor antagonist [38]. However, a recent randomized controlled trial with bosentan [63] did not show perfusion increment, probably because the relapsing-remitting Ms (RRMS) patients of that study had no baseline impaired CBF compared to healthy controls.

A recent DSC MRI study [15] showed a direct relationship between systemic and cerebral blood flow in primary-progressive MS (PPMS) patients, suggesting the failure in CVR mechanisms and insufficient perfusion control.

Also, the transit time was found to be altered in Ms subjects by previous studies [64,65] using various imaging techniques. In particular, prolonged transit time was described in Ms patients compared to controls: Mancini et al. [64] compared 103 Ms patients and 42 healthy subjects using contrast-enhanced ultrasound; Monti et al. [65] compared 80 Ms patients with RR, SP, and PP phenotypes and 44 healthy subjects using digital subtraction angiography; Ge et al. [66] using DSC MRI found a prolonged MTT in lesions and NAWM of 17 patients compared with WM of 17 healthy controls.

In the following sections of the chapter, studies have been reported separately based on their main focus: perfusion alterations in various Ms phenotypes, their relationship with disability, lesion formation, and atrophy.

Perfusion differences across multiple sclerosis phenotypes

Only a minor part of the perfusion studies tested differences across Ms phenotypes [23–25,40,43], with a cross-sectional design. However, those studies globally suggested a trend of perfusion impairment in the PPMS and secondary-progressive Ms (SPMS) compared to RRMS.

DSC studies at 3 T showed lower CBF in RRMS and PPMS compared to healthy controls [23–25], with wider altered regions in PPMS. Comparing PPMS and RRMS directly, lower CBF and CBV were found in the periventricular NAWM [23], thalamus and caudate head [25].

Conversely, comparable MTT was reported among Ms phenotypes [23–25].

However, it has to be highlighted that the three concordant studies [23–25] reported above were from the same research group.

Other studies on Ms phenotype differences used ASL with 1.5 T scanners [40,43]. Amann and colleagues reported cortical hypoperfusion in SPMS compared with RRMS; however, this significance did not survive correction for T2 lesion volume, age, gender, and disease duration [40]. Rashid et al. showed [43] no difference in RRMS compared to controls, wide hypoperfusion regions in PPMS compared to healthy controls, and both hypoperfusion and hyperperfusion in SPMS.

Physical disability

Studies investigating the relationships between perfusion and physical disability reported different results; however, most showed a correspondence between physical disability and perfusion [23,25–27,36,41,42,45,67]. Specifically, most of them found that disability worsening was associated with decreased CBF or CBV. Expanded-disability status scale (EDSS) was negatively correlated with CBF or CBV [23,25,36,41,42]; multiple sclerosis functional composite (MSFC) and 9-Hole Peg Test were positively correlated with CBF [42] and thalamic MTT [67]; multiple sclerosis severity score (MSSS) was negatively correlated with GM nCBF and hippocampal nCBV [67]. However, one study reported both negative and positive correlations between EDSS and CBF in diffuse GM areas [41]. Finally, a longitudinal study [36] reported negative correlations between CBF and EDSS, 9-Hole Peg Test, and Timed 25-Foot Walk, but lower CBF in frontal regions was related to a higher number of correct responses at the Symbol Digit Modalities Test that might indicate a maladaptive functional reorganization.

However, some studies using various MRI scanners, sequences and processing methods did not find a relationship between perfusion and physical disability scales [24,34,39,40,43].

Discordant results were also reported regarding the relationship between physical disability and transit time. Consistently with the prolonged transit time generally found in Ms subjects compared with healthy controls [64–66], prolonged transit time was also associated with higher EDSS [27,45]. A counterintuitive finding was reported in: [26] lower normalized MTT in RRMS patients with high versus low MSSS.

Since a trend for different perfusion patterns was shown among Ms phenotypes, the relationship between physical disability and perfusion should be investigated in Ms subgroups. Conversely, these correlations were tested by merging different Ms phenotypes by the previous studies [23–25,43].

Cognitive performance

Cognitive decline is a widely recognized Ms symptom, with strong impacts on daily activities and quality of life [68].

Consistent results reported a relationship between cognition and cerebral perfusion in Ms [17–22,24,37–39,69], and brain perfusion was suggested as a predictor for cognitive dysfunction in Ms [19,22,54].

In particular, most of the studies associated cognitive impairment with reduced cerebral perfusion [17–22,24,37–39].

This relationship was evidenced in specific brain areas and even at the global level, considering GM and WM as a whole, suggesting hypoperfusion in diffuse brain areas of cognitively impaired patients [19,22].

Many studies [18,21,22,37,39] specifically focused on frontal areas, due to their key role in high-level cognitive functions, such as working memory, executive functions, and control [70]. Most of the studies found hypoperfusion associated with cognitive impairment in RRMS and SPMS patients [21,22,24,39,71]. A recent study [36] reported the inverse relationship in patients with PPMS, explaining this result as a possible functional maladaptive mechanism.

Specifically, studies performing voxel-wise analyses showed frontal hypoperfusion in cognitively impaired versus cognitively preserved RRMS and healthy controls [18,37], and an association between hypoperfusion and memory [39]. The first study [18] used DSC data and reported lower CBF and CBV in the left middle frontal and left superior frontal gyri. The second one [37] used pCASL and described lower CBF in the left frontal and bilateral superior frontal lobes. The third study [39] used pCAS and showed a positive correlation between frontal and precentral gyri perfusion and memory, assessed with Brief Visuospatial Memory Test (BVMT) and California Verbal Learning Test (CVLT) in patients with RRMS. In patients with SPMS, perfusion of the frontal lobes was shown to predict the overall cognitive impairment [22]. Specifically, the study used DSC to estimate CBV in the left inferior frontal, middle frontal, superior frontal regions, and bilateral medial superior frontal regions, which were significant predictors of cognitive performances measured with the minimal assessment of cognitive functions. Another study by the same group [21] investigated perfusion in SPMS using DSC, showing lower CBF in medial frontal gyri and lower CBV in frontal gyri of cognitively impaired versus cognitively preserved patients. The perfusion indices of the resulting hypoperfused areas also significantly correlated with the scores of all Minimal Assessment of Cognitive Function in Ms (MACFIMS) tests (apart from the CVLT).

Besides perfusion of the frontal regions, that of the deep GM, in particular of the thalamus, seemed to be associated with cognition in people with Ms [24,39]. The basal ganglia and thalamus act as important hubs in information integration and modulation during complex attention and executive function tasks [71].

The previously cited studies by Vitorino et al. [18], Francis et al. [21], and Hojjat et al. [37] showed reduced thalamic CBF and CBV in cognitively impaired versus preserved patients with RRMS and SPMS [18,21,37], using DSC.

Another DSC study [24] showed an association between deep GM CBF and Rey Complex Figure Copy test and between deep GM CBV and Delis-Kaplan Executive Function System Sorting Test (Color Word Interference Test inhibition) in patients with PPMS and RRMS. A pCAS study [39] showed a positive correlation between thalamic CBF and memory, assessed with BVMT and CVLT.

Besides the thalamus, the caudate nucleus displayed altered perfusion in cognitively impaired RRMS and SPMS patients in two studies of the same group [21,37].

Finally, cognitive performance has been suggested to be associated with the perfusion of both WM lesions [17] and NAWM [20,21,72]. A significant correlation between WM lesion perfusion and Symbol Digit Modalities Test score was reported in patients with RRMS and SPMS [17]. CBF and CBV were reduced in cognitively impaired versus preserved patients with RRMS not only in WM lesions but also in the NAWM [20]. Interestingly, in a pCASL study, CBF in the normal-appearing semioval center, but not in the frontoparietal cortices, was significantly correlated with Paced Auditory Serial Addition Test scores [72].

The MTT obtained from DSC data was not significantly associated with a cognitive decline for most of the studies [18,21,22]. Two studies by Hojjatt [19,37] found that MTT was altered in cognitively impaired RRMS. The first one [19] reported that the MTT of the whole normal-appearing GM was higher in cognitively impaired versus preserved patients, and also in the NAWM for the cognitively impaired patients versus healthy controls. The second one [37] examined the MTT in cortical GM and WM, finding that the cognitively impaired RRMS patients had the highest MTT, but that also the cognitively preserved RRMS patients had higher MTT compared to healthy controls.

In conclusion, the association between cognitive dysfunction and cerebral hypoperfusion is suggested by most of the MRI studies, while only a few studies found prolonged MTT. But interestingly, a DSC study on clinically isolated syndrome patients [69], reported an inverse correlation between memory and CBV of the left frontal NAWM, bilateral thalami, right caudate and corpus callosum, regions involved in memory functions [69]. Also, a longitudinal pCASL study on PPMS reported an inverted relationship between perfusion and cognition, compared to the rest of the studies: specifically, CBF decreased with improved performance at the Symbol Digit Modalities Test. A functional maladaptive response to cognitive decline was hypothesized to be responsible for this CBF change.

Finally, promising results are shown by a recent ASL study [73] using the administration N-acetyl cysteine, an antioxidant and glutathione precursor: it was able not only to change CBF levels but also to improve self-reported cognition and attention scales.

Lesion evolution

A study using an atlas of single-photon emission-computed tomography [74] from healthy volunteers showed that chronic Ms lesions are preferentially located in regions with low perfusion, especially for SPMS patients. Almost two decades ago, it was shown that perfusion parameter alterations may precede the BBB breakdown [58], with CBF and CBV increment till 3 weeks before lesion enhancement. Various kinds of Gd-enhancing lesion evolution have been described, with an increased CBF and CBV that may reflect inflammatory activity, as shown by some DCE studies [33–35,58,75]. Different types of lesions (nodular or ring-like), as well as their development and expansion, have been described [35].

The perfusion values come back to baseline values till 20 weeks after their enhancement [58]. Regarding chronic lesions, the T1-hypointense ones have lower CBV than T1-isointense lesions [75], which have similar perfusion compared to the WM of the healthy subjects. The lower perfusion of the black holes (T1-hypointense lesions) is due to their axonal loss and lower metabolism.

Ge et al. [66] described two subsets of nonenhancing lesions: those with lower CBF and CBV compared to NAWM and those with increased CBV, probably in an inflammatory state. A similar pattern was also shown for cortical lesions in a DSC study [76], and more recently a CBF reduction in cortical lesions using ASL at 7 T [77].

Various studies investigated the relationship between perfusion and lesion load, with different findings. A direct correlation is reported by some of them [40,78], but GM hypoperfusion was found to be independent of lesions by several others [24,39,79].

Hypoperfusion and brain atrophy

Some studies suggested that hypoperfusion precedes or induces atrophy, and several others suggested the opposite mechanisms [39,42,80]. It is also reported that hypoperfusion and atrophy may occur in different cerebral areas, suggesting their independent alterations [44,52].

The first hypothesis suggests that chronic hypoperfusion, mediated by elevated levels of the vasospastic peptide endothelin-1, induces axonal degeneration in patients with Ms [62]. Cross-sectional studies described hypoperfusion in non-atrophic regions in RRMS compared to healthy controls [39,42] and in cognitively impaired versus preserved RRMS and SPMS patients [22,37]. The lack of correlation between GM hypoperfusion and atrophy found in another study [44] supports the hypothesis that factors different from axonal integrity disruption and neurodegeneration might be responsible for CBF reduction in Ms [81].

A multimodal perfusion-diffusion MRI study [60] showed a correlation between CBF and mean diffusivity, but not with fractional anisotropy in the corpus callosum of RRMS patients, an additional evidence that neurodegeneration cannot be the primary cause of hypoperfusion, given that the axonal Ms degeneration is typically associated with increased mean diffusivity and decreased fractional anisotropy. This kind of diffusion and perfusion status is the typical pattern of ischemia.

However, atrophy has also been suggested as one of the causes of hypoperfusion in Ms [80].

Hypoperfusion could be due to other factors, such as reduced axonal activity, astrocyte energy metabolism, vasoactive substances such as endothelin-1 [62,81–83], or lower cross-sectional area [84] in the main neck arteries. A greater reduction of the latter over 5 years was also recently shown [85].

Cerebrovascular reactivity

The neurovascular coupling integrity was recently evaluated using the CVR, an index of cerebral arterioles dilatory capacity in response to a vasomotor stimulus, representing the capacity of increasing the CBF after neural activation.

A map of CVR can be estimated as explained in the previous section of this chapter. A recent study [86] acquired ASL at 3 T and revealed lower CVR in wide areas of the GM of 19 Ms patients compared to 19 healthy subjects. CVR was negatively correlated with lesion volume and positively correlated with GM atrophy. A study by the same group [87] confirmed CVR reduction in 28 patients with Ms versus 28 healthy subjects, in particular in various functional networks (default mode, frontoparietal, somatomotor, and ventral attention networks). They also showed a correlation between lesion load and CVR in the default mode and ventral attention networks, and between GM atrophy and default mode network CVR. They hypothesized that the CVR decrement was due to the chronically elevated nitric oxide levels, a vasodilator constantly produced by inflammatory processes. The consequent habituation and desensitization of vascular endothelial and smooth muscle functions would cause a decreased vasodilatory capacity.

However, a similar study using 3D ASL at 1.5 T did not find CVR differences between 26 Ms patients and 26 healthy controls [8]. Other previous studies did not find CVR differences between Ms patients and healthy subjects, using MRI [88] and transcranial Doppler [89]. As regards the latter [89], breath-holding was used as a stimulus and no differences were found between RRMS patients in different disease activity stages compared to healthy subjects. As to the former study [88], CVR was not statistically different between 33 Ms patients and 22 healthy controls. However, cognitively impaired patients compared to cognitively normal ones had lower CVR in the whole brain and in eight regions of the Automated Anatomical Labeling atlas, with a significant correlation between CVR and education level.

In conclusion, only a few MRI studies estimated CVR with conflicting results, but they used various sequences, imaging and analysis methods.

Conclusions

All the perfusion MRI studies described earlier suggested that perfusion, estimates using MRI, might provide useful biomarkers in Ms. However, the studies used various clinical scales, such as EDSS, MSSS, and MSFC, as well as different neuropsychological batteries to assess cognitive status, including MACFIMS, Wechsler Memory Scale, and the Rey Complex Figure Copy test.

Furthermore, various acquisition and quantification approaches were used: different sequences and acquisition parameters, and perfusion indices were evaluated as averages in regions of interest [17,19,20,24,69,72] or at the voxel-level [18,21,37,39]. Moreover, most of the studies are cross-sectional, with different statistical approaches: some studies tested the correlation between physical disability and perfusion considering only one phenotype [26,27,34,39,41,42,45], while others mixed different phenotypes [23–25,40,43]; some tested the correlation between cognitive performance and CBF or CBV [17,21,24,36,38,39,53,69], while others tested differences between cognitively impaired and cognitively preserved Ms patients [18–22,37]. Since cognitive impairment is defined according to an arbitrary threshold, the classification as either impaired or preserved may change across studies.

Homogenizing the methods in future perfusion investigations would make the results comparable among studies. Voxel-wise analyses would prevent averaging hypo- and hyperperfusion regions, which could be contemporaneously present in the same Ms group [41,43]. Moreover, voxel-wise analyses or evaluations of perfusion-based indices in specific regions of interest would consider the different perfusion distributions across the brain. Indeed, a recent study [90] using ASL data from a large cohort of the Human Connectome Project-Aging showed that perfusion is spatially heterogeneous, with significant differences between cortical and subcortical GM and between juxtacortical and periventricular WM.

References

[1] Lycke J, Wikkelsö C, Bergh AC, Jacobsson L, Andersen O. Regional cerebral blood flow in multiple sclerosis measured by single photon emission tomography with technetium-99m hexamethylpropyleneamine oxime. Eur Neurol 1993;33(2):163–7.

[2] Sun X, Tanaka M, Kondo S, Okamoto K, Hirai S. Clinical significance of reduced cerebral metabolism in multiple sclerosis: a combined PET and MRI study. Ann Nucl Med 1998;12(2):89–94.

[3] Lapointe E, Li DKB, Traboulsee AL, Rauscher A. What have we learned from perfusion MRI in multiple sclerosis? AJNR Am J Neuroradiol 2018;39(6):994–1000.

[4] Clement P, Mutsaerts HJ, Vaclavu L, Ghariq E, Pizzini FB, Smits M, et al. Variability of physiological brain perfusion in healthy subjects – a systematic review of modifiers. Considerations for multi-center ASL studies. J Cereb Blood Flow Metab 2018;38(9):1418–37.

[5] Eskildsen SF, Gyldensted L, Nagenthiraja K, Nielsen RB, Hansen MB, Dalby RB, et al. Increased cortical capillary transit time heterogeneity in Alzheimer's disease: a DSC-MRI perfusion study. Neurobiol aging 2017;50:107–18.

[6] Xi YB, Kang XW, Wang N, Liu TT, Zhu YQ, Cheng G, et al. Differentiation of primary central nervous system lymphoma from high-grade glioma and brain metastasis using arterial spin labeling and dynamic contrast-enhanced magnetic resonance imaging. Eur J Radiol 2019;112:59–64.

[7] Corno S, Giani L, Lagana MM, Baglio F, Mariani C, Pantoni L, et al. The brain effect of the migraine attack: an ASL MRI study of the cerebral perfusion during a migraine attack. Neurol Sci 2018;39(Suppl 1):73–4.

[8] Pelizzari L, Laganà MM, Rossetto F, Bergsland N, Galli M, Baselli G, et al. Cerebral blood flow and cerebrovascular reactivity correlate with severity of motor symptoms in Parkinson's disease. Therapeutic Adv Neurological Disord 2019;.

[9] Laganà MM, Pirastru A, Pelizzari L, Rossetto F, Di Tella S, Bergsland N, et al. Multimodal evaluation of neurovascular functionality in early Parkinson's disease. Front Neurol 2020;11:831.

[10] Pelizzari L, Di Tella S, Rossetto F, Laganà MM, Bergsland N, Pirastru A, et al. Parietal perfusion alterations in Parkinson's disease patients without dementia. Front Neurol 2020;11:562.

[11] Pelizzari L, Laganà MM, Di Tella S, Rossetto F, Bergsland N, Nemni R, et al. Combined assessment of diffusion parameters and cerebral blood flow within basal ganglia in early Parkinson's disease. Front aging Neurosci 2019;11:134.

[12] Querzola G, Lovati C, Laganà MM, Pirastru A, Baglio F, Pantoni L. Incipient chronic traumatic encephalopathy in active American football players: neuropsychological assessment and brain perfusion measures. Neurological Sci 2022;43(9):5383–90.

[13] McGehee BE, Pollock JM, Maldjian JA. Brain perfusion imaging: how does it work and what should I use? J Magn Reson Imaging 2012;36(6):1257–72.

[14] Quarles CC, Bell LC, Stokes AM. Imaging vascular and hemodynamic features of the brain using dynamic susceptibility contrast and dynamic contrast enhanced MRI. Neuroimage 2019;187:32–55.

[15] Jakimovski D, Bergsland N, Dwyer MG, Choedun K, Marr K, Weinstock-Guttman B, et al. Cerebral blood flow dependency on systemic arterial circulation in progressive multiple sclerosis. Eur Radiol 2022;1–12.

[16] Jakimovski D, Zivadinov R, Dwyer MG, Bergsland N, Ramasamy DP, Browne RW, et al. High density lipoprotein cholesterol and apolipoprotein AI are associated with greater cerebral perfusion in multiple sclerosis. J Neurol Sci 2020;418:117120.

[17] Ma AY, Vitorino RC, Hojjat SP, Mulholland AD, Zhang L, Lee L, et al. The relationship between white matter fiber damage and gray matter perfusion in large-scale functionally defined networks in multiple sclerosis. Mult Scler 2017;23(14):1884–92.

[18] Vitorino R, Hojjat SP, Cantrell CG, Feinstein A, Zhang L, Lee L, et al. Regional frontal perfusion deficits in relapsing-remitting multiple sclerosis with cognitive decline. AJNR Am J Neuroradiol 2016;37(10):1800–7.

[19] Hojjat SP, Cantrell CG, Carroll TJ, Vitorino R, Feinstein A, Zhang L, et al. Perfusion reduction in the absence of structural differences in cognitively impaired versus unimpaired RRMS patients. Mult Scler 2016;22(13):1685–94.

[20] Hojjat SP, Kincal M, Vitorino R, Cantrell CG, Feinstein A, Zhang L, et al. Cortical perfusion alteration in normal-appearing gray matter is most sensitive to disease progression in relapsing-remitting multiple sclerosis. AJNR Am J Neuroradiol 2016;37(8):1454–61.

[21] Francis PL, Jakubovic R, O'Connor P, Zhang L, Eilaghi A, Lee L, et al. Robust perfusion deficits in cognitively impaired patients with secondary-progressive multiple sclerosis. AJNR Am J Neuroradiol 2013;34(1):62–7.

[22] Aviv RI, Francis PL, Tenenbein R, O'Connor P, Zhang L, Eilaghi A, et al. Decreased frontal lobe gray matter perfusion in cognitively impaired patients with secondary-progressive multiple sclerosis detected by the bookend technique. AJNR Am J Neuroradiol 2012;33(9):1779–85.

[23] Adhya S, Johnson G, Herbert J, Jaggi H, Babb JS, Grossman RI, et al. Pattern of hemodynamic impairment in multiple sclerosis: dynamic susceptibility contrast perfusion MR imaging at 3.0 T. Neuroimage 2006;33(4):1029–35.

[24] Inglese M, Adhya S, Johnson G, Babb JS, Miles L, Jaggi H, et al. Perfusion magnetic resonance imaging correlates of neuropsychological impairment in multiple sclerosis. J Cereb Blood Flow Metab 2008;28(1):164–71.

[25] Inglese M, Park SJ, Johnson G, Babb JS, Miles L, Jaggi H, et al. Deep gray matter perfusion in multiple sclerosis: dynamic susceptibility contrast perfusion magnetic resonance imaging at 3 T. Arch Neurol 2007;64(2):196–202.

[26] Sowa P, Nygaard GO, Bjornerud A, Celius EG, Harbo HF, Beyer MK. Magnetic resonance imaging perfusion is associated with disease severity and activity in multiple sclerosis. Neuroradiology 2017;59(7):655−64.
[27] Garaci FG, Marziali S, Meschini A, Fornari M, Rossi S, Melis M, et al. Brain hemodynamic changes associated with chronic cerebrospinal venous insufficiency are not specific to multiple sclerosis and do not increase its severity. Radiology 2012;265(1):233−9.
[28] Jakimovski D, Bergsland N, Dwyer MG, Ramasamy DP, Ramanathan M, Weinstock-Guttman B, et al. Serum neurofilament light chain levels are associated with lower thalamic perfusion in multiple sclerosis. Diagnostics 2020;10(9).
[29] Keston P, Murray AD, Jackson A. Cerebral perfusion imaging using contrast-enhanced MRI. Clin Radiol 2003;58(7):505−13.
[30] Willats L, Calamante F. The 39 steps: evading error and deciphering the secrets for accurate dynamic susceptibility contrast MRI. NMR Biomed 2013;26(8):913−31.
[31] Wintermark M, Sesay M, Barbier E, Borbely K, Dillon WP, Eastwood JD, et al. Comparative overview of brain perfusion imaging techniques. Stroke 2005;36(9):e83−99.
[32] Sourbron SP, Buckley DL. Classic models for dynamic contrast-enhanced MRI. NMR Biomed 2013;26(8):1004−27.
[33] Ingrisch M, Sourbron S, Morhard D, Ertl-Wagner B, Kumpfel T, Hohlfeld R, et al. Quantification of perfusion and permeability in multiple sclerosis: dynamic contrast-enhanced MRI in 3D at 3T. Invest Radiol 2012;47(4):252−8.
[34] Yin P, Xiong H, Liu Y, Sah SK, Zeng C, Wang J, et al. Measurement of the permeability, perfusion, and histogram characteristics in relapsing-remitting multiple sclerosis using dynamic contrast-enhanced MRI with extended Tofts linear model. Neurol India 2018;66(3):709−15.
[35] Gaitán MI, Shea CD, Evangelou IE, Stone RD, Fenton KM, Bielekova B, et al. Evolution of the blood−brain barrier in newly forming multiple sclerosis lesions. Ann Neurol 2011;70(1):22−9.
[36] Testud B, Delacour C, El Ahmadi AA, Brun G, Girard N, Duhamel G, et al. Brain grey matter perfusion in primary progressive multiple sclerosis: mild decrease over years and regional associations with cognition and hand function. Eur J Neurol 2022;29(6):1741−52.
[37] Hojjat SP, Cantrell CG, Vitorino R, Feinstein A, Shirzadi Z, MacIntosh BJ, et al. Regional reduction in cortical blood flow among cognitively impaired adults with relapsing-remitting multiple sclerosis patients. Mult Scler 2016;22(11):1421−8.
[38] D'Haeseleer M, Beelen R, Fierens Y, Cambron M, Vanbinst AM, Verborgh C, et al. Cerebral hypoperfusion in multiple sclerosis is reversible and mediated by endothelin-1. Proc Natl Acad Sci U S A 2013;110(14):5654−8.
[39] Debernard L, Melzer TR, Van Stockum S, Graham C, Wheeler-Kingshott CA, Dalrymple-Alford JC, et al. Reduced grey matter perfusion without volume loss in early relapsing-remitting multiple sclerosis. J Neurol Neurosurg Psychiatry 2014;85(5):544−51.
[40] Amann M, Achtnichts L, Hirsch JG, Naegelin Y, Gregori J, Weier K, et al. 3D GRASE arterial spin labelling reveals an inverse correlation of cortical perfusion with the white matter lesion volume in MS. Mult Scler 2012;18(11):1570−6.
[41] Zhang X, Guo X, Zhang N, Cai H, Sun J, Wang Q, et al. Cerebral blood flow changes in multiple sclerosis and neuromyelitis optica and their correlations with clinical disability. Front Neurol 2018;9:305.
[42] Doche E, Lecocq A, Maarouf A, Duhamel G, Soulier E, Confort-Gouny S, et al. Hypoperfusion of the thalamus is associated with disability in relapsing remitting multiple sclerosis. J Neuroradiol 2017;44(2):158−64.
[43] Rashid W, Parkes LM, Ingle GT, Chard DT, Toosy AT, Altmann DR, et al. Abnormalities of cerebral perfusion in multiple sclerosis. J Neurol Neurosurg Psychiatry 2004;75(9):1288−93.
[44] Lagana MM, Mendozzi L, Pelizzari L, Bergsland NP, Pugnetti L, Cecconi P, et al. Are cerebral perfusion and atrophy linked in multiple sclerosis? Evidence for a multifactorial approach to assess neurodegeneration. Curr Neurovasc Res 2018;15(4):282−91.
[45] Paling D, Thade Petersen E, Tozer DJ, Altmann DR, Wheeler-Kingshott CA, Kapoor R, et al. Cerebral arterial bolus arrival time is prolonged in multiple sclerosis and associated with disability. J Cereb Blood Flow Metab 2014;34(1):34−42.
[46] Chappell M, MacIntosh BJ, Okell T. Introduction to perfusion quantification using arterial spin labelling. Oxford: Oxford University Press; 2018.
[47] Hernandez-Garcia L, Lahiri A, Schollenberger J. Recent progress in ASL. Neuroimage 2019;187:3−16.

[48] Alsop DC, Detre JA, Golay X, Gunther M, Hendrikse J, Hernandez-Garcia L, et al. Recommended implementation of arterial spin-labeled perfusion MRI for clinical applications: a consensus of the ISMRM perfusion study group and the European consortium for ASL in dementia. Magn Reson Med 2015;73(1):102–16.

[49] Swank RL, Roth JG, Woody Jr DC. Cerebral blood flow and red cell delivery in normal subjects and in multiple sclerosis. Neurological Res 1983;5(1):37–59.

[50] Brooks D, Leenders K, Head G, Marshall J, Legg N, Jones T. Studies on regional cerebral oxygen utilisation and cognitive function in multiple sclerosis. J Neurol Neurosurg Psychiatry 1984;47(11):1182–91.

[51] de Paula Faria D, Copray S, Buchpiguel C, Dierckx R, de Vries E. PET imaging in multiple sclerosis. J Neuroimmune Pharmacol 2014;9(4):468–82.

[52] Shooli H, Nemati R, Chabi N, Larvie M, Jokar N, Dadgar H, et al. Multimodal assessment of regional gray matter integrity in early relapsing-remitting multiple sclerosis patients with normal cognition: a voxel-based structural and perfusion approach. Br J Radiol 2021;94(1127):20210308.

[53] Jakimovski D, Gibney BL, Marr K, Ramasamy DP, Dwyer MG, Bergsland N, et al. Lower cerebral arterial blood flow is associated with greater serum neurofilament light chain levels in multiple sclerosis patients. Eur J Neurol 2022;29(8):2299–308.

[54] Jakimovski D, Benedict RH, Marr K, Gandhi S, Bergsland N, Weinstock-Guttman B, et al. Lower total cerebral arterial flow contributes to cognitive performance in multiple sclerosis patients. Mult Scler 2019; 1352458518819608.

[55] Laganà MM, Pelizzari L, Baglio F. Relationship between MRI perfusion and clinical severity in multiple sclerosis. Neural Regeneration Res 2020;15(4):646.

[56] Jakimovski D, Topolski M, Genovese AV, Weinstock-Guttman B, Zivadinov R. Vascular aspects of multiple sclerosis: emphasis on perfusion and cardiovascular comorbidities. Expert Rev Neurother 2019;19(5):445–58.

[57] de la Peña MJ, Peña IC, García PG-P, Gavilán ML, Malpica N, Rubio M, et al. Early perfusion changes in multiple sclerosis patients as assessed by MRI using arterial spin labeling. Acta Radiol Open 2019;8(12), 2058460119894214.

[58] Wuerfel J, Bellmann-Strobl J, Brunecker P, Aktas O, McFarland H, Villringer A, et al. Changes in cerebral perfusion precede plaque formation in multiple sclerosis: a longitudinal perfusion MRI study. Brain 2004;127 (Pt 1):111–19.

[59] Jandric D, Lipp I, Paling D, Rog D, Castellazzi G, Haroon H, et al. Mechanisms of network changes in cognitive impairment in multiple sclerosis. Neurology 2021;97(19):e1886–97.

[60] Saindane AM, Law M, Ge Y, Johnson G, Babb JS, Grossman RI. Correlation of diffusion tensor and dynamic perfusion MR imaging metrics in normal-appearing corpus callosum: support for primary hypoperfusion in multiple sclerosis. AJNR Am J Neuroradiol 2007;28(4):767–72.

[61] Steen C, D'Haeseleer M, Hoogduin JM, Fierens Y, Cambron M, Mostert JP, et al. Cerebral white matter blood flow and energy metabolism in multiple sclerosis. Mult Scler 2013;19(10):1282–9.

[62] D'Haeseleer M, Hostenbach S, Peeters I, Sankari SE, Nagels G, De Keyser J, et al. Cerebral hypoperfusion: a new pathophysiologic concept in multiple sclerosis? J Cereb Blood Flow Metab 2015;35(9):1406–10.

[63] Hostenbach S, Raeymaekers H, Van Schuerbeek P, Vanbinst A-M, Cools W, De Keyser J, et al. The role of cerebral hypoperfusion in multiple sclerosis (ROCHIMS) trial in multiple sclerosis: insights from negative results. Front Neurol 2020;11:674.

[64] Mancini M, Morra VB, Di Donato O, Maglio V, Lanzillo R, Liuzzi R, et al. Multiple sclerosis: cerebral circulation time. Radiology 2012;262(3):947–55.

[65] Monti L, Donati D, Menci E, Cioni S, Bellini M, Grazzini I, et al. Cerebral circulation time is prolonged and not correlated with EDSS in multiple sclerosis patients: a study using digital subtracted angiography. PLoS One 2015;10(2):e0116681.

[66] Ge Y, Law M, Johnson G, Herbert J, Babb JS, Mannon LJ, et al. Dynamic susceptibility contrast perfusion MR imaging of multiple sclerosis lesions: characterizing hemodynamic impairment and inflammatory activity. AJNR Am J Neuroradiol 2005;26(6):1539–47.

[67] Jakimovski D, Bergsland N, Dwyer MG, Traversone J, Hagemeier J, Fuchs TA, et al. Cortical and deep gray matter perfusion associations with physical and cognitive performance in multiple sclerosis patients. Front Neurol 2020;11:700.

[68] Sumowski JF, Benedict R, Enzinger C, Filippi M, Geurts JJ, Hamalainen P, et al. Cognition in multiple sclerosis: state of the field and priorities for the future. Neurology 2018;90(6):278–88.

[69] Papadaki EZ, Simos PG, Mastorodemos VC, Panou T, Maris TG, Karantanas AH, et al. Regional MRI perfusion measures predict motor/executive function in patients with clinically isolated syndrome. Behav Neurol 2014;2014:252419.

[70] Badre D, Nee DE. Frontal cortex and the hierarchical control of behavior. Trends Cognit Sci 2018;22(2):170–88.

[71] Batista S, Zivadinov R, Hoogs M, Bergsland N, Heininen-Brown M, Dwyer MG, et al. Basal ganglia, thalamus and neocortical atrophy predicting slowed cognitive processing in multiple sclerosis. J Neurol 2012;259(1):139–46.

[72] D'Haeseleer M, Steen C, Hoogduin JM, van Osch MJ, Fierens Y, Cambron M, et al. Performance on Paced Auditory Serial Addition Test and cerebral blood flow in multiple sclerosis. Acta Neurol Scand 2013;128(5), e26-9.

[73] Shahrampour S, Heholt J, Wang A, Vedaei F, Mohamed FB, Alizadeh M, et al. N-acetyl cysteine administration affects cerebral blood flow as measured by arterial spin labeling MRI in patients with multiple sclerosis. Heliyon 2021;7(7):e07615.

[74] Holland CM, Charil A, Csapo I, Liptak Z, Ichise M, Khoury SJ, et al. The relationship between normal cerebral perfusion patterns and white matter lesion distribution in 1,249 patients with multiple sclerosis. J Neuroimaging 2012;22(2):129–36.

[75] Haselhorst R, Kappos L, Bilecen D, Scheffler K, Mori D, Radu EW, et al. Dynamic susceptibility contrast MR imaging of plaque development in multiple sclerosis: application of an extended blood-brain barrier leakage correction. J Magn Reson Imaging 2000;11(5):495–505.

[76] Peruzzo D, Castellaro M, Calabrese M, Veronese E, Rinaldi F, Bernardi V, et al. Heterogeneity of cortical lesions in multiple sclerosis: an MRI perfusion study. J Cereb Blood Flow Metab 2013;33(3):457–63.

[77] Dury RJ, Falah Y, Gowland PA, Evangelou N, Bright MG, Francis ST. Ultra-high-field arterial spin labelling MRI for non-contrast assessment of cortical lesion perfusion in multiple sclerosis. Eur Radiol 2018;.

[78] Ota M, Sato N, Nakata Y, Ito K, Kamiya K, Maikusa N, et al. Abnormalities of cerebral blood flow in multiple sclerosis: a pseudocontinuous arterial spin labeling MRI study. Magnetic Reson Imaging 2013;31(6):990–5.

[79] Bester M, Forkert ND, Stellmann JP, Sturner K, Aly L, Drabik A, et al. Increased perfusion in normal appearing white matter in high inflammatory multiple sclerosis patients. PLoS One 2015;10(3):e0119356.

[80] Debernard L, Melzer TR, Alla S, Eagle J, Van Stockum S, Graham C, et al. Deep grey matter MRI abnormalities and cognitive function in relapsing-remitting multiple sclerosis. Psychiatry Res 2015;234(3):352–61.

[81] D'Haeseleer M, Cambron M, Vanopdenbosch L, De Keyser J. Vascular aspects of multiple sclerosis. Lancet Neurol 2011;10(7):657–66.

[82] Monti L, Morbidelli L, Bazzani L, Rossi A. Influence of circulating endothelin-1 and asymmetric dimethylarginine on whole brain circulation time in multiple sclerosis. Biomarker Insights 2017;12, 1177271917712514.

[83] Monti L, Morbidelli L, Rossi A. Impaired Cerebral Perfusion in Multiple Sclerosis: Relevance of Endothelial Factors. Biomark Insights 2018;13, 1177271918774800.

[84] Belov P, Jakimovski D, Krawiecki J, Magnano C, Hagemeier J, Pelizzari L, et al. Lower arterial cross-sectional area of carotid and vertebral arteries and higher frequency of secondary neck vessels are associated with multiple sclerosis. Am J Neuroradiol 2018;39(1):123–30.

[85] Pelizzari L, Jakimovski D, Lagana MM, Bergsland N, Hagemeier J, Baselli G, et al. Five-year longitudinal study of neck vessel cross-sectional area in multiple sclerosis. AJNR Am J Neuroradiol 2018;39(9):1703–9.

[86] Marshall O, Lu H, Brisset JC, Xu F, Liu P, Herbert J, et al. Impaired cerebrovascular reactivity in multiple sclerosis. JAMA Neurol 2014;71(10):1275–81.

[87] Marshall O, Chawla S, Lu H, Pape L, Ge Y. Cerebral blood flow modulation insufficiency in brain networks in multiple sclerosis: a hypercapnia MRI study. J Cereb Blood Flow Metab 2016;36(12):2087–95.

[88] Metzger A, Le Bars E, Deverdun J, Molino F, Maréchal B, Picot M-C, et al. Is impaired cerebral vasoreactivity an early marker of cognitive decline in multiple sclerosis patients? Eur Radiology 2018;28(3):1204–14.

[89] Uzuner N, Ozkan S, Cinar N. Cerebrovascular reactivity in multiple sclerosis patients. Mult Scler 2007;13(6):737–41.

[90] Juttukonda MR, Li B, Almaktoum R, Stephens KA, Yochim KM, Yacoub E, et al. Characterizing cerebral hemodynamics across the adult lifespan with arterial spin labeling MRI data from the Human Connectome Project-Aging. Neuroimage 2021;230:117807.

CHAPTER

16

Magnetic resonance spectroscopy and myelin water fraction in multiple sclerosis

Cornelia Laule[1,2,3,4] and Irene M. Vavasour[1,4]

[1]Department of Radiology, University of British Columbia, Vancouver, Canada [2]Department of Pathology & Laboratory Medicine, University of British Columbia, Vancouver, Canada [3]Department of Physics & Astronomy, University of British Columbia, Vancouver, Canada [4]International Collaboration on Repair Discoveries (ICORD), University of British Columbia, Vancouver, Canada

OUTLINE

Magnetic resonance spectroscopy	288
Overview	288
Metabolites of interest for multiple sclerosis studies	289
N-acetylaspartate	289
Choline	289
Creatine	290
Myo-inositol	291
Glutamate and glutamine	291
Lactate	291
Magnetic resonance spectroscopy data acquisition	292
Magnetic resonance spectroscopy data analysis	293
Factors affecting reproducibility of magnetic resonance spectroscopy and consensus protcols	294
Magnetic resonance spectroscopy findings in multiple sclerosis lesions	295
Magnetic resonance spectroscopy findings in multiple sclerosis normal-appearing white matter and gray matter	298
Magnetic resonance spectroscopy and multiple sclerosis cognitive impairment	299
Magnetic resonance spectroscopy in multiple sclerosis clinical trials and disease modifying therapy evaluation	300

Limitations of magnetic resonance spectroscopy studies in multiple sclerosis 301	Myelin water fraction in multiple sclerosis subtypes 307
Future directions in magnetic resonance spectroscopy for multiple sclerosis research and clinical care 301	Myelin water fraction correlations with clinical measures 307
Myelin water fraction 302	Myelin water fraction and disease modifying treatments 307
Overview 302	Limitations 308
What is myelin water? 302	Future directions 308
Measurement of myelin water fraction 303	Summary 311
Myelin water fraction validation 305	References 311
Myelin water fraction in different multiple sclerosis tissues 305	

Multiple sclerosis (MS) is a chronic inflammatory disease of the central nervous system (CNS) that is characterized by demyelination, axonal loss, gliosis, and neurodegeneration in the brain and spinal cord. The clinical manifestations of MS vary widely, depending on the location and extent of the lesions, and include motor, sensory, cognitive, and visual symptoms. MS diagnosis is based on clinical criteria, supported by magnetic resonance imaging (MRI) findings. Clinical MRI scans are very sensitive to damage within the CNS; however, conventional MRI measures, such as lesion volume and brain atrophy, do not fully capture the complexity and heterogeneity of MS pathology. For example, MS lesions appear bright on proton density and T_2-weighted images, but the underlying pathology can include edema, inflammation, demyelination, axonal loss, and gliosis. In addition to conventional MRI's limited specificity for the underlying tissue changes, poor sensitivity for nonlesional pathology, and generally poor correlation with clinical outcomes have created a clinicoradiological paradox in MS [1]. Therefore there is a need for more advanced MRI techniques that can provide additional information on the metabolic and microstructural aspects of MS tissue pathology, clinical progression, and response to therapy.

Magnetic resonance spectroscopy

Overview

Magnetic resonance spectroscopy (MRS) is one such technique that allows for noninvasive measurement of the concentrations of various biochemicals in living CNS tissue. Proton, or hydrogen (^{1}H) MRS is the most common type of MRS used for CNS studies, whereby signals from hydrogen nuclei in brain metabolites are measured. MRS can reveal metabolic abnormalities related to key pathological processes of MS including inflammation, demyelination, axonal loss, gliosis, and neurodegeneration. MRS can also complement conventional MRI, which highlights anatomical abnormalities like lesions and atrophy, by detecting subtle changes in normal-appearing white matter (NAWM) and gray matter (NAGM) that are otherwise MRI-invisible.

The aim of this section is to review the current state of knowledge on MRS in MS. We will discuss the main metabolites that can be measured by ^{1}H MRS, the technical aspects of data acquisition and analysis, reproducibility and consensus protocols for brain MRS studies. We will also summarize MRS findings in MS lesions, NAWM, and NAGM, the relationship between MRS findings and clinical measures in MS, the use of MRS in MS clinical trials and for monitoring disease-modifying therapies, the limitations of MRS studies in MS, and future directions for MRS research and clinical applications in MS. For further details and more information about MRS and MS, the reader is pointed to a recent review by Swanberg et al. [2] and a book chapter by Harris and MacMillan [3].

Metabolites of interest for multiple sclerosis studies

The most commonly measured metabolites in MRS studies include N-acetylaspartate (NAA), creatine (Cr), choline (Cho), myo-inositol (mI), glutamate (Glu), glutamine (Gln), and lactate (Lac) (Fig. 16.1). These metabolites have mobile protons, are present in sufficient levels to allow for in vivo measurement, and can provide insights into neuronal viability, energy metabolism, membrane turnover, gliosis, oxidative stress, excitotoxicity, and anaerobic glycolysis. Brief overviews are provided below, and the reader is referred to references [4–8] for more detail.

N-acetylaspartate

NAA is the second most abundant metabolite in the CNS and is considered a marker of neuronal and axonal integrity, health, and function [3]. The NAA signal measured with MRS often includes signals from both NAA and N-acetylaspartaglutamate, and can be reported in the literature as total N-acetylaspartate (tNAA). NAA is synthesized in the mitochondria of neurons and transported to oligodendrocytes and axons. A number of metabolic pathways involve NAA, including myelin synthesis, energy production, neurotransmitter synthesis, osmoregulation, and axonal transport. A decrease in NAA reflects neuronal and axonal injury, dysfunction or loss, which is associated with neurodegeneration and disability in MS. NAA levels are reduced in MS lesions, NAWM, and NAGM, and NAA reduction correlates with disability progression, cognitive impairment and brain atrophy in MS [9–19].

Choline

Cho is a metabolite that reflects membrane turnover and myelin damage and synthesis in the CNS [3]. MRS-measured Cho contains signal from multiple Cho-containing compounds, in particular glycerophosphocholine, phosphocholine and, to a lesser extent, free

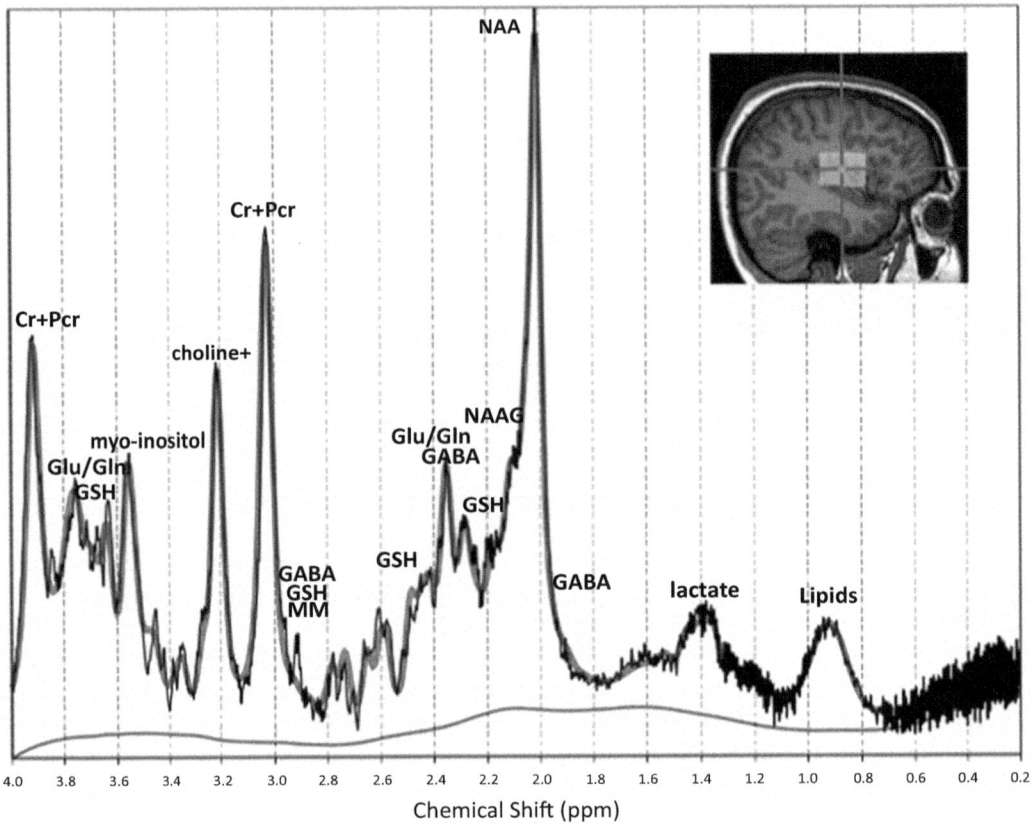

FIGURE 16.1 The ^{1}H MRS spectrum from a normal adult left temporal lobe at 3 T. The y-axis represents the detected concentration or intensity of the metabolite in moles per liter of tissue, or millimolar (mM). The x-axis is the frequency chemical shift in parts per million (ppm), upon which metabolites are specified. Note: choline +, total Choline; Cr + PCr, creatine + phosphocreatine; Gln, glutamine; Glu, glutamate; GSH, glutathione; MM, macromolecules; NAA, N-acetyl-aspartyl; NAAG, NAA-glutamic acid. Source: *Figure 1 from Ford TC, Crewther DP. A comprehensive review of the (1)H-MRS metabolite spectrum in autism spectrum disorder. Front Mol Neurosci 2016;9(9):14.*

Cho. Cho is a precursor for phosphatidylcholine, a major component of cell membranes, and also a product of phospholipase-mediated membrane breakdown. Cho levels are increased in MS lesions, NAWM, and NAGM due to increased membrane synthesis or degradation, and MRS studies have demonstrated that Cho levels correlate with lesion load, disease activity, and disability in MS [9,16,20–24].

Creatine

Cr is a marker of energy metabolism and cellular homeostasis [3]. MRS measured Cr arises from signals of Cr and phosphocreatine, which are part of the cellular system that provides energy for processes such as membrane transport and neurotransmission.

Cr levels are relatively stable in normal CNS tissue, and this metabolite is often used (misguidedly) as a reference in MS studies. However, Cr may change under pathological conditions that affect energy metabolism and cellular function, or even potentially gliosis [25]. Previous work has demonstrated that Cr may decrease in MS due to energy depletion or tissue atrophy; however, increased Cr levels in MS lesions, NAWM, and NAGM are also reported, which may be due to increased energy demand or decreased adenosine triphosphate synthesis. Cr levels also correlate with MS disability and lesion volume [26].

Myo-inositol

mI is a metabolite that reflects glial cell activity and osmoregulation [3]. mI is synthesized by astrocytes and MRS detectable mI has main sources as: (1) a constituent of phosphoglycerides forming biomembranes and (2) an osmolyte that maintains cell volume and water balance in the brain [5,7]. An increase in mI reflects increased glial cell proliferation or activation, which is associated with gliosis and inflammation in MS. mI levels are increased in MS lesions, NAWM, and NAGM, and mI levels correlate with disease duration, lesion load, disability, and cognitive impairment in MS [9,18,26–29].

Glutamate and glutamine

Glu and Gln are metabolite markers of excitatory neurotransmission and the Glu–Gln cycle [3]. Glu is the most abundant metabolite in the brain that can be measured with ^{1}H MRS and is the main excitatory neurotransmitter in the CNS, while Gln is its precursor. The MRS signals from Glu and Gln are very similar and overlap in the MRS spectrum, and many studies will report Glx which represents the combined signals: Glx = Glu + Gln. Glu and Gln are involved in synaptic plasticity, learning, memory, and neurotoxicity. Changes in Glu and Gln reflect altered neuronal activity and Glu homeostasis, which may be associated with cognitive impairment and neurodegeneration in MS. Altered Glu and Gln levels in MS lesions, NAWM, and gray matter (GM) have been reported, correlate with disease activity, disability, and cognitive function, and may be due to changes in neuronal activity, synaptic plasticity, astrocytic function, excitotoxicity mechanisms, or Glu metabolism [9,27,28,30,31].

Lactate

Lac is a metabolic marker of anaerobic glycolysis and tissue hypoxia in the CNS [3]. Healthy brain typically exhibits low concentrations of Lac, which is produced by the breakdown of glucose under low oxygen conditions. The MRS visibility of the Lac signal can be increased by modifying data acquisition parameters such as the echo time (TE). An increase in Lac reflects increased metabolic demand or reduced oxygen supply leading to anaerobic glycolysis, which may be associated with inflammation, hypoxia, acute demyelination, energy failure, metabolic stress, and/or mitochondrial dysfunction in MS. Increased Lac levels have been reported in MS lesions [32–37].

Magnetic resonance spectroscopy data acquisition

MRS pulse sequences are designed to selectively excite and detect the signals from specific metabolites in a defined volume of interest or voxel within the brain. There are a variety of different MRS data acquisition methods available, and a few common approaches will be summarized below [3]. The main parameters that affect the performance of MRS pulse sequences are:

Echo time—TE determines the signal-to-noise ratio (SNR) and the spectral resolution of MRS. Shorter TE increases SNR but reduces spectral resolution. The optimal TE depends on the relaxation properties of the metabolite(s) of interest. For example, NAA has a longer T_2 relaxation time than Cho or Cr, so it can be detected at longer TE with better resolution [38].

Repetition time (TR)—TR determines the recovery of magnetization and the saturation effect of MRS. Shorter TR increases scan time efficiency but reduces signal intensity. The optimal TR depends on the T_1 relaxation properties of the metabolites of interest. For example, NAA has a longer T_1 relaxation time than Cho or Cr, so it requires a longer TR to fully recover its magnetization [39].

Bandwidth (BW)—BW determines the sampling rate and the aliasing effect of MRS. Higher BW increases sampling rate but reduces SNR. The optimal BW depends on the spectral width and the chemical shift dispersion of the metabolites of interest. For example, Glu and Gln have a larger spectral width and a higher chemical shift dispersion than NAA, Cho, or Cr, so they require a higher BW to avoid aliasing and overlap [40].

Number of averages (NA)—NA determines the SNR and the scan time of MRS. Higher NA increases SNR but prolongs scan time. The optimal NA depends on the desired SNR and the available scan time for MRS.

Spatial localization—Spatial localization methods can be classified into single-voxel spectroscopy and multivoxel spectroscopy where spectra are acquired from multiple voxels within a slice or a volume of the brain. Single-voxel spectroscopy has higher SNR and spectral resolution than multivoxel spectroscopy, but it has lower spatial coverage. The choice of spatial localization method depends on the research question and/or the clinical application of MRS.

Shimming method—Manual shimming requires users to adjust the shim currents based on visual inspection of field maps or spectra while automatic shimming uses algorithms to calculate and apply the optimal shim currents based on field maps or spectra. The use of higher order shims can overcome magnetic field inhomogeneities that cannot be corrected for by linear shims [41]. Shimming methods are crucial for improving the quality and reliability of MRS.

Point resolved spectroscopy (PRESS)—PRESS is the most commonly used pulse sequence for single voxel MS MRS studies. With the PRESS acquisition method, signal is acquired at the intersection of the three orthogonal planes resulting from a 90 degree excitation pulse and two perpendicular 180 degree refocusing pulses to generate a spin echo [3]. PRESS has high SNR and spectral resolution, but it is sensitive to magnetic field inhomogeneities and chemical shift displacement errors, which can cause spectral distortions and spatial misregistration.

Stimulated echo acquisition mode (STEAM)—STEAM is another single-voxel MRS technique that uses three 90 degree pulses to select a volume of interest within the brain [3]. The volume of interest is defined by three orthogonal slice-selective gradients that are applied before each pulse. STEAM has lower SNR than PRESS, but is also less sensitive to magnetic field inhomogeneities and chemical shift displacement errors than PRESS, which results in less spectral distortions and spatial misregistration. However, STEAM uses a longer TE than PRESS, which reduces the detection of short T_2 metabolites such as Cr and Cho.

Semi−localization by adiabatic selective refocusing (semi-LASER)—Semi-LASER is a single-voxel MRS technique that uses two adiabatic 180 degree RF pulses instead of hard 180 degree RF pulses to refocus the spins within the volume of interest [3]. The volume of interest is defined by three orthogonal slice-selective gradients that are applied before each pulse. Semi-LASER has similar SNR as PRESS, but is less sensitive to magnetic field inhomogeneities and chemical shift displacement errors compared to PRESS and STEAM.

Two-dimensional chemical shift imaging (2D-CSI)—2D-CSI is a multivoxel approach where phase encoding gradients are used to acquire spectra from multiple voxels within a slice after an excitation pulse [3]. The slice thickness is defined by a slice-selective gradient that is applied before the excitation pulse. 2D-CSI has high spatial coverage and can provide information on metabolite distribution across the brain; however, limitations include long acquisition time, low spatial resolution, low SNR per voxel and sensitivity to motion artifacts.

Spectral editing is an MRS data acquisition method that aims to improve the measurement of metabolites with low concentrations and/or who's signals overlap with signals from other metabolites [3]. Common spectral editing techniques that enhance or suppress signals from certain metabolites include using different TE values or frequency-selective pulses. Metabolites such as gamma-aminobutyric acid (GABA), glutathione, Lac, and Glu can be probed using spectral editing techniques [42,43]; however, limitations include increases in acquisition time, lower SNR, and additional analysis.

Magnetic resonance spectroscopy data analysis

MRS data analysis is the process of extracting quantitative information about the metabolite ratios and/or concentrations from the acquired MRS spectra. Several steps are involved in MRS data analysis, including preprocessing, fitting, and postprocessing [3]. Different methods and software tools have been developed to perform MRS data analysis, each with their own advantages and limitations. Some of the most common methods and tools include:

Linear combination of model spectra (LCModel) performs MRS data analysis by fitting the measured spectrum to a linear combination of basis set spectra that reflect the expected metabolites in the tissue [44,45]. The basis set spectra used can either be simulated or obtained from in vitro measurements for each biochemical. LCModel estimates the metabolite concentrations, standard deviation, and also produces measures of fit quality like SNR, linewidth, and Cramer Rao lower bounds. LCModel is considered robust, reliable, and accurate, is the most commonly used MRS analysis method, and is now freely available. However, like all methods that use a basis set, LCModel is sensitive to the choice of basis set used for analysis.

Totally automatic robust quantitation in NMR (Tarquin) is an open-source MRS data analysis software tool that uses a time-domain nonlinear least squares algorithm based on a metabolite basis set [46,47]. Like other approaches, Tarquin can estimate the metabolite concentrations, uncertainties, and spectral quality measures such as Cramer Rao lower bounds and SNR.

FMRIB Software Library for MRS (FSL for MRS) is a newly developed MRS data analysis method that is integrated into the FSL neuroimaging software package [48]. FSL for MRS uses a Bayesian inference framework to fit the MRS spectra to a model that incorporates metabolite signals derived from a basis set of model spectra, chemical shifts, spectrum baseline, and noise. Metabolite concentrations and their uncertainties, as well as spectral quality measures like Cramer Rao lower bounds and SNR can be estimated. The integration of FSL for MRS into FSL neuroimaging software can be helpful for voxel segmentation analysis and image registration leading to absolute quantification.

Absolute quantification of metabolite concentrations is an MRS data analysis approach that converts metabolite ratios or relative concentrations obtained from spectral fitting to "absolute" concentration units expressed in mM or mol/g [3,49,50]. To incorporate absolute quantification analysis into an MRS analysis pipeline, additional information is needed including water content, tissue relaxation times, tissue density, tissue composition, coil loading, and receiver gain. Corrections for edema in MS longitudinal studies may be helpful [51]. Absolute quantification should lead to more comparable results across different scanners and studies, but accurate and reliable estimation of the various correction factors is needed, which requires additional data collection and analysis.

Factors affecting reproducibility of magnetic resonance spectroscopy and consensus protcols

Good reproducibility of MRS data and analysis is key for ensuring the validity and reliability of MS MRS findings, and crucial when comparing data over time or from different sources. Various factors can affect MRS reproducibility including scanner equipment, data acquisition, data analysis, and true biological variation. Important potential sources of variability in MRS studies to be aware of include:

Participant: age, sex, gender, disease state, therapy status, hydration levels, head motion, physiological noise.

Scanner: magnetic field inhomogeneity, magnetic field strength, coil, gradient strength, radiofrequency power, receiver gain, shims, drift.

Sequence: pulse sequence (type, scanning parameters), voxel size, voxel location, number of acquisitions, water suppression approach, shim options.

Analysis: preprocessing, choice of analysis method, basis set, correction factors, quality control criteria, quantification method chosen (if applicable).

To address the impact of the many possible factors that contribute to sources of variability in MRS studies, and provide guidelines and recommendations for best MRS data acquisition and analysis practices, efforts have been made to establish consensus protocols for brain MRS. These reports provide recommendations on choice of pulse sequences, acquisition parameters, spatial localization methods, spectral editing approaches, quality control measures and criteria, and analysis methods [52–62]. Numerous studies have attempted to characterize the reproducibility of MRS data in controls and MS [63–66].

Magnetic resonance spectroscopy findings in multiple sclerosis lesions

MS lesions are visible with proton density, and T_1 and T_2 weighted MRI scans, but the underlying biochemical and microstructural abnormalities present in these lesions cannot be ascertained with conventional MRI. MRS can provide complementary information to MRI by measuring the concentrations of various metabolites that reflect pathological processes and tissue damage, thereby yielding information about the heterogeneity and evolution of MS lesions.

The most consistent MRS finding in MS lesions is reduced NAA, which reflects neuronal and axonal injury, dysfunction, and/or loss. NAA decreases are observed in all types of MS lesions, with NAA levels being lower in chronic lesions than acute lesions [9,10]. NAA reductions are also more pronounced in T_1-hypointense lesions (also known as black holes) than in T_1-isointense lesions, suggestive of more severe axonal damage in the black hole subtype of lesion [67]. The NAA pattern in the acute phase of lesion development is highly variable, ranging from almost no change to significant decreases [68]. Initial NAA decreases in acute lesions may partly recover on timescale of weeks to months, or remain persistently low [21,34,69], and in some reports, acute lesions showed similar NAA levels pre- and post-formation, suggesting no long-term axonal damage (Fig. 16.2) [70]. Cervical cord MRS demonstrates reduced NAA/Cr in diffuse and focal lesions, which correlated with clinical progression during 2-year followup [71]. The severity of NAA reduction may also differ between MS subtypes, for example, NAA/Cr is significantly decreased in nonenhancing lesions in primary progressive multiple sclerosis (PPMS) compared to relapsing–remitting multiple sclerosis (RRMS) [72].

Increases in Cho are also commonly reported in MS lesions, reflecting increased membrane turnover from active myelin breakdown or remyelination. Cho increase is observed in both acute and chronic lesions [9,20,21], with higher levels in acute versus chronic plaques [9]. Elevations in Cho have also been reported in longitudinal studies of NAWM areas prior to lesion formation [21,30]. Case reports demonstrated increased Cho in Balo's Concentric Sclerosis, a rare subvariant of MS [33,73]. A positive correlation between longitudinal changes in Cho and lesion volume in moderately hypointense lesions could suggest that lesion size is mediated by ongoing membrane degradation [70]. Elevated Cho/Cr in RRMS cervical cord lesions, relative to matched controls, is also documented in the literature [74].

As mentioned earlier, Cr is typically used as a reference in MRS studies as it's thought to remain relatively stable in neurological disease like MS. However, abnormalities in Cr have been reported in MS, including increases in Cr in RRMS prior to new lesion formation [30,70], high Cr in focal and diffuse cervical cord lesions [71], and elevated Cr in secondary progressive multiple sclerosis (SPMS) lesions, suggestive of increased metabolic rate even in chronic lesions [20]. The severity of Cr increases may also vary between MS

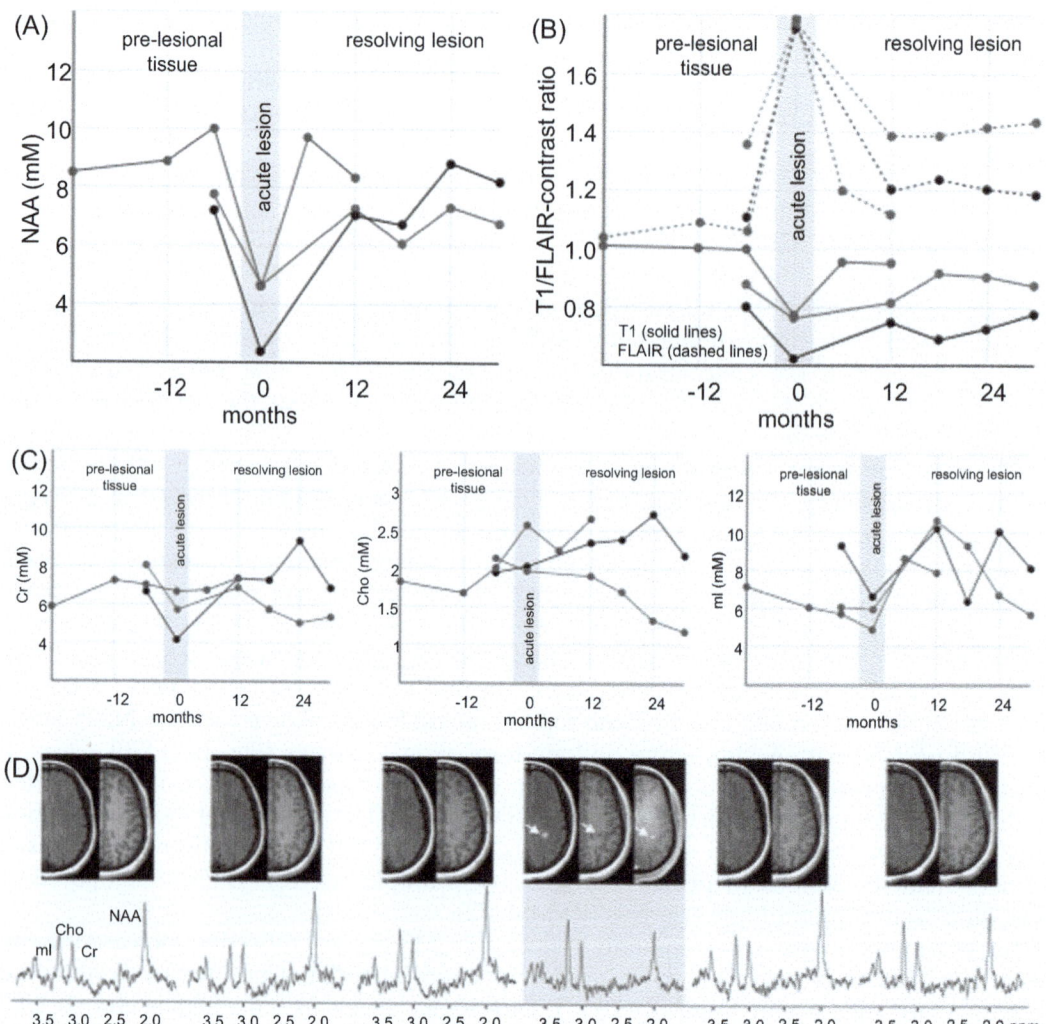

FIGURE 16.2 ^{1}H MRSI and MRI changes in resolving lesions. (A) Changes in NAA concentration and in (B) T_1/FLAIR-contrast ratios in the three tissues (green, blue, and black lines, corresponding to three lesions) undergoing transformation from a prelesional state to acute lesion (gray background) to a resolved lesion. (C) The corresponding changes in the concentrations of Cr, Cho, and mI. (D) Top: The FLAIR and MP-RAGE from which the values of the green lines in (B) were acquired. Note the appearance of the lesion at the fourth timepoint (white arrow), with contrast enhancement shown on the post-gadolinium chelate T_1-weighted image. Bottom: The spectra from which the metabolite values of the green lines in (A) and (C) were acquired. Note the similar NAA levels pre- and post-lesion formation. *Source: Figure 4 from Kirov II, Liu S, Tal A, Wu WE, Davitz MS, Babb JS, et al. Proton MR spectroscopy of lesion evolution in multiple sclerosis: steady-state metabolism and its relationship to conventional imaging. Hum Brain Mapp 2017;38(8):4047–63.*

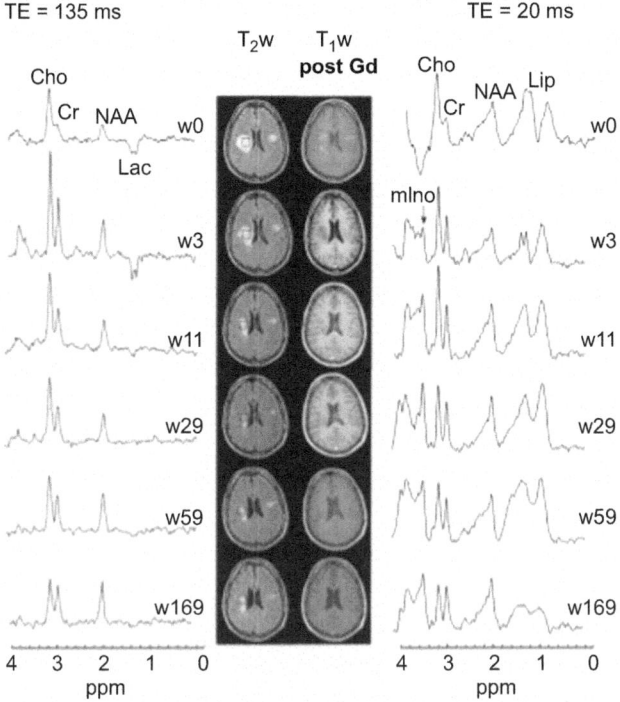

FIGURE 16.3 Metabolic pattern evolution occurring within the focal inflammatory brain lesion during the 3-year followup. Squares on the fluid-attenuated inversion recovery images represent location of the spectroscopic single voxel ($20 \times 20 \times 15$ mm^3) acquired at TE = 135 ms (left panel) and TE = 20 ms (right panel). At W0, low levels of NAA and creatine (Cr) and high levels of lactate (Lac) and lipids (Lip) were observed while myoinositol (mIno) appeared at W3 and increased over time. High lipid levels were present at the end of the followup while NAA and creatine recovered progressively over the whole period of the followup. T_1w post-Gd images illustrate that gadolinium enhancement was only present at W0 while lactate remained present until W29. Source: With permission from Figure 2 in Zaaraoui W, Rico A, Audoin B, Reuter F, Malikova I, Soulier E, et al. Unfolding the long-term pathophysiological processes following an acute inflammatory demyelinating lesion of multiple sclerosis. Magn Reson Imaging 2010;28(4):477–86.

subtypes, for example Cr was significantly increased in nonenhancing lesions in PPMS compared to RRMS. [72] Cr can also be reduced at lesion first appearance (Fig. 16.3) [35].

Elevations in mI, which reflects increased glial activation and/or proliferation arising from inflammation or gliosis, is also observed in MS lesions [27], including acute plaques [9]. In RRMS, chronic, moderately hypointense lesions, showed an increase in mI over 3 years, suggesting accumulating astrogliosis [70]. mI is also elevated in SPMS chronic lesions [20], and compared with isointense T_1 lesions, hypointense T_1 lesions exhibit higher mI [67]. Over 2 years, a significant increase in mI and mI/Cr was observed in both RRMS and SPMS white matter lesions, independently from the course of the disease, suggesting ongoing astrogliosis in nonenhancing, visually stable lesions [75]. Increased mI/Cr in RRMS cervical spinal cord lesions is also reported [74], while in neuromyelitis optica (NMO), lower mI/Cr is consistently found within the NMO cervical cord lesions when compared with healthy controls and MS, suggesting quantification of mI may help distinguish NMO from MS [76]. mI may also contribute to differentiating MS lesions from age-related white matter hyperintensities, as MS lesions show elevated mI while age-related lesions do not [77].

Apart from NAA, Cho, Cr, and mI, other metabolites in MS lesions are less frequently reported in the literature. Abnormalities in Glu and Gln (or Glx) are sometimes documented. In one study, Glu increased 12 and 6 months before new lesions appeared in RRMS [30], and another report found Glu concentrations are elevated in acute but not chronic lesions [9]. However, decreases in Glx/Cr are also described in RRMS lesions [27]. The presence of a lipid signal, signifying myelin breakdown, is evident in lesions, and in prelesional areas in MS as well as Balo's Concentric Sclerosis (Fig. 16.2) [13,21,35,78–80]. Such lipid peaks may remain elevated for 4–8 months in MS enhancing lesions [34]. Lactate is also found in MS plaques (Fig. 16.2) and Balo's Concentric Sclerosis [32,33], with variable timing after initial gadolinium enhancement [34–36], and less frequently in chronic lesions [37].

Magnetic resonance spectroscopy findings in multiple sclerosis normal-appearing white matter and gray matter

Many studies have described metabolic changes not only in focal MS lesions but also in NAWM and NAGM. In general, metabolite abnormalities in MS NAWM and NAGM mirror findings seen in MS lesions, but to a lesser degree of damage. Consequently, MRS can also provide valuable insight into the more global and diffuse disease processes in MS beyond what is visible on conventional MRI.

Akin to lesions, the most consistent finding from MRS studies of nonlesional tissue is a reduction of NAA in NAWM and NAGM in MS, as well as clinically isolated syndrome (CIS) [11–17]. NAWM NAA levels can vary between disease subtype [81] and may correlate with measures of disability, for example, NAA reduction correlates with motor deficit score [15], and in SPMS patients, NAWM NAA is negatively correlated with clinical disability, sometimes independent of the duration of the disease [13,18]. Longitudinal studies reveal that individuals with SPMS show significant reduction of tNAA in NAWM over 2 years, indicating possible diffuse neuroaxonal loss during this timeframe [75].

MRS measured Cho in MS NAWM and NAGM also consistently shows abnormal levels, with both increases and decreases reported. Cho increases [16,22–24] can be interpreted as ongoing membrane turnover (de- and remyelination) and Cho/Cr of NAWM is weakly correlated with expanded disability status scale (EDSS) [24]. Cho can also be decreased in NAWM of patients with and without lesions, and cortical gray matter [18,26,82,83]. Understanding the MS mechanisms responsible for both increases and decreases in Cho in nonlesional tissue requires further study.

Cr levels are also abnormal in MS NAWM and NAGM, with increases most commonly reported, including evidence of higher Cr levels in PPMS relative to RRMS, which is consistent with the notion that progressive disability in PPMS reflects increased gliosis

[22,84,85,23]. An increase in Cr can also be observed in CIS NAWM, which may be of prognostic significance [86]. However, reductions in Cr are also reported in MS NAWM and NAGM [81], and cortical NAWM Cr levels are correlated with multiple sclerosis functional composite score (MSFC) [26]. Like Cho, understanding the impact of MS on CNS Cr levels is still an active area of research.

Increases in NAWM and NAGM mI are also reported in MS and sometimes CIS [17,86]. EDSS and MSFC both show a positive correlation with mI concentration and the relationship between clinical impairment and mI may indicate that glial proliferation relates to function [18,26]. mI is negatively correlated with brain parenchymal fraction [28], and the ratio of mI:NAA in NAWM has consistent predictive power on brain atrophy and neurological disability evolution as measured by EDSS and MSFC [29].

Abnormalities in other metabolites in MS MRS studies are also documented. Glu concentrations are elevated in NAWM of RRMS, SPMS, and PPMS [9], and Glx is related to Multiple Sclerosis Severity Score, independent of number of lesions in the patient [28]. Furthermore, baseline Glu of NAWM is predictive of accelerated longitudinal decline in NAGM NAA, and an increase in NAWM Glu/NAA is associated with a loss of brain volume and MSFC score [31], providing further evidence that Glu and Glx may be important markers for pathology in nonlesional white matter in MS. Finally, strong lipid resonances, even in the absence of lesions, can be observed in both gray and white matter of PPMS [11].

Magnetic resonance spectroscopy and multiple sclerosis cognitive impairment

The ability of MRS to measure changes in brain metabolites offers a unique in vivo opportunity to assess functional changes happening in the CNS of people living with MS. While the impact of MS on motor and sensory function is well recognize and studied, cognitive impairment has received less attention in clinical and basic science research. The historical lack of focus on MS cognition is despite the fact that cognitive impairment is a common and disabling symptom of MS, affecting up to 70% of patients. Memory, processing speed, attention, executive function, and language can all be affected, leading to significant impacts on quality of life, social functioning, and ability to work [87].

Various metabolites show relationships with different cognitive measures in MS. For example, axonal damage, assessed using NAA/Cr of the right locus coeruleus correlates with selective attention impairment, measured by means of a dichotic listening paradigm, in early stage RRMS [88]. NAA/Cr correlates with a global cognitive measure and was able to distinguish between cognitively impaired and unimpaired early MS patients [19]. Left periventricular NAA concentration correlates with performance on verbal learning and memory, while right periventricular NAA correlates with mathematical performance (Fig. 16.4) [89]. Diverse groups of MS patients demonstrate correlations between NAA, mI, Cho and Glu/NAA, and a cognitive measure of auditory information processing speed and flexibility, as well as calculation ability in both NAWM and NAGM, suggesting metabolic markers of neuroaxonal integrity, neuronal activity, and astrogliosis in NAWM and membrane turnover in NAGM could be potential biomarkers of cognitive disability in progressive MS [31,90].

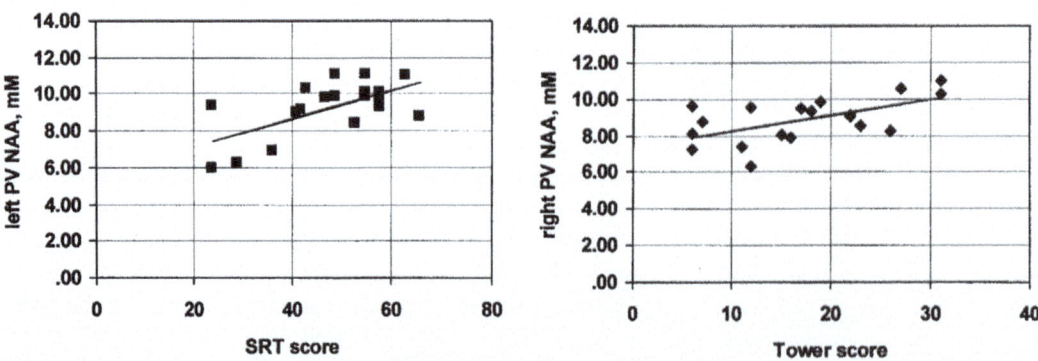

FIGURE 16.4 Left: Regression of selective reminding test (SRT, verbal memory) with measurements of left periventricular NAA from all the data. These data are significantly ($P < .005$) correlated with a Pearson correlation coefficient of $r = 0.63$. Right: Regression of performance on the Tower of Hanoi test (visuospatial conceptual planning) with measurements of right periventricular NAA. These data are significantly ($P < .02$) correlated with a Pearson correlation coefficient of $r = 0.58$. Source: *With permission from Figures 2 and 3 in Pan JW, Krupp LB, Elkins LE, Coyle PK. Cognitive dysfunction lateralizes with NAA in multiple sclerosis. Appl Neuropsychol. 2001;8(3):155–60.*

In RRMS, Cho/Cr- and mI/Cr of both the thalamus and hippocampus correlate with lower scores on the symbol digit modalities test, but not with EDSS, further highlighting the importance of studying subcortical gray matter in MS cognition research [91]. Finally, cognitive dysfunction is present very early in MS, including in CIS, where numerous correlations between NAA, Cho and mI, and a variety of cognitive domains (sustained attention, processing speed, visual scanning, motor speed, verbal learning), anxiety, and depression have also been reported [92]. Other metabolites have also been studied in relation to MS cognition. For example, lower GABA levels correlated with worse cognitive performance in patients with RRMS [93], and glycerophosphocholine correlated positively with multiple fatigue domains, tyrosine showed positive correlation with modified fatigue impact scale and cognitive fatigue was negatively correlated with total glutathione, while no correlations were found between lesion load or brain volumes and fatigue scores [94]. However, other studies report no relationship between MRS markers like NAA and mI with measures of fatigue or cognition in various subtypes of MS, quantified using the fatigue severity scale, and various tests assessing processing speed, verbal learning, delayed memory, verbal fluency, and executive function [95–97], so continued research is warranted to better understand the neurochemical changes that underlie cognitive impairment in MS.

Magnetic resonance spectroscopy in multiple sclerosis clinical trials and disease modifying therapy evaluation

MRS is well suited for use in evaluating MS disease-modifying therapies (DMT) that aim to prevent the development of new lesions, reduce the frequency and severity of MS relapses, slow down disability progression, and promote repair. By quantifying brain metabolites, MRS can potentially provide valuable information about how DMTs impact metabolic changes that occur with MS treatment. A number of studies and clinical trials have used MRS to evaluate safety and efficacy of DMTs, as well as explore their mechanism of action, and recommendations for the inclusion of MRS in trials have been published [98]. Some representative examples include studies which show increases or stability of NAA in MS lesions, NAWM, and/or NAGM with interferon treatment [99–103], increases of NAA in MS and CIS NAWM with glatiramer acetate [104–106], increases of NAA, Cr, and Glu in MS lesional white matter with natalizumab [107], stability of NAA, Cr, Cho, mI, and Glx in MS NAWM with alemtuzumab [108], and stability of Cr and Cho, and reductions in mI in MS NAWM with ocrelizumab [109].

In summary, when considering MRS in the context of DMTs, MRS can complement other imaging modalities, like MRI and fMRI, by providing additional information about the metabolic aspects of brain function in response to MS therapy, as well potentially stratifying patients for trial selection. More natural history studies are needed to characterize longitudinal MRS changes in untreated MS patients, so that any metabolite stability or change can indeed be attributed to DMT usage with confidence.

Limitations of magnetic resonance spectroscopy studies in multiple sclerosis

Although MRS has the potential to provide unique and quantitative information about metabolic changes occurring in MS CNS tissue, a number of challenges and limitations need to be kept in mind when conducting and interpreting MRS findings in the context of MS. The limited spatial resolution of MRS can pose challenges for a number of reasons including partial volume effects where an MS voxel may contain more than one type of tissue (CSF, WM, GM, lesion), so spectra from pure lesional tissue (or any one tissue) may be difficult to obtain. Reducing the size of the voxel enables one to capture signal from smaller regions; however, this will also reduce the SNR of the spectra; thus there is a constant trade-off between spatial resolution and SNR in MRS studies. In addition, the MRS signals being measured are ~10,000 times smaller than the signal from water, so excellent water suppression techniques, and in general careful optimization and standardization of data acquisition and analysis methods are needed to accurately measure metabolite signals. Noise and artifacts arising from motion, magnetic field inhomogeneity, eddy currents, and poor water suppression will all impact spectra quality, and thus care is needed to minimize these effects and improve reproducibility and reliability. When evaluating MRS studies in the literature it is also important to keep in mind that technical factors including field strength, hardware (i.e., coil type, gradients), pulse sequences, acquisition parameters like TE and TR, and choice of analysis software and basis sets will all potentially impact metabolite quantification, and therefore direct comparisons between studies may not be appropriate or valid.

Future directions in magnetic resonance spectroscopy for multiple sclerosis research and clinical care

MRS is clearly established as an MR method that is capable of providing valuable information about the metabolic changes that occur in MS. Despite some of the above-mentioned challenges with MRS studies, metabolite quantification can complement other MR approaches to yield biochemical information that is not otherwise available. Strategies are being developed which will help alleviate some of these limitations, including more studies at higher field, the development of new shim elements, the development of prospective frequency and motion corrections, and the increasing adoption of semi-LASER [3]. "High" field typically refers to magnetic field strengths above 3 T, where advantages such as increased SNR, increased spectral resolution, reduced scan time, better sensitivity to low concentration metabolites can be realized and visibility of metabolites not discernable at lower fields like glutathione, taurine, and glycine made possible. However, with higher field, also comes greater field inhomogeneity, increased susceptibility artifacts and more complicated hardware, corrections for which are all active areas of research. More sophisticated techniques are increasing in popularity such as 31-phosphorus MRS, which can probe mitochondrial function and energy metabolism [110], and diffusion-MRS which measures metabolite diffusion to yield information about the microstructural environment and organization of metabolites in the CNS, like cell size, shape, density, and intracellular/extracellular distribution [111,112]. Alongside such technical advances, the creation of normative atlases will be needed such that abnormalities can be highlighted in a single person (as opposed to group-level differences), thereby moving MRS closer to daily clinical use for individual patient management.

Myelin water fraction

Overview

Myelin damage is the pathological hallmark of MS, leading to a wide range of symptoms as nerve conduction velocity is diminished and axons lose trophic support. Demyelination, in particular, is a pathology that may be reversible with naturally occurring remyelination as well as the promise of remyelination therapies [113–115]. Therefore techniques that can quantify myelin in vivo are particularly useful when monitoring MS.

The aim of this section is to review the current state of knowledge on myelin water fraction (MWF) in MS. We will discuss the biological basis of myelin water, measurement of MWF, human and animal MWF validation studies, MWF findings in lesional and nonlesional MS tissue, the relationship between MWF findings and clinical measures in MS, the use of MWF in MS clinical trials, the limitations of MWF studies in MS, and future directions for MWF research and clinical applications in MS.

What is myelin water?

Myelin water imaging (MWI) is an advanced MRI method that uses the difference in T_2 relaxation time of water in various environments to extract the signal from water located between myelin bilayers (Fig. 16.5) [116,117]. This water pool, labeled myelin water, has a shorter T_2 ($\sim 10-20$ ms) than other water due to its close interaction with the macromolecules of myelin. To quantify the myelin water, the ratio of myelin water to all the water within a voxel is calculated and called the MWF.

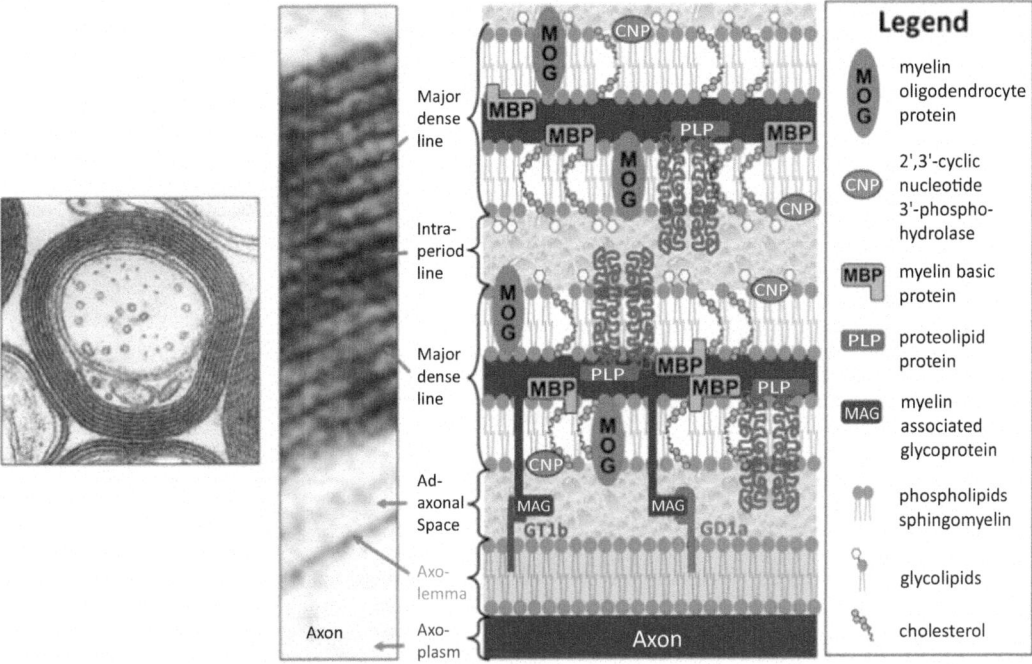

FIGURE 16.5 Electron micrograph of myelinated central nervous system (CNS) tissue at low and high magnifications depicting the major dense line, which represents the fusion of the cytoplasmic aspects of the oligodendrocyte cell membrane, and the intraperiod line, a potential extracellular space formed by the apposition of the extracellular faces of adjacent oligodendrocyte cell membranes. As shown in the accompanying schematic, the intraperiod line forms a restricted water reservoir, and, thus, is thought to give rise to the short-T_2 component, the signal of which can be displayed anatomically as the myelin water map (see Figs. 16.6-16.8). The oligodendrocyte cell membrane is a bilayer of lipids in which are embedded the major myelin proteins, which include myelin basic protein (MBP), proteolipid protein (PLP), 2′,3′-cyclic nucleotide 3′-phosphodiesterase (CNP), myelin oligodendrocyte protein (MOG), and myelin-associated glycoprotein (MAG). Note, however, that on the inner aspect of the myelin sheath MAG is restricted to the membrane adjacent to the adaxonal space, which it spans to bind the myelin sheath to its axolemmal ganglioside receptors, GD1a and GT1b. The exact position of some of the components of myelin shown in this schematic has not been determined. Source: *Low magnification (left): adapted from Figures 4–7, originally by Dr. W.T. Norton and Dr. C. S. Raine in Morell P, Quarles RH. Myelin formation, structure, and biochemistry. In Siegel GJ, Agranoff BW, Albers RW, Fisher SK, Uhler MD, editors. Basic neurochemistry. 6th ed. Philadelphia: Lippincott-Raven; 1999. ISBN 0-397-51820-X with permission; high magnification (middle): adapted from Peters A, Palay SL, Webster H deF. Fine structure of the nervous system: the cells and their processes. 1st ed. p. 89, Figure 33, New York: Paul B. Hoeber Inc; 1970, with permission from Dr. Alan Peters. From Laule C, Moore GRW. Myelin water imaging to detect demyelination and remyelination and its validation in pathology. Brain Pathol 2018;28(5):750–64.*

Measurement of myelin water fraction

The first in vivo human MWF measurement was published in 1994 and required ~25 minutes to acquire a single 5 mm thick slice of data [116]. The Carr-Purcell-Meiboom-Gill (CPMG) sequence used in this pioneering work consisted of the collection of images from 32 spin echoes at 10 ms spacing flanked by descending crusher gradients. The signal from each echo was fit using a regularized nonnegative least-squares (NNLS) method to

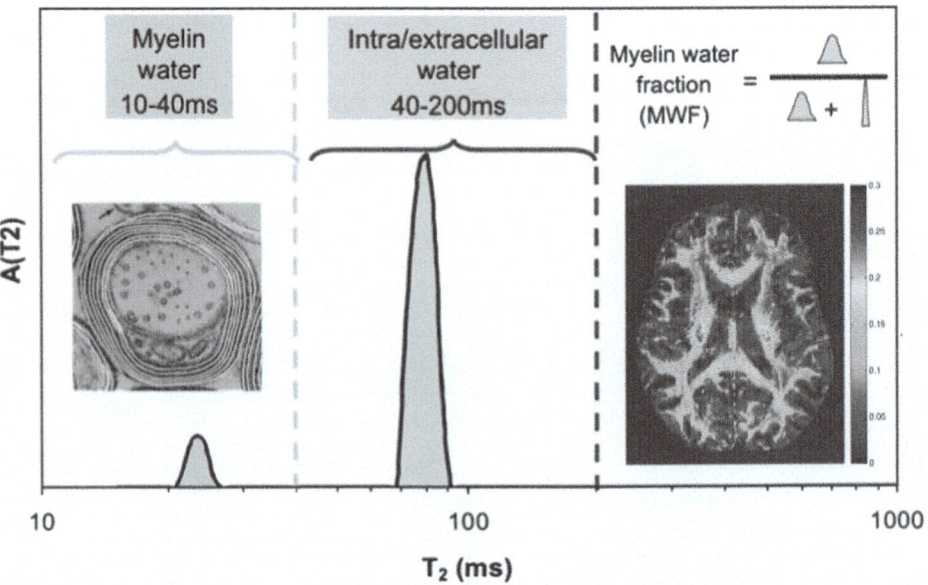

FIGURE 16.6 T_2 distribution from normal white matter and corresponding myelin water fraction image. Source: *Adapted from MacKay AL, Laule C. Magnetic resonance of myelin water: an in vivo marker for myelin. Brain Plast 2016;2(1):71–91.*

extract multiple exponential components; each component is associated with a separate water environment [118]. In normal brain and spinal cord tissue, three exponential T_2 components can be found; the component with $T_2 > 1$ s is attributed to cerebrospinal fluid, the component with T_2 around 80 ms is attributed to intra/extracellular water and the shortest T_2 component ($T_2 \sim 20$ ms) is assigned to myelin water. Additional flip angle optimization was later added to the NNLS analysis method to account for stimulated echoes [119]. When each imaging voxel is analyzed separately, a MWF image can be created (Fig. 16.6). Publicly available analysis resources can be accessed using https://mriresearch.med.ubc.ca/news-projects/myelin-water-fraction/ or DECAES [120]. Within normal brain white matter, MWF from different tracts range by over a factor of 2 indicating that myelin content is variable across the white matter with small values in the frontal area and larger values in the posterior internal capsules [121].

Improvement in MWI sequences allowed for faster data collection which in turn permitted multislice acquisitions. For example, the addition of flanking gradient echoes to each spin echo (GRASE) resulted in a 3× faster acquisition without significant image distortion or artefacts [122,123]. Further acceleration can be obtained by using compressed sensing; multispin echo T_2 relaxation imaging with compressed sensing results in whole brain imaging within ∼5 minutes [124]. Other techniques such as T_2^*, multicomponent-driven equilibrium single pulse observation of T_1 and T_2 (mcDESPOT), FAST-T_2, and ViSTA, have advantages in speed and brain coverage; however, their correlation with myelin is less strong [125–132]. For comprehensive technical overviews of MWI the reader is referred to excellent reviews by Alonso-Ortiz et al. [128] and Lee et al. [129].

In terms of reproducibility, the coefficient of variation for intrasite MWF determined using CPMG, GRASE, and FAST-T_2 methods ranges from 3% to 6% in normal brain white matter [130,133–136], and ~6% in normal cervical cord [137]. For between sites, MWF coefficients of variation ranges from 3% to 5% although different scanning techniques will result in different reproducibility measures [133,134].

Myelin water fraction validation

MWF within MS brain and spinal cord tissue samples has been compared to histological staining for myelin using Luxol Fast Blue (LFB), a stain for phospholipids. Good visual correspondence between lesional areas with less myelin stain and areas of decreased MWF is seen at both 1.5 and 7 T (Fig. 16.7) [138–145]. Strong correlation coefficients between the LFB optical density and MWF are also found [139,141].

Animal models of demyelination and MS also shown good correlation between MWF and stained myelin. In guinea pigs with experimental autoimmune encephalomyelitis, regions of reduced MWF correspond to myelin loss detected with histopathology [146], and a remyelinating mouse model shows increasing MWF with remyelination [147]. In rat spinal cord a strong correlation between MWF and histological staining for myelin is also observed [148], and in a longitudinal study characterizing white matter damage in the rat spinal cord after dorsal column transection, changes in MWF correspond to changes in myelin detected with staining [149]. Recent work in fetal guinea pig demonstrates a correlation between mcDESPOT-derived myelin water and myelin basic protein staining [150].

Myelin water fraction in different multiple sclerosis tissues

Lesions were one of the initial focuses of MWF studies in MS. MS lesions show greatly reduced MWF when compared to healthy white matter and NAWM [121]. The reduction in

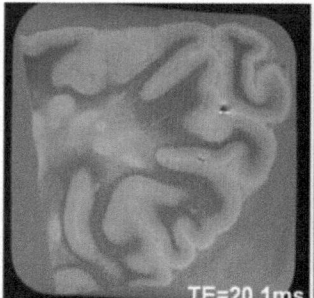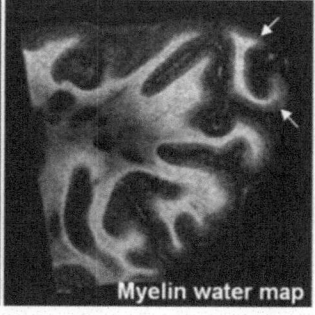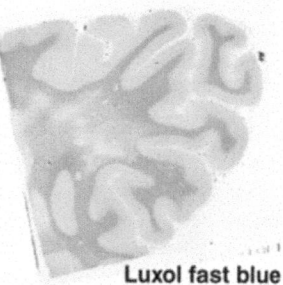

FIGURE 16.7 Example of a 7 T (A) TE = 20.1 ms image and (B) myelin water map, and (C) corresponding luxol fast blue histology image of the parieto-occipital region of an MS patient. A good qualitative correspondence is observed between the myelin water map and histology stain for myelin. The normal prominent myelination of the deeper cortical layers (arrows) is also visible on the myelin water image. Source: *With permission from Figure 2 in Laule C, Kozlowski P, Leung E, Li DK, Mackay AL, Moore GR. Myelin water imaging of multiple sclerosis at 7 T: correlations with histopathology. Neuroimage. 2008;40:1575–80.*

lesional MWF is variable with some lesions showing complete lack of myelin water signal whereas others have only slightly reduced MWF (Fig. 16.8) [151]. In a study of 35 MS participants, Laule et al. found that the MWF in lesions was reduced by on average 52% compared to healthy white matter [151]. In another study, contrast enhancing lesions showed a 26% reduction and T_2 lesions a 29% reduction in MWF compared to healthy white matter [152]. Finally, in a study comparing numerous different myelin sensitive measures from 105 MS participants, MWF was the best at separating T_2 lesions from NAWM [153]. When comparing between different lesion types, periventricular lesions show reduced MWF compared to juxtacortical lesions [154], contrast enhancing lesions have higher MWF than T_2 lesions and T_1 black holes [155], but T_2 lesions and T_1 black holes show no MWF difference [156]. Finally, chronic active lesions, as measured by a paramagnetic rim, have lower MWF than nonparamagnetic rim lesions [157]. The longitudinal progress of MWF in lesions is inconclusive as one study showed that contrast enhancing lesions have increased MWF over 6 months [158], whereas another study showed no change over 1 year [159]. T_2 lesions tend to show either no change or decreased MWF over time [158,160]. Overall, MWF can vary greatly within and between MS cohorts [161].

NAWM generally shows decreased MWF compared to healthy white matter and these decreases are diffuse throughout the tissue. Different studies have found a range of reduction (11%–37%) in NAWM MWF compared to healthy white matter [151,155,162,163], although there are also studies that find no difference between the MWF in NAWM and healthy white matter. Over the course of 6 months, NAWM MWF in MS is stable [164], but over 5 years, NAWM MWF decreases by ~8% whereas healthy control normal white matter shows no change [165].

Diffusely abnormal white matter (DAWM) are areas of diffusely increase signal intensity on proton density- and T_2-weighted images in 25%–50% of people living with CIS and MS [166–168]. MWF within DAWM is reduced compared to NAWM and healthy

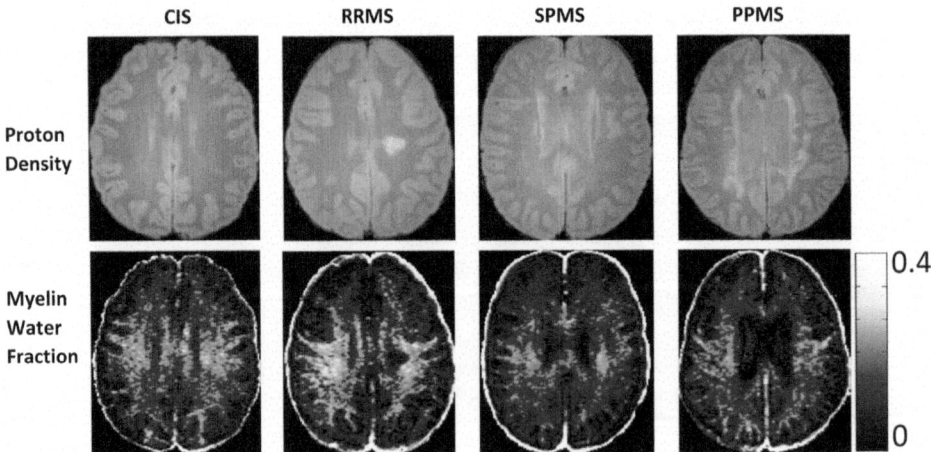

FIGURE 16.8 Example proton density and myelin water fraction images for each subtype of MS (clinically isolated syndrome [CIS], relapsing-remitting multiple sclerosis [RRMS], secondary progressive multiple sclerosis [SPMS], and primary progressive multiple sclerosis [PPMS]). *Source: Adapted from Figure 1 in Vavasour IM, Sun P, Graf C, Yik JT, Kolind SH, Li DK, et al. Characterization of multiple sclerosis neuroinflammation and neurodegeneration with relaxation and diffusion basis spectrum imaging. Mult Scler 2022;28(3):418–28..*

white matter [142,169,170]. Histological studies on myelin in areas of DAWM show a selective loss of myelin phospholipids with myelin proteins being relatively unaffected [142].

Spinal cord MWF in MS participants is lower than in healthy cord by 19% on average [162]. MWF decreases can be detected in both lesioned and nonlesioned cord segments indicating the presence of diffuse white matter myelin water loss even in the absence of lesions, and MWF continues to decrease over 1 year [171]. MWF in PPMS cervical cord also shows a 10.5% decrease over 2 years [172].

Myelin water fraction in multiple sclerosis subtypes

A number of studies show differences in white matter MWF between MS subtypes, CIS and healthy controls (HC) [152,161,170,173], although there are also studies showing no difference [170]. Progressive multiple sclerosis (PMS) generally has the lowest MWF, RRMS MWF is larger than PMS but lower than HC and CIS has a MWF in between RRMS and HC [161,173–175]. Early MS (including RRMS and CIS participants with <5 years disease duration) has a higher MWF than late MS (including RRMS and SPMS participants with >5 years disease duration) [152]. PPMS has a 6%–11% decrease in NAWM and 19% decrease in cord MWF compared to HC [162]. PPMS also shows a ∼10% decrease in cord MWF over 2 years [172]. Although Kitzler et al. found no significant difference between mean NAWM MWF in PMS and RRMS, PMS and CIS or CIS and HC, if the number of myelin deficient voxels (calculated as the number of voxels within the NAWM that had a z-score lower than −4 compared to a control MWF atlas) was used, CIS had more deficient voxels than HC and SPMS had more deficient voxels than RRMS [170]. An increased number of myelin deficient voxels in CIS also predicts conversion to MS [176]. Over 1 year, the mean MWF in NAWM did not change; however, the number of myelin deficient voxels did increase.

Myelin water fraction correlations with clinical measures

The significance of results from comparison of MWF and clinical measures is mixed. Some studies show no correlation between NAWM or lesion MWF and age, sex, EDSS, or disease duration [155]. Others find significant correlations between MWF and disease duration and EDSS [163,175,177,178], as well as the number of myelin deficient voxels and EDSS [170]. Lesion, whole brain and NAWM MWF is correlated with serum neurofilament light chain (a proposed marker of axonal damage) [154,179]. Within the cord, MWF and the myelin heterogeneity index (MHI = MWF standard deviation/mean MWF) correlates with nine hole peg test (a measure of manual dexterity) and timed 25 foot walk [180]. Finally, patterns of MWF decreases correlate with different aspects of cognitive impairment [181–183].

Myelin water fraction and disease modifying treatments

MWF has only been measured in a couple of studies involving disease modifying treatments. In a population of highly active MS treated with alemtuzumab, no change in MWF was detected [108]. In a substudy of the OPERA trial, the change in MWF over 2 years shows a favorable outcome in participants treated with ocrelizumab compared to participants treated with interferon β-1a (Fig. 16.9) [184]. Furthermore, after 4 years, participants originally randomized to interferon β-1a have lower lesion MWF than participants randomized to ocrelizumab [184].

Limitations

There are a few limitations associated with MWI that should be considered when collecting data, conducting analysis and/or developing interpretations about pathology. The measurement of the different water pools assumes that there is no exchange between them. There are studies that suggest that water exchanges fast enough that the measured MWF will be artificially smaller [185]; however, other studies suggest that exchange will have a minimal effect [186,187]. When myelin is damaged and broken down, myelin debris may remain within the tissue for some time before removal by macrophages. This myelin will not be functional but may be included within the MWF measurement [148]. Since myelin is attached to axons which are a very directional structure, orientation of the axon within the MRI scanner may effect the measure MWF; however, the possibility of this effect is still under investigation [135,188]. Iron may also affect the measured MWF since the presence of iron can decrease the T_2 relaxation time causing the protons to become part of the short T_2 component. In an experiment that removed the iron from white matter, the MWF decreased by 26%–28% [189], although it should be noted that removal of iron may disrupt the underlying tissue microstructure confounding the MWF interpretation. Imperfect flip angle estimates as well as noise can lead to underestimates of the MWF; optimization of regularized NNLS parameters can lead to more stable measurements [190]. Finally, it should be noted that the multiecho T_2 relaxation approach to MWF measurement does not work well within gray matter as the SNR is poor so other techniques are needed.

Future directions

Traditionally, the MWI sequence has been lengthy so new acquisition schemes for faster data collection are being developed. FAST-T_2 uses a smaller number of echoes and a spiral readout to speed up data collection to ~4 minutes [130]. mcDESPOT combines two fast gradient echo sequences with a limited number of flip angles to fit a model and extract MWF as well as several other potentially useful parameters [191]. The Constrained, Adaptive, Low-dimensional, Intrinsically Precise Reconstruction (CALIPR) framework harnesses the advancements in compressed sensing to collect multiecho T_2 data with ~15-fold faster acquisition times [192]. In addition to faster data acquisition, research is also being done to speed up MWF analysis. Machine learning is being used to obtain MWF images within under a minute as opposed to the several minutes to several hours that it usually takes [193,194]. Additional regularization has also been added to the NNLS method such as spatial regularization or joint sparsity constraint [195,196]. All of these methods have the advantage of being more robust to noise.

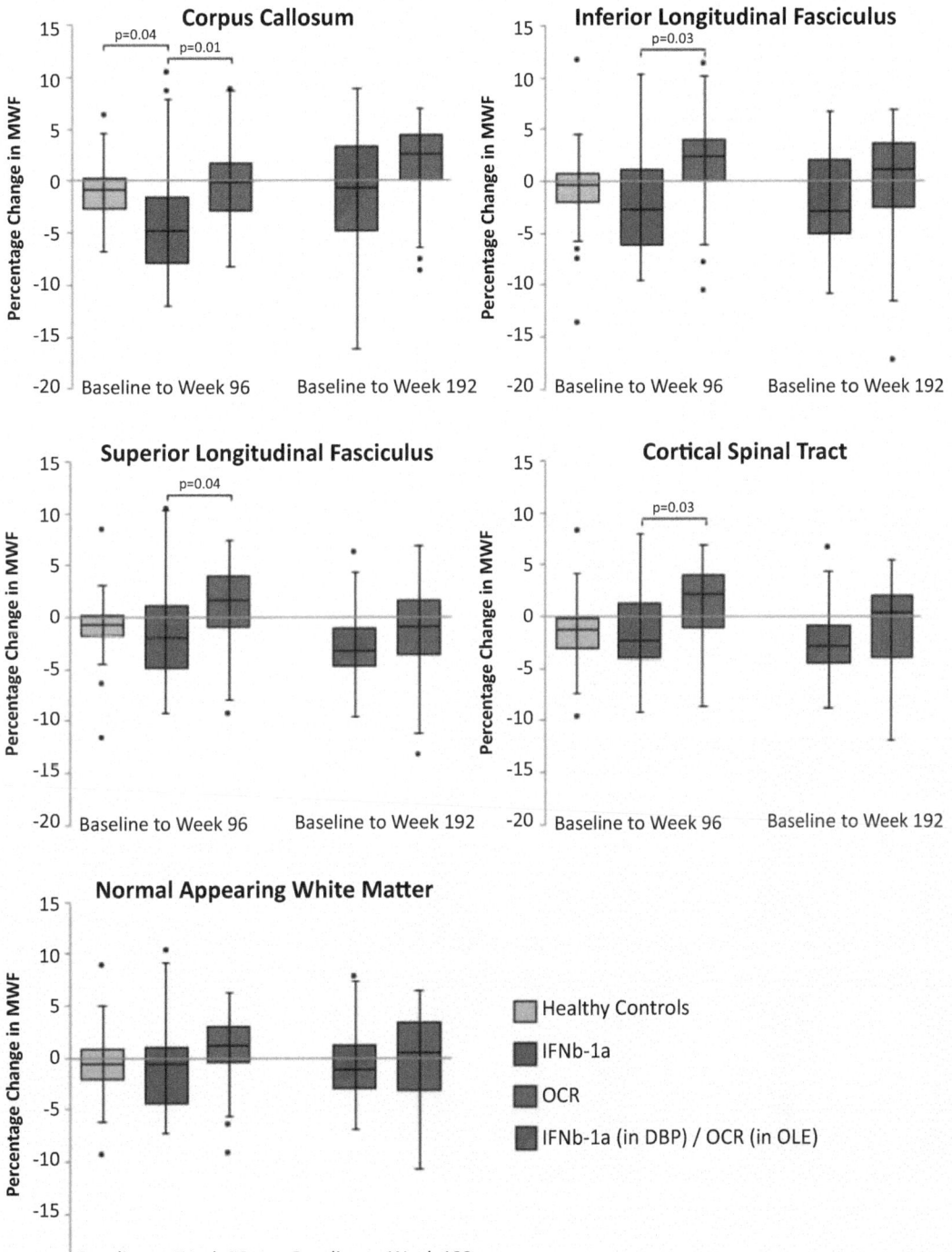

FIGURE 16.9 Boxplots showing the percent change in MWF in different white matter regions of interest from baseline to week 96 and baseline to week 192 for healthy controls and two treatment arms: (1) interferon beta 1a (IFNb-1a) from baseline to week 96 switching to ocrelizumab (OCR) from week 96 to week 192 and (2) OCR from baseline to week 192. For clarity, only p-values < 0.05 between treatment using an ANOVA are indicated. Source: in Kolind S, Abel S, Taylor C, Tam R, Laule C, Li DKB, et al. Myelin water imaging in relapsing multiple sclerosis treated with ocrelizumab and interferon beta-1a. Neuroimage Clin. 2022;35:103109.

To allow for even more data collection as well as harmonization of data across centers, consensus protocols will need to be developed and deployed on each of the vendor scanners. MWFs from the GRASE sequence have been compared on a Philips and Siemens scanner with quite reproducible results [134]. The newer CALIPR sequence is being developed for Philips, Siemens, and GE scanners [192]. Most of the current human in vivo MWF studies have been carried out at either 1.5 or 3 T; however, as more and more centers invest in higher field scanners, MWI may also move to 7 T and above. One study at 7 T showed that the results were similar to 3 T; however, there were a few extra considerations needed such as changing the myelin water cut-off window and using shorter TEs [197].

With more and more MWF data being collected, it is now possible to create healthy control MWF atlases. Patient data can then be compared to the atlas to obtain individualized results. Several MWF atlas are currently available (Fig. 16.10) and the continued expansion of such atlases will help move forward the use of MWF towards individual patient studies [198–201].

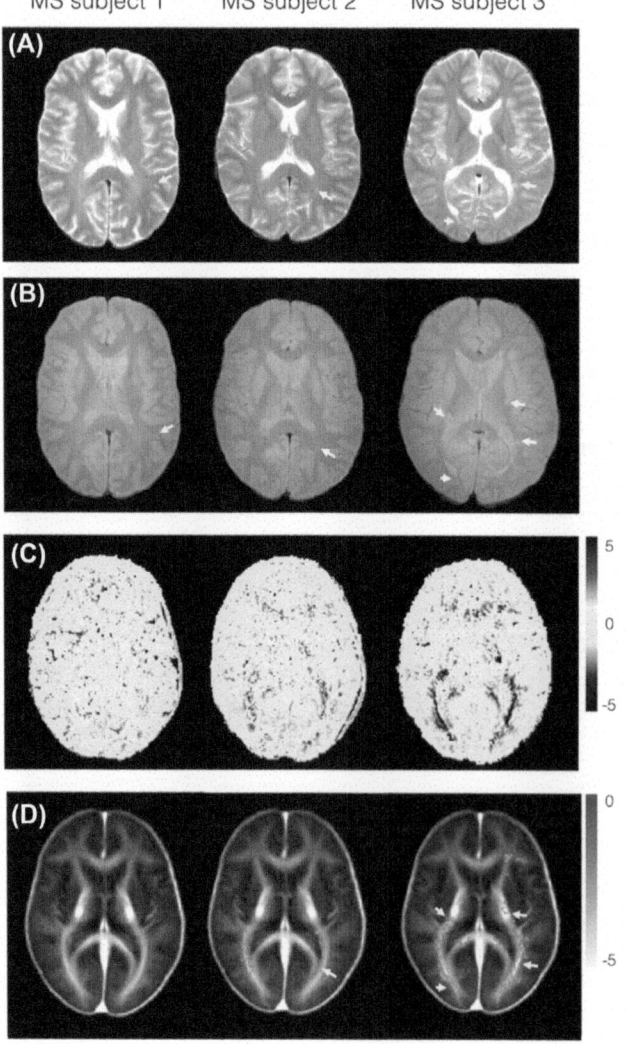

FIGURE 16.10 Z-score maps for three individuals with multiple sclerosis. (A) T_2-weighted images in individual space of three RRMS participants with lesions indicated by arrows; (B) proton density weighted images in individual space; (C) corresponding z-score maps (color scale: −5 to 5); and (D) z-score maps thresholded for significant reduction ($z < -1.96$, uncorrected, color scale: −5 to 5). Regions with coefficient of variation (mean divided by standard deviation) larger than 0.75 were excluded. MS subject 1 was a 25-year-old female with expanded disability status score (EDSS) = 0. MS subjects 2 and 3 were 26-year-old females with EDSS = 2.0. MS subject 3 demonstrates the most extensive global demyelination. Source: *With permission from Figure 4 in Liu H, Rubino C, Dvorak AV, Jarrett M, Ljungberg E, Vavasour IM, et al. Myelin water atlas: a template for myelin distribution in the brain. J Neuroimaging 2019;29(6):699–706.*

Summary

Both MRS and MWF offer quantitative and specific information about MS pathology beyond what is available with conventional MRI, which is not specific to the type of pathology underlying tissue changes, and not sensitive to abnormalities in normal-appearing tissue. MRS allows for noninvasive measurement of the concentrations of various biochemicals, which can provide insights into neuronal viability, energy metabolism, membrane turnover, gliosis, oxidative stress, excitotoxicity, and anaerobic glycolysis. MWF provides information about myelin content. MR signals giving rise to both MRS and MWF are small, but mighty, with great potential to provide quantitative and specific information about focal and diffuse damage in MS.

References

[1] Mollison D, Sellar R, Bastin M, Mollison D, Chandran S, Wardlaw J, et al. The clinico-radiological paradox of cognitive function and MRI burden of white matter lesions in people with multiple sclerosis: a systematic review and meta-analysis. PLoS ONE 2017;12(5):e0177727.

[2] Swanberg KM, Landheer K, Pitt D, Juchem C. Quantifying the metabolic signature of multiple sclerosis by in vivo proton magnetic resonance spectroscopy: current challenges and future outlook in the translation from proton signal to diagnostic biomarker. Front Neurol 2019;10:1173.

[3] Harris AD, MacMillian FL. MRS in neuroinflammation. In: Laule C, Port JD, editors. Imaging neuroinflammation. editors Elsevier; 2023.

[4] Govindaraju V, Young K, Maudsley AA. Proton NMR chemical shifts and coupling constants for brain metabolites. NMR Biomed 2000;13(3):129−53.

[5] Rae CD. A guide to the metabolic pathways and function of metabolites observed in human brain 1H magnetic resonance spectra. Neurochem Res 2014;39(1):1−36.

[6] Govind V, Young K, Maudsley AA. Corrigendum: Proton NMR chemical shifts and coupling constants for brain metabolites. Govindaraju V, Young K, Maudsley AA, NMR Biomed. 2000; 13: 129-153. NMR Biomed 2015;28(7):923−4.

[7] Bottomley PA, Griffiths JR. Handbook of magnetic resonance spectroscopy in vivo: MRS theory, practice and applications. John Wiley & Sons; 2016.

[8] Harris AD, Saleh MG, Edden RA. Edited (1) H magnetic resonance spectroscopy in vivo: methods and metabolites. Magn Reson Med 2017;77(4):1377−89.

[9] Srinivasan R, Sailasuta N, Hurd R, Nelson S, Pelletier D. Evidence of elevated glutamate in multiple sclerosis using magnetic resonance spectroscopy at 3 T. Brain 2005;128(Pt 5):1016−25.

[10] Simone IL, Tortorella C, Federico F. The contribution of (1)H-magnetic resonance spectroscopy in defining the pathophysiology of multiple sclerosis. Ital J Neurol Sci 1999;20(5 Suppl):S241−5.

[11] Narayana PA, Wolinsky JS, Rao SB, He R, Mehta M. Multicentre proton magnetic resonance spectroscopy imaging of primary progressive multiple sclerosis. Mult Scler 2004;10(Suppl 1):S73−8.

[12] Tiberio M, Chard DT, Altmann DR, Davies G, Griffin CM, McLean MA, et al. Metabolite changes in early relapsing-remitting multiple sclerosis. a two year follow-up study. J Neurol 2006;253(2):224−30.

[13] Tourbah A, Stievenart JL, Gout O, Fontaine B, Liblau R, Lubetzki C, et al. Localized proton magnetic resonance spectroscopy in relapsing remitting versus secondary progressive multiple sclerosis. Neurology 1999;53(5):1091−7.

[14] van Walderveen MA, Barkhof F, Pouwels PJ, van Schijndel RA, Polman CH, Castelijns JA. Neuronal damage in T1-hypointense multiple sclerosis lesions demonstrated in vivo using proton magnetic resonance spectroscopy. Ann Neurol 1999;46(1):79−87.

[15] Pendlebury ST, Lee MA, Blamire AM, Styles P, Matthews PM. Correlating magnetic resonance imaging markers of axonal injury and demyelination in motor impairment secondary to stroke and multiple sclerosis. Magn Reson Imaging 2000;18(4):369−78.

[16] Inglese M, Liu S, Babb JS, Mannon LJ, Grossman RI, Gonen O. Three-dimensional proton spectroscopy of deep gray matter nuclei in relapsing-remitting MS. Neurology 2004;63(1):170−2.
[17] Wattjes MP, Harzheim M, Lutterbey GG, Klotz L, Schild HH, Traber F. Axonal damage but no increased glial cell activity in the normal-appearing white matter of patients with clinically isolated syndromes suggestive of multiple sclerosis using high-field magnetic resonance spectroscopy. AJNR Am J Neuroradiol 2007;28(8):1517−22.
[18] Gustafsson MC, Dahlqvist O, Jaworski J, Lundberg P, Landtblom AM. Low choline concentrations in normal-appearing white matter of patients with multiple sclerosis and normal MR imaging brain scans. AJNR Am J Neuroradiol 2007;28(7):1306−12.
[19] Mathiesen HK, Jonsson A, Tscherning T, Hanson LG, Andresen J, Blinkenberg M, et al. Correlation of global N-acetyl aspartate with cognitive impairment in multiple sclerosis. Arch Neurol 2006;63(4):533−6.
[20] Marshall I, Thrippleton MJ, Bastin ME, Mollison D, Dickie DA, Chappell FM, et al. Characterisation of tissue-type metabolic content in secondary progressive multiple sclerosis: a magnetic resonance spectroscopic imaging study. J Neurol 2018;265(8):1795−802.
[21] Narayana PA, Doyle TJ, Lai D, Wolinsky JS. Serial proton magnetic resonance spectroscopic imaging, contrast-enhanced magnetic resonance imaging, and quantitative lesion volumetry in multiple sclerosis. Ann Neurol 1998;43(1):56−71.
[22] He J, Inglese M, Li BS, Babb JS, Grossman RI, Gonen O. Relapsing-remitting multiple sclerosis: metabolic abnormality in nonenhancing lesions and normal-appearing white matter at MR imaging: initial experience. Radiology 2005;234(1):211−17.
[23] Inglese M, Li BS, Rusinek H, Babb JS, Grossman RI, Gonen O. Diffusely elevated cerebral choline and creatine in relapsing-remitting multiple sclerosis. Magn Reson Med 2003;50(1):190−5.
[24] Anik Y, Demirci A, Efendi H, Bulut SS, Celebi I, Komsuoglu S. Evaluation of normal appearing white matter in multiple sclerosis: comparison of diffusion magnetic resonance, magnetization transfer imaging and multi-voxel magnetic resonance spectroscopy findings with expanded disability status scale. Clin Neuroradiol 2011;21(4):207−15.
[25] Hattingen E, Magerkurth J, Pilatus U, Hubers A, Wahl M, Ziemann U. Combined (1)H and (31)P spectroscopy provides new insights into the pathobiochemistry of brain damage in multiple sclerosis. NMR Biomed 2011;24(5):536−46.
[26] Chard DT, Griffin CM, McLean MA, Kapeller P, Kapoor R, Thompson AJ, et al. Brain metabolite changes in cortical grey and normal-appearing white matter in clinically early relapsing-remitting multiple sclerosis. Brain 2002;125(Pt 10):2342−52.
[27] Al-Iedani O, Ribbons K, Gholizadeh N, Lechner-Scott J, Quadrelli S, Lea R, et al. Spiral MRSI and tissue segmentation of normal-appearing white matter and white matter lesions in relapsing remitting multiple sclerosis patients(☆). Magn Reson Imaging 2020;74:21−30.
[28] Tisell A, Leinhard OD, Warntjes JB, Aalto A, Smedby O, Landtblom AM, et al. Increased concentrations of glutamate and glutamine in normal-appearing white matter of patients with multiple sclerosis and normal MR imaging brain scans. PLoS ONE 2013;8(4):e61817.
[29] Llufriu S, Kornak J, Ratiney H, Oh J, Brenneman D, Cree BA, et al. Magnetic resonance spectroscopy markers of disease progression in multiple sclerosis. JAMA Neurol 2014;71(7):840−7.
[30] Klauser AM, Wiebenga OT, Eijlers AJ, Schoonheim MM, Uitdehaag BM, Barkhof F, et al. Metabolites predict lesion formation and severity in relapsing-remitting multiple sclerosis. Mult Scler 2018;24(4):491−500.
[31] Azevedo CJ, Kornak J, Chu P, Sampat M, Okuda DT, Cree BA, et al. In vivo evidence of glutamate toxicity in multiple sclerosis. Ann Neurol 2014;76(2):269−78.
[32] Confort-Gouny S, Vion-Dury J, Nicoli F, Dano P, Donnet A, Grazziani N, et al. A multiparametric data analysis showing the potential of localized proton MR spectroscopy of the brain in the metabolic characterization of neurological diseases. J Neurol Sci 1993;118(2):123−33.
[33] Al Ashi AK, Meray V, Aziz AM. A rare case of Balo concentric sclerosis, a subtype of tumefactive multiple sclerosis, in a 40-year-old male: case report. Cureus 2022;14(4):e24033.
[34] Davie CA, Hawkins CP, Barker GJ, Brennan A, Tofts PS, Miller DH, et al. Serial proton magnetic resonance spectroscopy in acute multiple sclerosis lesions. Brain 1994;117(Pt 1):49−58.
[35] Zaaraoui W, Rico A, Audoin B, Reuter F, Malikova I, Soulier E, et al. Unfolding the long-term pathophysiological processes following an acute inflammatory demyelinating lesion of multiple sclerosis. Magn Reson Imaging 2010;28(4):477−86.

[36] Kocevar G, Stamile C, Hannoun S, Roch JA, Durand-Dubief F, Vukusic S, et al. Weekly follow up of acute lesions in three early multiple sclerosis patients using MR spectroscopy and diffusion. J Neuroradiol 2018;45(2):108—13.

[37] Miller DH, Austin SJ, Connelly A, Youl BD, Gadian DG, McDonald WI. Proton magnetic resonance spectroscopy of an acute and chronic lesion in multiple sclerosis. Lancet 1991;337(8732):58—9.

[38] Zaaraoui W, Fleysher L, Fleysher R, Liu S, Soher BJ, Gonen O. Human brain-structure resolved T(2) relaxation times of proton metabolites at 3 Tesla. Magn Reson Med 2007;57(6):983—9.

[39] Gasparovic C, Chen H, Mullins PG. Errors in (1) H-MRS estimates of brain metabolite concentrations caused by failing to take into account tissue-specific signal relaxation. NMR Biomed 2018;31(6):e3914.

[40] Ramadan S, Lin A, Stanwell P. Glutamate and glutamine: a review of in vivo MRS in the human brain. NMR Biomed 2013;26(12):1630—46.

[41] Spielman DM, Adalsteinsson E, Lim KO. Quantitative assessment of improved homogeneity using higher-order shims for spectroscopic imaging of the brain. Magn Reson Med 1998;40(3):376—82.

[42] Chan KL, Snoussi K, Edden RAE, Barker PB. Simultaneous detection of glutathione and lactate using spectral editing at 3 T. NMR Biomed 2017;30(12).

[43] Wijtenburg SA, Rowland LM, Edden RA, Barker PB. Reproducibility of brain spectroscopy at 7T using conventional localization and spectral editing techniques. J Magn Reson Imaging 2013;38(2):460—7.

[44] Provencher S. LCModel and LCMGui user's manual. Stephen W. Provencher; 2005.

[45] Provencher SW. Estimation of metabolite concentrations from localized in vivo proton NMR spectra. Magn Reson Med 1993;30(6):672—9.

[46] Wilson M, Reynolds G, Kauppinen RA, Arvanitis TN, Peet AC. A constrained least-squares approach to the automated quantitation of in vivo (1)H magnetic resonance spectroscopy data. Magn Reson Med 2011;65(1):1—12.

[47] Reynolds G, Wilson M, Peet A, Arvanitis TN. An algorithm for the automated quantitation of metabolites in in vitro NMR signals. Magn Reson Med 2006;56(6):1211—19.

[48] Clarke WT, Stagg CJ, Jbabdi S. FSL-MRS: an end-to-end spectroscopy analysis package. Magn Reson Med 2021;85(6):2950—64.

[49] Bagory M, Durand-Dubief F, Ibarrola D, Confavreux C, Sappey-Marinier D. "Absolute" quantification in magnetic resonance spectroscopy: validation of a clinical protocol in multiple sclerosis. Annu Int Conf IEEE Eng Med Biol Soc 2007;2007:3458—61.

[50] Sarchielli P, Presciutti O, Pelliccioli GP, Tarducci R, Gobbi G, Chiarini P, et al. Absolute quantification of brain metabolites by proton magnetic resonance spectroscopy in normal-appearing white matter of multiple sclerosis patients. Brain 1999;122(Pt 3):513—21.

[51] Helms G. Volume correction for edema in single-volume proton MR spectroscopy of contrast-enhancing multiple sclerosis lesions. Magn Reson Med 2001;46(2):256—63.

[52] Wilson M, Andronesi O, Barker PB, Bartha R, Bizzi A, Bolan PJ, et al. Methodological consensus on clinical proton MRS of the brain: review and recommendations. Magn Reson Med 2019;82(2):527—50.

[53] Oz G, Cocozza S, Henry PG, Lenglet C, Deistung A, Faber J, et al. MR imaging in Ataxias: consensus recommendations by the Ataxia Global Initiative Working Group on MRI biomarkers. Cerebellum 2023;23(3):931—45.

[54] Peek AL, Rebbeck TJ, Leaver AM, Foster SL, Refshauge KM, Puts NA, et al. A comprehensive guide to MEGA-PRESS for GABA measurement. Anal Biochem 2023;669:115113.

[55] Oz G, Deelchand DK, Wijnen JP, Mlynarik V, Xin L, Mekle R, et al. Advanced single voxel (1) H magnetic resonance spectroscopy techniques in humans: Experts' consensus recommendations. NMR Biomed 2020;e4236.

[56] Andronesi OC, Bhattacharyya PK, Bogner W, Choi IY, Hess AT, Lee P, et al. Motion correction methods for MRS: experts' consensus recommendations. NMR Biomed 2021;34(5):e4364.

[57] Cudalbu C, Behar KL, Bhattacharyya PK, Bogner W, Borbath T, de Graaf RA, et al. Contribution of macromolecules to brain (1) H MR spectra: experts' consensus recommendations. NMR Biomed 2021;34(5):e4393.

[58] Juchem C, Cudalbu C, de Graaf RA, Gruetter R, Henning A, Hetherington HP, et al. B(0) shimming for in vivo magnetic resonance spectroscopy: experts' consensus recommendations. NMR Biomed 2021;34(5):e4350.

[59] Lin A, Andronesi O, Bogner W, Choi IY, Coello E, Cudalbu C, et al. Minimum reporting standards for in vivo magnetic resonance spectroscopy (MRSinMRS): experts' consensus recommendations. NMR Biomed 2021;34(5):e4484.

[60] Maudsley AA, Andronesi OC, Barker PB, Bizzi A, Bogner W, Henning A, et al. Advanced magnetic resonance spectroscopic neuroimaging: experts' consensus recommendations. NMR Biomed 2021;34(5):e4309.
[61] Near J, Harris AD, Juchem C, Kreis R, Marjanska M, Oz G, et al. Preprocessing, analysis and quantification in single-voxel magnetic resonance spectroscopy: experts' consensus recommendations. NMR Biomed 2021;34(5):e4257.
[62] Tkac I, Deelchand D, Dreher W, Hetherington H, Kreis R, Kumaragamage C, et al. Water and lipid suppression techniques for advanced (1) H MRS and MRSI of the human brain: hxperts' consensus recommendations. NMR Biomed 2021;34(5):e4459.
[63] Busch MH, Vollmann W, Mateiescu S, Stolze M, Deli M, Garmer M, et al. Reproducibility of brain metabolite concentration measurements in lesion free white matter at 1.5 T. BMC Med Imaging 2015;15:40.
[64] Graf C, MacMillan EL, Fu E, Harris T, Traboulsee A, Vavasour IM, et al. Intra- and inter-site reproducibility of human brain single-voxel proton MRS at 3 T. NMR Biomed 2019;32(6):e4083.
[65] Zhang Y, Taub E, Salibi N, Uswatte G, Maudsley AA, Sheriff S, et al. Comparison of reproducibility of single voxel spectroscopy and whole-brain magnetic resonance spectroscopy imaging at 3T. NMR Biomed 2018;31(4):e3898.
[66] Vafaeyan H, Ebrahimzadeh SA, Rahimian N, Alavijeh SK, Madadi A, Faeghi F, et al. Quantification of diagnostic biomarkers to detect multiple sclerosis lesions employing (1)H-MRSI at 3T. Australas Phys Eng Sci Med 2015;38(4):611−18.
[67] Brex PA, Parker GJ, Leary SM, Molyneux PD, Barker GJ, Davie CA, et al. Lesion heterogeneity in multiple sclerosis: a study of the relations between appearances on T1 weighted images, T1 relaxation times, and metabolite concentrations. J Neurol Neurosurg Psychiatry 2000;68(5):627−32.
[68] Rovira A, Auger C, Alonso J. Magnetic resonance monitoring of lesion evolution in multiple sclerosis. Ther Adv Neurol Disord 2013;6(5):298−310.
[69] De Stefano N, Matthews PM, Antel JP, Preul M, Francis G, Arnold DL. Chemical pathology of acute demyelinating lesions and its correlation with disability. Ann Neurol 1995;38(6):901−9.
[70] Kirov II, Liu S, Tal A, Wu WE, Davitz MS, Babb JS, et al. Proton MR spectroscopy of lesion evolution in multiple sclerosis: steady-state metabolism and its relationship to conventional imaging. Hum Brain Mapp 2017;38(8):4047−63.
[71] Bellenberg B, Busch M, Trampe N, Gold R, Chan A, Lukas C. 1H-magnetic resonance spectroscopy in diffuse and focal cervical cord lesions in multiple sclerosis. Eur Radiol 2013;23(12):3379−92.
[72] Rahimian N, Saligheh Rad H, Firouznia K, Ebrahimzadeh SA, Meysamie A, Vafaiean H, et al. Magnetic resonance spectroscopic findings of chronic lesions in two subtypes of multiple sclerosis: primary progressive versus relapsing remitting. Iran J Radiol 2013;10(3):128−32.
[73] Yeo CJJ, Hutton GJ, Fung SH. Advanced neuroimaging in Balo's concentric sclerosis: MRI, MRS, DTI, and ASL perfusion imaging over 1 year. Radiol Case Rep 2018;13(5):1030−5.
[74] Marliani AF, Clementi V, Albini Riccioli L, Agati R, Carpenzano M, Salvi F, et al. Quantitative cervical spinal cord 3T proton MR spectroscopy in multiple sclerosis. AJNR Am J Neuroradiol 2010;31(1):180−4.
[75] Obert D, Helms G, Sattler MB, Jung K, Kretzschmar B, Bahr M, et al. Brain metabolite changes in patients with relapsing-remitting and secondary progressive multiple sclerosis: a two-year follow-up study. PLoS ONE 2016;11(9):e0162583.
[76] Ciccarelli O, Thomas DL, De Vita E, Wheeler-Kingshott CA, Kachramanoglou C, Kapoor R, et al. Low myoinositol indicating astrocytic damage in a case series of neuromyelitis optica. Ann Neurol 2013;74(2):301−5.
[77] Kapeller P, Ropele S, Enzinger C, Lahousen T, Strasser-Fuchs S, Schmidt R, et al. Discrimination of white matter lesions and multiple sclerosis plaques by short echo quantitative 1H-magnetic resonance spectroscopy. J Neurol 2005;252(10):1229−34.
[78] Larsson HB, Christiansen P, Jensen M, Frederiksen J, Heltberg A, Olesen J, et al. Localized in vivo proton spectroscopy in the brain of patients with multiple sclerosis. Magn Reson Med 1991;22(1):23−31.
[79] Kim MO, Lee SA, Choi CG, Huh JR, Lee MC. Balo's concentric sclerosis: a clinical case study of brain MRI, biopsy, and proton magnetic resonance spectroscopic findings. J Neurol Neurosurg Psychiatry 1997;62(6):655−8.
[80] Tourbah A, Stievenart JL, Edan G, Abanou A, Dormont D, Lyon-Caen O. Acute demyelination: an insight into the effect of mitoxantrone on CNS lesions. J Neuroradiol 2005;32(1):63−6.
[81] Sarchielli P, Presciutti O, Tarducci R, Gobbi G, Alberti A, Pellicciolli GP, et al. Localized (1)H magnetic resonance spectroscopy in mainly cortical gray matter of patients with multiple sclerosis. J Neurol 2002;249(7):902−10.

[82] Tedeschi G, Bonavita S, McFarland HF, Richert N, Duyn JH, Frank JA. Proton MR spectroscopic imaging in multiple sclerosis. Neuroradiology 2002;44(1):37–42.

[83] Oh J, Pelletier D, Nelson SJ. Corpus callosum axonal injury in multiple sclerosis measured by proton magnetic resonance spectroscopic imaging. Arch Neurol 2004;61(7):1081–6.

[84] Rooney WD, Goodkin DE, Schuff N, Meyerhoff DJ, Norman D, Weiner MW. 1H MRSI of normal appearing white matter in multiple sclerosis. Mult Scler 1997;3(4):231–7.

[85] Suhy J, Rooney WD, Goodkin DE, Capizzano AA, Soher BJ, Maudsley AA, et al. 1H MRSI comparison of white matter and lesions in primary progressive and relapsing-remitting MS. Mult Scler 2000;6(3):148–55.

[86] Fernando KT, McLean MA, Chard DT, MacManus DG, Dalton CM, Miszkiel KA, et al. Elevated white matter myo-inositol in clinically isolated syndromes suggestive of multiple sclerosis. Brain 2004;127(Pt 6):1361–9.

[87] Morrow SA, Baldwin C, Alkabie S. Importance of identifying cognitive impairment in multiple sclerosis. Can J Neurol Sci 2022;1–7.

[88] Gadea M, Martinez-Bisbal MC, Marti-Bonmati L, Espert R, Casanova B, Coret F, et al. Spectroscopic axonal damage of the right locus coeruleus relates to selective attention impairment in early stage relapsing-remitting multiple sclerosis. Brain 2004;127(Pt 1):89–98.

[89] Pan JW, Krupp LB, Elkins LE, Coyle PK. Cognitive dysfunction lateralizes with NAA in multiple sclerosis. Appl Neuropsychol 2001;8(3):155–60.

[90] Solanky BS, John NA, DeAngelis F, Stutters J, Prados F, Schneider T, et al. NAA is a marker of disability in secondary-progressive MS: a proton MR spectroscopic imaging study. AJNR Am J Neuroradiol 2020;41(12):2209–18.

[91] Kantorova E, Hnilicova P, Bogner W, Grendar M, Grossmann J, Kovacova S, et al. Neurocognitive performance in relapsing-remitting multiple sclerosis patients is associated with metabolic abnormalities of the thalamus but not the hippocampus- GABA-edited 1H MRS study. Neurol Res 2022;44(1):57–64.

[92] Guenter W, Bielinski M, Bonek R, Borkowska A. Neurochemical changes in the brain and neuropsychiatric symptoms in clinically isolated syndrome. J Clin Med 2020;9(12).

[93] Cao G, Edden RAE, Gao F, Li H, Gong T, Chen W, et al. Reduced GABA levels correlate with cognitive impairment in patients with relapsing-remitting multiple sclerosis. Eur Radiol 2018;28(3):1140–8.

[94] Arm J, Al-Iedani O, Ribbons K, Lea R, Lechner-Scott J, Ramadan S. Biochemical correlations with fatigue in multiple sclerosis detected by MR 2D localized correlated spectroscopy. J Neuroimaging 2021;31(3):508–16.

[95] Pokryszko-Dragan A, Bladowska J, Zimny A, Slotwinski K, Zagrajek M, Gruszka E, et al. Magnetic resonance spectroscopy findings as related to fatigue and cognitive performance in multiple sclerosis patients with mild disability. J Neurol Sci 2014;339(1-2):35–40.

[96] Cox D, Pelletier D, Genain C, Majumdar S, Lu Y, Nelson S, et al. The unique impact of changes in normal appearing brain tissue on cognitive dysfunction in secondary progressive multiple sclerosis patients. Mult Scler 2004;10(6):626–9.

[97] Blinkenberg M, Mathiesen HK, Tscherning T, Jonsson A, Svarer C, Holm S, et al. Cerebral metabolism, magnetic resonance spectroscopy and cognitive dysfunction in early multiple sclerosis: an exploratory study. Neurol Res 2012;34(1):52–8.

[98] De Stefano N, Filippi M, Miller D, Pouwels PJ, Rovira A, Gass A, et al. Guidelines for using proton MR spectroscopy in multicenter clinical MS studies. Neurology 2007;69(20):1942–52.

[99] Zacharzewska-Gondek A, Pokryszko-Dragan A, Sasiadek M, Zimny A, Bladowska J. Magnetic resonance spectroscopy of the normal appearing grey matter in the posterior cingulate gyrus in the prognosis and monitoring of disease activity in MS patients treated with interferon-beta in a 3-year follow-up. J Clin Neurosci 2020;79:205–14.

[100] Yetkin MF, Mirza M, Donmez H. Monitoring interferon beta treatment response with magnetic resonance spectroscopy in relapsing remitting multiple sclerosis. Med (Baltim) 2016;95(36):e4782.

[101] Takeuchi C, Ota K, Ono Y, Iwata M. Interferon Beta-1b may reverse axonal dysfunction in multiple sclerosis. Neuroradiol J 2007;20(5):531–40.

[102] Schubert F, Seifert F, Elster C, Link A, Walzel M, Mientus S, et al. Serial 1H-MRS in relapsing-remitting multiple sclerosis: effects of interferon-beta therapy on absolute metabolite concentrations. MAGMA 2002;14(3):213–22.

[103] Narayanan S, De Stefano N, Francis GS, Arnaoutelis R, Caramanos Z, Collins DL, et al. Axonal metabolic recovery in multiple sclerosis patients treated with interferon beta-1b. J Neurol 2001;248(11):979–86.

[104] Khan O, Shen Y, Caon C, Bao F, Ching W, Reznar M, et al. Axonal metabolic recovery and potential neuroprotective effect of glatiramer acetate in relapsing-remitting multiple sclerosis. Mult Scler 2005;11(6):646–51.

[105] Khan O, Shen Y, Bao F, Caon C, Tselis A, Latif Z, et al. Long-term study of brain 1H-MRS study in multiple sclerosis: effect of glatiramer acetate therapy on axonal metabolic function and feasibility of long-Term H-MRS monitoring in multiple sclerosis. J Neuroimaging 2008;18(3):314–19.

[106] Arnold DL, Narayanan S, Antel S. Neuroprotection with glatiramer acetate: evidence from the PreCISe trial. J Neurol 2013;260(7):1901–6.

[107] Wiebenga OT, Klauser AM, Schoonheim MM, Nagtegaal GJ, Steenwijk MD, van Rossum JA, et al. Enhanced axonal metabolism during early natalizumab treatment in relapsing-remitting multiple sclerosis. AJNR Am J Neuroradiol 2015;36(6):1116–23.

[108] Vavasour IM, Tam R, Li DK, Laule C, Taylor C, Kolind SH, et al. A 24-month advanced magnetic resonance imaging study of multiple sclerosis patients treated with alemtuzumab. Mult Scler 2019;25(6):811–18.

[109] MacMillan EL, Schubert JJ, Vavasour IM, Tam R, Rauscher A, Taylor C, et al. Magnetic resonance spectroscopy evidence for declining gliosis in MS patients treated with ocrelizumab versus interferon beta-1a. Mult Scler J Exp Transl Clin 2019;5(4) 2055217319879952.

[110] Kauv P, Chalah MA, Creange A, Lefaucheur JP, Ayache SS, Hodel J. Phosphorus magnetic resonance spectroscopy and fatigue in multiple sclerosis. J Neural Transm (Vienna) 2020;127(8):1177–83.

[111] Ricigliano VA, Tonietto M, Palladino R, Poirion E, De Luca A, Branzoli F, et al. Thalamic energy dysfunction is associated with thalamo-cortical tract damage in multiple sclerosis: a diffusion spectroscopy study. Mult Scler 2021;27(4):528–38.

[112] Bodini B, Branzoli F, Poirion E, Garcia-Lorenzo D, Didier M, Maillart E, et al. Dysregulation of energy metabolism in multiple sclerosis measured in vivo with diffusion-weighted spectroscopy. Mult Scler 2018;24(3):313–21.

[113] Oh J, Bar-Or A. Emerging therapies to target CNS pathophysiology in multiple sclerosis. Nat Rev Neurol 2022;18(8):466–75.

[114] Caprariello AV, Adams DJ. The landscape of targets and lead molecules for remyelination. Nat Chem Biol 2022;18(9):925–33.

[115] Borda M, Aquino JB, Mazzone GL. Cell-based experimental strategies for myelin repair in multiple sclerosis. J Neurosci Res 2023;101(1):86–111.

[116] MacKay A, Whittall K, Adler J, Li D, Paty D, Graeb D. In vivo visualization of myelin water in brain by magnetic resonance. Magnetic Reson Med 1994;31(6):673–7.

[117] Laule C, Moore GRW. Myelin water imaging to detect demyelination and remyelination and its validation in pathology. Brain Pathol 2018;28(5):750–64.

[118] Whittall KP, MacKay AL. Quantitative interpretation of nmr relaxation data. J Magn Reson 1989;84:134–52.

[119] Prasloski T, Madler B, Xiang QS, MacKay A, Jones C. Applications of stimulated echo correction to multicomponent T2 analysis. Magn Reson Med 2012;67(6):1803–14.

[120] Doucette J, Kames C, Rauscher A. DECAES – decomposition and component analysis of exponential signals. Z Med Phys 2020;30(4):271–8.

[121] Vavasour IM, Whittall KP, MacKay AL, Li DK, Vorobeychik G, Paty DW. A comparison between magnetization transfer ratios and myelin water percentages in normals and multiple sclerosis patients. Magnetic Reson Med 1998;40(5):763–8.

[122] Prasloski T, Rauscher A, MacKay AL, Hodgson M, Vavasour IM, Laule C, et al. Rapid whole cerebrum myelin water imaging using a 3D GRASE sequence. Neuroimage 2012;63(1):533–9.

[123] Zhang J, Vavasour I, Kolind SH, Baumeister TR, Rauscher A, MacKay A. Advanced myelin water imaging techniques for rapid data acquisition and long T_2 component measurements. Proc Int Soc Magn Reson Med 2015;23:0824.

[124] Dvorak AV, Wiggermann V, Gilbert G, Vavasour IM, MacMillan EL, Barlow L, et al. Multi-spin echo T(2) relaxation imaging with compressed sensing (METRICS) for rapid myelin water imaging. Magn Reson Med 2020;84(3):1264–79.

[125] Oh SH, Bilello M, Schindler M, Markowitz CE, Detre JA, Lee J. Direct visualization of short transverse relaxation time component (ViSTa). Neuroimage 2013;83:485–92.

[126] Deoni SC, Kolind SH. Investigating the stability of mcDESPOT myelin water fraction values derived using a stochastic region contraction approach. Magn Reson Med 2015;73(1):161–9.
[127] Bouhrara M, Spencer RG. Improved determination of the myelin water fraction in human brain using magnetic resonance imaging through Bayesian analysis of mcDESPOT. Neuroimage 2016;127:456–71.
[128] Alonso-Ortiz E, Levesque IR, Pike GB. MRI-based myelin water imaging: a technical review. Magn Reson Med 2015;73:70–81.
[129] Lee J, Hyun JW, Lee J, Choi EJ, Shin HG, Min K, et al. So you want to image myelin using MRI: an overview and practical guide for myelin water imaging. J Magn Reson Imaging 2021;53(2):360–73.
[130] Nguyen TD, Deh K, Monohan E, Pandya S, Spincemaille P, Raj A, et al. Feasibility and reproducibility of whole brain myelin water mapping in 4 minutes using fast acquisition with spiral trajectory and adiabatic T2prep (FAST-T2) at 3T. Magn Reson Med 2015;.
[131] Nguyen TD, Spincemaille P, Gauthier SA, Wang Y. Rapid whole brain myelin water content mapping without an external water standard at 1.5T. Magn Reson Imaging 2017;39:82–8.
[132] Hwang D, Kim DH, Du YP. In vivo multi-slice mapping of myelin water content using T2* decay. Neuroimage 2010;52(1):198–204.
[133] Meyers SM, Vavasour IM, Madler B, Harris T, Fu E, Li DK, et al. Multicenter measurements of myelin water fraction and geometric mean T2: intra- and intersite reproducibility. J Magn Reson Imaging 2013;38(6):1445–53.
[134] Lee LE, Ljungberg E, Shin D, Figley CR, Vavasour IM, Rauscher A, et al. Inter-vendor reproducibility of myelin water imaging using a 3D gradient and spin echo sequence. Front Neurosci 2018;12:854.
[135] Morris SR, Vavasour IM, Smolina A, MacMillan EL, Gilbert G, Lam M, et al. Myelin biomarkers in the healthy adult brain: correlation, reproducibility, and the effect of fiber orientation. Magn Reson Med 2023;89(5):1809–24.
[136] Drenthen GS, Backes WH, Aldenkamp AP, Jansen JFA. Applicability and reproducibility of 2D multi-slice GRASE myelin water fraction with varying acquisition acceleration. Neuroimage 2019;195:333–9.
[137] Ljungberg E, Vavasour I, Tam R, Yoo Y, Rauscher A, Li DKB, et al. Rapid myelin water imaging in human cervical spinal cord. Magn Reson Med 2017;78(4):1482–7.
[138] Moore GRW, Leung E, MacKay AL, Vavasour IM, Whittall KP, Cover KS, et al. A pathology-MRI study of the short-T2 component in formalin-fixed multiple sclerosis brain. Neurology 2000;55(10):1506–10.
[139] Laule C, Leung E, Lis DK, Traboulsee AL, Paty DW, MacKay AL, et al. Myelin water imaging in multiple sclerosis: quantitative correlations with histopathology. Mult Scler 2006;12(6):747–53.
[140] Laule C, Yung A, Pavlova V, Bohnet B, Kozlowski P, Hashimoto SA, et al. High-resolution myelin water imaging in post-mortem multiple sclerosis spinal cord: a case report. Mult Scler 2016;22(11):1485–9.
[141] Laule C, Kozlowski P, Leung E, Li DK, Mackay AL, Moore GR. Myelin water imaging of multiple sclerosis at 7 T: correlations with histopathology. Neuroimage 2008;40:1575–80.
[142] Laule C, Pavlova V, Leung E, Zhao G, MacKay AL, Kozlowski P, et al. Diffusely abnormal white matter in multiple sclerosis: further histologic studies provide evidence for a primary lipid abnormality with neurodegeneration. J Neuropathol Exp Neurol 2013;72(1):42–52.
[143] Moore GR, Laule C, Mackay A, Leung E, Li DK, Zhao G, et al. Dirty-appearing white matter in multiple sclerosis: preliminary observations of myelin phospholipid and axonal loss. J Neurol 2008;255(11):1802–11.
[144] Galbusera R, Bahn E, Weigel M, Schaedelin S, Franz J, Lu PJ, et al. Postmortem quantitative MRI disentangles histological lesion types in multiple sclerosis. Brain Pathol 2022;e13136.
[145] McDowell AR, Petrova N, Carassiti D, Miquel ME, Thomas DL, Barker GJ, et al. High-resolution quantitative MRI of multiple sclerosis spinal cord lesions. Magn Reson Med 2022;87(6):2914–21.
[146] Stewart WA, MacKay AL, Whittall KP, Moore GR, Paty DW. Spin-spin relaxation in experimental allergic encephalomyelitis. Analysis of CPMG data using a non-linear least squares method and linear inverse theory. Magnetic Reson Med 1993;29(6):767–75.
[147] McCreary CR, Bjarnason TA, Skihar V, Mitchell JR, Yong VW, Dunn JF. Multiexponential T2 and magnetization transfer MRI of demyelination and remyelination in murine spinal cord. Neuroimage 2009;45(4):1173–82.
[148] Kozlowski P, Raj D, Liu J, Lam C, Yung AC, Tetzlaff W. Characterizing white matter damage in rat spinal cord with quantitative MRI and histology. J Neurotrauma 2008;25(6):653–76.
[149] Kozlowski P, Rosicka P, Liu J, Yung AC, Tetzlaff W. In vivo longitudinal myelin water imaging in rat spinal cord following dorsal column transection injury. Magn Reson Imaging 2014;32(3):250–8.

[150] Sethi S, Friesen-Waldner LJ, Wade TP, Courchesne M, Nygard K, Sarr O, et al. Feasibility of MRI quantification of myelin water fraction in the fetal guinea pig brain. J Magn Reson Imaging 2023;57(6):1856—64.

[151] Laule C, Vavasour IM, Moore GRW, Oger J, Li DKB, Paty DW, et al. Water content and myelin water fraction in multiple sclerosis: a T2 relaxation study. J Neurol 2004;251(3):284—93.

[152] Oh J, Han ET, Lee MC, Nelson SJ, Pelletier D. Multislice brain myelin water fractions at 3T in multiple sclerosis. J Neuroimaging 2007;17(2):156—63.

[153] Lipp I, Jones DK, Bells S, Sgarlata E, Foster C, Stickland R, et al. Comparing MRI metrics to quantify white matter microstructural damage in multiple sclerosis. Hum Brain Mapp 2019;.

[154] Rahmanzadeh R, Weigel M, Lu PJ, Melie-Garcia L, Nguyen TD, Cagol A, et al. A comparative assessment of myelin-sensitive measures in multiple sclerosis patients and healthy subjects. Neuroimage Clin 2022;36:103177.

[155] Faizy TD, Thaler C, Kumar D, Sedlacik J, Broocks G, Grosser M, et al. Heterogeneity of multiple sclerosis lesions in multislice myelin water imaging. PLoS ONE 2016;11(3):e0151496.

[156] Vavasour IM, Li DK, Laule C, Traboulsee AL, Moore GR, Mackay AL. Multi-parametric MR assessment of T(1) black holes in multiple sclerosis: evidence that myelin loss is not greater in hypointense versus isointense T(1) lesions. J Neurol 2007;254(12):1653—9.

[157] Yao Y, Nguyen TD, Pandya S, Zhang Y, Hurtado Rua S, Kovanlikaya I, et al. Combining quantitative susceptibility mapping with automatic zero reference (QSM0) and myelin water fraction imaging to quantify iron-related myelin damage in chronic active MS lesions. AJNR Am J Neuroradiol 2018;39(2):303—10.

[158] Vargas WS, Monohan E, Pandya S, Raj A, Vartanian T, Nguyen TD, et al. Measuring longitudinal myelin water fraction in new multiple sclerosis lesions. Neuroimage Clin 2015;9:369—75.

[159] Levesque IR, Giacomini PS, Narayanan S, Ribeiro LT, Sled JG, Arnold DL, et al. Quantitative magnetization transfer and myelin water imaging of the evolution of acute multiple sclerosis lesions. Magn Reson Med 2010;63(3):633—40.

[160] Pandya S, Kaunzner UW, Hurtado Rua SM, Nealon N, Perumal J, Vartanian T, et al. Impact of lesion location on longitudinal myelin water fraction change in chronic multiple sclerosis lesions. J Neuroimaging 2020;30(4):537—43.

[161] Vavasour IM, Sun P, Graf C, Yik JT, Kolind SH, Li DK, et al. Characterization of multiple sclerosis neuroinflammation and neurodegeneration with relaxation and diffusion basis spectrum imaging. Mult Scler 2022;28(3):418—28.

[162] Kolind S, Seddigh A, Combes A, Russell-Schulz B, Tam R, Yogendrakumar V, et al. Brain and cord myelin water imaging: a progressive multiple sclerosis biomarker. Neuroimage Clin 2015;9:574—80.

[163] Kolind S, Matthews L, Johansen-Berg H, Leite MI, Williams SC, Deoni S, et al. Myelin water imaging reflects clinical variability in multiple sclerosis. Neuroimage 2012;60(1):263—70.

[164] Liang AL, Vavasour IM, Madler B, Traboulsee AL, Lang DJ, Li DK, et al. Short-term stability of T1 and T2 relaxation measures in multiple sclerosis normal appearing white matter. J Neurol 2012;259(6):1151—8.

[165] Vavasour IM, Huijskens SC, Li DK, Traboulsee AL, Madler B, Kolind SH, et al. Global loss of myelin water over 5 years in multiple sclerosis normal-appearing white matter. Mult Scler 2018;24(12):1557—68.

[166] Holmes RD, Vavasour IM, Greenfield J, Zhao G, Lee JS, Moore GRW, et al. Nonlesional diffusely abnormal appearing white matter in clinically isolated syndrome: prevalence, association with clinical and MRI features, and risk for conversion to multiple sclerosis. J Neuroimaging 2021;.

[167] Cairns J, Vavasour IM, Traboulsee A, Carruthers R, Kolind SH, Li DKB, et al. Diffusely abnormal white matter in multiple sclerosis. J Neuroimaging 2022;32(1):5—16.

[168] Vertinsky AT, Li DKB, Vavasour IM, Miropolsky V, Zhao G, Zhao Y, et al. Diffusely abnormal white matter, T(2) burden of disease, and brain volume in relapsing-remitting multiple sclerosis. J Neuroimaging 2019;29(1):151—9.

[169] Papadaki E, Mastorodemos V, Panou T, Pouli S, Spyridaki E, Kavroulakis E, et al. T2 Relaxometry Evidence of Microstructural Changes in Diffusely Abnormal White Matter in Relapsing-Remitting Multiple Sclerosis and Clinically Isolated Syndrome: Impact on Visuomotor Performance. J Magn Reson Imaging 2021;54:1077—87.

[170] Kitzler HH, Su J, Zeineh M, Harper-Little C, Leung A, Kremenchutzky M, et al. Deficient MWF mapping in multiple sclerosis using 3D whole-brain multi-component relaxation MRI. Neuroimage 2012;59(3):2670—7.

[171] Combes AJE, Matthews L, Lee JS, Li DKB, Carruthers R, Traboulsee AL, et al. Cervical cord myelin water imaging shows degenerative changes over one year in multiple sclerosis but not neuromyelitis optica spectrum disorder. Neuroimage Clin 2017;16:17−22.
[172] Laule C, Vavasour IM, Zhao Y, Traboulsee AL, Oger J, Vavasour JD, et al. Two-year study of cervical cord volume and myelin water in primary progressive multiple sclerosis. Mult Scler 2010;16(6):670−7.
[173] Dayan M, Hurtado Rua SM, Monohan E, Fujimoto K, Pandya S, LoCastro EM, et al. MRI analysis of white matter myelin water content in multiple sclerosis: a novel approach applied to finding correlates of cortical thinning. Front Neurosci 2017;11:284.
[174] Rahmanzadeh R, Lu PJ, Barakovic M, Weigel M, Maggi P, Nguyen TD, et al. Myelin and axon pathology in multiple sclerosis assessed by myelin water and multi-shell diffusion imaging. Brain 2021;144(6):1684−96.
[175] Panou T, Kavroulakis E, Mastorodemos V, Pouli S, Kalaitzakis G, Spyridaki E, et al. Myelin content changes in clinically isolated syndrome and relapsing-remitting multiple sclerosis: associations with lesion type and severity of visuomotor impairment. Mult Scler Relat Disord 2021;54:103108.
[176] Kitzler HH, Wahl H, Eisele JC, Kuhn M, Schmitz-Peiffer H, Kern S, et al. Multi-component relaxation in clinically isolated syndrome: lesion myelination may predict multiple sclerosis conversion. Neuroimage Clin 2018;20:61−70.
[177] Choi JY, Jeong IH, Oh SH, Oh CH, Park NY, Kim HJ, et al. Evaluation of normal-appearing white matter in multiple sclerosis using direct visualization of short transverse relaxation time component (ViSTa) myelin water imaging and gradient echo and spin echo (GRASE) myelin water imaging. J Magn Reson Imaging 2019;49(4):1091−8.
[178] Hurtado Rua SM, Kaunzner UW, Pandya S, Sweeney E, Tozlu C, Kuceyeski A, et al. Lesion features on magnetic resonance imaging discriminate multiple sclerosis patients. Eur J Neurol 2022;29(1):237−46.
[179] Yik JT, Becquart P, Gill J, Petkau J, Traboulsee A, Carruthers R, et al. Serum neurofilament light chain correlates with myelin and axonal magnetic resonance imaging markers in multiple sclerosis. Mult Scler Relat Disord 2022;57:103366.
[180] Lee LE, Vavasour IM, Dvorak A, Liu H, Abel S, Johnson P, et al. Cervical cord myelin abnormality is associated with clinical disability in multiple sclerosis. Mult Scler 2021;27(14):2191−8.
[181] Baumeister TR, Kolind SH, MacKay AL, McKeown MJ. Inherent spatial structure in myelin water fraction maps. Magn Reson Imaging 2020;67:33−42.
[182] Abel S, Vavasour I, Lee LE, Johnson P, Ackermans N, Chan J, et al. Myelin damage in normal appearing white matter contributes to impaired cognitive processing speed in multiple sclerosis. J Neuroimaging 2020;30(2):205−11.
[183] Abel S, Vavasour I, Lee LE, Johnson P, Ristow S, Ackermans N, et al. Associations between findings from myelin water imaging and cognitive performance among individuals with multiple sclerosis. JAMA Netw Open 2020;3(9):e2014220.
[184] Kolind S, Abel S, Taylor C, Tam R, Laule C, Li DKB, et al. Myelin water imaging in relapsing multiple sclerosis treated with ocrelizumab and interferon beta-1a. Neuroimage Clin 2022;35:103109.
[185] Dula AN, Gochberg DF, Valentine HL, Valentine WM, Does MD. Multiexponential T2, magnetization transfer, and quantitative histology in white matter tracts of rat spinal cord. Magn Reson Med 2010;63(4):902−9.
[186] Kalantari S, Laule C, Bjarnason TA, Vavasour IM, Mackay AL. Insight into in vivo magnetization exchange in human white matter regions. Magn Reson Med 2011;66(4):1142−51.
[187] Stanisz GJ, Kecojevic A, Bronskill MJ, Henkelman RM. Characterizing white matter with magnetization transfer and T(2). Magnetic Reson Med 1999;42(6):1128−36.
[188] Birkl C, Doucette J, Fan M, Hernandez-Torres E, Rauscher A. Myelin water imaging depends on white matter fiber orientation in the human brain. Magn Reson Med 2021;85(4):2221−31.
[189] Birkl C, Birkl-Toeglhofer AM, Endmayr V, Hoftberger R, Kasprian G, Krebs C, et al. The influence of brain iron on myelin water imaging. Neuroimage 2019;199:545−52.
[190] Wiggermann V, Vavasour IM, Kolind SH, MacKay AL, Helms G, Rauscher A. Non-negative least squares computation for in vivo myelin mapping using simulated multi-echo spin-echo T(2) decay data. NMR Biomed 2020;33(12):e4277.
[191] Deoni SC, Rutt BK, Arun T, Pierpaoli C, Jones DK. Gleaning multicomponent T1 and T2 information from steady-state imaging data. Magn Reson Med 2008;60(6):1372−87.

[192] Dvorak A., Kumar D., Gilbert G., C L., G.R.W. M., AL M., et al. The CALIPR framework comprehensively improves acquisition, reconstruction & analysis of multi-component relaxation imaging. In: Proceedings joint annual meeting ISMRM-ESMRMB. 07-12 May 2022, London England. 2022.

[193] Liu H, Joseph TS, Xiang QS, Tam R, Kozlowski P, Li DKB, et al. A data-driven T2 relaxation analysis approach for myelin water imaging: spectrum analysis for multiple exponentials via experimental condition oriented simulation (SAME-ECOS). Magn Reson Med 2022;87(2):915–31.

[194] Liu H, Xiang QS, Tam R, Dvorak AV, MacKay AL, Kolind SH, et al. Myelin water imaging data analysis in less than one minute. Neuroimage 2020;210:116551.

[195] Nagtegaal M, Koken P, Amthor T, de Bresser J, Madler B, Vos F, et al. Myelin water imaging from multi-echo T(2) MR relaxometry data using a joint sparsity constraint. Neuroimage 2020;219:117014.

[196] Kumar D, Hariharan H, Faizy TD, Borchert P, Siemonsen S, Fiehler J, et al. Using 3D spatial correlations to improve the noise robustness of multi component analysis of 3D multi echo quantitative T2 relaxometry data. Neuroimage 2018;178:583–601.

[197] Wiggermann V, MacKay AL, Rauscher A, Helms G. In vivo investigation of the multi-exponential T(2) decay in human white matter at 7 T: implications for myelin water imaging at UHF. NMR Biomed 2021;34 (2):e4429.

[198] Liu H, Ljungberg E, Dvorak AV, Lee LE, Yik JT, MacMillan EL, et al. Myelin water fraction and intra/extra-cellular water geometric mean T2 normative atlases for the cervical spinal cord from 3T MRI. J Neuroimaging 2020;30(1):50–7.

[199] Liu H, Rubino C, Dvorak AV, Jarrett M, Ljungberg E, Vavasour IM, et al. Myelin water atlas: a template for myelin distribution in the brain. J Neuroimaging 2019;29(6):699–706.

[200] Morris SR, Holmes RD, Dvorak AV, Liu H, Yoo Y, Vavasour IM, et al. Brain myelin water fraction and diffusion tensor imaging atlases for 9-10 year-old children. J Neuroimaging 2020;30(2):150–60.

[201] Dvorak AV, Swift-LaPointe T, Vavasour IM, Lee LE, Abel S, Russell-Schulz B, et al. An atlas for human brain myelin content throughout the adult life span. Sci Rep 2021;11(1):269.

CHAPTER 17

High-field imaging in multiple sclerosis

Francesca Bagnato[1,2], Kelsey Barter[1,3,4], Chloe Cho[4], Carynn Koch[1], Zachery Rohm[1] and Colin McKnight[5]

[1]Neuroimaging Unit, Neuro-immunology Division, Department of Neurology, Vanderbilt University Medical Center, Nashville, TN, United States [2]Department of Neurology, VA Hospital, TN Valley Healthcare Center, Nashville, TN, United States [3]Division of Pediatric and Developmental Neurology, Washington University School of Medicine, St. Louis, MO, United States [4]Vanderbilt University School of Medicine, Nashville, TN, United States [5]Department of Radiology and Radiological Sciences, Vanderbilt University Medical Center, Nashville, TN, United States

OUTLINE

Introduction	322
Multiple sclerosis—induced disease under the microscope of high-field imaging	323
Improved visibility of white matter lesions	324
Paramagnetic rim lesions	325
The central vein sign	327
Cortical pathology	330
Leptomeningeal enhancement	332
Clinical application of high-field imaging	333
Technical challenges associated with the use of high-field imaging	334
Summary and conclusions	335
Acknowledgments	335
Conflict of interest	335
References	335

Introduction

Clinical magnetic resonance imaging (MRI) is a cornerstone in diagnosing and monitoring multiple sclerosis (MS). A clinical MRI protocol includes T_1-weighted (T_1-w) and T_2-w sequences, the former obtained before and after the intravenous injection of the contrast agent gadolinium diethylenetriamine penta-acetic acid (Gd-DTPA). Clinical MRI is typically performed using 1.0, 1.2, 1.5, or 3.0 T scanners. T_1-w and T_2-w sequences are sensitive to MS disease-induced changes but lack pathological specificity and offer limited spatial resolution. Both factors preclude accurate tissue injury characterization and quantification [1].

In recent years, the advent of high-field imaging (HFI), herein defined as 7 T MRI and above, allowed for overcoming some of the limitations of clinical MRI. HFI offered the ground truth for novel radionomics subsequently translated to lower field MRI, leading to significant advancement of our understanding of MS and more precise disease phenotyping.

The use of HFI in MS has two main advantages. The first gain is the increase of the high signal-to-noise ratio (SNR) that rises linearly as a function of the static magnetic field strength [2]. The use of multichannel detectors for HFI can increase signal sensitivity in the brain up to three times (for a 32-channel coil [2]) and maximize the benefits of parallel imaging using a higher acceleration factor [3], further increasing SNR. The combination of these elements yields images with $\leq 200\,\mu m$ in-plane resolution, which amplifies the detection of pathological changes in people with MS (pwMS) even when these changes are in small structures of the central nervous system (CNS), such as the cortical ribbon [4] and the spinal cord [5]. Accordingly, also, HFI improves white matter (WM) lesion detectability in the brain of pwMS [6–8].

The second advantage of HFI is the enhancement of the susceptibility contrast which leads to the detection of pathological processes or anatomical structures such as vessels, due to the presence of paramagnetic metals such as iron [9], calcium, and magnesium [2]. This mechanism of contrast represents a significant change from that offered by hydrogen-based nuclear magnetic resonance and opens a new window on the disease characterization and treatment targets discovery.

Notwithstanding these considerations, HFI has not made it to day-to-day clinical practice for two main reasons. First, while HFI has substantially improved sensitivity and specificity to MS-induced disease, MRI characterization and detection remain far from ideal. Second, HFI presents several unique technical challenges and safety requirements that are currently viewed as obstacles to clinical translation.

In this chapter, we will discuss the specific areas where HFI improved our understanding of MS pathophysiology and related disability (Section "Multiple sclerosis—induced disease under the microscope of high-field imaging"). When reviewing clinical-MRI associations, we will focus only on those based on 7 T data given the scope of this chapter. Nonetheless, we note that most of the results generated in smaller pwMS cohorts imaged at 7 T have been reproduced and expanded in larger cohorts scanned at 3 T. We will then address the role of HFI in both clinical and research settings (Section "Clinical application of high-field imaging") and will present the major technical challenges currently preventing clinical translation of HFI clinical (Section "Technical challenges associated with the use of high-field imaging").

Multiple sclerosis—induced disease under the microscope of high-field imaging

MS is an autoimmune disease of the CNS featured by a complex interplay of inflammatory, demyelinating, and neurodegenerative processes, driven by both the innate and adaptive immune systems. The MS-induced disease can be seen in the WM and the gray matter (GM) tissue of the brain and the spinal cord as well as in the leptomeningeal space [10]. Major pathological mechanisms and sites of injury vary from person to person and within the same person over time.

WM lesions encompass an acute and a chronic phase (Fig. 17.1). The acute stage is primarily driven by the adaptative immune system with CNS influx of autoreactive T and B cells from the bloodstream. This process leads to inflammation and blood—brain barrier (BBB) breakdown and is accompanied by macrophages and microglia-driven demyelination. Clinical T_1-w MRI obtained upon the injection of the contrast agent Gd-DTPA allows the detection and characterization of acutely inflamed lesions where the BBB is disrupted. Upon the resolution of the BBB leakage, acutely inflamed WM lesions may evolve into chronic inactive or chronic active lesions (CAL). A subset of them may remyelinate. Irrespective of their pathological outcome, all chronic WM lesions appear as hyperintense on T_2-w MRI. Fig. 17.1

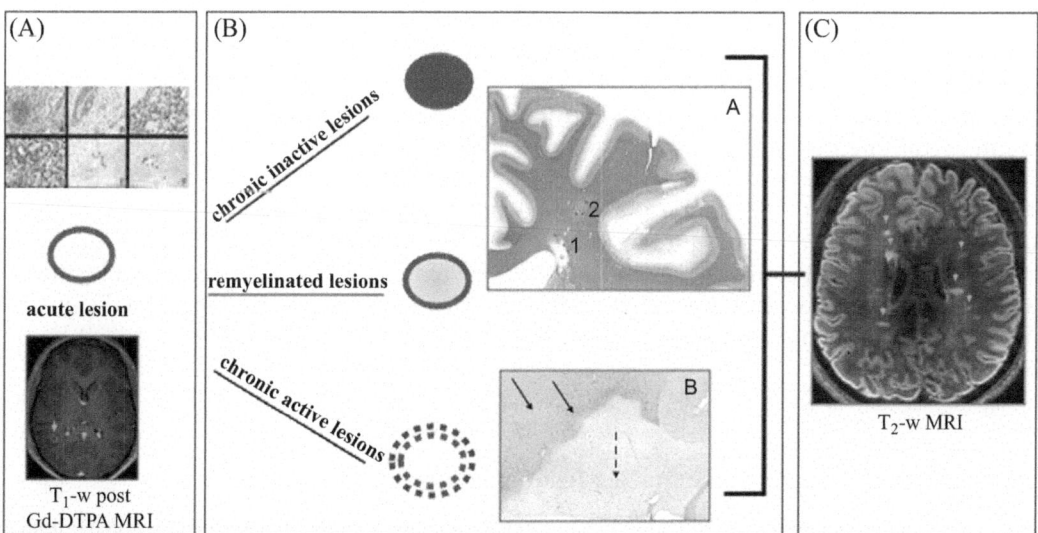

FIGURE 17.1 **Multiple sclerosis lesion evolution.** Acutely active multiple sclerosis lesions (panel A) are featured by focal perivascular inflammatory infiltrate of lymphocytes, macrophages, and reactive astrocytes (A in panel A); Periodic acid-Schiff—positive granules in macrophages (B in panel A); rarefied axonal structures (C in panel A); microglial/macrophagocytic infiltrates (D in panel A); B cells (E) and T cells (F). Due to the breakdown of the blood—brain barrier, lesions so formed appear hyperintense on T_1-weighted postgadolinium diethylenetriamine penta-acetic acid clinical scans. Upon the resolution of the acute phase, acute lesions may evolve into chronic inactive (lesion #1 in figure A of panel B), remyelinated (lesion #2 in figure A of panel B) or chronic active lesions featured by microglia activation (solid black arrow) surrounding demyelinated plaques (dotted black arrow) and extending far from it (solid black arrow), (figure B of panel B). All these lesions appear hyperintense on T_2-weighted magnetic resonance imaging (panel C).

2. Non-conventional MRI use in multiple sclerosis

shows all the pathological and imaging aspects of these processes. The ability of T_2-w and T_1-w Gd-DTPA-enhanced MRI to visualize WM lesions has revolutionized the way we diagnose and monitor MS. However, the inability of clinical sequences to identify all chronic lesions, define their subtype, and distinguish them from mimickers has been a significant limiting factor for our understanding of disease outcome and disability accumulation in individual people as well as for an accurate diagnosis of less clinically defined cases. HFI filled some of this knowledge gap improving the detection of brain and spinal cord WM lesions and allowing the characterization of CAL and the identification of the central vein sign (CVS), two novel biomarkers of disease indicative of specific lesion state and stage.

Cortical GM lesions, for example, cortical lesions (CL), are also an important component of the MS pathological axis. Unlike WM lesions, CL are sustained by less acute inflammation, and BBB breakdown is rarely seen [11]. CL are difficult to detect on clinical MRI due to their small size, volume averaging from the adjacent cerebrospinal fluid, and the lower level of myelination of the cortex. The latter factor contributes to the small contrast-to-noise ratio (CNR) between CL and the surrounding normal-appearing gray matter (NAGM) when demyelination occurs [12]. The advent of HFI has made it possible to overcome some of the limitations of clinical MRI and has allowed a more accurate identification of cortical pathology in vivo.

Lastly, while neglected for a long time, the potential role of leptomeningeal inflammation as a contributor to disability progression has been the focus of attention after the identification of foci of leptomeningeal enhancement (LME) using high-resolution postcontrast T_2-w fluid-attenuated inversion recovery (FLAIR) sequences. While there is still controversy on the clinical utility and pathological specificity of this biomarker to MS, the role of HFI in permitting better identification remains important.

In this chapter, we will detail how the use of HFI improved our understanding of the above-mentioned pathological processes. We note that MS spreads through other sites, such as the NAGM and the normal-appearing white matter (NAWM) of both the brain and the spinal cord. Furthermore, neurodegeneration along with acute and chronic inflammation plays an important role in this disease. However, the advantage of HFI on our understanding of neurodegenerative processes is still unclear and will therefore not be the focus of this chapter.

Improved visibility of white matter lesions

A few comparative studies have been performed to assess differences in WM lesion conspicuity between HFI and clinical scanners. The use of spin echo (SE) sequences at 7 T is substantially hindered by the increased specific absorption rates (SAR) related to SE refocusing pulses in the high-field environment [13]. On the contrary, T_2-w SE images are the gold standard, along with T_2-w FLAIR, for WM lesion detection at lower field strength MRI. Likely due to this limitation of HFI, only one study [6] compared WM lesion detectability between 1.5 T/7 T proton density and T_2-w SE images in 12 pwMS. The authors counted 97 brain WM lesions at 1.5 T and 126 at 7 T.

In 14 pwMS, Mistry et al. [7] identified 812 brain WM lesions on 3 T T_2-w FLAIR, 186 additional lesions on 3 T T_1-w magnetization-prepared rapid-gradient-echo (MPRAGE), and 231 additional (relative to the 3 T T_2-w FLAIR) lesions on 7 T T_1-w MPRAGE. Similarly, in 18 pwMS, Sinnecker et al. [8] found that 7 T T_1-w MP$_2$RAGE detected 728 WM lesions relative to 399 seen on T_1-w MP$_2$RAGE at 1.5 T. Notably, at 7 T, T_2-w fast low-angle shot (FLASH) sequence detected only 584 WM lesions.

Only one study [5] compared spinal cord lesion detectability in 9 pwMS using T2-turbo spin echo images at 3 T and T_2^*-w gradient echo (GRE) sequences at 7 T. The authors counted 28 WM lesions on the former and 42 lesions on the latter sequences.

Paramagnetic rim lesions

Iron localization in proximity to MS demyelinating lesions provides insight into MS pathophysiology. The presence of iron deposits surrounding MS demyelinating plaques was first characterized in 1982 [14]. The origin of these deposits is not fully understood. As oligodendrocytes are the primary reservoir of iron in the brain [15], it is speculated that oligodendrocyte apoptosis at the lesion periphery results in iron liberalization to the extracellular space, which may augment demyelination and promote neurodegeneration via increased oxidative stress [16]. Microglia and macrophages, activated by oligodendrocyte destruction, take up extracellular iron [16], promote further axonal injury, and are the hallmark of CAL [17]. Low-grade inflammation propagated by microglia/macrophages in CAL is thought to contribute to MS disease progression [18]. A portion of CAL also shows signs of active demyelination which likely promotes their slow radial enlargement over time. This subset of CAL is defined as slowly expanding lesions [18] or smoldering lesions [19].

Visualizing chronic active lesions on magnetic resonance imaging

The advent of HFI has made it possible to identify CAL in the brains of pwMS [20]. As discussed above, iron-laden microglia/macrophages at the lesion peripheries in a subset of CAL contribute to chronic inflammation and lesion expansion. This iron rim is readily detectable by susceptibility-based MRI as a rim of paramagnetic signal surrounding select MS lesions visible on T_2-w MRI [21]. These lesions are termed paramagnetic rim lesions (PRL). Multiple ex vivo histopathological and MRI studies have demonstrated that PRL accurately reflect iron-rich phagocytes surrounding a subset of chronic MS lesions at the tissue level [20,22,23]. PRL have been validated with multiple susceptibility-based imaging methods, including multiecho-derived R2*/T2* and phase maps [20,22], susceptibility-weighted imaging (SWI, [23]), and quantitative susceptibility mapping[24]. Fig. 17.2 displays PRL visibility on each of these MRI methods. Although prospective studies comparing PRL visibility among different MRI methods are lacking, retrospective metanalyses have shown no significant differences in sensitivity across different methods of susceptibility-based imaging when analyzing pooled data from 7 T studies [25]. Nonetheless, the reason why not all PRL are visible on all susceptibility-based MRI methods remains to be investigated (Fig. 17.2). Technical differences among with variability in iron content are likely the main factors. Although 7 T allowed for the discovery of PRL, PRL are now routinely identified at lower field strengths. Whether 7 T outperforms lower field strength MRI remains to be seen. In a 2018 study with a cohort of 20 pwMS, investigators found that nearly all 7 T PRL can be visualized at 3 T with phase imaging [26].

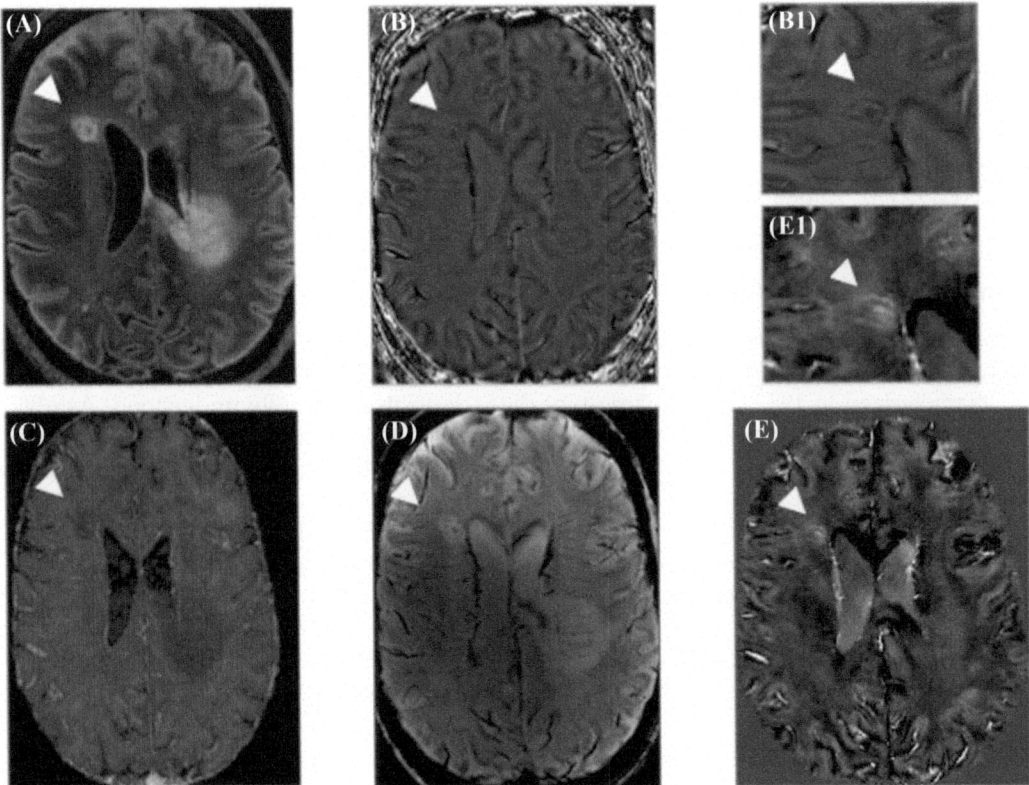

FIGURE 17.2 **Paramagnetic rim lesions.** A paramagnetic rim lesion (PRL) seen as a hyperintense lesion on a T_2-weighted fluid-attenuated inversion recovery image (white arrow in A). The PRL is detectable on phase (white arrow in B and B1 for higher magnification), susceptibility-weighted images (white arrow in D) and quantitative susceptibility mapping (white arrow in E and E1 for higher magnification) but not on the R2* map (white arrow in C).

However, investigators have also found reduced subject-level PRL sensitivity at 3 T compared to 7 T (24.4% vs 57.6%, respectively; [25]). More studies are needed to compare the sensitivity of 3 T versus 7 T in detecting PRL. More recently, a 2022 study utilizing susceptibility-weighted angiography suggested that 1.5 T is comparable to 3 T in identifying PRL [27].

It is also important to highlight that microglia activation can reach far into the NAWM (Fig. 17.1) [28] surrounding PRL, but HFI fails to disclose it.

Lastly, although no comparative studies are available, only 7 T GRE images likely allow for the visibility of PRL in the spinal cord [29].

Clinical significance of paramagnetic rim lesions

HFI studies indicate that PRL indicate more advanced disease at the lesion and subject levels.

On a lesion level, it is well established that PRL evolve differently than rimless lesions. PRL are larger, have lower levels of macromolecular pool size ratio measured on quantitative magnetization transfer imaging [30], and have longer T_1 values [31] relative to rimless lesions. PRL tend to slowly expand over time, whereas rimless lesions tend to shrink [23]. Furthermore, persisting PRL have a darker T_1 signal than lesions which have only a transient PRL at the time of the Gd-DTPA enhancement [22]. Altogether these findings are indicative of PRL as sites encompassing larger areas with a higher degree of demyelination and diminished WM integrity.

PwMS with PRL tend to have higher levels of disability and disease radiological progression on MRI. PwMS with PRL perform worse on cognitive batteries [32] and have a higher likelihood of accumulating disability over time [31]. PwMS with PRL also have larger T_2-lesion volume [30], greater GM atrophy [32], and higher levels of neurofilaments [33]. Using machine-learning algorithms, investigators have demonstrated that PRL volume at 7 T is a major predictor of MS disability progression [34].

Although not necessarily demonstrated at 7 T, it is important to underscore that the presence of PRL is shown to increase the accuracy of an MS diagnosis. Diagnostic accuracy is of paramount importance, as MS misdiagnosis is common and may lead to patient morbidity from unnecessary immunomodulatory treatment and financial hardship due to the high cost of disease-modifying therapy. In a 2022 study, the magnetic resonance imaging in multiple sclerosis (MAGNIMS) group found that the presence of PRL at 3 T is highly specific for MS/clinically isolated syndrome (CIS) with a specificity of 99.7% and positive predictive value of 98.4%, although sensitivity was limited at 24% [35]. Another study reported 93% PRL specificity in differentiating between MS and non-MS at 3 T [36]. PRL had a higher specificity than the CVS, another emerging MS imaging biomarker, which had a specificity of 88.3% [35]. This data suggests that the presence of PRL substantially increases the likelihood of an MS/CIS diagnosis, whereas the absence of PRL does not appreciably reduce the likelihood of an MS/CIS diagnosis.

The central vein sign

The generally accepted pathophysiology of MS is based upon the notion of a vasocentric WM lesion distribution. As stated previously, WM lesion formation depends on the entry of lymphocytes and other inflammatory cells from systemic circulation, across the BBB, into the CNS. The perivascular space is thought to be a privileged site where immune cells interact with antigen-presenting cells, leading to an inflammatory cascade and formation of demyelinating lesions around a central vein.

The finding of a central vein was first identified in ex vivo studies in 1820 when many MS lesions were noted to have small parenchymal veins running through them [37]. In early 2000s, Tan and collaborators were the first to study this concept in vivo. Using T_2^*-w imaging at 1.5 T and SWI they demonstrated this relationship between lesions and blood vessels, with 94/95 detected lesions possessing a central vein. However, it was only upon the advent of HFI that a substantial improvement in the visualization of the CVS occurred. In 2008, indeed, the use of 7 T T_2^*-w GRE sequences and SWI made it possible to further improve the visualization of MS lesions and their central veins and led to the definition of

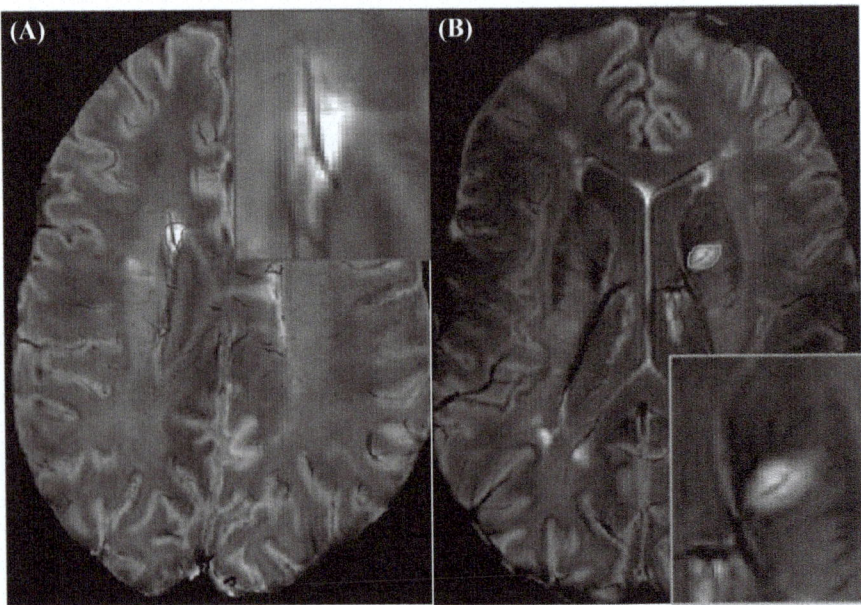

FIGURE 17.3 **The central vein sign.** T_2-weighted fluid-attenuated inversion recovery images fused with susceptibility-weighted images at 7.0 Tesla in a person with newly diagnosed multiple sclerosis. Two white matter lesions (purple and blue contours in A and B, respectively) crossed by a central vein are shown.

the CVS [38–40]. Radiologically, the CVS is defined by the appearance of a thin hypointense line or small dot, with a diameter of less than 2 mm, running partially or entirely through the center of a lesion regardless of lesion shape, that is visualized in at least two perpendicular MRI planes ([41], Fig. 17.3).

Visualizing central vein sign using high-field imaging and 3 T magnetic resonance imaging

An early 7 T study showed that a central vein was visualized in 87% of lesions in pwMS, compared to only 8% in healthy volunteers [42]. This raised the question of its clinical significance, and whether it could hold diagnostic utility. Subsequently, Tallantyre et al. [43] demonstrated that by using T_2^*-w MRI at 7 T, demyelinating MS lesions in pwMS could be distinguished from nonspecific WM lesions in healthy volunteers and those with vascular risk factors with 100% specificity based on the presence of CVS in more than 40% of lesions. Additional studies sought to contextualize this finding in MS [44,45] showed that in 1004 brain lesions from 33 people with varying MS phenotypes, for example, 19 with relapsing–remitting multiple sclerosis (RRMS), 9 with primary progressive (PP) MS, and 5 with secondary progressive (SP) MS, 78% of were located around a central vein. This preference for a vasocentric distribution of lesions regardless of MS clinic phenotype was supported by a followup 7 T study [46]. The likelihood of identifying a central vein also varied based on location, with the highest proportion in periventricular lesions (94%) and deep WM lesions (84%).

As HFI is not yet widely available, several investigations have been performed to assess the role of 3 T MRI in CVS detection. One study used a variety of T2* sequences on 3 T MRI to question whether CVS was able to be identified in 10 pwMS and 10 people with small vessel disease. They found that the CVS could be identified on 3 T MRI, irrespective of specific imaging methods. Compared to T2* GRE and susceptibility-weighted angiography, T2* with high echo planar imaging was the sequence with the highest sensitivity, identifying the CVS in 69.6% of lesions in pwMS and 6.1% of lesions in people with small vessel disease [47].

Deemed to be a capable resource, head-to-head 3 T–7 T studies were pursued. An early study of 358 lesions (7 pwMS and 7 controls), reported the identification of a central vessel in 45% of visible lesions at 3 T, compared to 87% of visible lesions at 7 T in T2* sequences [42]. In a small cohort of 3 pwMS and two controls, Dixon et al. [48] showed that while T_2^*-w at 7 T has the highest sensitivity for detection of CVS, optimized T_2^*-w 3 T protocols can improve the detection of the CVS to >80%. More recently, Okromelidze et al. [49] conducted a retrospective study to compare 3 T SWI, 7 T SWI, and T_2^*-w sequences in detecting CVS. This has been the largest cohort study, with 61 pwMS (903 lesions) and 39 controls (1088 lesions). SWI and T_2^*-w at 7 T detected 73% and 87% of CVS, respectively, compared to 31% on 3 T SWI. The 7 T sequences could distinguish MS lesions from WM lesions of presumed vascular origin with 100% accuracy [49].

Clinical significance of central vein sign

The CVS has been proposed as a sensitive and specific biomarker for MS and may serve as a way to differentiate MS common MS mimics.

1. *Migraines:* A 2015 study based on 3 T T_2-w FLAIR* acquired on 10 pwMS and 10 people with migraine, identified the CVS in 84% of MS lesions compared to 22% of migraine-related lesions. Notably, when considering subcortical and deep WM regions, this difference was even greater at 88% of lesions in pwMS patients and 19% of lesions in people with migraine [50].
2. *Cerebral small vessel disease:* It is well known that cerebral small vessel disease can cause WM lesions that often mimic MS. Not only can this cause diagnostic query, but as pwMS age or accumulate vascular comorbidities, there can be progression of WM disease that possibly represents new demyelination or chronic small vessel disease. Using a T_2-w FLAIR* at 7 T MRI, Kilsdonk et al. [44,45] looked at a total of 433 brain lesions from 16 pwMS and 86 brain lesions from people with vascular risk factors. They found that a central vein was present in 74% of MS lesions and only 47% of vascular lesions. When considering deep WM lesions, this difference was more marked, with 81% of MS lesions and only 44% of vascular lesions. The utility of the CVS to differentiate MS lesions from small vessel disease has been supported in similar studies using T_2^*-w imaging at both 7 T [43] and 3 T [51].
3. *Neuromyelitis optica spectrum disorder:* The neuromyelitis optica spectrum disorder (NMOSD) is an umbrella term that comprehends several autoimmune diseases of the CNS predominantly affecting the optic nerves and spinal cord. There can be significant overlap in clinical and radiologic features of NMOSD and MS, and tools for differentiating these conditions early in the disease course can often prevent costly

relapse. To this end, Kister et al. [52] used 7 T MRI to look at a total of 92 lesions in 10 people with aquaporin-4 autoantibody-positive neuromyelitis optica (NMO) and found that only eight lesions (9%) were traversed by a central vein. In a prospective study of pwMS and antibody-positive NMO people, 533 WM lesions were counted in 18 pwMS and 140 WM lesions were seen in those with NMO using T2*-w imaging on 7T MRI. Of these, 92% of MS lesions were perivascular, compared to 35% of NMO lesions [8].

Cortical pathology

Cortical damage is an important component of MS disease pathology. It may be secondary to WM disease via retrograde degeneration or independent from it and mediated by a primary GM inflammatory-demyelination-neurodegenerative process. Cortical demyelination presents as focally demarcated regions of tissue loss, that is, CL, or as wider spread areas of cortical demyelination. These two processes may coexist and are not mutually exclusive.

Histopathology studies classify CL into three subtypes based on their anatomical localization relative to the cortex and the adjacent WM [46,53,54]. Type-I (leukocortical) mixed WM and GM CL are featured by demyelination which spreads entirely or partially through the cortex and also in the adjacent WM; type-II (purely intracortical) demyelinating intra-CL span exclusively over the cortex and typically evolves around a vessel; and type-III (subpial) CL extend from the pial surface with variable depth into the cortex and are more frequently encountered with histopathologic studies (Fig. 17.4).

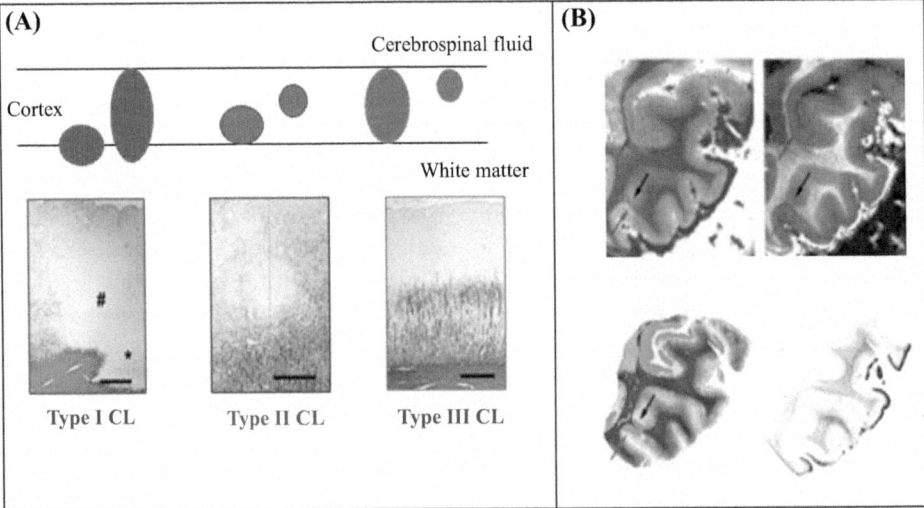

FIGURE 17.4 **Cortical disease in multiple sclerosis.** In panel A, the three types of cortical lesions are represented. In panel B multiecho gradient, echo magnitude, and R2* images show a normal-appearing cortex which is instead featured by complete myelin loss (pink masks in the proteolipid staining underneath) and iron loss.

Evidence from ex vivo studies

Ex vivo studies at HFI offer an opportunity to improve the identification of CL due to high image resolution with less penalization of the SNR. This is made possible by fewer constraints on scanning time. Accordingly, 7 T MRI detected 225% more CL than 3 T and 7 T T_2^*-w detected 200% more CL than 3 T T_2^* [55].

A few combined MRI-histology ex vivo imaging studies have been performed to assess the degree of sensitivity of various MRI methods and field strengths in detecting CL, both prospectively, for example, before knowing about CL existence on histology, and retrospectively, for example, knowing about CL existence on histology. Schmierer et al. [56] found only 78% of CL seen on histology by using T_2-w MRI at 9.4 T. Pitt et al. [57] used a three-dimensional (3D) T_2^*-w GRE and a WM-attenuated inversion recovery turbo-field-echo (TFE) sequences at 7 T and prospectively detected 46% (T_2^*-w GRE) and 42% (TFE) CL. Retrospectively, sensitivity increased to 93% and 82%, respectively. Notably, HFI here provided the first identification of PRL in the cortex of pwMS. Yao et al. [58] reported prospective and retrospective CL sensitivities of 48% and 67%, respectively, using T_2^*-w multiple-echo recombined gradient echo MRI at 7 T. More importantly, the authors showed that large areas of extensive subpial demyelination and remyelination escape HFI visibility. Thus, despite the improvement, HFI remains far from disclosing the amount of cortical pathology seen in histology.

Using high-field imaging in vivo improves cortical lesions visibility and allows the detection of subpial disease

In vivo HFI studies parallel the ex vivo evidence showing improved CL visibility and enabling the identification of subpial pathology. Although detecting subpial CL remains a challenge, HFI T_2^*-w at 7 T allowed the differentiation of various subtypes of CL and detected more CL [59] and subpial CL compared with 3 T [60]. Similarly, 7 T T_2-w FLASH detected subpial CL which double inversion recovery at 3 T failed [61]. This is especially important because subpial cortical lesions in MS are specific pathological features compared to cortical damage in other diseases [53]. Sinnecker et al. [62] also reported that 7 T T_1-w MPRAGE detected CL twice more than 1.5 T MPRAGE and found that 7 T T_1-w MPRAGE improves the sensitivity to CL compared to T_2-w FLASH. Maranzano et al. [63] compared CL counts between 7 T and 3 T 3D-MP$_2$RAGE/2D-GRE and reported a higher incidence of leukocortical and subpial CL at 7 T in people with secondary progressive form of multiple sclerosis (pwSPMS).

Lastly, as seen in ex vivo studies [58] HIF allowed the detection of cortical disease outside CL. Cohen-Adad et al. [64] showed subpial T_2^* relaxation time changes in large cortical areas using the T_2^* mapping technique at 7 T. Widespread cortical injury tended to be localized in the subpial space early in the disease course and extended through the deeper layers later [65]. Similarly, Barletta et al. [66] assessed the health of the normal-appearing cortex using a combined myelin estimation (CME) from quantitative T_2^* and T_1 maps. They found CME alterations in several areas of the seemingly healthy cortex and noted that the cortical CME correlated with measurements of cognitive impairment in pwMS.

Clinical relevance of cortical lesions

Quantitative MRI studies show that CL have various degrees of and heterogenous patterns of demyelination [66] and iron loss [66,67]. Opposite to WM lesions which are the disease hallmark of RRMS, CL tend to form in the progressive phase of MS [68,69] and independently from the formation of new WM lesions [70]. Cross-sectional HFI studies showed that CL have a strong relationship to physical and cognitive disability [61] as well as with thalamic lesions [68]. In a longitudinal 7T-based study of 2.8 ± 0.7 years, Beck et al. [70] found that CL volume is associated with changes in physical disability measured using the 25-foot timed walk, the 9-hole peg test but not with change in brain atrophy. Baseline CL volume was also higher in people who converted to SPMS and showed progression at the expanded disability status scale (EDSS).

Leptomeningeal enhancement

Meningeal lymphatics are thought to represent a port of entry to the CNS for T- and B-cells [71]. Ectopic B-cell follicles with germinal centers were first detected in the meninges of pwSPMS and were found to be associated with the early onset of disease and increased cortical pathology, including adjacent CL [72]. Subsequently, these structures were found to correlate with a quantitative increase in meningeal inflammation [73], measured by microglial activation and GM cortical demyelination detected with immunocytochemistry [73]. Histopathologic studies suggest meningeal inflammation is part of the pathology of MS.

In vivo visualization of meningeal inflammation

Inflammation of the meninges facilitates the breakdown of the BBB, which consequently allows for visualization of LME on Gd-DTPA—enhanced MRI. Pre- and postcontrast 3D T_2-w FLAIR is superior to T_1-w images for detecting meningeal inflammation because of the former sequence's longer relaxation time, which increases sensitivity to very small amounts of leakage of gadolinium through the BBB [74]. Three main patterns of LME have been described in pwMS including nodular, spread/fill sulcal, and spread/fill gyral areas of enhancement within the leptomeningeal space [75].

Studies assessing the incidence of LME at 3 and 7 T have varying results [75–82]. Metanalytic data show that a higher number of LME foci are detected at 7 T relative to 3 and 1.5 T, a factor that is important to consider in the design of studies assessing the clinical significance of LME in MS and potential medication efficacy.

Histopathologic and clinical significance of leptomeningeal enhancement

Histopathologic validation of LME foci is limited. Absinta et al. [22] reported that the LME areas correspond to sites of perivascular monocellular and lymphocytic infiltrates in the vicinity of subpial demyelination. However, it remains to be assessed if other areas with similar histopathological features are present and lack LME correlates.

The clinical significance of LME is still controversial. First, it should be noted that LME is not unique to MS, although with differences in morphology, it can be seen at a similar rate in some neoplastic, infectious, and neuroinflammatory diseases as well as in heathy controls [77]. LME is detectable in neurosarcoidosis (highest proportions), NMO and NMOSD (lowest proportions), anti−myelin oligodendrocyte glycoprotein and anti−N-methyl-D-aspartate -antibody mediated diseases as well as autoimmune glial fibrillary acidic protein astrocytopathy.

Concerning the association between LME and measures of clinical impairment and radiological progression seen in pwMS, we here focus only on evidence gathered from HFI-based studies. Initial work from Harrison et al. [77] used a 3D T_2-w magnetization-prepared fluid-attenuated inversion recovery (MPFLAIR) along with a T_1-w MPRAGE obtained before, 20 (T_2-w MPFLAIR) and 3 (T_1-w MP$_2$RAGE) minutes after the injection of Gd-DTPA. They detected the presence of LME in 90% of the 29 studied pwMS and noted that LME was associated with a reduction in cortical GM volume.

Jonas and collaborators imaged 31 pwMS (21 pwRRMS, 7 pwSPMS, and 3 people with primary progressive multiple sclerosis [pwPPMS]) using the same approach [77] and detected LME in 81% of the enrolled people [94]. Over 2 years, they found more progressors at the EDSS score in the LME + group relative to the LME − group.

Zurawski et al. [81] imaged 30 pwRRMS using a 3D T_2-w FLAIR obtained before and 10 minutes after the injection of Gd-DTPA. They detected LME in two-third of the pwRRMS and found that LME + people had longer disease duration, larger WM lesions, and CL volume. They also detected a correlation between the number of LME foci and that of CL and thalamic lesions. Contemporaneously, Ighani et al. [75] imaged 41 pwMS (31 pwRRMS, 5 pwSPMS, and 5 pwPPMS) using the same imaging methods as in Ref. [77] and detected LME in 81 pwMS and 60% of the five healthy controls. They did not confirm the association between LME and CL volume seen in [82] but detected an association between LME and hippocampus and WM lesions burden as well as cortical thinning.

Subsequent work from Mizell et al. [80] found an association between LME and retinal thinning in 40 pwMS and confirmed the association between LME and cortical thinning.

Thus, clinical evidence parallels the histopathologic concept suggesting that meningeal inflammation, potentially visualizable with LME, is associated with more widespread cortical injury.

It is noteworthy that B-cell depletion is not effective in reducing the number and volume of LME foci when administered either intrathecally [83] or endogenously [84].

Clinical application of high-field imaging

7 T MRI was approved for clinical use in the brain, upper extremities, and lower extremities in 2017 [85]. Nevertheless, as of today, 7 T MRI has not made its use for routine clinical operations, apart from a few sites in the world. A few obstacles still impede clinical translation. First, implants with a safety record at 1.5 and 3 T cannot be assumed to be suitable for the high-field magnetic resonance environment. In one study, 5 out of 39 implants failed safety testing at 7 T, including two stents, one heel implant, and one fibular implant [86]. Such results prompt safety-based restrictions to many people who might otherwise benefit from HFI. Other factors associated with HFI, while not strictly safety

concerns, could present a detriment to patient experience. For example, dizziness, possibly related to vestibular effects has been noted in a substantial minority of patients who move in and out of the high-field environment [87]. At 7 T, peripheral nerve stimulation has been noted to occur in approximately 63% of patients, with more than 11% of patients describing it as "moderately to very uncomfortable" [88].

Finally, the cost of 7 T MRI systems is higher than that of lower field magnets and siting requirements are more restrictive, limiting HFI to a subset of people only. While the expense of 7 T MRI may decline with increased commercial availability, greater financial expenditure associated with 7 T MRI will have to be justified by clinical and or research utility.

From a research standpoint, the advantage of 7 T over lower field strength MRI is undeniable though likely limited to the fields indicated in previous sections. Incorporating 7 T into clinical trials will certainly prove advantageous when assessing the effect of experimental treatments in reducing CL or PRL formation. Designing such studies may prove more complex but will certainly enrich the field.

Technical challenges associated with the use of high-field imaging

Despite the undeniable advantages, HFI presents a few unique challenges. Perhaps most importantly, B_1 field heterogeneities are more pronounced at 7 T, which results in increased image inhomogeneities [89]. Most significantly, signal dropout at 7 T challenges imaging the posterior fossa and the identification of lesions located in this region. This is a significant drawback as brainstem and cerebellar lesions are prominent in MS and contribute substantially to disability. This has been partially mitigated with the use of the MP$_2$RAGE technique which corrects for B1 field inhomogeneities [90], though further work is needed.

Susceptibility effects are increased in the HFI. While these effects have been harnessed to assess for differences in iron content and venous structures with deoxygenated hemoglobin, they also lead to increased susceptibility artifacts. These artifacts can obscure anatomy at 7 T, particularly in regions where there is variance of the local magnetic field, such as along the skull base or in the spinal canal. Spinal imaging is also particularly challenging at 7 T due to the augmented artifacts related to respiratory and cardiac motion. Such challenges may explain why HFI studies focusing on the spinal cord in MS are scarce.

7 T evaluation is largely GRE-based, as this technique is associated with substantially less SAR. SE sequences are relatively insensitive to magnetic susceptibility artifacts. The relative lack of availability of SE sequences in the 7 T environment due to SAR considerations can lead to frequent occurrences of these artifacts. Future progress in 7 T imaging about both clinical utility and improved disease characterization will depend on the development of sequences with low SAR as well as the development of other mitigating strategies such as parallel imaging and multislice imaging [91].

Spinal cord imaging is also significantly hindered by the lack of coils to study the thoracic and lumbar portions of the spine [5]. Strategies to mitigate cardiac and respiratory movements are also warranted.

Summary and conclusions

The use of HFI has led to the discovery of novel biomarkers of disease. PRL, the CVS sign, and CL may all be visualized at multiple field strengths (1.5, 3, and 7 T) although with likely compromised sensitivity. When detected, PRL, CVS, and CL can increase MS diagnostic certainty due to the high level of specificity. Incorporating this imaging biomarker into routine clinical practice and clinical guidelines may increase the accuracy of MS diagnoses [92], stratify an individual patient's risk for disease progression, and inform decisions regarding disease-modifying therapy.

Against these undeniable advancements, HFI remains limited and far from clinical translation.

Technical challenges and costs are the main factors hindering its advancement to clinical practice.

For the near future, the use of HFI in pwMS will likely remain confined to select research institutions. Notwithstanding these considerations, work at HFI remains to be supported. The unique improvement in the accuracy of disease detection is key to novel biomarkers discovery and the advancement of our understanding of MS [93].

Acknowledgments

This work was supported by the generosity of the National MS Society (RG1901-33190: FB) the National MS Society Medical Student Mentorship Program (KB).

Conflict of interest

None of the authors has any conflict relevant to this manuscript.

References

[1] Bagnato F, Gauthier SA, Laule C, Moore GRW, Bove R, Cai Z, et al.NAIMS Cooperative Imaging mechanisms of disease progression in multiple sclerosis: beyond brain atrophy. J Neuroimaging 2020;30:251–66.
[2] Duyn JH. The future of ultra-high field MRI and fMRI for study of the human brain. Neuroimage 2012;62:1241–8.
[3] Sodickson DK, Manning WJ. Simultaneous acquisition of spatial harmonics (SMASH): fast imaging with radiofrequency coil arrays. Magn Reson Med 1997;38:591–603.
[4] Mainero C, Benner T, Radding A, van der Kouwe A, Jensen R, Rosen BR, et al. In vivo imaging of cortical pathology in multiple sclerosis using ultra-high field MRI. Neurology 2009;73:941–8.
[5] Dula AN, Pawate S, Dortch RD, Barry RL, George-Durrett KM, Lyttle BD, et al. Magnetic resonance imaging of the cervical spinal cord in multiple sclerosis at 7T. Mult Scler 2016;22:320–8.
[6] Kollia K, Maderwald S, Putzki N, Schlamann M, Theysohn JM, Kraff O, et al. First clinical study on ultra-high-field MR imaging in patients with multiple sclerosis: comparison of 1.5T and 7T. AJNR Am J Neuroradiol 2009;30:699–702.
[7] Mistry N, Tallantyre EC, Dixon JE, Galazis N, Jaspan T, Morgan PS, et al. Focal multiple sclerosis lesions abound in 'normal appearing white matter. Mult Scler 2011;17:1313–23.
[8] Sinnecker T, Dörr J, Pfueller CF, Harms L, Ruprecht K, Jarius S, et al. Distinct lesion morphology at 7-T MRI differentiates neuromyelitis optica from multiple sclerosis. Neurology 2012;79:708–14.
[9] Yao B, Bagnato F, Matsuura E, Merkle H, van Gelderen P, Cantor FK, et al. Chronic multiple sclerosis lesions: characterization with high-field-strength MR imaging. Radiology 2012;262:206–15.

[10] Jakimovski D, Bittner S, Zivadinov R, Morrow SA, Benedict RH, Zipp F, et al. Multiple sclerosis. Lancet 2023;S0140-6736(23) 01473.
[11] Mainero C, Treaba CA, Barbuti E. Imaging cortical lesions in multiple sclerosis. Curr Opin Neurol 2023;36:222–8.
[12] Bruschi N, Boffa G, Inglese M. Ultra-high-field 7-T MRI in multiple sclerosis and other demyelinating diseases: from pathology to clinical practice. Eur Radiol Exp 2020;4:59–71.
[13] Schindler MK, Sati P, Reich DS. Insights from ultrahigh field imaging in multiple sclerosis. Neuroimaging Clin N Am 2017;27:357–66.
[14] Craelius W, Migdal MW, Luessenhop CP, Sugar A, Mihalakis I. Iron deposits surrounding multiple sclerosis plaques. Arch Pathol Lab Med 1982;106:397–9.
[15] Connor JR, Menzies S,L. Cellular management of iron in the brain. J Neurol Sci 1995;134(Suppl):33–44.
[16] Hametner S, Wimmer I, Haider L, Pfeifenbring S, Brück W, Lassmann H. Iron and neurodegeneration in the multiple sclerosis brain. Ann Neurol 2013;74:848–61.
[17] Lassmann H, Brück W, Lucchinetti CF. The immunopathology of multiple sclerosis: an overview. Brain Pathol 2007;17:210–18.
[18] Prineas JW, Kwon EE, Cho ES, Sharer LR, Barnett MH, Oleszak EL, et al. Immunopathology of secondary-progressive multiple sclerosis. Ann Neurol 2001;50:646–57.
[19] Kuhlmann T, Ludwin S, Prat A, Antel J, Brück W, Lassmann H. An updated histological classification system for multiple sclerosis lesions. Acta Neuropathol 2017;133:13–24.
[20] Bagnato F, Hametner S, Yao B, van Gelderen P, Merkle H, Cantor FK, et al. Tracking iron in multiple sclerosis: a combined imaging and histopathological study at 7 Tesla. Brain 2011;134:3602–15.
[21] Bagnato F, Sati P, Hemond CC, Elliott C, Gauthier SA, Harrison DM, et al.on behalf of the NAIMS Cooperative Imaging chronic active lesions in multiple sclerosis: a consensus statement. Brain 2024;147(9):2913–33.
[22] Absinta M, Sati P, Schindler M, Leibovitch EC, Ohayon J, Wu T, et al. Persistent 7-tesla phase rim predicts poor outcome in new multiple sclerosis patient lesions. J Clin Invest 2016;126:2597–609.
[23] Dal-Bianco A, Grabner G, Kronnerwetter C, Weber M, Höftberger R, Berger T, et al. Slow expansion of multiple sclerosis iron rim lesions: pathology and 7 T magnetic resonance imaging. Acta Neuropathol 2017;133:25–42.
[24] Gillen KM, Mubarak M, Park C, Ponath G, Zhang S, Dimov A, et al. QSM is an imaging biomarker for chronic glial activation in multiple sclerosis lesions. Ann Clin Transl Neurol 2021;8:877–86.
[25] Ng Kee Kwong KC, Mollison D, Meijboom R, York EN, Kampaite A, Thrippleton MJ, et al. The prevalence of paramagnetic rim lesions in multiple sclerosis: a systematic review and meta-analysis. PLoS One 2021;16(9):e0256845.
[26] Absinta M, Sati P, Fechner A, Schindler MK, Nair G, Reich DS. Identification of chronic active multiple sclerosis lesions on 3T MRI. AJNR Am J Neuroradiol 2018;39:1233–8.
[27] Hemond CC, Reich DS, Dundamadappa SK. Paramagnetic rim lesions in multiple sclerosis: comparison of visualization at 1.5-T and 3-T MRI. AJR Am J Roentgenol 2022;219:120–31.
[28] Airas L, Yong VW. Microglia in multiple sclerosis - pathogenesis and imaging. Curr Opin Neurol 2022;35:299–306.
[29] Clarke MA, Witt AA, Robison RK, Fleishman S, Combes AJE, Houston D, et al. Cervical spinal cord susceptibility-weighted MRI at 7T: application to multiple sclerosis. Neuroimage 2023;284:120460.
[30] Clarke MA, Cheek R, Kazimuddin HF, Hernandez B, Clarke R, McKnight CD, et al. Paramagnetic rim lesions and the central vein sign: characterizing multiple sclerosis imaging markers. J Neuroimaging 2023;34(1):86–94.
[31] Choi S, Lake S, Harrison DM. Evaluation of the blood-brain barrier, demyelination, and neurodegeneration in paramagnetic rim lesions in multiple sclerosis on 7 Tesla MRI. J Magn Reson Imaging 2024;59(3):941–51.
[32] Kazimuddin H.F., Wang J., Hernandez B., Sun L., Eaton J.A., Taylor S., et al. Paramagnetic rim lesions and their relationship with neurodegeneration and clinical disability at the time of multiple sclerosis diagnosis. In: ACTRIMS poster present; 2024.
[33] Dal-Bianco A, Schranzer R, Grabner G, Lanzinger M, Kolbrink S, Pusswald G, et al. Iron rims in patients with multiple sclerosis as neurodegenerative marker? A 7-Tesla magnetic resonance study. Front Neurol 2021;12(2):632749.
[34] Treaba CA, Conti A, Klawiter EC, Barletta VT, Herranz E, Mehndiratta A, et al. Cortical and phase rim lesions on 7 T MRI as markers of multiple sclerosis disease progression. Brain Commun 2021;3(3) fcab134.

[35] Meaton I, Altokhis A, Allen CM, Clarke MA, Sinnecker T, Meier D, et al. Paramagnetic rims are a promising diagnostic imaging biomarker in multiple sclerosis. Mult Scler 2022;28:2212–20.
[36] Maggi P, Sati P, Nair G, Cortese ICM, Jacobson S, Smith BR, et al. Paramagnetic rim lesions are specific to multiple sclerosis: an international multicenter 3T MRI study. Ann Neurol 2020;88:1034–42.
[37] Rae-Grant AD, Wong C, Bernatowicz R, Fox RJ. Observations on the brain vasculature in multiple sclerosis: a historical perspective. Mult Scler Relat Disord 2014;3:156–62.
[38] Ge Y, Zohrabian VM, Grossman RI. Seven-Tesla magnetic resonance imaging: new vision of microvascular abnormalities in multiple sclerosis. Arch Neurol 2008;65:812–16.
[39] Hammond KE, Metcalf M, Carvajal L, Okuda DT, Srinivasan R, Vigneron D, et al. Quantitative in vivo magnetic resonance imaging of multiple sclerosis at 7 Tesla with sensitivity to iron. Ann Neurol 2008;64:707–13.
[40] Tallantyre EC, Brookes MJ, Dixon JE, Morgan PS, Evangelou N, Morris PG. Demonstrating the perivascular distribution of MS lesions in vivo with 7-Tesla MRI. Neurology 2008;70:2076–8.
[41] Sati P, Oh J, Constable RT, Evangelou N, Guttmann CR, Henry RG, et al.NAIMS Cooperative The central vein sign and its clinical evaluation for the diagnosis of multiple sclerosis: a consensus statement from the North American Imaging in Multiple Sclerosis Cooperative. Nat Rev Neurol 2016;12:714–22.
[42] Tallantyre EC, Morgan PS, Dixon JE, Al-Radaideh A, Brookes MJ, Evangelou N, et al. A comparison of 3T and 7T in the detection of small parenchymal veins within MS lesions. Invest Radiol 2009;44:491–4.
[43] Tallantyre EC, Dixon JE, Donaldson I, Owens T, Morgan PS, Morris PG, et al. Ultra-high-field imaging distinguishes MS lesions from asymptomatic white matter lesions. Neurology 2011;76:534–9.
[44] Kilsdonk ID, Wattjes MP, Lopez-Soriano A, Kuijer JP, de Jong MC, de Graaf WL, et al. Improved differentiation between MS and vascular brain lesions using FLAIR* at 7 Tesla. Eur Radiol 2014;24:841–9.
[45] Kilsdonk ID, Lopez-Soriano A, Kuijer JP, de Graaf WL, Castelijns JA, Polman CH, et al. Morphological features of MS lesions on FLAIR* at 7 T and their relation to patient characteristics. J Neurol 2014;261:1356–64.
[46] Kuchling J, Ramien C, Bozin I, Dörr J, Harms L, Rosche B, et al. Identical lesion morphology in primary progressive and relapsing-remitting MS—an ultrahigh field MRI study. Mult Scler (Houndmills, Basingstoke, Engl) 2014;20:1866–71.
[47] Samaraweera AP, Clarke MA, Whitehead A, Falah Y, Driver ID, Dineen RA, et al. The central vein sign in multiple sclerosis lesions is present irrespective of the T2* sequence at 3 T. J Neuroimaging 2017;27:114–21.
[48] Dixon JE, Simpson A, Mistry N, Evangelou N, Morris PG. Optimisation of T_2*-weighted MRI for the detection of small veins in multiple sclerosis at 3 T and 7 T. Eur J Radiol 2013;82:719–27.
[49] Okromelidze L, Patel V, Singh RB, Lopez Chiriboga AS, Tao S, Zhou X, et al. Central vein sign in multiple sclerosis: a comparison study of the diagnostic performance of 3T versus 7T MRI. AJNR Am J Neuroradiol 2023;10:3174.
[50] Solomon AJ, Schindler MK, Howard DB, Watts R, Sati P, Nickerson JP, et al. "Central vessel sign" on 3T FLAIR* MRI for the differentiation of multiple sclerosis from migraine. Ann Clin Transl Neurol 2015;3:82–7.
[51] Mistry N, Abdel-Fahim R, Samaraweera A, Mougin O, Tallantyre E, Tench C, et al. Imaging central veins in brain lesions with 3-T T2*-weighted magnetic resonance imaging differentiates multiple sclerosis from microangiopathic brain lesions. Mult Scler 2016;22:1289–96.
[52] Kister I, Herbert J, Zhou Y, Ge Y. Ultrahigh-field MR (7 T) imaging of brain lesions in neuromyelitis optica. Mult Scler Int 2013;2013:398259.
[53] Lassmann H. Multiple sclerosis pathology. Cold Spring Harb Perspect Med 2018;8:1–16.
[54] Peterson JW, Bo L, Mork S, Chang A, Trapp BD. Transected neurites, apoptotic neurons, and reduced inflammation in cortical multiple sclerosis lesions. Ann Neurol 2001;50:389–400.
[55] Kilsdonk ID, Jonkman LE, Klaver R, van Veluw SJ, Zwanenburg JJM, Kuijer JPA, et al. Increased cortical grey matter lesion detection in multiple sclerosis with 7T MRI: a postmortem verification study. Brain 2016;139:1472–82.
[56] Schmierer K, Parkes HG, So PW, An SF, Brandner S, Ordidge RJ, et al. High field (9.4 Tesla) magnetic resonance imaging of cortical grey matter lesions in multiple sclerosis. Brain 2010;133:858–67.
[57] Pitt D, Boster A, Pei W, Wohleb E, Jasne A, Zachariah CR, et al. Imaging cortical lesions in multiple sclerosis with ultra-high-field magnetic resonance imaging. Arch Neurol 2010;67:812–18.
[58] Yao B, Hametner S, van Gelderen P, Merkle H, Chen C, Lassmann H, et al. 7 Tesla magnetic resonance imaging to detect cortical pathology in multiple sclerosis. PLoS One 2014;9(10):e108863.
[59] de Graaf WL, Kilsdonk ID, Lopez-Soriano A, Zwanenburg JJ, Visser F, Polman CH, et al. Clinical application of multi-contrast 7-T MR imaging in multiple sclerosis: increased lesion detection compared to 3 T confined to grey matter. Eur Radiol 2013;23:528–40.

[60] Trattnig S, Bogner W, Gruber S, Szomolanyi P, Juras V, Robinson S, et al. Clinical applications at ultrahigh field (7 T). Where does it make the difference? NMR Biomed 2016;29:1316−34.
[61] Nielsen AS, Kinkel RP, Tinelli E, Benner T, Cohen-Adad J, Mainero C. Focal cortical lesion detection in multiple sclerosis: 3 Tesla DIR versus 7 Tesla FLASH-T2. J Magn Reson Imaging 2012;35:537−42.
[62] Sinnecker T, Mittelstaedt P, Dörr J, Pfueller CF, Harms L, Niendorf T, et al. Multiple sclerosis lesions and irreversible brain tissue damage: a comparative ultra-high-field strength magnetic resonance imaging study. Arch Neurol 2012;69:739−45.
[63] Maranzano J, Dadar M, Rudko DA, De Nigris D, Elliott C, Gati JS, et al. Comparison of multiple sclerosis cortical lesion types detected by multicontrast 3T and 7T MRI. Am J Neuroradiol 2019;40:1162−9.
[64] Cohen-Adad J, Benner T, Greve D, Kinkel RP, Radding A, Fischl B, et al. In vivo evidence of disseminated subpial T2* signal changes in multiple sclerosis at 7 T: a surface-based analysis. Neuroimage 2011;57:55−62.
[65] Mainero C, Louapre C, Govindarajan S, Gianni C, Nielsen AS, Cohen-Adad J, et al. A gradient in cortical pathology in multiple sclerosis by in vivo quantitative 7T imaging. Brain 2015;138:932−45.
[66] Barletta V, Herranz E, Treaba CA, Mehndiratta A, Ouellette R, Mangeat G, et al. Quantitative 7-Tesla imaging of cortical myelin changes in early multiple sclerosis. Front Neurol 2021;3(12):714820.
[67] Louapre C, Govindarajan ST, Giannì C, Langkammer C, Sloane JA, Kinkel RP, et al. Beyond focal cortical lesions in MS: an in vivo quantitative and spatial imaging study at 7T. Neurology 2015;85:1702−9.
[68] Harrison DM, Oh J, Roy S, Wood ET, Whetstone A, Seigo MA, et al. Thalamic lesions in multiple sclerosis by 7T MRI: clinical implications and relationship to cortical pathology. Mult Scler 2015;21:1139−50.
[69] Treaba CA, Granberg TE, Sormani MP, Herranz E, Ouellette RA, Louapre C, et al. Longitudinal characterization of cortical lesion development and evolution in multiple sclerosis with 7.0-T MRI. Radiology 2019;291:740−9.
[70] Beck ES, Mullins WA, Dos Santos Silva J, Filippini S, Parvathaneni P, Maranzano J, et al. Cortical lesions uniquely predict motor disability accrual and form rarely in the absence of new white matter lesions in multiple sclerosis. medRxiv [Prepr] 2023; 2023.09.22.23295974.
[71] Louveau A, Herz J, Alme MN, Salvador AF, Dong MQ, Viar KE, et al. CNS lymphatic drainage and neuroinflammation are regulated by meningeal lymphatic vasculature. Nat Neurosci 2018;21:1380−91.
[72] Serafini B, Rosicarelli B, Magliozzi R, Stigliano E, Aloisi F. Detection of ectopic B-cell follicles with germinal centers in the meninges of patients with secondary progressive multiple sclerosis. Brain Pathol 2004;14:164−74.
[73] Magliozzi R, Howell O, Vora A, Serafini B, Nicholas R, Puopolo M, et al. Meningeal B-cell follicles in secondary progressive multiple sclerosis associate with early onset of disease and severe cortical pathology. Brain 2007;130:1089−104.
[74] Bagnato F. Uncovering and characterizing multiple sclerosis lesions: the aid of fluid-attenuated inversion recovery images in the presence of gadolinium contrast agent. J Neuroimaging 2009;19:201−4.
[75] Ighani M, Jonas S, Izbudak I, Choi S, Lema-Dopico A, Hua J, et al. No association between cortical lesions and leptomeningeal enhancement on 7-Tesla MRI in multiple sclerosis. Mult Scler 2020;26:165−76.
[76] Absinta M, Cortese IC, Vuolo L, Nair G, de Alwis MP, Ohayon J, et al. Leptomeningeal gadolinium enhancement across the spectrum of chronic neuroinflammatory diseases. Neurology 2017;88:1439−44.
[77] Harrison DM, Wang KY, Fiol J, Naunton K, Royal 3rd. W, Hua J, et al. Leptomeningeal enhancement at 7T in multiple sclerosis: frequency, morphology, and relationship to cortical volume. J Neuroimaging 2017;27:461−8.
[78] Hildesheim FE, Ramasamy DP, Bergsland N. Leptomeningeal, dura mater and meningeal vessel wall enhancements in multiple sclerosis. Mult Scler Relat Disord 2021;47:102653.
[79] Makshakov G, Magonov E, Totolyan N, Nazarov V, Lapin S, Mazing A, et al. Leptomeningeal contrast enhancement is associated with disability progression and grey matter atrophy in multiple sclerosis. Neurol Res Int 2017;2017:8652463.
[80] Mizell R, Chen H, Lambe J, Saidha S, Harrison DM. Association of retinal atrophy with cortical lesions and leptomeningeal enhancement in multiple sclerosis on 7T MRI. Mult Scler 2022;28:393−405.
[81] Zivadinov R, Ramasamy DP, Vaneckova M, Gandhi S, Chandra A, Hagemeier J, et al. Leptomeningeal contrast enhancement is associated with progression of cortical atrophy in MS: a retrospective, pilot, observational longitudinal study. Mult Scler 2017;23:1336−45.
[82] Zurawski J, Tauhid S, Chu R, Khalid F, Healy BC, Weiner HL, et al. 7T MRI cerebral leptomeningeal enhancement is common in relapsing-remitting multiple sclerosis and is associated with cortical and thalamic lesions. Mult Scler 2020;26:177−87.

[83] Bhargava P, Wicken C, Smith MD, Strowd RE, Cortese I, Reich DS, et al. Trial of intrathecal rituximab in progressive multiple sclerosis patients with evidence of leptomeningeal contrast enhancement. Mult Scler Relat Disord 2019;30:136–40.
[84] Dahal S, Allette YM, Naunton K, Harrison DM. A pilot trial of ocrelizumab for modulation of meningeal enhancement in multiple sclerosis. Mult Scler Relat Disord 2023;81:105344.
[85] Barisano G, Sepehrband F, Ma S, Jann K, Cabeen R, Wang DJ, et al. Clinical 7 T MRI: are we there yet? A review about magnetic resonance imaging at ultra-high field. Br J Radiol 2019;92:20180492.
[86] Feng DX, McCauley JP, Morgan–Curtis FK, Salam RA, Pennell DR, Loveless ME, et al. Evaluation of 39 medical implants at 7.0 T. Br J Radiol 2015;88:20150633.
[87] Versluis MJ, Teeuwisse WM, Kan HE, van Buchem MA, Webb AG, van Osch MJ. Subject tolerance of 7 T MRI examinations. J Magn Reson Imaging 2013;38:722–5.
[88] Hansson B, Markenroth Bloch K, Owman T, Nilsson M, Lätt J, Olsrud J, et al. Subjectively reported effects experienced in an actively shielded 7T MRI: a large-scale study. J Magn Reson Imaging 2020;52:1265–76.
[89] Sati P. Diagnosis of multiple sclerosis through the lens of ultra-high-field MRI. J Magn Reson Imaging 2018;291:101–9.
[90] Marques JP, Kober T, Krueger G, van der Zwaag W, Van de Moortele PF, Gruetter R. MP2RAGE, a self bias-field corrected sequence for improved segmentation and T1-mapping at high field. Neuroimage 2010;49:1271–81.
[91] Ineichen BV, Beck ES, Piccirelli M, Reich DS. New prospects for ultra-high-field magnetic resonance imaging in multiple sclerosis. Invest Radiol 2021;56:773–84.
[92] Cagol A, Cortese R, Barakovic M, Schaedelin S, Ruberte E, Absinta M, et al.MAGNIMS Study Group Diagnostic performance of cortical lesions and the central vein sign in multiple sclerosis. JAMA Neurol 2023;81(2):143–53.
[93] Bagnato F, Gore JC. Ultra-high-field (7.0 Tesla and above) MRI is now necessary to make the next step forward in understanding MS pathophysiology – YES. Mult Scler 2017;23:372–3.
[94] Jonas SN, Izbudak I, Frazier AA, Harrison DM. Longitudinal Persistence of Meningeal Enhancement on Postcontrast 7T 3D-FLAIR MRI in Multiple Sclerosis. American Journal of Neuroradiology 2018;39(10):1799–805. Available from: https://doi.org/10.3174/ajnr.A5796 In press.

PART 3

Use of other imaging acquisition and analysis modalities in multiple sclerosis

CHAPTER 18

Positron emission tomography imaging in multiple sclerosis

Steven Cicero, Caleb Hansel, Eero Rissanen and Tarun Singhal

Department of Neurology, Ann Romney Center for Neurologic Diseases, Brigham and Women's Hospital, Harvard Medical School, Boston, MA, United States

OUTLINE

Introduction	344
Positron emission tomography imaging	344
Microglial activation and translocator protein–positron emission tomography	345
Translocator protein–positron emission tomography ligands	346
Translocator protein–positron emission tomography and gray matter pathology in multiple sclerosis	347
Translocator protein–positron emission tomography and white matter	347
Translocator protein–positron emission tomography and symptom pathogenesis	350
Translocator protein–positron emission tomography and prognostication in multiple sclerosis	351
Translocator protein–positron emission tomography and treatment effects	353
Beyond translocator protein–positron emission tomography: other glial imaging targets and PET ligands	356
Beyond translocator protein–positron emission tomography: PET imaging of nonimmune mechanisms in MS	356
Future directions and conclusion	356
References	357

Introduction

Multiple sclerosis (MS) is a neuroinflammatory and neurodegenerative disease of the central nervous system (CNS) [1]. About 90% of MS patients are initially diagnosed with a relapsing–remitting course of disease, experiencing periods of symptom "flare ups" and progression followed by improvement and stability. About half of relapsing–remitting patients eventually progress to a steady state of symptom development within 10–20 years of disease onset. These patients are said to have secondary-progressive multiple sclerosis (SPMS). This contrasts from a small population of people (about 10%–15%) who initially experience a gradual progression of symptoms, without steady remission. This population is diagnosed with primary-progressive multiple sclerosis (PPMS) and rarely experiences a remission period. If a relapse does occur, their condition has been historically labeled as progressive relapsing MS [1]. There have been several recent updates to MS diagnostic criteria and to the clinical descriptors of MS disease course [2–4]. Further, there has been increasing realization of MS disease progression independent of clinical relapses or apparent magnetic resonance imaging (MRI) disease activity [5].

While MRI has revolutionized the diagnosis and clinical monitoring of MS, there are several limitations to MRI in relation to identifying and tracking progressive MS. Conventional assessments such as MRI lesion load are weak predictors of clinical disability progression in longitudinal MS studies [6]. Further, it is common clinical experience to take care of MS patients who are getting worse clinically without their MRI scans demonstrating any apparent changes in terms of lesion load or contrast enhancement. Moreover, even more advanced measures such as volumetric assessments on MRI have limited predictive ability for clinical disability and disease progression in MS [6,7]. In addition, conventional MRI measures do not provide an explanation for some of the "hidden" symptoms of MS, such as depression and fatigue. These observations emphasize the presence of an additional pathology that may account for disease progression and symptom pathogenesis in MS [6].

Positron emission tomography imaging

Positron emission tomography (PET) is a molecular imaging technique that uses radioisotopes to evaluate biological processes, in vivo [8]. Positrons are positively charged electron antimatter particles that are emitted by radioisotopes that have an excess of protons as compared to neutrons in the atomic nucleus and attain nuclear stability by emission of the positron. The most commonly used positron emitting radioisotopes in medical applications are carbon-11, oxygen-15, fluorine-18, and nitrogen-13. These isotopes have a physical half-life of approximately 20, 2, 110, and 10 minutes, respectively. These radionucleotides are isotopes of natural building blocks of biomolecules, and thereby, can be used to trace biological processes in vivo without affecting the biochemical processes themselves. PET is based on the tracer kinetic principle which states that the tracers are administered in such small amounts, often picomolar ranges, that they do not interfere with the primary biological process. On radioactive decay, the emitted positrons interact with electrons in their environment, and thereby, there is an interaction between the antimatter positron and the matter electron particle, which leads to their annihilation and conversion of their mass into energy according to the $E = mc^2$ equation. The energy of the emitted gamma particles is 511 keV and two such gamma rays are emitted in almost opposite directions, which are then detected by the PET scanners and the information obtained is converted into a special distribution map of radioactivity and converted into an image for biomedical purposes. PET is a quantitative imaging tool that can allow quantification of biological processes, using various mathematical approaches, including semiquantitative evaluation and tracer kinetic modeling. [F-18]Fluorodeoxyglucose-PET has been widely used in oncology and cardiology and is Food and Drug Administration (FDA) approved for use in neurology for dementia and refractory epilepsy evaluation. Similarly, new tracers targeting amyloid and tau deposition in the brain have also been FDA approved. From a practical standpoint, fluorine-18 labeled PET radiopharmaceuticals are most clinically relevant as they have a long half-life, that is, 110 minutes, are more stable and provide a greater likelihood of a higher signal-to-noise ratio, better image quality, and logistical feasibility. These radiopharmaceuticals can be produced in a central radiopharmacy and can be transported to various institutions because of their longer half-life. On the other hand, for a carbon-11 labeled radiopharmaceutical, an on-site cyclotron is necessary because of the short half-life of carbon-11, that is, 20 minutes. Moreover, images obtained using a short half-life isotope may lead to more noisy images given lower count statistics and poor signal-to-noise ratio by the time equilibrium is achieved in vivo following tracer administration.

Microglial activation and translocator protein—positron emission tomography

Several lines of evidence suggest that microglial activation may play a critical role in pathogenesis of MS symptoms and disease progression [6,9—11]. It is critical to be able to measure microglial activation in vivo in MS patients in order to overcome the limitations of conventional MRI in MS, which were delineated above. Noninvasive assessment of microglial activation can be accomplished using PET in MS, and has also been employed in various other neurologic and neuropsychiatric diseases [12].

One of the most common markers used to quantitatively measure innate immune activity in the CNS is the expression of the 18 kDa-translocator protein (TSPO). The TSPO molecule is located on the outer mitochondrial membrane and is composed of five transmembrane domains with a primarily helical structure, which functions to transport cholesterol from the cytosol into the mitochondria for potential steroid synthesis [13,14]. Historically, TSPO was identified as a peripheral binding site for diazepam (or a "peripheral benzodiazepine receptor"), but was soon recognized as being overexpressed at the tissue level under disease conditions in the CNS [15,16]. Initial conclusions drawn from studying rodents suggested a correlation between TSPO expression and proinflammatory microglial activation [16]. This observation was supported by early studies, which found an elevated TSPO presence throughout diseased brains compared to healthy controls, although distribution varied based on stage and type of disease [12]. In MS patients, TSPO binding has been shown to be increased in active lesions and perilesional sites, as well as in normal-appearing white matter (NAWM) and the thalamus when compared to healthy controls [6]. In animal models, there is a known overexpression of TSPO in pathologic microglia; however, studies of MS lesions in humans have suggested that the number of TSPO molecules is not elevated on a "per-cell" level [17,18]. On the other hand, the increased TSPO-PET signal seen in MS patients reflects an increased density of glial cells at the tissue level. In the perilesional MS tissues, the majority (55%–60%) of TSPO + ve cells were found to be CD68 + (microglia/macrophages) while approximately, 20%–25% cells were GFAP + ve (astrocytes) [17]. Microglia and astrocytes synergistically contribute to MS pathology and hence, although it has been criticized, the ability of TSPO-PET to demonstrate increased density of these cells is actually a strength of the technology rather than a limitation, in the clinical context [17,18].

Translocator protein—positron emission tomography ligands

Several PET radioligands have been developed targeting TSPO, of which [C-11]PK11195 was one of the first, and has been most widely utilized and contributed significantly to our understanding of the role of innate immune activation in MS. Several studies using [C-11]PK11195 have found an increased uptake in active lesions, perilesional area, NAWM, and the thalamus as compared to healthy controls [19–23]. However, [C-11] PK11195 suffers from several disadvantages. The C-11 isotope has a 20-minute half-life and requires an on-site cyclotron to produce the radiopharmaceutical. This is not feasible at many centers, thereby limiting the practical utility of this PET tracer. Moreover, [C-11] PK11195 has a low binding affinity for TSPO ($K_i = 9.4$ nM) [24] and poor signal-to-noise ratio. Hence, a range of second-generation TSPO-PET radioligands was developed to overcome these limitations. [C-11]PBR28, an aryloxyanilide analog, is a second-generation TSPO-PET radioligand that has a higher TSPO-binding affinity ($K_i = 2.5$ nM) [25] than [C-11]PK11195 and has been widely used in studying various neurological diseases [12]. Tissue binding of [C-11]PBR28 and [C-11]PK11195 has been shown to correlate strongly with TSPO expression in MS brain specimens [18]. [F-18]PBR06 is an aryloxyanilide analog targeting TSPO and has the advantages of a longer half-life radioisotope (half-life of F-18 = 110 minutes), higher binding affinity to TSPO ($K_i = 1.0$ nM) and high lipophilicity (log D = 4.1), facilitating increased blood–brain barrier penetration [26]. [F-18]PBR06 has been validated as a marker of TSPO expression in various disease models and wild-type mice [27,28]. Several other TSPO-PET ligands including [F-18]DPA714, [C-11]DPA713, and [F-18]PBR111 have also reportedly been studied in MS [29,30]. Similar to [C-11]PK11195, the second-generation TSPO-PET radioligands have also provided evidence for widespread innate immune activation in MS that is present beyond conventionally measurable white matter lesions, as will be discussed in detail below [29–33].

While second-generation TSPO radioligands have higher specificity and improved signal-to-noise ratio, these PET ligands have demonstrated variable binding across individuals in the human population. Human subjects may be classified as "high-affinity," "medium-affinity," or "low-affinity" binders based on a genetic polymorphism [34]. A genetic polymorphism in the TSPO gene (rs6971) has been shown to be the cause of this heterogeneity in binding of these PET ligands. TSPO-PET studies involving human subjects predominantly focus on high- and medium-affinity binders and exclude low-affinity binders [34].

Translocator protein—positron emission tomography and gray matter pathology in multiple sclerosis

Although MS is considered to be a predominantly white matter disease, pathological studies have shown that there is gray matter inflammation in MS. In fact, gray matter atrophy has been shown to correlate more strongly with disease progression, as compared to overall atrophy, in MS [7]. TSPO-PET studies using [C-11]PK11195 have shown increased gray matter binding in clinically isolated syndrome and progressive MS patients as compared to healthy controls [21,35]. In progressive MS, gray matter binding has been shown to correlate with physical disability in progressive MS patients using [C-11]PK11195 [21] (Table 18.1). Subsequently, [C-11]PBR28 has also been shown to demonstrate higher cortical binding in MS patients [31]. Using [F-18]PBR06, our group has shown increased gray matter inflammation in progressive MS as compared to relapsing MS patients [32] (Table 18.1). Interestingly, many of the cortical regions affected by increased TSPO binding were in the limbic and paralimbic cortical areas. Using [C-11]PBR28 and [F-18]PBR06, it has been shown that cortical TSPO binding is linked with cortical atrophy in MS patients [31,40].

Interestingly, several studies have shown that thalamic binding of TSPO-PET ligands correlates strongly with physical disability as measured by expanded disability status scale (EDSS) and timed 25-foot walk (T25-FW) in MS patients [22,31,32]. We have shown that thalamic [F-18] PBR06 uptake predicts overall brain atrophy in MS patients [32]. This is consistent with widespread thalamic pathology that has been reported on neuropathological and radiological assessment in MS tissue specimens and clinical studies [41,42]. In terms of other subcortical regions, we have also shown increased [F-18]PBR06 binding in putamen in progressive versus relapsing MS patients [32]. Putaminal TSPO binding was also shown to correlate with disability measures [32].

Translocator protein—positron emission tomography and white matter

While some of the studies with TSPO-PET have demonstrated increased PET radioligand uptake in contrast enhancing lesions [43], the value of PET does not lie in identifying inflammation in lesions that have already clearly demonstrated active blood—brain barrier injury on routine MRI. In fact, pathological evaluation has clearly demonstrated that there is widespread microglial activation in the otherwise NAWM [44,45] that is not picked up by routine MRI evaluations.

Using [F-18]PBR06, we have observed widespread abnormalities in the otherwise NAWM in MS patients [46,47]. The NAWM TSPO-PET abnormality has been linked with magnetization transfer ratio and diffusion tensor imaging changes in MS [36,48].

We have also demonstrated that there is increased microglial activation in the NAWM as compared to T1 black holes and T2 lesions [46]. Similarly, using [C-11]PK11195,

TABLE 18.1 Translocator protein–positron emission tomography (TSPO-PET) studies on progressive multiple sclerosis (MS).

Study	Tracer	Cohort	Duration	Findings
Politis et al. 2012 [21]	[11C]-R-PK11195	10 RRMS; 8 SPMS; 8 HC	Cross-sectional	– Mean cortical BP_{ND} higher in SPMS vs HC, and in RRMS vs HC – As compared to HC, >100% increases in BP_{ND} in postcentral, middle frontal, anterior orbital, fusiform, and parahippocampal gyri in RRMS. In addition, >100% increases in precentral, superior parietal, lingual and anterior superior, medial, and inferior temporal gyri in SPMS. – Mean cortical BP_{ND} correlates with EDSS and MSIS-29 in SPMS, but not in RRMS – Mean WM BP_{ND} higher in SPMS vs HC and RRMS vs HC, but no correlation with disability
Rissanen et al. 2014 [22]	[11C]-R-PK11195	10 SPMS; 8 HC	Cross-sectional	– DVR higher in NAWM and thalamus of SPMS patients vs HC – Average DVR in FLAIR hyperintense lesions lower than in NAWM of SPMS patients – 57% of all lesions in the 10 patients had increased DVR in perilesional area of T1 hypointense lesions, as compared to surrounding NAWM (visual evaluation)
Herranz et al. 2016 [31]	[11C] PBR28	15 SPMS; 12 RRMS; 14 HC	Cross-sectional	– SUVR higher in SPMS vs HC in whole cortex, intra/ and leukocortical lesions, thalamus, hippocampus, basal ganglia, and NAWM, but among RRMS higher only in intracortical lesions and whole cortex – Cortical thinning associated with increased SUVR in thalamus in all MS, SPMS, and RRMS – In surface-based analyses, increase in cortical SUVR more diffuse in SPMS vs HC, and more localized in RRMS vs HC
Rissanen et al. 2018 [36]	[11C]-R-PK11195	10 RRMS, 10 SPMS; 17 HC	Cross-sectional	– DVR higher in NAWM in SPMS vs HC and SPMS vs RRMS, and higher in thalamus in SPMS vs HC. – DVR higher in cortex in RRMS vs HC, but not in SPMS. – DVR in Gd- T1 perilesional 3–6 mm rim from lesion edge higher in SPMS vs RRMS. – DVR lower within T1 lesions than in NAWM in SPMS, but not in RRMS.
Singhal et al. 2019 [32]	[18F] PBR06	7 RRMS; 5 SPMS; 5 HC	Cross-sectional	– Cortical SUVR higher in rolandic operculum, hippocampus, midcingulate, parahippocampus, olfactory, insula, and middle temporal gyrus in SPMS vs HC, – Cortical SUVR higher in supramarginal gyrus in SPMS vs RRMS – Subcortical GM (thalamus, caudate, putamen, and pallidum) SUVR higher in SPMS vs RRMS – In subcortical GM subregions, thalamic SUVR higher in SPMS vs RRMS and SPMS vs HC.

(Continued)

TABLE 18.1 (Continued)

Study	Tracer	Cohort	Duration	Findings
Bezukladova et al. 2020 [37]	[11C]-R-PK11195	40 RRMS; 15 SPMS; 15 HC	Cross-sectional	– DVR higher in whole NAWM in SPMS vs HC and RRMS vs HC – Of NAWM subregions, frontal DVR higher in SPMS vs RRMS, SPMS vs HC, and RRMS vs HC; and cingulate and deep (mainly periventricular) NAWM in SPMS vs HC and SPMS vs RRMS.
Herranz et al. 2020 [38]	[11C] PBR28	10 SPMS; 9 RRMS; 14 HC	Cross-sectional	– SUVR higher in active cortical lesions and abnormal q-T2 ROIs in RRMS and SPMS when compared to the HC cortical ROI – SUVR in NACGM higher in SPMS vs HC cortical ROI
Kang et al. 2021 [39]	[11C]-R-PK11195	9 SPMS; 6 PPMS; 16 HC	Longitudinal (2 years followup PET scans)	– DVR higher in NAWM, thalamus, putamen, hippocampus, and amygdala in PMS vs HC at baseline. WM lesional DVR in PMS lower than in HC WM. – No change in DVR in any region from baseline to 6 months. – From baseline to 12 months, increases in NAWM, cortical, thalamus, and putamen DVR – No further increases at 24 months – No change in EDSS

BP_{ND}, binding potential; *FLAIR*, fluid-attenuated inversion recovery; *Gd*, gadolinium; *MSIS-29*, multiple sclerosis impact scale; *NACGM*, normal-appearing cortical gray matter*q-T2*, quantitative T2; *ROI*, region of interest; *SUVR*, standardized uptake value ratio.

NAWM has been shown to have increased TSPO binding in MS patients, which was even higher than the uptake in the perilesional white matter, [22] and higher in NAWM in SPMS than in relapsing–remitting multiple sclerosis (RRMS) (Table 18.1) [21,36,37]. [C-11] PBR28 has also demonstrated increased glial activity in NAWM as compared to lesions in MS volunteers [29]. In a head-to-head comparison study, [F-18]PBR06 demonstrated a lower coefficient of variation as compared to [C-11]PBR28 [46].

Neuropathological studies have demonstrated that MS lesions can be classified as active, inactive, and mixed active/inactive lesions [45]. Active lesions have inflammatory cells both in the core and at the periphery of the lesion; inactive lesions do not have any concentration of inflammatory cells in or around the lesion; while mixed active/inactive lesions have presence of a rim of inflammatory cells at the periphery of the lesion but the core of the lesion does not show increased inflammatory cells [45]. On qualitative evaluation, lesional heterogeneity in terms of microglial activation can be seen using TSPO-PET [49].

In terms of quantitation of lesional binding, most non–contrast-enhancing T2 lesions in treated MS patients show decreased TSPO uptake as compared to surrounding white matter [29,46]. Our finding of increased TSPO uptake in T1 hypointense lesions as compared to T2 lesions underscores the destructive nature of T1 hypointense lesions [46]. Several studies have shown that TSPO-PET demonstrates rim of increased inflammation around MS lesions. Perilesional increased glial activation can be observed using various TSPO-PET ligands such as [C-11]PK11195, [F-18]DPA714, and [F-18]PBR06 [23,47,50].

In a study using [C-11]PK11195, iron positive rim lesions had higher tracer uptake than iron negative rim lesions [17]. Various other threshold base approaches have been reported to demonstrate increased perilesional TSPO-PET signal and individualized classification of lesions based on arbitrary thresholds has been attempted [50,51]. Our group has developed individualized, voxel-wise Z score approaches that can demonstrate increased widespread and perilesional TSPO uptake that can be evaluated in longitudinal studies [47,52]. In a case of progressive MS, we showed that although conventional MRI images demonstrated only two T2 bright lesions, [F-18]PBR06-PET was able to demonstrate widespread neuroinflammation in NAWM with additional precise localization of TSPO binding in the perilesional area around a brainstem T2 bright lesion (Fig. 18.1) [47].

Translocator protein—positron emission tomography and symptom pathogenesis

Fatigue, defined as an overwhelming sense of tiredness, lack of energy, or feeling of exhaustion has a lifetime prevalence of 80% in MS patients and has been shown to be the most disabling symptom in up to 60% of MS patients. Fatigue is also a leading cause of absence from work in MS [53]. However, pathogenesis or fatigue in MS has been unclear. Various hypotheses, such as neurochemical abnormalities, neuroendocrine, and functional disconnection hypothesis have been proposed to explain fatigue in MS. However, there is a lack of a unifying mechanism to account for fatigue pathogenesis in MS patients. Using [F-18]PBR06, our group has shown that fatigue scores in MS patients are strongly correlated with widespread microglial activation, particularly in the substantia nigra (Figs. 18.2 and 18.3) in MS patients after adjusting for physical disability and age [54]. MS patients who had fatigue had higher thalamic and substantia nigra radiotracer uptake as compared to those who did not have fatigue and in comparison with healthy controls [54]. Previous studies have shown a relationship between fatigue in MS and hypometabolism on fluorodeoxyglucose (FDG)-PET (Fig. 18.4) [55].

Similarly, lifetime prevalence of depression is up to 50% in MS. Depression is underreported and undertreated in MS. There is increased risk of suicide in MS patients. Depression is an important determinant of quality of life and although it has been known since Charcot's initial description in 1868, its pathogenesis in MS is not clear. Depression has shown poor correlation with overall lesion load and there have been mixed results from few MRI studies. Our preliminary data demonstrates that there is increased glial activity in anterior cingulate gyrus that correlates with depression scores [56]. In fact, increased [F-18]PBR06 uptake in the vagal nucleus tractus solitarius correlated with depressive symptoms in MS patients, after adjusting for disability and other covariates in our pilot evaluation [57]. Another study has shown increased TSPO binding in the hippocampus that correlated with depression scores and was associated with functional disconnection in relation to a subgenual cingulate gyrus hub on functional connectivity magnetic resonance imaging (fcMRI) in MS patients [58].

Hence, TSPO-PET can lead to critical and unique insights into the pathogenesis of some of the hidden symptoms of MS such as fatigue and depression.

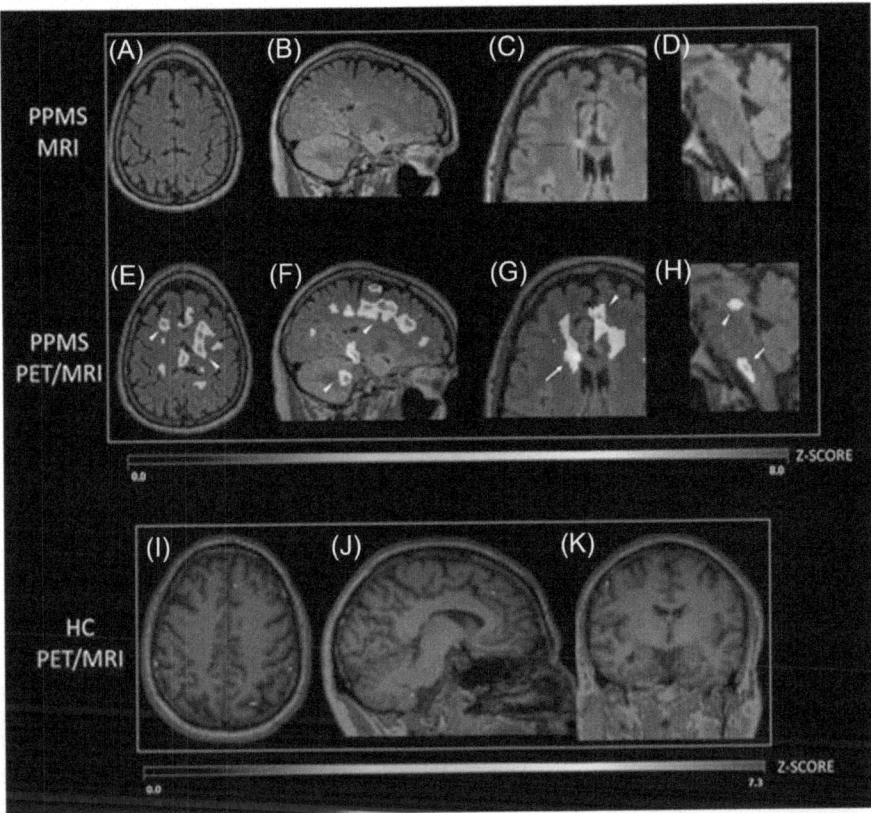

FIGURE 18.1 Widespread glial activation in a 64-year-old PPMS patient who is a medium affinity binder with few lesions on MRI. Transverse and sagittal fluid-attenuated inversion recovery MRI slices show normal-appearing white matter (NAWM) (A, B) except for a few fluid-attenuated inversion recovery bright lesions (C, D, crosshairs), which are seen in right periventricular white matter (WM) and medulla. Z-score maps of increased 18F-PBR06 PET uptake (thresholded at $z > 4.0$) superimposed on MRI (E–G) reveal widespread increased radiotracer uptake in NAWM (arrowheads). Additionally, there is focal increased uptake corresponding to a right periventricular WM lesion (G, arrow) and in a perilesional area in the medulla (H, arrow). A focal area of increased 18F-PBR06 uptake in the midbrain (H, arrowhead) has no corresponding MRI abnormality (D). In contrast, while aging can be associated with increased glial activity, the age- and genotype-matched HC (I–K) demonstrated no significant clusters of voxels with increased 18F-PBR06 uptake. This case demonstrates that short-duration, static 18F-PBR06 PET imaging can demonstrate widespread increased glial activation in PPMS using a clinically feasible, individualized z-score mapping approach, similar to the approaches used in 18F-FDG PET literature. Source: *From Singhal T, Rissanen E, Ficke J, Cicero S, Carter K, Weiner HL. Widespread glial activation in primary progressive multiple sclerosis revealed by 18F-PBR06 PET: a clinically feasible, individualized approach. Clin Nucl Med 2021;46:136–37.*

Translocator protein–positron emission tomography and prognostication in multiple sclerosis

A few longitudinal studies have reported on the ability of TSPO-PET to predict long-term outcomes in MS patients (Table 18.2). Increased NAWM PET uptake at baseline was

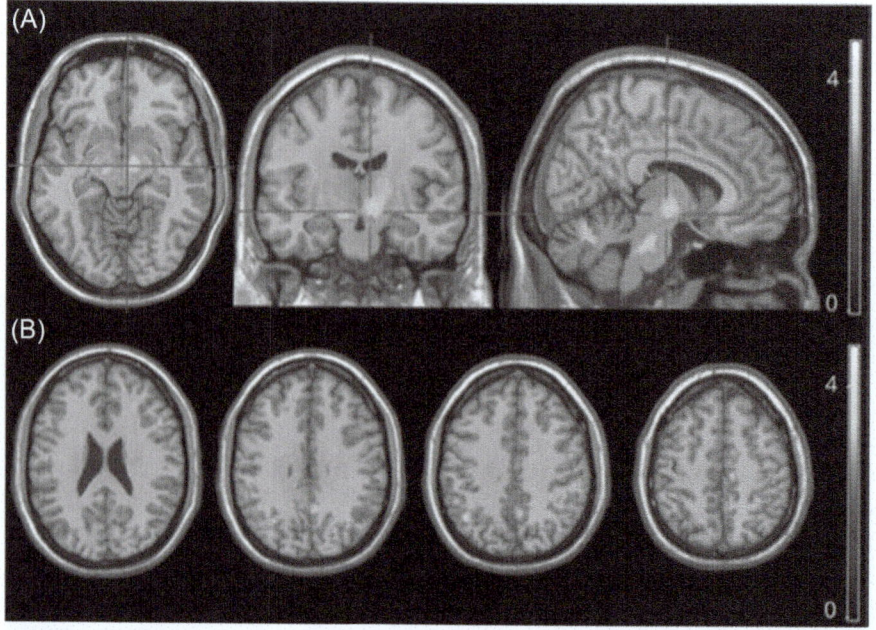

FIGURE 18.2 Statistical parametric mapping (SPM) analysis. (A) SPM-based, voxel-by-voxel maps of correlation between modified fatigue impact scale (MFIS) and PET uptake. The strongest positive correlation was seen in the right substantia nigra (cross-bars). Additional regions of significant correlations were seen in the left substantia nigra, periaqueductal gray, pons, medulla, precuneus, midcingulate, cerebellar vermis, and insular cortex regions. (B) SPM-based, voxel-by-voxel maps demonstrate widespread positive correlation in juxtacortical white matter between MFIS scores and PET uptake. Source: *From Singhal T, Cicero S, Pan H, et al. Regional microglial activation in the substantia nigra is linked with fatigue in MS. Neurol Neuroimmunol Neuroinflamm 2020;7.*

associated with higher T2 lesion volume changes after 1 year of followup [60]. Interestingly, baseline white matter lesional PET uptake was associated with enlarging white matter lesion volume load, whole brain, and gray matter atrophy at 1 year of followup [60].

A recent study, using [C-11]PK11195, has shown that at baseline, increased PET uptake in NAWM and perilesional areas was associated with an increased risk of clinical progression without evidence of relapse activity in MS patients, at 2 years of followup [23]. Notably, baseline T1 lesional volume and brain volume were not predictive of clinical progression in this study. Similarly, increased thalamic radiotracer uptake was predictive of disease progression at a mean followup of 3 years [20]. In this study, patients with disability progression had reduced thalamic, caudate, and putaminal volumes at baseline as compared to patients without progression of disease. However, thalamic uptake was independently associated with disability progression after adjustment for other imaging variables [20].

In summary, initial evidence suggests that TSPO-PET can predict disease progression in MS. If confirmed in larger studies, these findings can facilitate significantly improved risk stratification for clinical trials and in clinical management of individual MS patients.

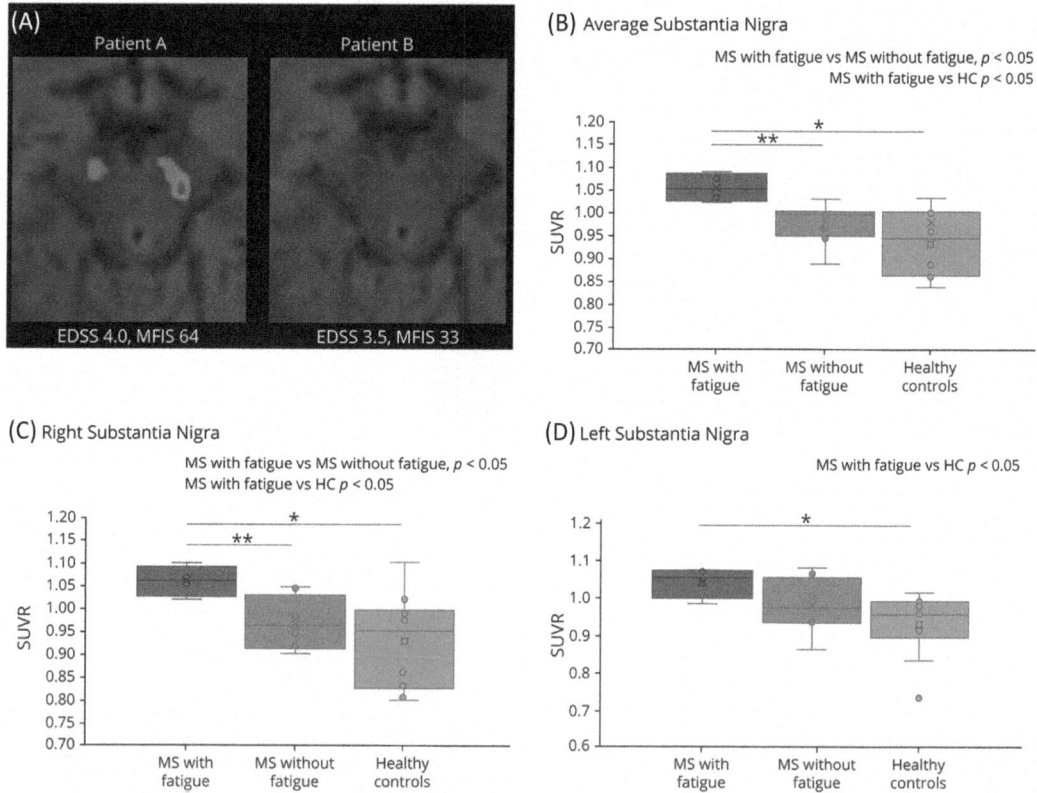

FIGURE 18.3 Comparison of substantia nigra PET uptake between subjects with MS with fatigue and subjects with MS without fatigue and healthy participants. (A) Individualized z-score maps showing increased [F-18] PBR06 PET uptake in the bilateral substantia nigra in a patient with MS with fatigue with a high total MFIS score (total MFIS score = 64) compared with a patient with MS without fatigue with a comparable EDSS score (3.5 vs 4) and a low total MFIS score (MFIS score = 33). For the latter patient, the ROIs for the substantia nigra are delineated but do not demonstrate an increased z-score of >2 compared with a healthy control group. (B) Increased average substantia nigra standardized uptake value ratio (SUVR) in patients with MS with fatigue compared with patients with MS without fatigue and healthy participants. (C) Increased right substantia nigra SUVR in patients with MS with fatigue compared with patients with MS without fatigue and healthy participants. (D) Increased left substantia nigra SUVR in patients with MS with fatigue compared with healthy controls. *$P < .05$; **$P < .01$. Source: *From Singhal T, Cicero S, Pan H, et al. Regional microglial activation in the substantia nigra is linked with fatigue in MS. Neurol Neuroimmunol Neuroinflamm 2020;7.*

Translocator protein—positron emission tomography and treatment effects

TSPO-PET imaging offers a promising method to monitor the progression of disease, by assessing longitudinal changes in microglial activation in patients who are not on any disease-modifying therapy (DMT) and can also assess response to current and emerging DMTs. Several recent studies have trialed this approach, using it to gauge the efficacy of multiple MS medications, using various TSPO-PET ligands. Two such studies [61,62]

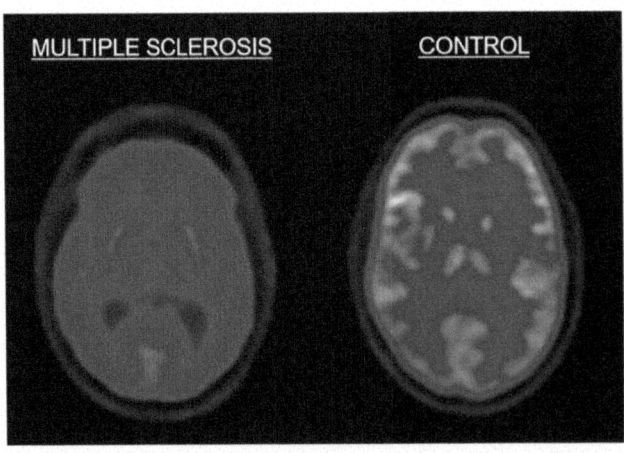

FIGURE 18.4 FDG-PET in MS. FDG-PET showing diffuse hypometabolism in an MS patient (left panel). In comparison, there is preserved 18F-FDG uptake in a healthy control individual (right panel). *Source: From Singhal lab, Brigham and Women's Hospital, Boston, MA.*

TABLE 18.2 Translocator protein—positron emission tomography (TSPO-PET) and prognostication in multiple sclerosis (MS).

Giannetti et al. 2014 [59]	[11C]-R-PK11195	10 RRMS; 8 SPMS; 1 PPMS	2 years (clinical followup)	— Of a total of 1242 T1 black holes within all MS, 76% of the black holes presented with TSPO-binding (defined with a threshold of mean $BP_{ND} > 0$). — Total T1 black holes PET signal predictive of clinical disability outcome at 2 years
Giannetti et al. 2015 [35]	[11C]-R-PK11195	18 CIS; 8 HC	2 years (MRI and clinical follow-up)	— NAWM and central GM BP_{ND} higher in CIS vs HC — NAWM BP_{ND} correlates with EDSS in CIS — Higher NAWM BP_{ND} at baseline in patients developing CDMS ($n = 12$) by 2 years vs those who did not ($n = 4$)
Datta et al. 2017 [60]	[11C] PBR28	14 RRMS; 7 SPMS	1 year (MRI and clinical followup)	— Higher NAWM and T2 lesional DVR at baseline associated with enlarging T2 lesion volume at 1 year in RRMS — Higher T2 lesional DVR at baseline associated with progression of whole brain and GM atrophy at 1 year in SPMS. Similar, nonsignificant, trend for baseline NAWM DVR, and whole brain volume change in SPMS
Sucksdorff et al. 2020 [23]	[11C]-R-PK11195	49 RRMS; 24 SPMS; 18 HC	4 years (clinical followup)	— Both among all MS and in those without relapses, DVR in NAWM, and T1 perilesional ROI higher in those with later progression vs those without later progression — Multivariate logistic regression among all MS: higher NAWM DVR at baseline predicts progression at followup (OR 4.26) — Logistic regression in patients with progression independent of relapses: higher T1 perilesional DVR at baseline associated with disease progression at followup (OR 4.57)

BP_{ND}, binding potential; *CDMS*, clinically definite MS; *CIS*, clinically isolated syndrome; *ROI*, region of interest.

analyzed the effect of the monoclonal antibody medication natalizumab, which works to decrease the influx of lymphocytes to the CNS and ultimately reduce the effects of the adaptive immune response, regarded as a contributing factor to the pathology of MS. The first study, published in 2017, used data from 18 patients with RRMS or SPMS over the course of 6 months, using MRI and PET imaging from baseline, 3 months, and 6 months into treatment [61]. The study found that there was reduction of PK11195 uptake in enhancing chronic lesions after the start of treatment, mostly within the first 3 months [61]. The authors noted that conclusions were complicated by the high nonspecific binding that is associated with PK11195, yet their data was supported by trends found in the next study, conducted in 2019 by another group [62], which similarly followed the progression of 10 RRMS and SPMS patients beginning treatment with natalizumab. This study was expanded by gathering MRI and PET imaging from before and after 1 year of treatment and comparing results to 11 MS patients (10 SPMS and 1 RRMS) without treatment and 8 healthy controls. The previous conclusions were similar, as it concluded that overall, natalizumab treatment reduced microglial activation in NAWM and perilesional areas, with suggested reduction in focal inflammatory lesions as well. In contrast, it also found that there was an overall increase in microglial activation during the 1 year followup with the untreated MS patients, which was later associated with more rapid disability progression, suggesting the specificity of the findings seen in the natalizumab-treated arm [62].

The effects of other MS medications have also been assessed with TSPO imaging. Fingolimod was the first oral DMT approved for the treatment of RRMS, which works also to block the migration of lymphocytes from periphery to CNS. In a study also done in 2017, 10 RRMS patients, who were switched to fingolimod therapy due to safety reasons or therapy escalation, were imaged at baseline and after 24 weeks of treatment [63]. MRI and PET images were compared to the parallel data of 8 healthy individuals. It initially found that microglial activation, measured as [C-11]PK11195 distribution volume ratio (DVR), was higher in NAWM and the thalamus in MS patients compared to healthy controls. After 6 months of treatment, the study concluded that [C-11]PK11195 binding was reduced specifically in the combined T2 lesion area after 6 months of treatment, but not in other brain areas such as NAWM and GM [63].

Another study in eight RRMS patients undergoing treatment with glatiramer acetate, using [C-11]PK11195-PET imaging performed before and after 1 year of treatment found that the whole brain [C-11]PK11195 binding potential per unit volume decreased by 3.17%, consistent with a reduction in inflammation following treatment [64]. Significant decrease of binding potential was noted in cortical GM and cerebral WM, with a decreasing trend also identified in the thalamus and putamen [64].

wFurther, in another study using [F-18]PBR06-PET, we employed PET and MRI imaging to assess the effects of Ofatumumab, a B-cell depleting therapy, in 10 RMS patients. PET scans were planned to be obtained at baseline, 5, 28, 90, and 270 days, approximately, after treatment. On interim analysis of 3-month data in the first five patients, we reported that there was significantly decreased microglial activation in the cortical gray matter, which was preceded by B-cell depletion within the first week of treatment initiation, suggesting a downstream effect of B-cell depletion on cortical microglial activity [52].

3. Use of other imaging acquisition and analysis modalities in multiple sclerosis

Beyond translocator protein—positron emission tomography: other glial imaging targets and PET ligands

Glial imaging targets, beyond TSPO, are being evaluated for future PET studies [65]. Macrophage colony stimulating factor 1 receptor, CSF-1R, has been studied in animal models using a ligand, [C-11]CPPC [66]. In this study, brain PET uptake was shown to correlate with experimental allergic encephalomyelitis severity [66]. A P2X7 purinergic receptor ligand, [C-11]SMW139 has been evaluated in relapsing MS patients and was shown to have increased gray matter and white matter binding in MS patients as compared to healthy controls [67]. Despite its initial promise, it appears that P2X7 receptor may be expressed on oligodendrocyte precursor cells and astrocytes, in addition to microglia, and therefore, may not demonstrate the desired specificity for microglia. Monoamine oxidase B is a promising target for astrocyte imaging and a new F-18 labeled PET radioligand has been used in other disease conditions to assess astrocytic activation in neurodegenerative diseases [68] and have the potential to be employed in MS. Several other PET radioligands are being developed for improved imaging of neuroinflammation and are reviewed in detail elsewhere [65].

Beyond translocator protein—positron emission tomography: PET imaging of nonimmune mechanisms in MS

PET is a versatile technique that can be used to image various components of MS pathogenesis [69]. PET has been used to image neuronal degeneration, myelin integrity, neurochemical alterations, and cerebral metabolism and blood flow in MS [69]. For example, several PET ligands that have been otherwise developed for imaging amyloid pathology have been investigated to assess demyelination in MS [70–72]. C–11 labeled Pittsburgh compound B ([C-11]PIB) has been used to study myelin integrity in MS and was found to be decreased in demyelinating lesions [70]. [C-11]PIB was also able to demonstrate a gradient of demyelination along the lesion edges, and demonstrate patterns of remyelination in MS lesions [70]. Similarly, F-18 florbetaben and flutemetamol have been evaluated for assessing demyelination in MS in tandem with other MRI techniques [71,72].

C-11 labeled flumazenil has demonstrated decreased synaptic integrity in gray matter in MS patients [73]. Novel synaptic density markers targeting synaptic vesicle 2a are being used to measure and track synaptic pathology in various neurodegenerative conditions [74] and can be employed in MS.

Widespread glucose hypometabolism has been seen in MS patients that has been linked with clinical disability, brain atrophy, and fatigue scores [55,75]. Various neurochemical abnormalities, such as changes in serotonin transporter, have been reported using neurochemical PET in MS [76].

Future directions and conclusion

In conclusion, PET is a unique tool that can truly enhance our understanding of MS pathogenesis, its diagnosis and treatment, and aid novel drug development. Future studies are needed to evaluate the longitudinal predictive value of PET in MS. Each novel tracer needs to be validated in the appropriate clinical context, and in tandem with advanced MRI approaches. Studies to evaluate the effects of existing and novel therapeutic molecules, using PET as an imaging endpoint, are urgently needed. Moreover, PET imaging can have a crucial role in evaluation of progressive MS, and needs to be extensively and systematically studied in this area of high unmet need in MS. We expect to see a potentially transformative role of PET imaging in MS care and its understanding in the coming years.

References

[1] Reich DS, Lucchinetti CF, Calabresi PA. Multiple sclerosis. N Engl J Med 2018;378:169–80.
[2] Thompson AJ, Banwell BL, Barkhof F, et al. Diagnosis of multiple sclerosis: 2017 revisions of the McDonald criteria. Lancet Neurol 2018;17:162–73.
[3] Lublin FD, Reingold SC, Cohen JA, et al. Defining the clinical course of multiple sclerosis: the 2013 revisions. Neurology 2014;83:278–86.
[4] Lublin FD, Coetzee T, Cohen JA, Marrie RA, Thompson AJ. International Advisory Committee on Clinical Trials in MS. The 2013 clinical course descriptors for multiple sclerosis: a clarification. Neurology 2020;94:1088–92.
[5] Tur C, Carbonell-Mirabent P, Cobo-Calvo A, et al. Association of early progression independent of relapse activity with long-term disability after a first demyelinating event in multiple sclerosis. JAMA Neurol 2023;80:151–60.
[6] Singhal T, Weiner HL, Bakshi R. TSPO-PET imaging to assess cerebral microglial activation in multiple sclerosis. Semin Neurol 2017;37:546–57.
[7] Fisher E, Lee JC, Nakamura K, Rudick RA. Gray matter atrophy in multiple sclerosis: a longitudinal study. Ann Neurol 2008;64:255–65.
[8] Phelps ME. PET: molecular imaging and its biological applications. New York: Springer-Verlag; 2012.
[9] Gandhi R, Laroni A, Weiner HL. Role of the innate immune system in the pathogenesis of multiple sclerosis. J neuroimmunology 2010;221:7–14.
[10] Guerrero BL, Sicotte NL. Microglia in multiple sclerosis: friend or foe? Front Immunol 2020;11:374.
[11] Prineas JW, Parratt JDE. Multiple sclerosis: microglia, monocytes, and macrophage-mediated demyelination. J Neuropathol Exp Neurol 2021;80:975–96.
[12] Kreisl WC, Kim MJ, Coughlin JM, Henter ID, Owen DR, Innis RB. PET imaging of neuroinflammation in neurological disorders. Lancet Neurol 2020;19:940–50.
[13] Jaremko L, Jaremko M, Giller K, Becker S, Zweckstetter M. Structure of the mitochondrial translocator protein in complex with a diagnostic ligand. Science 2014;343:1363–6.
[14] Scarf AM, Ittner LM, Kassiou M. The translocator protein (18 kDa): central nervous system disease and drug design. J Med Chem 2009;52:581–92.
[15] Braestrup C, Squires RF. Specific benzodiazepine receptors in rat brain characterized by high-affinity (3H) diazepam binding. Proc Natl Acad Sci U S A 1977;74:3805–9.
[16] Schoemaker H, Morelli M, Deshmukh P, Yamamura HI. 3H]Ro5-4864 benzodiazepine binding in the kainate lesioned striatum and Huntington's diseased basal ganglia. Brain Res 1982;248:396–401.
[17] Kaunzner UW, Kang Y, Zhang S, et al. Quantitative susceptibility mapping identifies inflammation in a subset of chronic multiple sclerosis lesions. Brain 2019;142:133–45.
[18] Nutma E, Stephenson JA, Gorter RP, et al. A quantitative neuropathological assessment of translocator protein expression in multiple sclerosis. Brain 2019;142:3440–55.
[19] Banati RB, Newcombe J, Gunn RN, et al. The peripheral benzodiazepine binding site in the brain in multiple sclerosis: quantitative in vivo imaging of microglia as a measure of disease activity. Brain 2000;123(Pt 11):2321–37.
[20] Misin O, Matilainen M, Nylund M, et al. Innate immune cell-related pathology in the thalamus signals a risk for disability progression in multiple sclerosis. Neurol Neuroimmunol Neuroinflamm 2022;9.

[21] Politis M, Giannetti P, Su P, et al. Increased PK11195 PET binding in the cortex of patients with MS correlates with disability. Neurology 2012;79:523–30.
[22] Rissanen E, Tuisku J, Rokka J, et al. In vivo detection of diffuse inflammation in secondary progressive multiple sclerosis using PET imaging and the radioligand ^{11}C-PK11195. J Nucl Med 2014;55:939–44.
[23] Sucksdorff M, Matilainen M, Tuisku J, et al. Brain TSPO-PET predicts later disease progression independent of relapses in multiple sclerosis. Brain 2020;143:3318–30.
[24] Winkeler A, Boisgard R, Awde AR, et al. The translocator protein ligand [^{18}F]DPA-714 images glioma and activated microglia in vivo. Eur J Nucl Med Mol Imaging 2012;39:811–23.
[25] Fujita M, Imaizumi M, Zoghbi SS, et al. Kinetic analysis in healthy humans of a novel positron emission tomography radioligand to image the peripheral benzodiazepine receptor, a potential biomarker for inflammation. Neuroimage 2008;40:43–52.
[26] Luu TG, Kim HK. (18)F-Radiolabeled translocator protein (TSPO) PET tracers: recent development of TSPO radioligands and their application to PET study. Pharmaceutics 2022;14.
[27] James ML, Belichenko NP, Nguyen TV, et al. PET imaging of translocator protein (18 kDa) in a mouse model of Alzheimer's disease using N-(2,5-dimethoxybenzyl)-2-18F-fluoro-N-(2-phenoxyphenyl)acetamide. J Nucl Med 2015;56:311–16.
[28] Lartey FM, Ahn GO, Shen B, et al. PET imaging of stroke-induced neuroinflammation in mice using [^{18}F]PBR06. Mol Imaging Biol: 2014;16:109–17.
[29] Datta G, Colasanti A, Kalk N, et al. ^{11}C-PBR28 and ^{18}F-PBR111 detect white matter inflammatory heterogeneity in multiple sclerosis. J Nucl Med 2017;58:1477–82.
[30] Hamzaoui M, Garcia J, Boffa G, et al. Positron emission tomography with [^{18}F]-DPA-714 unveils a smoldering component in most multiple sclerosis lesions which drives disease progression. Ann Neurol 2023;94(2):366–83.
[31] Herranz E, Gianni C, Louapre C, et al. Neuroinflammatory component of gray matter pathology in multiple sclerosis. Ann Neurol 2016;80:776–90.
[32] Singhal T, O'Connor K, Dubey S, et al. Gray matter microglial activation in relapsing vs progressive MS: A [F-18]PBR06-PET study. Neurol Neuroimmunol Neuroinflamm 2019;6:e587.
[33] Singhal T, Pan H, Dubey S, Cicero S, Hurwitz S, Tauhid S, et al. The relationship of microglial activation and multiple sclerosis-associated fatigue: a [F-18]PBR06 PET study. Mult Scler J 2018;24(S2):1008.
[34] Owen DR, Yeo AJ, Gunn RN, et al. An 18-kDa translocator protein (TSPO) polymorphism explains differences in binding affinity of the PET radioligand PBR28. J Cereb Blood Flow Metab 2012;32:1–5.
[35] Giannetti P, Politis M, Su P, et al. Increased PK11195-PET binding in normal-appearing white matter in clinically isolated syndrome. Brain 2015;138:110–19.
[36] Rissanen E, Tuisku J, Vahlberg T, et al. Microglial activation, white matter tract damage, and disability in MS. Neurol Neuroimmunol Neuroinflamm 2018;5:e443.
[37] Bezukladova S, Tuisku J, Matilainen M, et al. Insights into disseminated MS brain pathology with multimodal diffusion tensor and PET imaging. Neurol Neuroimmunol Neuroinflamm 2020;7.
[38] Herranz E, Louapre C, Treaba CA, et al. Profiles of cortical inflammation in multiple sclerosis by (11)C-PBR28 MR-PET and 7 Tesla imaging. Mult Scler 2020;26:1497–509.
[39] Kang Y, Pandya S, Zinger N, Michaelson N, Gauthier SA. Longitudinal change in TSPO PET imaging in progressive multiple sclerosis. Ann Clin Transl Neurol 2021;8:1755–9.
[40] Singhal T., S. Cicero, E. Rissanen, J.H. Ficke, P. Kukreja, S. Vaquerano, et al. Microglial activation in multiple sclerosis patients treated with current high-efficacy disease modifying treatments: Individualized [F-18]PBR06-PET analysis. In: ECTRIMS, Amsterdam, Netherlands; 2022.
[41] Houtchens MK, Benedict RH, Killiany R, et al. Thalamic atrophy and cognition in multiple sclerosis. Neurology 2007;69:1213–23.
[42] Zivadinov R, Bergsland N, Jakimovski D, et al. Thalamic atrophy measured by artificial intelligence in a multicentre clinical routine real-word study is associated with disability progression. J Neurol Neurosurg Psychiatry 2022. Available from: https://doi.org/10.1136/jnnp-2022-329333.
[43] Unterrainer M, Mahler C, Vomacka L, et al. TSPO PET with [(18)F]GE-180 sensitively detects focal neuroinflammation in patients with relapsing-remitting multiple sclerosis. Eur J Nucl Med Mol Imaging 2018;45:1423–31.

[44] Frischer JM, Bramow S, Dal-Bianco A, et al. The relation between inflammation and neurodegeneration in multiple sclerosis brains. Brain 2009;132:1175–89.
[45] Kuhlmann T, Ludwin S, Prat A, Antel J, Bruck W, Lassmann H. An updated histological classification system for multiple sclerosis lesions. Acta Neuropathol 2017;133:13–24.
[46] Singhal T, O'Connor K, Dubey S, et al. 18F-PBR06 Versus 11C-PBR28 PET for assessing white matter translocator protein binding in multiple sclerosis. Clin Nucl Med 2018;43:e289–95.
[47] Singhal T, Rissanen E, Ficke J, Cicero S, Carter K, Weiner HL. Widespread glial activation in primary progressive multiple sclerosis revealed by 18F-PBR06 PET: a clinically feasible, individualized approach. Clin Nucl Med 2021;46:136–7.
[48] Colasanti A, Guo Q, Muhlert N, et al. In vivo assessment of brain white matter inflammation in multiple sclerosis with (18)F-PBR111 PET. J Nucl Med 2014;55:1112–18.
[49] Tavazzi E, Zivadinov R, Dwyer MG, et al. MRI biomarkers of disease progression and conversion to secondary-progressive multiple sclerosis. Expert Rev Neurother 2020;20:821–34.
[50] Bodini B, Poirion E, Tonietto M, et al. Individual mapping of innate immune cell activation is a candidate marker of patient-specific trajectories of worsening disability in multiple sclerosis. J Nucl Med 2020;61:1043–9.
[51] Nylund M, Sucksdorff M, Matilainen M, Polvinen E, Tuisku J, Airas L. Phenotyping of multiple sclerosis lesions according to innate immune cell activation using 18 kDa translocator protein-PET. Brain Commun 2022;4:fcab301.
[52] Singhal T., Carter K., J.H. Ficke, P. Kukreja, E. Rissanen GB, B. Glanz, J. Zurawski, et al. Efficacy of ofatumumab on microglial activity in patients with relapsing forms of multiple sclerosis: interim analysis. In: Presentation at the American Academy of Neurology (AAN) 2022, April 2–7, 2022.
[53] Glanz BI, Degano IR, Rintell DJ, Chitnis T, Weiner HL, Healy BC. Work productivity in relapsing multiple sclerosis: associations with disability, depression, fatigue, anxiety, cognition, and health-related quality of life. Value Health 2012;15:1029–35.
[54] Singhal T, Cicero S, Pan H, et al. Regional microglial activation in the substantia nigra is linked with fatigue in MS. Neurol Neuroimmunol Neuroinflamm 2020;7.
[55] Roelcke U, Kappos L, Lechner-Scott J, et al. Reduced glucose metabolism in the frontal cortex and basal ganglia of multiple sclerosis patients with fatigue: a 18F-fluorodeoxyglucose positron emission tomography study. Neurology 1997;48:1566–71.
[56] Cicero S. CK, Dubey S., Chu R., Tauhid S., Pan H., et al. Depression is linked with abnormal regional grey matter microglial activation in multiple sclerosis: a [F-18]PBR06-PET study. In: ACTRIMS. West Palm Beach, FL. 2020.
[57] Singhal T., Pan H., Cicero S., Rissanen E., Vaquerano S., Kukreja P., et al. Regional microglial activation in vagal afferent network is linked with depression in multiple sclerosis: a statistical parametric mapping analysis study using [F-18]PBR06-PET. In: ACTRIMS. West Palm Beach, FL, 2022.
[58] Colasanti A, Guo Q, Giannetti P, et al. Hippocampal neuroinflammation, functional connectivity, and depressive symptoms in multiple sclerosis. Biol Psychiatry 2016;80:62–72.
[59] Giannetti P, Politis M, Su P, et al. Microglia activation in multiple sclerosis black holes predicts outcome in progressive patients: an in vivo [(11)C](R)-PK11195-PET pilot study. Neurobiol Dis 2014;65:203–10.
[60] Datta G, Colasanti A, Rabiner EA, et al. Neuroinflammation and its relationship to changes in brain volume and white matter lesions in multiple sclerosis. Brain 2017;140:2927–38.
[61] Kaunzner UW, Kang Y, Monohan E, et al. Reduction of PK11195 uptake observed in multiple sclerosis lesions after natalizumab initiation. Mult Scler Relat Disord 2017;15:27–33.
[62] Sucksdorff M, Tuisku J, Matilainen M, et al. Natalizumab treatment reduces microglial activation in the white matter of the MS brain. Neurol Neuroimmunol Neuroinflamm 2019;6:e574.
[63] Sucksdorff M, Rissanen E, Tuisku J, et al. Evaluation of the effect of fingolimod treatment on microglial activation using serial PET imaging in multiple sclerosis. J Nucl Med 2017;58:1646–51.
[64] Ratchford JN, Endres CJ, Hammoud DA, et al. Decreased microglial activation in MS patients treated with glatiramer acetate. J Neurol 2012;259:1199–205.
[65] Narayanaswami V, Dahl K, Bernard-Gauthier V, Josephson L, Cumming P, Vasdev N. Emerging PET radiotracers and targets for imaging of neuroinflammation in neurodegenerative diseases: outlook beyond TSPO. Mol Imaging 2018;17:1536012118792317.

[66] Horti AG, Naik R, Foss CA, et al. PET imaging of microglia by targeting macrophage colony-stimulating factor 1 receptor (CSF1R). Proc Natl Acad Sci U S A 2019;116:1686–91.

[67] Hagens MHJ, Golla SSV, Janssen B, et al. The P2X(7) receptor tracer [(11)C]SMW139 as an in vivo marker of neuroinflammation in multiple sclerosis: a first-in man study. Eur J Nucl Med Mol Imaging 2020;47:379–89.

[68] Villemagne VL, Harada R, Dore V, et al. First-in-humans evaluation of (18)F-SMBT-1, a novel (18)F-labeled monoamine oxidase-B PET tracer for imaging reactive astrogliosis. J Nucl Med 2022;63:1551–9.

[69] Niccolini F, Su P, Politis M. PET in multiple sclerosis. Clin Nucl Med 2015;40:e46–52.

[70] Bodini B, Veronese M, Garcia-Lorenzo D, et al. Dynamic imaging of individual remyelination profiles in multiple sclerosis. Ann Neurol 2016;79:726–38.

[71] Carotenuto A, Giordano B, Dervenoulas G, et al. [(18)F]Florbetapir PET/MR imaging to assess demyelination in multiple sclerosis. Eur J Nucl Med Mol Imaging 2020;47:366–78.

[72] Matias-Guiu JA, Cabrera-Martin MN, Matias-Guiu J, et al. Amyloid PET imaging in multiple sclerosis: an (18)F-florbetaben study. BMC Neurol 2015;15:243.

[73] Freeman L, Garcia-Lorenzo D, Bottin L, et al. The neuronal component of gray matter damage in multiple sclerosis: a [(11) C]flumazenil positron emission tomography study. Ann Neurol 2015;78:554–67.

[74] Carson RE, Naganawa M, Toyonaga T, et al. Imaging of synaptic density in neurodegenerative disorders. J Nucl Med 2022;63:60S–7S.

[75] Bakshi R, Miletich RS, Kinkel PR, Emmet ML, Kinkel WR. High-resolution fluorodeoxyglucose positron emission tomography shows both global and regional cerebral hypometabolism in multiple sclerosis. J Neuroimaging 1998;8:228–34.

[76] Carotenuto A, Valsasina P, Preziosa P, Mistri D, Filippi M, Rocca MA. Monoaminergic network abnormalities: a marker for multiple sclerosis-related fatigue and depression. J Neurol Neurosurg Psychiatry 2023;94:94–101.

CHAPTER 19

Optical coherence tomography in multiple sclerosis

Nik Krajnc[1,2] and Gabriel Bsteh[1,2]

[1]Department of Neurology, Medical University of Vienna, Vienna, Austria [2]Comprehensive Center for Clinical Neurosciences and Mental Health, Medical University of Vienna, Vienna, Austria

OUTLINE

Introduction	361
Technical principles of optical coherence tomography	362
Retina as a window to the brain	363
Optical coherence tomography in optic neuritis	365
Optical coherence tomography is a marker of multiple sclerosis—associated neuroaxonal damage	367
Cross-sectional optical coherence tomography—retinal thickness indicates neuroaxonal reserve	368
Longitudinal optical coherence tomography—retinal thinning indicates ongoing neuroaxonal damage	368
Optical coherence tomography in progressive multiple sclerosis	368
Optical coherence tomography in treatment monitoring	369
Practical issues and limitations affecting clinical application of optical coherence tomography in multiple sclerosis	370
Conclusion	371
References	372

Introduction

Optical coherence tomography (OCT) is an inexpensive, noninvasive, repeatable, and accessible technique that uses near-infrared light to create tomographic, two-dimensional structural in vivo images of biological tissues, including the retina [1]. It is based on the physical concept of time-of-flight delay and intensity of backscattered light from microscopic particles within the tissue [2], and is considered an optical analog of ultrasound-based tomographic imaging with yet much higher resolution [3,4].

In the last two decades, the traditional application of OCT in ophthalmological diseases has been introduced into neurology, with a variety of studies demonstrating retinal pathology in different disease models, including multiple sclerosis (MS) [5—7]. Neuroaxonal degeneration as a consequence of inflammation and demyelination is regarded as the principal substrate of disability progression in MS [8,9], which can also be reliably detected by OCT.

Herewith we summarize recent advances in clinical applications of OCT in MS. First, we provide an overview of the basic technological characteristics of OCT, followed by a discussion of retinal anatomy. Furthermore, we elaborate on its application in case of optic neuritis (ON) and its value as a biomarker of neuroaxonal damage in MS. In the end, we discuss the relevant issue of OCT use in treatment monitoring.

Technical principles of optical coherence tomography

OCT is an optical imaging modality that uses near-infrared light to perform high-resolution, cross-sectional images of the retina. It is often compared to ultrasound because they both direct waves to the tissue where the waves are scattered and reflected back, with their delay measuring the depth in which the reflection occurred [2]. It offers higher resolution than other medical imaging technologies ($\sim$ 1—2 μm) with yet limited penetration ($\sim$ 1 mm) [10]. However, due to its performance, it is particularly well-suited for diagnostic applications in which the multilayered tissue, for example, retina, may be disrupted by pathology with each layer having distinct optic features in terms of reflection, backscattering, and absorption. The light directed onto the retina crosses the transparent structures of the eye and is partially absorbed, whereas a larger fraction is backscattered by the different retinal layers [2].

The functional principle behind OCT is interferometry (Fig. 19.1). Light from one arm is reflected or scattered off the retina and interferes with light from the reference arm in a spectrometer. By using the time-delay information contained in the light waves which have been reflected from different depths, a depth-profile reconstruction of the sample structure can be made [11]. Three-dimensional images can be created by scanning the light beam laterally across the sample surface. While the lateral resolution is determined by the spot size of the light beam, the depth resolution depends primarily on the optical bandwidth of the light source. In that way, OCT may combine high axial resolutions with large depths of field [4].

OCT first entered the market in 1996 as a time domain OCT (TD-OCT), which required acquisition of a depth scan for every location, offering very slow imaging speed and low image quality [3]. Since TD-OCT has been supplemented by spectral domain OCT (SD-OCT), which provides a better image resolution and enables the use of segmentation algorithms, its validity has increased so much that small changes in the micrometer spectrum can be reliable reproduced, improving its signal-to-noise ratio and increasing tissue contrast [12,13]. Contrary to TD-OCT, in SD-OCT, the interference signal is a function of different wavelengths and not different echo time delays [11]. The wavelength spectrum is converted into time delay signals by Fourier transformation, which enables a simultaneous analysis of all echoes from different retinal layers [14].

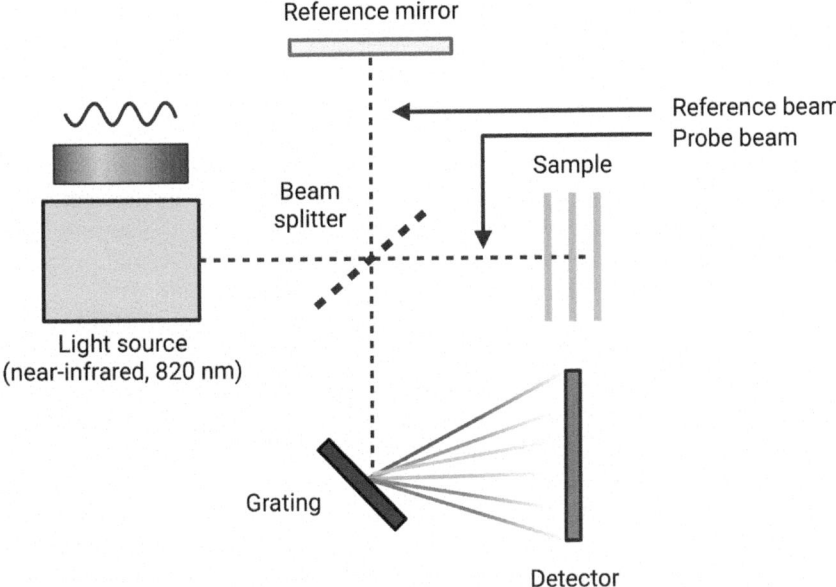

FIGURE 19.1 Basic principles of SD-OCT. Light from one arm is reflected off the retina and interferes with the light from the reference arm in a spectrometer. In SD-OCT, a complete spectrum of interference signals is processed simultaneously, allowing faster scan acquisition. *Source: Created with BioRender.com.*

OCT images are analyzed using manual or automated segmentation of retinal layers. However, as manual segmentation leads to potential bias [15], automated segmentation is preferred. Among those, two commonly utilized techniques are Cirrus and Spectralis [16]. They have both proven high reproducibility and repeatability in retinal layer measurements, especially when eye tracking and averaging of multiple images are used [17–19]. To obtain a cross-section image of the retina, the light beam is moved horizontally across the retina. For volumetric scans, the cross-sections are repeated in parallel or ring-shaped patterns depending on the region of interest and the program used. The currently fastest commercially available SD-OCT has an imaging speed of about 40,000 axial scans per second, which is approximately 100-time faster than TD-OCT.

In order to improve comparability between studies that use different devices and segmentation techniques, quality and reporting criteria have been developed. Currently, quality control criteria such as OSCAR-IB, and reporting guidelines APOSTEL 2.0 are being used [15,20].

Retina as a window to the brain

During embryonic development, the retina and optic nerve extend from the diencephalon and are therefore considered part of the central nervous system (CNS) [21]. The developmental process of the retina is similar to that of the cerebral cortex, where multipotent neuroblastic precursors proliferate, differentiate, and migrate to its laminar destination producing multilayered tissue [22].

The retinal thickness varies from 0.56 mm at the optic disk to 0.1 mm at the most peripheral areas [23]. It is composed of layers of specialized neurons that are connected through synapses, with light entering the eye, passing its structures and being captured by photoreceptors in the outermost layer of the retina. In photoreceptors, a cascade of neuronal signals is initiated that in the end reaches the retinal ganglion cells, whose axons then form the optic nerve. These axons extend to the lateral geniculate nucleus in the thalamus and the superior colliculus in the midbrain, from which information is further relayed to the higher visual processing centers mostly in the occipital cortex that finally enable us to perceive an image.

Retinal layers include, among other, the retinal nerve fiber layer (RNFL), the ganglion cell layer (GCL), the inner plexiform layer (IPL), the inner nuclear layer (INL), and the outer nuclear layer (ONL) (Fig. 19.2). Owing to low contrast between GCL and IPL in OCT images, those are frequently summarized as the ganglion cell-inner plexiform layer (GCIPL) [15]. The pRNFL consists of unmyelinated axons of retinal ganglion cells (about 1.2 million per eye) that stem from their cell bodies located in the GCIPL. The absence of myelin sheath leads to its uniqueness, with its thickness allowing monitoring of the actual axonal injury unaffected by the thickness of myelin sheath. Ganglion cells are the final cells (third-order neuron) in the process of transmitting visual signals from the retina to

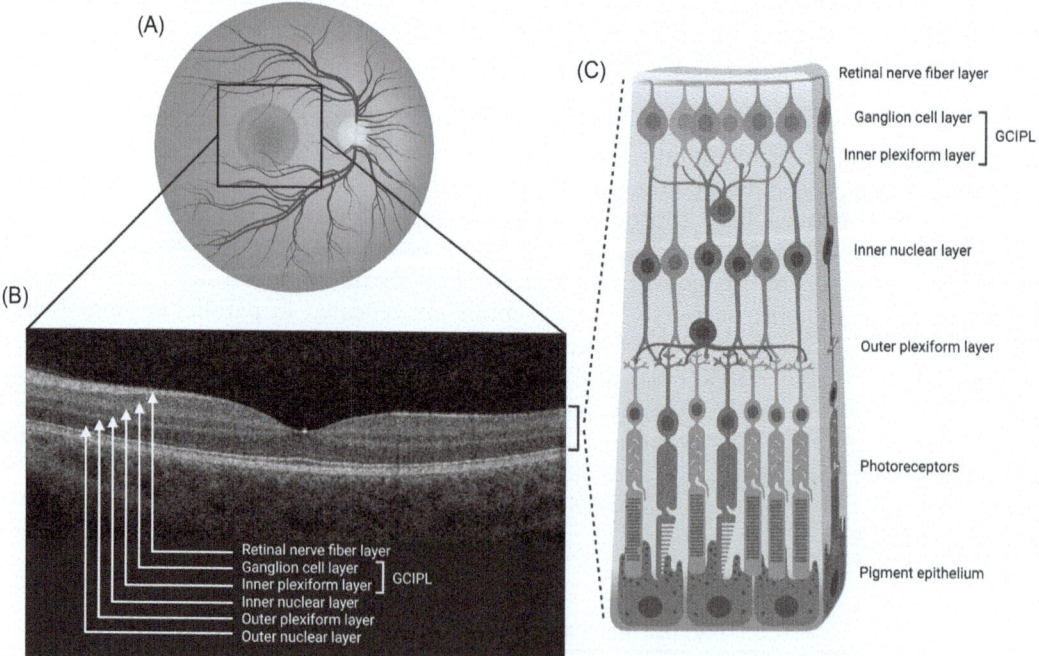

FIGURE 19.2 A fundus (A) with individual retinal layers in the fovea (B) generated by Spectralis OCT from the macular region. Panel (C) illustrates the cellular composition of retinal layers depicted in panel (B). *GCIPL*, ganglion cell-inner plexiform layer. Source: *Created with BioRender.com.*

the brain. When exiting the globe, those axons are covered with oligodendrocyte-produced myelin and are ensheathed in all three meningeal layers.

On the other hand, the INL contains cell bodies and nuclei of different types of interneurons: the horizontal cells, the bipolar cells, the amacrine cells, the Müller cells, and the interplexiform cells. These form complex connections via their processes in the inner and outer plexiform layers which are considered to connect photoreceptors (first-order neuron) with the ganglion cells (third-order neuron).

Optical coherence tomography in optic neuritis

Acute ON occurs in as many as 30%–70% of patients with MS [24,25]; even more, optic nerve lesions were discovered in 94%–99% of post mortem samples of patients with MS regardless of prior clinical ON [26,27]. Changes of retinal layers during and after an acute ON can be used to investigate pathophysiological mechanisms of ON and provide additional information regarding its location, severity, and prognosis.

The pathophysiology of acute ON resembles that of MS brain lesions, where inflammatory demyelination is mediated by autoreactive T cells, with involvement of B cells, microglia, and antibodies [28]. In one-third of ON, acute inflammation is also present in the retina, with initial, yet reversible increase in pRNFL thickness compared to the unaffected side. However, as optic nerve suffers from axonal damage during demyelination, its thinning is already visible 2–4 months after the episode of ON and increases further with time compared to the healthy eye, reaching its floor within 3–12 months from ON [29]. A metaanalysis of cross-sectional studies involving patients with MS who had a previous episode of ON showed that a mean pRNFL of up to 30 μm can be expected after ON [5]. Unless an adequate improvement of sight in patients with more significant pRNFL thinning occurs within 3–6 months from the development of ON, subsequent improvement of sight cannot be expected.

On the other hand, GCIPL does not swell during initial stages of ON which allows more accurate quantification of neurodegeneration following ON [30–32]. As its thinning manifests as early as one month after symptom onset [30], GCIPL is the more sensitive measurement for early detection of neuronal loss. Acute changes in GCIPL within 1 month of ON onset predict visual acuity after 6 months [31]. Its reduction is believed to be the result of retrograde axonal degeneration of the retinal nerve fibers. The absence of GCIPL swelling during ON as well as minimal astroglial influence on its thickness measures (retinal astrocytes are mainly located in the RNFL) may contribute to better reproducibility and lower variance of GCIPL in comparison to pRNFL thickness.

Most recently, retinal layer thinning after ON has been shown to reliably predict future non-ON relapse remission in patients with relapsing MS [33]. Patients with incomplete relapse remission displayed significantly more thinning of both pRNFL (30.4 μm vs 22.1 μm, $P = .002$) and GCIPL (16.3 μm vs. 11.3 μm, $P < .001$) from baseline (within 12 months before ON) to follow-up (3–6 months after ON) [33]. Incomplete remission of a non-ON relapse was also associated with change in pRNFL and GCIPL thickness from baseline to follow-up (odds ratio [OR] 1.9 per 10 μm and 2.4 per 5 μm, $P < .001$, respectively) [33]. Incomplete recovery, the clinical correlate of neuroaxonal damage, may result from a more severe initial injury or limited repair and/or functional compensation processes [34]. In younger patients with less severe relapses and better repair and compensation capacities [34,35], complete clinical recovery after a relapse might mask the accumulation of neuroaxonal damage below the clinical threshold, stressing the importance of OCT and its prognostic value for determining the likelihood of incomplete recovery from future relapses outside the visual system.

ON appears to affect outer retinal layers, too, namely the INL, OPL, and ONL [36], which exhibit thickening rather than thinning, indicating outer retinal inflammation. INL thickening occurs in the first 3 months after any MS relapse, including ON, and returns to baseline level within 1 year [37,38]. The plexiform layers surrounding the INL contain primary networks of retinal microglia [39], and act as diffusion barrier, making the INL susceptible for fluid accumulation during acute inflammation. It is assumed that an injury of Müller cells leads to increased vascular permeability within the retina, followed by accumulation of retinal fluid and INL thickening [38].

OCT can also serve as a marker to help differentiate between MS and other disorders that can lead to ON, for example, neuromyelitis optica spectrum disorder (NMOSD) and myelin oligodendrocyte glycoprotein antibody associated disease (MOGAD). In MS-associated ON, pRNFL atrophy is frequently observed only in the temporal area or with an temporal accentuation [27], whereas the pattern of pRNFL loss in NMOSD/MOGAD is more global, typically greater (about 10–20 μm worse in NMOSD) and shows less predilection for the temporal quadrants (Fig. 19.3)

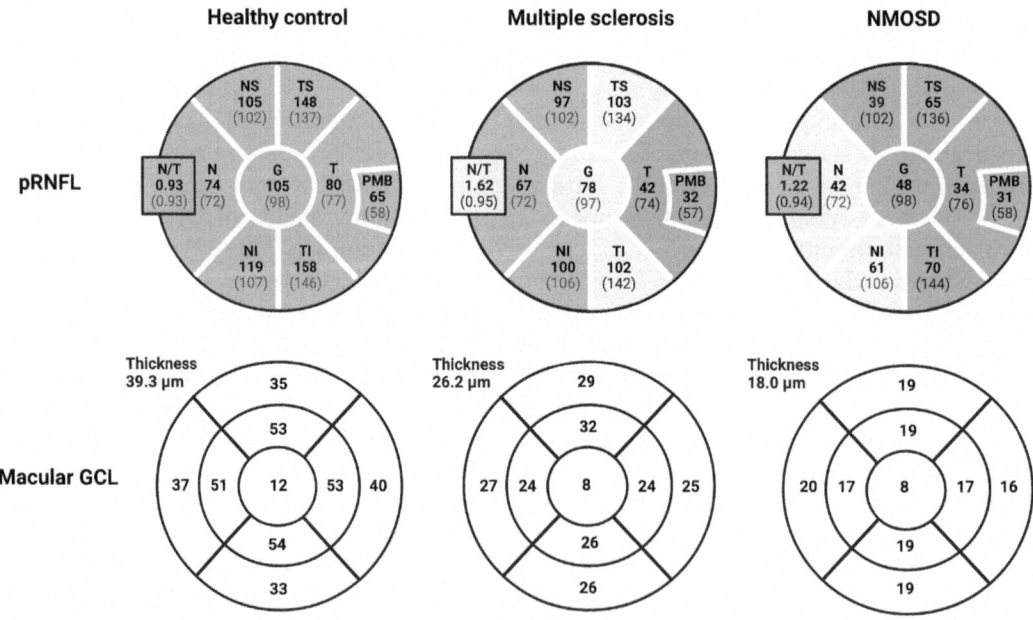

FIGURE 19.3 OCT-derived pRNFL and macular GCL thicknesses by quadrants for the left eye of a healthy control, and patients with MS and NMOSD. Colors in pRNFL quadrants represent the thickness' percentile of a built-in normative database (green: between the 95th and 5th percentile; yellow: between the 5th and 1st percentile; red: bellow the 1st percentile). In MS, pRNFL atrophy is most frequently observed in the temporal area, whereas the pRNFL loss in NMOSD is typically greater and more global. Source: *Created with BioRender.com.*

[40,41]. Besides, patients with NMOSD also exhibit more substantial intereye pRNFL differences irrespective of ON history (−31.0 μm vs −9.9 μm) [42], allowing a relatively reliable distinction of MS from NMOSD in case of diagnostic uncertainty. Microcystic macular edema (MME), a phenomenon typically located in the INL likely caused by a disturbance of the blood−retina barrier and indicative of acute intraretinal inflammation, is another diagnostic feature, which can be used in differential diagnosis of ON: MME is most commonly associated with chronic relapsing inflammatory ON und is much less often caused by NMOSD (up to 25% of eyes following ON) [43], and only rarely by MOGAD or MS (up to 5% of patients) [41,44].

OCT may also serve as a reliable tool to detect prior ON episodes which is of great importance as patients with an initial demyelinating event and previous symptoms consistent with ON can be a challenge in clinical practice. Providing proof of past ON episodes can change the diagnosis and treatment options, especially in patients with clinically isolated syndrome (CIS). Validated cut-off values of intereye asymmetry (≥ 4 μm/4% for GCIPL and ≥ 5 μm/5% for RNFL) discriminate between healthy eyes and eyes with previous ON [45,46].

Optical coherence tomography is a marker of multiple sclerosis−associated neuroaxonal damage

Apart from depicting presence, pattern and degree of neuroaxonal damage caused by ON, OCT has also been demonstrated to reflect global disease processes in MS. Atrophy of pRNFL and GCIPL correlates with clinical and radiological disease progression, manifested by worsening disability measured by Expanded Disability Status Scale (EDSS) [47−51], progressive disease course [52,53], new/enlarging T2- and contrast-enhancing lesions on MRI [54], brain and spinal cord atrophy [55−57], cognitive performance and executive function [58−60], and higher serum neurofilament levels [61,62].

Normal RNFL thickness has been reported to reach mean values of 104.4 ± 7.6 μm in the third life decade, and 89.5 ± 7.5 μm by the seventh life decade [63], indicating a physiologic loss of approximately 0.017% per year in retinal thickness (10−20 μm loss over 60 years) in healthy individuals. On the other hand, patients with MS display on average 20 and 12 μm lower pRNFL and GCIPL thicknesses, respectively [64−67]. Although axonal and neuronal degeneration is primarily thought to occur as sequelae of direct focal inflammatory demyelination, that is, ON in case of retinal neuroaxonal damage, other contributory pathophysiological processes play a role in the disease process, too [68,69]. The effect of pRNFL/GCIPL thinning on disability appears to be greatest early in the disease process [54,67,70], probably due to a greater vulnerability of retinal ganglion cells for neurodegeneration or a greater tendency for inflammatory disease activity earlier in the disease process, and seems to reach a plateau (flooring) effect with longer disease duration [70].

The pathophysiology underlying retinal thinning occurring independent of ON is still not entirely clear and may have several contributors. Clinically silent ON episodes may play a role in some patients. However, the bulk of retinal thinning is thought to be caused by MS lesions located in the visual pathway, which lead to retrograde axonal degeneration and/or trans-synaptic degeneration of retinal cells, resulting in pRNFL and GCIPL thinning. In that way, retinal thinning is understood as a surrogate of neurodegeneration and brain atrophy.

3. Use of other imaging acquisition and analysis modalities in multiple sclerosis

Cross-sectional optical coherence tomography—retinal thickness indicates neuroaxonal reserve

Based on these observations, pRNFL and GCIPL thicknesses are viewed as a measure of neuroaxonal reserve in patients with MS. Several studies using a stratification of patients into tiers of retinal thickness based on a one-time OCT of eyes without a history of ON have shown that reduced thickness of pRNFL and GCIPL indicates a higher risk of future disability worsening in MS. The lowest tier, defined by a GCIPL thickness <77 μm and a pRNFL thickness <88 μm, was robustly and reproducibly associated with a two to three increased risk of disability progression (EDSS and/or cognition measured by Symbol Digit Modalities Test) within the next 3 years [49,71,72]. In another study, GCIPL thickness <70 μm was associated with a fourfold increased risk of disease progression over a median of 10 years later [73]. Patients with lower retinal thickness have likely suffered a higher degree of neuroaxonal damage and therefore may have less neuroaxonal reserve remaining, which may in turn give them a lower capacity to compensate further neuroaxonal damage, leading to a higher likelihood of clinical worsening in subsequent years. Thus OCT can be used to stratify MS patients informing prognosis and treatment strategy.

Longitudinal optical coherence tomography—retinal thinning indicates ongoing neuroaxonal damage

Repeatedly performing OCT enables measuring the retinal thinning over time. Retinal thinning in patients with MS occurs in a slowly progressing pattern at mean rates reported to range from 1–2 μm/year in pRNFL and 0.6–1.8 μm/year in GCIPL, exceeding the mean rate of age-related average pRNFL thinning of about 0.2–0.5 and 0.1–0.3 μm/year observed in healthy individuals [54,74–77].

Numerous studies have investigated longitudinal OCT in relapsing-remitting multiple sclerosis (RRMS) patients. Thinning of pRNFL and GCIPL is faster in patients with disability worsening, new T2 and/or Gd-enhancing lesions, more pronounced brain atrophy and elevated sNfL [54,61]. These findings indicate that the degree of retinal changes in MS reflects global CNS processes, having the utility as an outcome measure to assess neuroprotective agents, especially in early, active MS. Both loss of pRNFL and GCIPL can predict occurrence of clinical events related to neurodegeneration such as EDSS progression and cognitive decline with good specificity and moderate sensitivity [71,72]. In multivariate models, both aLmGCIPL $\geq$ 1.0 μm and aLpRNFL $\geq$ 1.5 μm were strong predictors of clinical progression (OR 18.3 and 15.1, respectively) [72]. Of note, GCIPL thinning (87% sensitivity and 90% specificity) significantly exceeds the strength of association found with pRNFL thinning (76% sensitivity and 90% specificity) [72]. Importantly, both retinal and cortical volume loss rates seem to be age-dependent with faster rates occurring in younger patients, which indicates that the period of greatest adaptive-immune-mediated inflammatory activity is also the period of the greatest neuroaxonal loss [78].

Optical coherence tomography in progressive multiple sclerosis

Data on OCT in progressive MS are scarce compared to RRMS. However, patients with progressive MS show reduced retinal layer thickness and more pronounced retinal layer thinning compared to healthy persons and also patients with RRMS [52,53]. This might occur as those patients show more pronounced neuroaxonal loss based on a more severe and also different disease pathology than patients with RRMS [79], and due to longer disease duration, which leads to a greater damage accumulation [80,81]. The degree of pRNFL reduction is similar in patients with primary progressive MS, who by definition do not suffer from relapsing ON, and SPMS patients without previous ON (pRNFL: -0.69%, GCIPL: -0.67% to -0.54%), but faster than in RRMS [82]. Patients that are on the verge of conversion to SPMS also show higher annual RNFL and GCIPL thinning compared to RRMS without progression (RNFL: $-2.62\,\mu m$ vs $-0.59\,\mu m$; GCIPL $-1.11\,\mu m$ vs $-0.19\,\mu m$) [83], which is also associated with higher sNfL levels [84].

Optical coherence tomography in treatment monitoring

Given the sensitivity and reliability of OCT measures, OCT is also an attractive candidate to be also used in treatment monitoring. Most clinical trials routinely examine impact of disease-modifying therapy (DMT) on MRI measures, for example, new/expanding T2 lesions, Gd-enhancing lesions, and brain atrophy, with only few studies assessing their effect on retinal layer thickness. Defining response to DMT (i.e., recognizing treatment failure) is one of the biggest challenges in managing patients with MS, as it is the key for tailoring treatment according to individual disease activity.

One of current trends as a treatment outcome is known as no evidence of disease activity (NEDA), most commonly being used in the form of NEDA-3 (the absence of relapses, EDSS progression, new/enlarging, and/or Gd-enhancing lesions on MRI). Patients who do not meet NEDA-3 criteria show higher rates of pRNFL thinning ($-2.8 \pm 2.0\,\mu m$) compared to patients fulfilling the criteria ($-0.9 \pm 1.4\,\mu m$) over 2 years [85]. In a study by Zimmermann et al., patients with CIS were grouped into tertiles according to pRNFL, GCIPL, and INL thickness, with patients in the thinnest GCIPL tertile having a significantly higher probability of not meeting NEDA-3 criteria (hazard ratio [HR] = 3.33; $P<.001$) and increased rates of MS diagnosis during follow-up (HR = 4.05; $P<.001$) compared to patients from the thickest tertile [86]. Increased INL and ONL volumes were reported to be associated with progression to MS in patients with CIS [87], reflecting direct effects of subclinical inflammatory processes within the retina of patients with MS. The combination of sNfL and GCIPL thickness seems to be promising for predicting future disease activity on a group level; patients with abnormal sNfL and thin GCIPL have an elevated risk for NEDA-3 violation (HR = 3.61) compared to abnormal sNfL alone (HR = 2.28) [88]. A cut-off of $-1.25\,\mu m$ of pRNFL thinning appeared to classify NEDA patients with 81.4% specificity and 80% sensitivity (AUC 0.80; $P<.001$) [85].

Although some studies reported no significant difference in rates of retinal layer atrophy in MS patients treated with first-line DMTs compared to untreated patients over a 1-year period [89–91], 1 year seems to be too short to reliably assess therapeutic effect. Study by Zivadinov et al. showed that over a 24-month period, patients with MS treated with glatiramer acetate show similar evolution of OCT measures compared to healthy controls (HC) [92]. Moreover, another study confirmed greater RNFL thinning in untreated patients with MS [93], and comparable GCIPL atrophy in patients receiving second-line therapies, for example, natalizumab and HC [94]. Those results are not surprising as DMTs reduce the relapse rate and are therefore expected to lead to reduced axonal damage, pointing out the importance of early treatment in order to prevent future axonal loss. Siponimod has been shown to preserve retinal thickness in patients with secondary progressive MS, yet with a delayed effect which is demarcated not before 24 months of treatment [95]. Also, rituximab-treated patients with MS exhibit similar rates of GCIPL thinning compared to natalizumab-treated patients with MS or HCs, with greater attenuation of retinal atrophy occurring after 12 months of treatment [96]. Recently, alemtuzumab-treated patients with MS demonstrated no significant loss of RNFL thickness and GCIPL volume in comparison to patients with MS receiving first-line DMTs over a 5-year period [97].

DMT failure, defined as physical and/or cognitive disability worsening, can be predicted by measuring GCIPL thinning over 12 months (specificity 85% and sensitivity 78%) with only a minor additional gain in accuracy when extending measurement to 24 months (specificity 91% and sensitivity 81%) [98]. aLGCIPL of more than 0.5 μm at 2 years seems to be the strongest predictor of disability worsening (HR 4.5), closely followed by the same measurement at year 1 (HR 3.9) [98]. On the other hand, measuring pRNFL thinning is considerably less accurate and necessitates 24 months of observation (specificity 84% and sensitivity 69%) [98]. While GCIPL thickness is the most promising marker to assess the neuroprotective effect of DMTs [94], INL measures could potentially be used to assess their impact on inflammatory activity [99]. INL thickness has been reported to be globally reduced within 6–9 months of treatment initiation of different DMTs, whereas higher INL thickness predicts the development of future relapses, new T2 lesions and/or Gd-enhancing lesions, and disability progression [100]. However, treatment with fingolimod is associated with slightly higher MME incidence compared to other DMTs (approximately 1% of fingolimod-treated patients with MS) [101], which can lead to enhanced macular volumes and thinner INL in those patients [102].

As history of ON does not influence further loss of retinal thickness, but DMT does [94], it is suggested that OCT measuring both pRNFL and GCIPL thickness should be performed within 3–6 months after initiating DMT as a rebaseline and should then be repeated annually in routine monitoring.

Practical issues and limitations affecting clinical application of optical coherence tomography in multiple sclerosis

Despite the growing evidence supporting the potential utility of OCT in management of patients with MS, there is a need for more comprehensive longitudinal studies to validate OCT parameters for monitoring neurodegeneration and especially treatment monitoring. Besides, changes in retinal layer thickness are rather unspecific and occur in several other diseases apart from MS, including Alzheimer's and Parkinson's disease, amyotrophic lateral sclerosis, frontotemporal dementia, and Huntington's disease. Also, numerous primary ocular diseases may affect MME emergence in INL of patients with MS, which must be acknowledged when making conclusions about INL assessment and disease control.

There are also multiple confounders that affect longitudinal measurements of retinal thickness including race and the presence of other neurological and ophthalmological conditions [103–105]. OCT is also not applicable in case of high myopy (>4–6 dpt) or retinal comorbidities (glaucoma, optic drusen, diabetes mellitus, etc.). Patients with shorter disease duration show the highest annual atrophy rates, suggesting a subclinical neuroaxonal loss occurring early in MS due to more intact neuroaxonal substrate in comparison to patients with a preexisting injury to afferent visual pathways. After 15–20 years, its value is much lower due to the flooring effect, significantly hampering detection of longitudinal changes regardless of disease status [70]. Automatic segmentation also requires quality control which is partly overcome by the implementation of OSCAR-IB criteria. And last but not least, its availability is still limited in some centers around the world.

Conclusion

OCT has emerged as an inexpensive, noninvasive, high-resolution imaging technique that provides a window into global CNS inflammatory and neurodegenerative processes. In recent years, the evolution of SD-OCT has facilitated the incorporation of OCT into

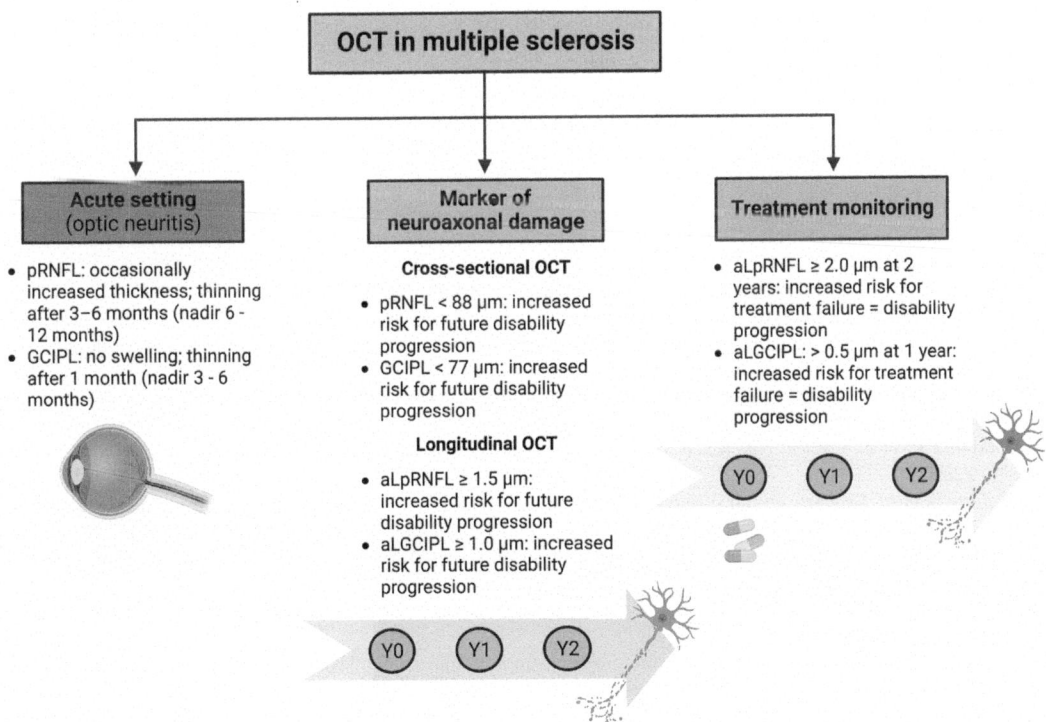

FIGURE 19.4 OCT presents a reliable and reproducible imaging biomarker of neuroaxonal with various applicable in acute optic neuritis, as a surrogate to determine neuroaxonal damage disease and monitor treatment efficacy. Source: *Created with BioRender.com.*

routine clinical practice as a reliable and reproducible imaging biomarker (Fig. 19.4). Given that within the retina, axons of the retinal ganglion cells are not myelinated, axonal damage can be directly quantified, rendering OCT an excellent tool to evaluate neuroaxonal damage in vivo. Moreover, as different diseases may show distinct patterns of retinal pathology, OCT may be helpful in the differential diagnostic workup in difficult situations. Most importantly, in the context of MS, it is now well established that at least in the absence of ON, retinal neuroaxonal degeneration determined by OCT reflects global neurodegeneration in the CNS, suggesting especially GCIPL, but also pRNFL, as surrogates for measuring CNS neurodegeneration. Although the etiology of INL changes in MS remains to be elucidated, its increased thickness at baseline has been shown to predict inflammatory disease activity and disability progression in MS.

From a scientific point of view, OCT has a growing role as a secondary endpoint in Phase 2 and Phase 3 clinical trials of neuroprotective and potentially remyelinating therapies. The evidence about its pathological correlate (axonal loss) is undeniable, and it also has a high clinical relevance, correlating with clinical measures and therefore capturing a pivotal feature associated with disability progression.

While advances in OCT have led to the development of new measures for tracking neurodegeneration and potentially also assessing neuroprotection and remyelination in patients with MS, much further work is needed. However, the results to date are exciting, and its clinical application will definitely continue to grow, shaping novel therapeutic avenues in MS.

References

[1] Britze J, Frederiksen JL. Optical coherence tomography in multiple sclerosis. Eye (Lond) 2018;32(5):884–8.
[2] Frohman EM, et al. Optical coherence tomography: a window into the mechanisms of multiple sclerosis. Nat Clin Pract Neurol 2008;4(12):664–75.
[3] Huang D, et al. Optical coherence tomography. Science 1991;254(5035):1178–81.
[4] Aumann S, et al. Optical coherence tomography (OCT): principle and technical realization. In: Bille JF, editor. High resolution imaging in microscopy and ophthalmology: new frontiers in biomedical optics. Cham: Springer International Publishing; 2019. p. 59–85.
[5] Petzold A, et al. Optical coherence tomography in multiple sclerosis: a systematic review and meta-analysis. Lancet Neurol 2010;9(9):921–32.
[6] Jindahra P, et al. Optical coherence tomography of the retina: applications in neurology. Curr Opin Neurol 2010;23(1):16–23.
[7] Saidha S, et al. Primary retinal pathology in multiple sclerosis as detected by optical coherence tomography. Brain 2011;134(Pt 2):518–33.
[8] Friese MA, et al. Mechanisms of neurodegeneration and axonal dysfunction in multiple sclerosis. Nat Rev Neurol 2014;10(4):225–38.
[9] Correale J, et al. Mechanisms of neurodegeneration and axonal dysfunction in progressive multiple sclerosis. Biomedicines 2019;7(1).
[10] Popescu DP, et al. Optical coherence tomography: fundamental principles, instrumental designs and biomedical applications. Biophys Rev 2011;3(3):155.
[11] Forte R, et al. Comparison of time domain Stratus OCT and spectral domain SLO/OCT for assessment of macular thickness and volume. Eye (Lond) 2009;23(11):2071–8.
[12] Petzold A, et al. Retinal layer segmentation in multiple sclerosis: a systematic review and meta-analysis. Lancet Neurol 2017;16(10):797–812.

[13] Kiernan DF, et al. Spectral-domain optical coherence tomography: a comparison of modern high-resolution retinal imaging systems. Am J Ophthalmol 2010;149(1):18–31.
[14] Schippling S. Basic principles of optical coherence tomography. In: Calabresi PA, et al., editors. Optical coherence tomography in neurologic diseases. Cambridge: Cambridge University Press; 2015. p. 4–13.
[15] Aytulun A, et al. APOSTEL 2.0 recommendations for reporting quantitative optical coherence tomography studies. Neurology 2021;97(2):68–79.
[16] Seigo MA, et al. In vivo assessment of retinal neuronal layers in multiple sclerosis with manual and automated optical coherence tomography segmentation techniques. J Neurol 2012;259(10):2119–30.
[17] Wu H, et al. Reproducibility of retinal nerve fiber layer thickness measurements using spectral domain optical coherence tomography. J Glaucoma 2011;20(8):470–6.
[18] Pemp B, et al. Effectiveness of averaging strategies to reduce variance in retinal nerve fibre layer thickness measurements using spectral-domain optical coherence tomography. Graefes Arch Clin Exp Ophthalmol 2013;251(7):1841–8.
[19] Wadhwani M, et al. Test-retest variability of retinal nerve fiber layer thickness and macular ganglion cell-inner plexiform layer thickness measurements using spectral-domain optical coherence tomography. J Glaucoma 2015;24(5):e109–15.
[20] Tewarie P, et al. The OSCAR-IB consensus criteria for retinal OCT quality assessment. PLoS One 2012;7(4):e34823.
[21] London A, et al. The retina as a window to the brain-from eye research to CNS disorders. Nat Rev Neurol 2013;9(1):44–53.
[22] Balcer LJ, et al. Vision and vision-related outcome measures in multiple sclerosis. Brain 2015;138(Pt 1):11–27.
[23] Kremser B, et al. Retinal thickness analysis in subjects with different refractive conditions. Ophthalmologica 1999;213(6):376–9.
[24] Balcer LJ. Clinical practice. Optic neuritis. N Engl J Med 2006;354(12):1273–80.
[25] Frohman EM, et al. The neuro-ophthalmology of multiple sclerosis. Lancet Neurol 2005;4(2):111–21.
[26] Toussaint D, et al. Clinicopathological study of the visual pathways, eyes, and cerebral hemispheres in 32 cases of disseminated sclerosis. J Clin Neuroophthalmol 1983;3(3):211–20.
[27] Ikuta F, Zimmerman HM. Distribution of plaques in seventy autopsy cases of multiple sclerosis in the United States. Neurology 1976;26(6 PT 2):26–8.
[28] Roed H, et al. Systemic T-cell activation in acute clinically isolated optic neuritis. J Neuroimmunol 2005;162(1-2):165–72.
[29] Costello F, et al. Tracking retinal nerve fiber layer loss after optic neuritis: a prospective study using optical coherence tomography. Mult Scler 2008;14(7):893–905.
[30] Kupersmith MJ, et al. Retinal ganglion cell layer thinning within one month of presentation for optic neuritis. Mult Scler 2016;22(5):641–8.
[31] Gabilondo I, et al. Dynamics of retinal injury after acute optic neuritis. Ann Neurol 2015;77(3):517–28.
[32] Syc SB, et al. Optical coherence tomography segmentation reveals ganglion cell layer pathology after optic neuritis. Brain 2012;135(Pt 2):521–33.
[33] Bsteh G, et al. Retinal layer thinning after optic neuritis is associated with future relapse remission in relapsing multiple sclerosis. Neurology 2022;.
[34] Trapp BD, et al. Axonal transection in the lesions of multiple sclerosis. N Engl J Med 1998;338(5):278–85.
[35] Caggiula M, et al. Neurotrophic factors and clinical recovery in relapsing-remitting multiple sclerosis. Scand J Immunol 2005;62(2):176–82.
[36] Al-Louzi OA, et al. Outer retinal changes following acute optic neuritis. Mult Scler 2016;22(3):362–72.
[37] Schurz N, et al. Evaluation of retinal layer thickness parameters as biomarkers in a real-world multiple sclerosis cohort. Eye Brain 2021;13:59–69.
[38] Pisa M, et al. Subclinical anterior optic pathway involvement in early multiple sclerosis and clinically isolated syndromes. Brain 2021;144(3):848–62.
[39] Hume DA, et al. Immunohistochemical localization of a macrophage-specific antigen in developing mouse retina: phagocytosis of dying neurons and differentiation of microglial cells to form a regular array in the plexiform layers. J Cell Biol 1983;97(1):253–7.
[40] Schneider E, et al. Optical coherence tomography reveals distinct patterns of retinal damage in neuromyelitis optica and multiple sclerosis. PLoS One 2013;8(6):e66151.

[41] Bennett JL, et al. Neuromyelitis optica and multiple sclerosis: seeing differences through optical coherence tomography. Mult Scler 2015;21(6):678–88.
[42] Peng A, et al. Evaluation of the retinal nerve fiber layer in neuromyelitis optica spectrum disorders: a systematic review and meta-analysis. J Neurol Sci 2017;383:108–13.
[43] Sotirchos ES, et al. In vivo identification of morphologic retinal abnormalities in neuromyelitis optica. Neurology 2013;80(15):1406–14.
[44] Bennett JL. Optic neuritis. Contin (Minneap Minn 2019;25(5):1236–64.
[45] Nolan-Kenney RC, et al. Optimal intereye difference thresholds by optical coherence tomography in multiple sclerosis: an international study. Ann Neurol 2019;85(5):618–29.
[46] Bsteh G, et al. Validation of inter-eye difference thresholds in optical coherence tomography for identification of optic neuritis in multiple sclerosis. Mult Scler Relat Disord 2020;45:102403.
[47] Talman LS, et al. Longitudinal study of vision and retinal nerve fiber layer thickness in multiple sclerosis. Ann Neurol 2010;67(6):749–60.
[48] Saidha S, et al. Optical coherence tomography reflects brain atrophy in multiple sclerosis: a four-year study. Ann Neurol 2015;78(5):801–13.
[49] Martinez-Lapiscina EH, et al. Retinal thickness measured with optical coherence tomography and risk of disability worsening in multiple sclerosis: a cohort study. Lancet Neurol 2016;15(6):574–84.
[50] Berek K, et al. Retinal layer thinning as a biomarker of long-term disability progression in multiple sclerosis. Mult Scler 2022;28(12):1871–80.
[51] Bsteh G, et al. Retinal layer thinning is reflecting disability progression independent of relapse activity in multiple sclerosis. Mult Scler J Exp Transl Clin 2020;6(4) 2055217320966344.
[52] Green AJ, et al. Ocular pathology in multiple sclerosis: retinal atrophy and inflammation irrespective of disease duration. Brain 2010;133(Pt 6):1591–601.
[53] Abalo-Lojo JM, et al. Retinal nerve fiber layer thickness, brain atrophy, and disability in multiple sclerosis patients. J Neuroophthalmol 2014;34(1):23–8.
[54] Ratchford JN, et al. Active MS is associated with accelerated retinal ganglion cell/inner plexiform layer thinning. Neurology 2013;80(1):47–54.
[55] Vidal-Jordana A, et al. Optical coherence tomography measures correlate with brain and spinal cord atrophy and multiple sclerosis disease-related disability. Eur J Neurol 2020;27(11):2225–32.
[56] Sepulcre J, et al. Diagnostic accuracy of retinal abnormalities in predicting disease activity in MS. Neurology 2007;68(18):1488–94.
[57] Saidha S, et al. Relationships between retinal axonal and neuronal measures and global central nervous system pathology in multiple sclerosis. JAMA Neurol 2013;70(1):34–43.
[58] Jakimovski D, et al. Visual deficits and cognitive assessment of multiple sclerosis: confounder, correlate, or both? J Neurol 2021;268(7):2578–88.
[59] Baetge SJ, et al. Association of retinal layer thickness with cognition in patients with multiple sclerosis. Neurol Neuroimmunol Neuroinflamm 2021;8(4).
[60] Esmael A, et al. Retinal thickness as a potential biomarker of neurodegeneration and a predictor of early cognitive impairment in patients with multiple sclerosis. Neurol Res 2020;42(7):564–74.
[61] Bsteh G, et al. Serum neurofilament levels correlate with retinal nerve fiber layer thinning in multiple sclerosis. Mult Scler 2020;26(13):1682–90.
[62] Uzunkopru C, et al. Retinal nerve fiber layer thickness correlates with serum and cerebrospinal fluid neurofilament levels and is associated with current disability in multiple sclerosis. Noro Psikiyatr Ars 2021;58(1):34–40.
[63] Nolan RC, et al. Optical coherence tomography for the neurologist. Semin Neurol 2015;35(5):564–77.
[64] Puliken M, et al. Optical coherence tomography and disease subtype in multiple sclerosis. Neurology 2007;69(22):2085–92.
[65] Gundogan FC, et al. Is optical coherence tomography really a new biomarker candidate in multiple sclerosis? A structural and functional evaluation. Invest Ophthalmol Vis Sci 2007;48(12):5773–81.
[66] Pueyo V, et al. Sub-clinical atrophy of the retinal nerve fibre layer in multiple sclerosis. Acta Ophthalmol 2010;88(7):748–52.
[67] Knier B, et al. Association of retinal architecture, intrathecal immunity, and clinical course in multiple sclerosis. JAMA Neurol 2017;74(7):847–56.

[68] Shindler KS, et al. Inflammatory demyelination induces axonal injury and retinal ganglion cell apoptosis in experimental optic neuritis. Exp Eye Res 2008;87(3):208–13.
[69] Frischer JM, et al. The relation between inflammation and neurodegeneration in multiple sclerosis brains. Brain 2009;132(Pt 5):1175–89.
[70] Balk LJ, et al. Timing of retinal neuronal and axonal loss in MS: a longitudinal OCT study. J Neurol 2016;263(7):1323–31.
[71] Bsteh G, et al. Peripapillary retinal nerve fibre layer thinning rate as a biomarker discriminating stable and progressing relapsing-remitting multiple sclerosis. Eur J Neurol 2019;26(6):865–71.
[72] Bsteh G, et al. Macular ganglion cell-inner plexiform layer thinning as a biomarker of disability progression in relapsing multiple sclerosis. Mult Scler 2021;27(5):684–94.
[73] Lambe J, et al. Association of spectral-domain OCT with long-term disability worsening in multiple sclerosis. Neurology 2021;96(16):e2058–69.
[74] Graham EC, et al. Progressive loss of retinal ganglion cells and axons in nonoptic neuritis eyes in multiple sclerosis: a longitudinal optical coherence tomography study. Invest Ophthalmol Vis Sci 2016;57(4):2311–17.
[75] Bsteh G, et al. Peripapillary retinal nerve fibre layer as measured by optical coherence tomography is a prognostic biomarker not only for physical but also for cognitive disability progression in multiple sclerosis. Mult Scler 2019;25(2):196–203.
[76] Leung CK, et al. Retinal nerve fiber layer imaging with spectral-domain optical coherence tomography: interpreting the RNFL maps in healthy myopic eyes. Invest Ophthalmol Vis Sci 2012;53(11):7194–200.
[77] Mansoori T, Balakrishna N. Effect of aging on retinal nerve fiber layer thickness in normal Asian Indian eyes: a longitudinal study. Ophthalmic Epidemiol 2017;24(1):24–8.
[78] Cordano C, et al. Differences in age-related retinal and cortical atrophy rates in multiple sclerosis. Neurology 2022;.
[79] Saidha S, et al. Visual dysfunction in multiple sclerosis correlates better with optical coherence tomography derived estimates of macular ganglion cell layer thickness than peripapillary retinal nerve fiber layer thickness. Mult Scler 2011;17(12):1449–63.
[80] Gelfand JM, et al. Retinal axonal loss begins early in the course of multiple sclerosis and is similar between progressive phenotypes. PLoS One 2012;7(5):e36847.
[81] Oberwahrenbrock T, et al. Retinal damage in multiple sclerosis disease subtypes measured by high-resolution optical coherence tomography. Mult Scler Int 2012;2012:530305.
[82] Rothman A, et al. Retinal measurements predict 10-year disability in multiple sclerosis. Ann Clin Transl Neurol 2019;6(2):222–32.
[83] Sabbagh H, et al. Retinal OCT used as a clinical biomarker of conversion to SPMS. ECTRIMS, 2021. Vienna: ECTRIMS; 2021.
[84] Seitz CB, et al. Serum neurofilament levels reflect outer retinal layer changes in multiple sclerosis. Ther Adv Neurol Disord 2021;14 17562864211003478.
[85] Pisa M, et al. No evidence of disease activity is associated with reduced rate of axonal retinal atrophy in MS. Neurology 2017;89(24):2469–75.
[86] Zimmermann HG, et al. Association of retinal ganglion cell layer thickness with future disease activity in patients with clinically isolated syndrome. JAMA Neurol 2018;75(9):1071–9.
[87] Knier B, et al. Optical coherence tomography indicates disease activity prior to clinical onset of central nervous system demyelination. Mult Scler 2016;22(7):893–900.
[88] Lin TY, et al. Increased serum neurofilament light and thin ganglion cell-inner plexiform layer are additive risk factors for disease activity in early multiple sclerosis. Neurol Neuroimmunol Neuroinflamm 2021;8(5).
[89] Chilinska A, et al. Analysis of retinal nerve fibre layer, visual evoked potentials and relative afferent pupillary defect in multiple sclerosis patients. Clin Neurophysiol 2016;127(1):821–6.
[90] Pul R, et al. Longitudinal time-domain optic coherence study of retinal nerve fiber layer in IFNbeta-treated and untreated multiple sclerosis patients.. Exp Ther Med 2016;12(1):190–200.
[91] Garcia-Martin E, et al. Effect of treatment in loss of retinal nerve fibre layer in multiple sclerosis patients. Arch Soc Esp Oftalmol 2010;85(6):209–14.
[92] Zivadinov R, et al. The effect of glatiramer acetate on retinal nerve fiber layer thickness in patients with relapsing-remitting multiple sclerosis: a longitudinal optical coherence tomography study. CNS Drugs 2018;32(8):763–70.

[93] Herrero R, et al. Progressive degeneration of the retinal nerve fiber layer in patients with multiple sclerosis. Invest Ophthalmol Vis Sci 2012;53(13):8344–9.
[94] Button J, et al. Disease-modifying therapies modulate retinal atrophy in multiple sclerosis: a retrospective study. Neurology 2017;88(6):525–32.
[95] Vermesch P, et al. 147 Siponimod preserves retinal thickness: findings from the EXPAND OCT substudy.. J Neurol Neurosurg Psychiatry 2022;93(9):e2.
[96] Lambe J, et al. Modulation of retinal atrophy with rituximab in multiple sclerosis. Neurology 2021;96(20): e2525–33.
[97] Chan JK, et al. Long-term stability of neuroaxonal structure in alemtuzumab-treated relapsing-remitting multiple sclerosis patients. J Neuroophthalmol 2020;40(1):37–43.
[98] Bsteh G, et al. Retinal layer thinning predicts treatment failure in relapsing multiple sclerosis. Eur J Neurol 2021;28(6):2037–45.
[99] Knier B, et al. Retinal inner nuclear layer volume reflects response to immunotherapy in multiple sclerosis. Brain 2016;139(11):2855–63.
[100] Saidha S, et al. Microcystic macular oedema, thickness of the inner nuclear layer of the retina, and disease characteristics in multiple sclerosis: a retrospective study. Lancet Neurol 2012;11(11):963–72.
[101] Nolan R, et al. Fingolimod treatment in multiple sclerosis leads to increased macular volume. Neurology 2013;80(2):139–44.
[102] Dinkin M, Paul F. Higher macular volume in patients with MS receiving fingolimod: positive outcome or side effect? Neurology 2013;80(2):128–9.
[103] Jenkins TM, Toosy AT. Optical coherence tomography should be part of the routine monitoring of patients with multiple sclerosis: no. Mult Scler 2014;20(10):1299–301.
[104] Gao L, et al. Abnormal retinal nerve fiber layer thickness and macula lutea in patients with mild cognitive impairment and Alzheimer's disease. Arch Gerontol Geriatr 2015;60(1):162–7.
[105] Ringelstein M, et al. Subtle retinal pathology in amyotrophic lateral sclerosis. Ann Clin Transl Neurol 2014;1(4):290–7.

CHAPTER 20

Imaging of multiple sclerosis in resource-poor settings

Avinash Chandra[1,2]

[1]Consultant Neurologist Neurology NAMS, Bir Hospital, Kathmandu, Bagmati, Nepal
[2]Consultant Neurologist Neurology Annapurna Neurological Institute and Allied Sciences, Kathmandu, Bagmati, Nepal

OUTLINE

Optimization strategies	379	References	380
Cost-reduction strategies	379		
Conclusion	380		

Multiple sclerosis (MS) is characterized pathologically by the development of inflammatory demyelinating white matter lesions in multifocal areas in the central nervous system.

While the prevalence magnetic resonance spectroscopy (MRS); of MS has increased substantially in the past two decades particularly in Europe and North America, in most low- and middle-income countries (LMICs), it is still considered an orphan/rare disease.

The public health programs are run poorly with little to no support by the government in LMICs and most of the health cost is covered out of the pocket by the individual on their own. Personal funding for diagnostic investigations and therapy for chronic disorders including MS is a severe strain on the family. Many resources limited areas, LMICs lack both neurologists and neuroimaging and diagnostic facilities. Therefore precise and modern diagnosis of MS remains elusive for a substantial number of people with MS worldwide. Imaging has helped clinicians to make decisions for their patients when and whether to start treatment for MS or clinically isolated syndrome (CIS). MS can now be diagnosed in some patients after CIS using new magnetic resonance imaging (MRI) diagnostic criteria. MRI is considered the gold standard imaging technique for the identification of demyelinating lesions, which can be used to support a clinical diagnosis of MS.

The imaging protocol that is usually followed are (detail is discussed elsewhere in other chapters):

1. **MRI:** MRI is the gold standard for diagnosing and monitoring MS. It provides high-resolution images of the brain and spinal cord, aiding in identifying lesions, inflammation, and atrophy. Despite its efficacy, MRI equipment, maintenance, and interpretation can be cost-prohibitive for resource-poor regions.
2. **Magnetic resonance spectroscopy (MRS):** Mrs supplements MRI by measuring chemical composition in tissues, helping distinguish active lesions from scar tissue. Its implementation, however, requires specialized hardware and expertise.
3. **Computed tomography (CT):** In areas where MRI is unavailable, CT scans can provide some information about brain atrophy and white matter changes [1]. However, CT is less sensitive in detecting early MS lesions and lacks the specificity of MRI.
4. **Optical coherence tomography (OCT):** OCT is used to assess retinal nerve fiber layer thickness, which can reflect axonal damage in MS [2]. It is a noninvasive and cost-effective option to monitor disease progression [3].

Imaging in resource-poor setting is a very daunting task. The cost of imaging along with poor availability of the imaging facility adds the difficulty to such a level that many of the times the imaging is either skipped or not done and thus patients remain undiagnosed.

This chapter explores the unique obstacles faced in these regions, discusses the impact of limited accesses, and suggests strategies to overcome these challenges.

The challenges that are frequently faced are:

1. **Inadequate access to advanced imaging technologies**

    In resource-limited nations, the availability of advanced imaging technologies, MRI, is often limited. The high cost of MRI machines, maintenance, and consumables makes them unaffordable for many healthcare facilities. This scarcity leads to long waiting times, delayed diagnoses, and hindered disease monitoring for MS patients. As a result, MS patients in these regions often have limited access to these essential diagnostic tools, leading to delays in diagnosis and suboptimal disease management.

2. **Limited access to contrast agents**

    Having MRI machines will not only be enough for MS diagnosis and characterization as well as management [4]. Contrast agents play a vital role in enhancing the sensitivity of MRI for detecting MS lesions [5,6]. However, resource-limited nations often have limited access to contrast agents due to their high cost and storage requirements. The lack of availability hampers the ability to perform contrast-enhanced MRI scans, limiting the diagnostic accuracy and comprehensive assessment of MS patients.

3. **Shortage of trained radiologists and technicians**

    Even when imaging technologies are available, resource-limited nations frequently face a shortage of trained radiologists and technicians who can accurately interpret the images. The complex nature of MS lesions requires expertise in distinguishing MS-related abnormalities from other conditions. The lack of trained professionals often leads to misinterpretation or underutilization of available imaging resources, compromising patient care and leading to suboptimal patient care.

4. Infrastructure and technological limitations

   The inadequate infrastructure in resource-limited nations poses additional challenges for MS imaging. MRI needs continuous electrical supply, better cooling system. Power outages (which is not uncommon in resource-limited nations), unreliable and slow internet connectivity, and poor maintenance of imaging equipment result in frequent interruptions in imaging services. Moreover, the absence of dedicated IT systems for image storage and retrieval hampers the long-term monitoring and follow-up of MS patients.

5. Lack of long-term follow-up and image storage systems

   Long-term follow-up and monitoring of MS patients are crucial for assessing disease progression and treatment response. However, resource-limited nations often lack dedicated image storage systems and electronic medical record infrastructure to store and retrieve MRI images efficiently. This limits the ability to compare images over time, hindering accurate assessment, and making it challenging to evaluate treatment efficacy.

Optimization strategies

The limitations bring opportunities and new ideas as well. There are few adaptations that can be made and used at centers in resource limited setting and optimal results can be obtained.

Cost-reduction strategies

1. Making cost-effective imaging

   Protocol optimization: Tailoring imaging protocols to focus on critical MS-related findings can reduce scan times, leading to lower costs and increased patient throughput. Limited healthcare budgets often prioritize other urgent needs, leaving MRI imaging for MS as a lower priority and thereby causing diagnostic delays and dilemmas. To address the limitations of expensive imaging technologies, several low-cost alternatives have been explored in resource-limited nations. Potential solutions for initial screenings and monitoring MS patients can be done by CT scan of head [1] (thin cut slices with good exposure CT with contrast enhancement). These affordable options can be deployed in remote areas where access to advanced imaging technologies is severely limited. However, their accuracy and reliability compared to standard imaging modalities need further evaluation and monitoring, but this can at least help in screening purposes. Some other few modifications that can make easier and cheaper imaging using MRI can be that we use the fluid attenuated inversion recovery and T2-weighted imaging only as a screening tool to monitor the MS patients during follow ups as these will show the progression of the disease. Mobile MRI units that can be used in some regions that employ mobile MRI units that can travel to remote areas, increasing access to imaging services.

2. Telemedicine and telemonitoring:

   Telemedicine has emerged as a valuable tool for overcoming imaging challenges in resource-limited nations [7]. Collaborative efforts with distant specialists through telemedicine can aid in accurate image interpretation, reducing the need for local experts and potentially cutting costs. Through telemedicine platforms, healthcare providers can

3. Use of other imaging acquisition and analysis modalities in multiple sclerosis

remotely connect with expert radiologists in more developed regions. This allows for the interpretation of imaging results and facilitates accurate diagnosis and treatment planning for MS patients [8]. Through telemedicine, neurologists can also be available for making the clinical diagnosis. Additionally, telemedicine enables the exchange of knowledge and expertise, empowering local healthcare professionals to enhance their skills in interpreting MS-related imaging findings. It was best implemented during COVID pandemic period [9]. Cloud-based solutions can be an effective method for storing and sharing images, allowing experts to remotely review scans and provide feedback.

3. Training and education initiatives

    Investing in training and education initiatives is essential for building local capacity in MS imaging. Scholarships, fellowships, and workshops can provide opportunities for healthcare professionals in resource-limited nations to acquire the necessary skills in interpreting MS-related imaging findings. Additionally, educational programs should focus on raising awareness among healthcare providers about the importance of early MS diagnosis and the proper utilization of available imaging resources. Machine learning and artificial intelligence algorithms can aid in preliminary image analysis [10], assisting less-experienced personnel in identifying potential MS-related findings [11].

4. International collaborations and support

    International collaborations and supports play a crucial role in addressing imaging challenges in resource-limited nations. Partnering with private entities for equipment sharing, maintenance, and expertise can help in reducing costs. By pooling resources and sharing knowledge, it becomes possible to improve access to advanced imaging technologies, support capacity building, and establish sustainable imaging services in those resource limited regions. Global organizations, academic institutions, and nonprofit entities should join forces to provide financial assistance, technical expertise, and training programs.

5. Task shifting: Training nonradiologist healthcare workers to perform basic MRI scans and assessments under supervision can alleviate the shortage of radiologists.

Conclusion

Imaging challenges and limitations in resource-limited nations significantly impact the diagnosis and management of MS. However, innovative strategies such as low-cost imaging alternatives, telemedicine, international collaborations, and training initiatives offer potential solutions. By addressing these challenges and providing equitable access to advanced imaging technologies, healthcare systems in resource-limited nations can improve the care and outcomes of MS patients.

References

[1] Loizou LA, Rolfe EB, Hewazy H. Cranial computed tomography in the diagnosis of multiple sclerosis. J Neurol Neurosurg Psychiatry 1982;45(10):905−12. Pubmed Central PMID: 6292371, PMCID: 491595.

[2] Britze J, Frederiksen JL. Optical coherence tomography in multiple sclerosis. Eye 2018;32(5):884−8. Pubmed Central PMID: 29391574, PMCID: 5944645.

[3] Frohman EM, Fujimoto JG, Frohman TC, Calabresi PA, Cutter G, Balcer LJ. Optical coherence tomography: a window into the mechanisms of multiple sclerosis. Nat Clin Pract Neurol 2008;4(12):664–75. Pubmed Central PMID: 19043423, PMCID: 2743162.

[4] Campbell Z, Sahm D, Donohue K, Jamison J, Davis M, Pellicano C, et al. Characterizing contrast-enhancing and re-enhancing lesions in multiple sclerosis. Neurology 2012;78(19):1493–9. Pubmed Central PMID: 22539575, PMCID: 3345616.

[5] Maravilla KR. Enhancing our understanding of multiple sclerosis: tracking contrast-enhancing plaques with MR imaging. AJNR Am J Neuroradiol 2001;22(4):601–3. Pubmed Central PMID: PMCID: 11290465 7976026.

[6] He J, Grossman RI, Ge Y, Mannon LJ. Enhancing patterns in multiple sclerosis: evolution and persistence. AJNR Am J Neuroradiol 2001;22(4):664–9. Pubmed Central PMID: 11290475, PMCID: 7976005.

[7] Wootton R, Bonnardot L. Telemedicine in low-resource settings. Front Public Health 2015;3:3. Pubmed Central PMID: 25654074, PMCID: 4300819.

[8] Roth EG, Minden SL, Maloni HW, Miles ZJ, Wallin MT. A qualitative, multiperspective inquiry of multiple sclerosis telemedicine in the United States. Int J MS Care 2022;24(6):275–81. Pubmed Central PMID: 36545645, PMCID: 9749833.

[9] Keszler P, Maloni H, Miles Z, Jin S, Wallin M. Telemedicine and multiple sclerosis: a survey of health care providers before and during the COVID-19 pandemic. Int J MS Care 2022;24(6):266–70. Pubmed Central PMID: 36545646, PMCID: 9749831.

[10] Schlaeger S, Shit S, Eichinger P, Hamann M, Opfer R, Kruger J, et al. AI-based detection of contrast-enhancing MRI lesions in patients with multiple sclerosis. Insights Imaging 2023;14(1):123. Pubmed Central PMID: 37454342, PMCID: 10350445.

[11] Bonacchi R, Filippi M, Rocca MA. Role of artificial intelligence in MS clinical practice. NeuroImage Clin 2022;35103065. Pubmed Central PMID: 35661470, PMCID: 9163993.

CHAPTER 21

Use of artificial intelligence in multiple sclerosis imaging

Ceren Tozlu[1], Amy Kuceyeski[1] and Michael G. Dwyer[2,3]

[1]Department of Radiology, Weill Cornell Medicine, New York, NY, United States [2]Buffalo Neuroimaging Analysis Center (BNAC), Department of Neurology, Jacobs School of Medicine and Biomedical Sciences, University at Buffalo, State University of New York, Buffalo, NY, United States [3]Center for Biomedical Imaging at Clinical and Translational Science Institute, University of Buffalo, State University of New York, New York, NY, United States

OUTLINE

Introduction	384
Basics of AI in medical imaging	**385**
Supervised vs. unsupervised ML	386
Deep learning architectures for image analysis	387
Assessment of model performance	390
Potential benefits of using AI and neuroimaging in MS	392
AI in MS neuroimaging: survey of current applications	**393**
Improving MRI acquisition protocols and image quality	393
Tissue segmentation/volumetrics	396
Lesion segmentation	397
Diagnosis, prognosis, and clustering	403
AI pitfalls and ethical concerns	**408**
Generalization issues	408
Biases in the data	408
Quality of study description	409
Ethical concerns	409
Explainability of AI algorithms	409
Practical challenges of AI in clinical practice	410
Future directions	**410**
Training healthcare professionals	411
Federated learning	411
Foundational models	411
Personalized treatment plans and digital twins	412
Regulatory landscape	412
Conclusions	**413**
References	**413**

Introduction

Multiple Sclerosis (MS) is a chronic autoimmune, demyelinating, and inflammatory disease affecting the central nervous system. Conventional and advanced magnetic resonance imaging (MRI) techniques play a crucial role in MS diagnosis, prognosis, and monitoring of treatment response. In addition, MRI can be used to better understand the complex relationship between the brain and disability, which is not yet characterized fully [1]. While MRI has significantly advanced our understanding of MS, there are still many challenges associated with MRI use in MS research and clinical care. In the near term, we still need more accurate and automated lesion segmentation approaches, improved image acquisition and reconstruction protocols, shorter scan duration, and a lower amount of contrast agent usage to identify active lesions. More broadly, we still lack highly accurate image-based guidance for diagnosis, prognosis, and treatment decisions. Addressing these challenges with novel, cutting-edge technologies that allow the use of multi-modal neuroimaging datasets is essential to improve the overall management of MS.

Artificial intelligence (AI) has shown to be a promising tool in the field of neuroimaging because of its ability to accomplish specific tasks using large amounts of high-dimensional complex imaging data, which would not be possible with traditional statistical approaches. AI is largely used for both "downstream" and "upstream" tasks (relative to neuroimaging acquisition) in MS and other neurological disorders. Downstream applications include the segmentation of anatomical structures [2–4], differentiating patients from healthy individuals [5,6], prediction of disease severity [7–9], as well as identification of tissue abnormalities such as MS lesions [10,11] or white matter hyperintensities [12]. AI's ability to identify abnormalities helps neuroradiologists save time and focus on more complex and nuanced aspects of the disease, decreases inter- and intra-rater variability, and improves the segmentation of the tissue which is not easily detectable by the human eye. Upstream applications include image enhancement such as creating high-resolution images from low-resolution images [13], removing artifacts [14], lowering contrast dosage [15], and creating synthetic brain images with better contrast that can better classify certain tissue abnormalities such double inversion recovery (DIR) [16]. AI's upstream applications for improving image acquisition can be used to improve clinical care in centers where a high-field MRI is not available, to shorten scanning time which decreases cost and motion artifacts, and to improve the quality of existing low-resolution clinical images. Overall, both downstream and upstream applications of AI provide an opportunity for enhanced patient outcomes and optimized healthcare workflows in the clinical care of individuals with MS.

While the integration of AI into the neuroimaging field has the potential to significantly improve clinical care, there are several practical challenges that need careful consideration to better navigate the ethical implications and avoid potential biases. AI models are most robust and useful when they are trained on large and diverse datasets, as this helps avoid overfitting (i.e. performance decreases when the model is applied to new data) and minimize the effects of bias (i.e. when the model is biased toward the majority subgroup in the data). While data-sharing between institutions is one avenue to creating large datasets, it is crucial to establish trusted data ecosystems to avoid unauthorized access to highly sensitive and potentially identifying neuroimaging data. In addition to creating trusted data ecosystems, increasing the AI models' transparency and interpretability is important to foster trust in the predictions performed with AI algorithms, which is an essential step to facilitate the integration of AI into clinical care. To ensure safe, transparent, and unbiased AI applications, several professional organizations including the US Food and Drug Administration (FDA) (https://www.fda.gov/medical-devices/software-medical-device-samd/artificial-intelligence-and-machine-learning-software-medical-device) and European Medicines Agency (EMA) (https://www.ema.europa.eu/en/about-us/how-we-work/regulatory-science-strategy#regulatory-science-strategy-to-2025-section) have published guidelines on how to improve the real-world applications of AI in medicine. The FDA has reviewed and authorized a growing number of AI-assisted technologies in medicine (155 AI-assisted devices as of June 2023), including AI applications in neuroimaging. One such application was from a startup called Vuno which created an AI-powered brain imaging device that improves MRI image quality, segments the brain into 100 regions, and identifies the regional and global brain atrophy for characterizing neurological disorders such as dementia and Alzheimer's disease. Another FDA-approved AI tool is BrainSee (https://brainsee.ai/) which predicts the likelihood of developing Alzheimer's disorder in people with mild cognitive impairment using conventional MRI scans such as T1 weighted MRI, demographics such as age and sex, and cognitive scores. Specific to MS, Pixyl.Neuro and Imeka both have FDA-approved software that applies AI to MRI data to detect and monitor MS disease activity. More specifically, Imeka applies AI to advanced neuroimaging techniques such as diffusion MRI to detect microstructural changes in white matter bundles of individuals with neurological disorders including MS; it provides the user with a PDF report that highlights the bundles most affected by the disease. While the approved AI-assisted tools provide an opportunity to shape the future of clinical care for more accurate diagnosis, prognosis, and treatment decisions, careful post-approval monitoring is needed to ensure the safe, accurate, and ethical use of AI in clinical care.

The overall goal of this chapter is to provide an overview of AI techniques and their application in neuroimaging individuals with MS. In the first section, we introduce supervised and unsupervised AI techniques, the most frequently used AI algorithms, and the metrics used to evaluate AI algorithm performance. In the second section, we will review how AI is used to (1) improve MRI protocol and image quality, (2) tissue and lesion segmentation, and (3) diagnosis, prognosis, and identify disease subtypes in individuals with MS. In the third section, we talk about the pitfalls of AI including generalization issues, perpetuation of biases, study and data quality, ethical concerns, and (un)explainability of AI algorithms. Finally, the fourth section is a discussion of future directions for promoting the integration of AI into clinical practice, data-sharing policies, and the regulatory landscape.

Basics of AI in medical imaging

AI refers to computer systems that mimic human intelligence in performing specific tasks such as analysis, perception, and decision-making. Machine learning (ML) is a subset of AI that focuses on teaching machines to learn patterns from data and make decisions or predictions. One of the most popular ML algorithms is the artificial neural network (ANN) which is inspired by the human neural network and consists of multiple layers of interconnected artificial neurons that process data in a hierarchical manner. Deep learning (DL) is a specialized version of ANN that employs deep (many layered) neural networks and can be used for more complex and high dimensional problems than many other ANN and other ML algorithms, although it often incurs a need for more data for training. One key advantage of DL is that DL performs better than ML techniques when the raw data (for example structural MRI images) is used. Using the raw MRI data rather than features selected from the MRI images (for example cortical/subcortical thickness/volume) is advantageous to decrease the variability in feature selection (for example variability in brain atlas). In this section, we introduce supervised and unsupervised ML techniques, key DL architectures, the metrics to evaluate the performance of ML/DL algorithms, key general applications of AI, and the benefits of using AI in MS neuroimaging.

Supervised vs. unsupervised ML

ML techniques can be categorized broadly into two groups: (1) unsupervised learning which is used for data-driven applications and (2) supervised learning which is used for task-driven applications. In unsupervised learning, the algorithm uses unlabeled data and tries to learn the underlying pattern and structure of the data itself without any supervision or feedback mechanism. Unsupervised learning can be categorized into two groups: clustering and dimension reduction. Clustering refers to creating groups that consist of similar data points based on certain features and characteristics, for example, clustering MS patients and controls separately using T2 fluid attenuated inversion recovery (FLAIR) imaging without any label regarding their diagnosis status. The most commonly used clustering approaches include k-means clustering which partitions the data into a predefined number (k) of clusters. K-means clustering iteratively assigns data points to clusters based on their similarity (e.g., Euclidian distance) to the cluster centroids and adjusts the centroids until convergence. Another commonly used unsupervised clustering technique is hierarchical clustering which builds a hierarchy of clusters, creating a tree-like structure.

Supervised learning is used for task optimization and consists of two main task categories: (i) regression where the outcome we aim to predict is a continuous measurement such as disability and cognitive scores and (ii) classification analysis where the outcome we aim to predict consists of binary or categorical measurements such as healthy vs disease groups or MS clinical subtypes (relapsing—remitting MS (RRMS), secondary progressive MS (SPMS), and primary progressive MS (PPMS)). The most widely used examples of supervised learning are tree-based algorithms (e.g., classification and regression trees [CART] and random forests), support vector machines (SVM), k-nearest neighbors (k-NN), and artificial neural networks (ANN).

Tree-based algorithms consist of decision trees that break down a dataset into smaller subsets while progressively making decisions based on input features. In a decision tree, each node represents a feature, each branch represents a decision rule based on that feature, and each leaf node represents the outcome or the final decision. While one type of decision tree called CART is easier to interpret, the approach is not stable as the predictions are highly dependent on the hyperparameter choice (e.g., the number of features to select the best feature in each node). Therefore, multiple decision tree algorithms such as random forest that bootstrap the data (i.e. create multiple subsamples of the original data) and build a decision tree for each subsample may be an alternative solution for more robust predictions. The final decision of the random forest is calculated as the majority vote of the decision trees for the classification problems and the average of the prediction for the regression problems.

Another frequently used machine learning algorithm is support vector machines (SVM) which aims to find the optimal boundary, known as the hyperplane, that best separates different classes in a dataset. In cases where the data is linearly separable, SVM aims to find a hyperplane that minimizes the classification error and maximizes the margin, which is defined as the distance between the hyperplane and the nearest data points (support vectors) of different classes. When the data is not linearly separable in the original feature space (for example in two-dimensional space), SVM maps the data into a higher-dimensional space (for example into three-dimensional space) where the data might become linearly separable using a hyperplane. SVM is effective in handling high-dimensional linear and non-linear data.

In k-NN, the classification of a new data point is computed using the class labels of its k nearest neighbors, for example, seven nearest observations when $k = 7$. The term "nearest" is defined based on a chosen distance metric, often Euclidean distance. K-NN predicts the output by averaging (in the case of regression) or taking the majority vote (in the case of classification) of the values from the k nearest neighbors. The advantages of k-NN are ease of implementation and not requiring any assumptions about the underlying data distribution; however, k-NN tends to overfit when applied to high-dimensional and complex datasets.

ANNs, which have driven much of the recent exponential expansion of AI, are inspired by the structure and function of biological neurons. ANNs consist of interconnected nodes called artificial neurons that are organized in layers: an input layer that receives the data, one or more hidden layers that process the results of the input layer by applying weights, and an output layer that combines the results of the hidden layer to produce output via an activation function. The weights on the connections between neurons are adjusted during the training process to enable the network to learn from data, using a process called back-propagation, where the network learns by iteratively adjusting the weights to minimize the difference between predicted and observed outputs. Traditional, fully connected ANNs are not suitable for images such as MRIs where the spatial information between each voxel and the shape/texture of the image needs to be modeled. Deep neural network architectures such as convolutional neural networks (CNNs), encoder−decoder models, and transformers are developed for the implementation of high-dimensional images.

Deep learning architectures for image analysis

CNNs are one of the DL techniques that are primarily used for image recognition and processing due to its ability to recognize patterns in images. CNN is inspired by the human visual system which is organized as hierarchical layer structures that extract simple local features such as edges and lines in early layers of the network and progressively build more complex representations such as size, color, shape, and textures in deeper layers. CNN consists of three key components: (A) convolutional layers, which are inspired by the receptive fields of neurons in the human visual cortex, and capture local patterns, textures, and spatial hierarchies in an image, (B) pooling layers, which can simulate the way the human visual system focuses on essential information while discarding unnecessary details, and (C) fully connected layers that perform the final prediction based on the extracted features from the convolutional and pooling layers (see Fig. 21.1). To explain each layer in detail: (A) Convolutional layers apply filters or kernels to input images, extracting features by sliding the same filter across the image. Different kernels can be applied concurrently to learn different features in the image; for example, one kernel can be applied across the image to learn vertical lines and then another kernel can be applied across the image to learn the horizontal lines. The application of kernels in the CNN helps to identify patterns, edges, and textures in different parts of the image (see Fig. 21.1). This process is an important part of the CNN algorithm, which does not exist in the traditional fully-connected ANN, where each voxel needs to be assigned to the node in the input layer, therefore spatial information in the image is not used in the traditional fully-connected ANNs. (B) Pooling layers downsample the feature maps generated by convolutional layers, reducing their dimensionality. Max pooling or average pooling operations help retain essential information while reducing computational complexity. For example, the max pooling approach chooses the maximum number in each grid and average pooling takes the average of the numbers in the grid. (C) Fully connected layers flatten the output of the pooling layer, process the extracted features, and make predictions.

Another most frequently used DL approach is the encoder-decoder model, also called an autoencoder model. This model is particularly suited for transformation purposes such as transforming sequences from one language to another or image-to-image translation for denoising. The encoder-decoder model first compresses the input data into a condensed, meaningful representation through the encoder (see Fig. 21.2A). The decoder then takes this condensed representation and reconstructs or generates the desired output. The weights between the layers are optimized by minimizing the difference between the

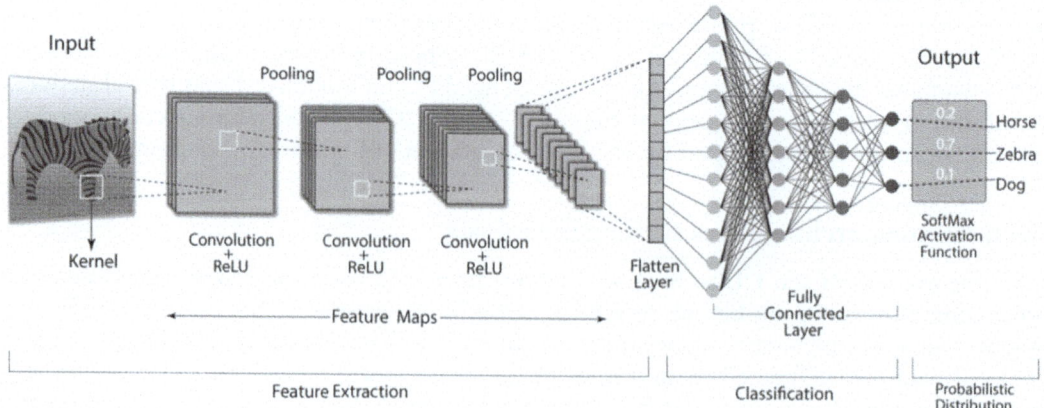

FIGURE 21.1 The convolutional neural network (CNN) and its components including convolutional, pooling, and fully connected layers. Image source from https://www.linkedin.com/pulse/what-convolutional-neural-network-cnn-deep-learning-nafiz-shahriar/.

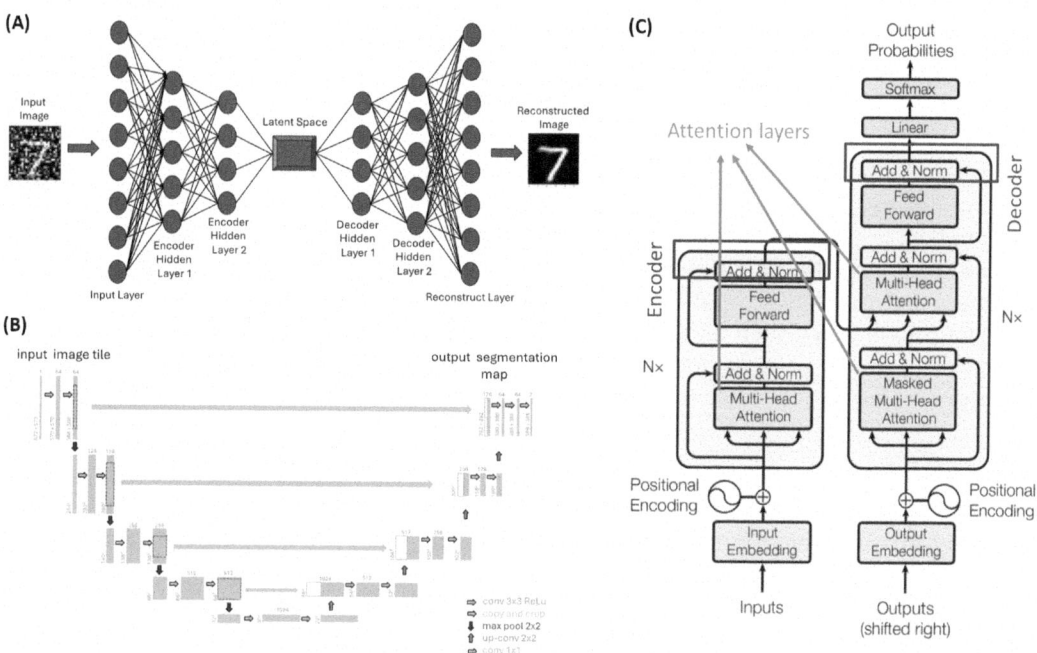

FIGURE 21.2 The encoder–decoder models. (A) The traditional encoder-decoder model with encoder, decoder, and latent space layers, (B) U-Net, and (C) the transformer. Image source from https://wikidocs.net/193827, https://machinelearningmastery.com/the-transformer-model/, and [17].

prediction and observed outcome, while the bottleneck of the condensed representation ensures that high-level representations are learned rather than simple memorization. U-Net is a specific version of this encoder-decoder model that includes skip connections allowing to connect the encoder's output directly to the corresponding layers in the decoder (see Fig. 21.2B). This helps to preserve spatial information during the encoding and decoding process while still providing the context of higher-level (more abstract) features, and is particularly helpful for image segmentation tasks where precise pixelwise localization is crucial. Another encoder–decoder model that is used in neuroimaging is the transformer which was initially developed for language processing [17] (see Fig. 21.2C). The attention mechanism of the transformers teaches the model the important part of the input, conditioned on the input itself. For example, in a language transformation input, some words (e.g., subject and verb) in a sentence have greater importance compared to other words (e.g., adjective) making the transformation easier and more robust. One of the largely used transformers is the large language models (LLM), which is also used for ChatGPT. The potential applications of transformers in neuroimaging are (1) to create patient reports using the patient's multi-modal data including neuroimaging, clinical, and demographic data, (2) to create imaging protocol based on the patient's medical history, (3) to convert the patient's report in a more structured way and suggest relevant differential diagnosis to the clinician to decrease the risk of misdiagnosis, and (4) to summarize and simplify the clinical and radiological reports for the patients [18].

3. Use of other imaging acquisition and analysis modalities in multiple sclerosis

Assessment of model performance

In addition to the choice of AI algorithm based on the best fit to the data and output type, the assessment of model performance is a crucial step to ensure the effectiveness and reliability of the model in a new dataset. AI methods, including DL ANN's, have tremendous potential to learn abstract representations, but because they have so many parameters they can also frequently overfit (i.e., memorize) on training data without developing useful or generalizable internal representations. Cross-validation (CV) is commonly employed to evaluate how well the model generalizes to new, unseen data, providing insights into its robustness and predictive capabilities. Depending on the output type (categorical vs continuous), various metrics such as accuracy, sensitivity, specificity, Dice coefficient, and mean squared error can be used to find the best hyperparameters of the model during CV as well as to measure the performance of the model in the new test dataset.

Cross-validation

To ensure generalization of the ML and DL techniques in a new dataset, CV is performed by dividing the original data into (1) a training dataset where the hyperparameters and parameters of ML/DL models are optimized and (2) a test dataset where the optimal ML/DL model is tested to assess how well the model performs in a new dataset (see Fig. 21.3). The training dataset is used to train the model by finding the best hyperparameter(s) of the model (e.g., number of trees in the random forest algorithm) and then once those are identified, the parameters of the model. To be able to find the best hyperparameter value from a list of candidate values (e.g., between 10 and 100 trees with a step of 10 trees), the training dataset is divided into training and validation sets where the training set is fitted using one of the candidate hyperparameter value (e.g., 20 trees) and validation set is used to test how well the model performs using this candidate hyperparameter. This approach is repeated for each candidate hyperparameter value and validation set is used. This internal loop provides the best hyperparameter(s) of the model which is used

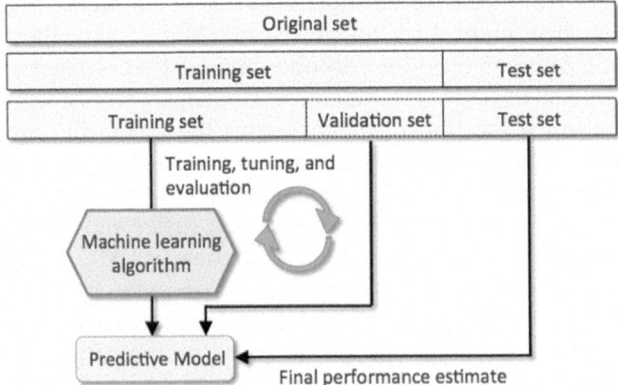

FIGURE 21.3 Cross validation structure used in the ML algorithms. Image source from https://scikit-learn.org/stable/modules/cross_validation.html and [19].

later to optimize the model parameters using the training set and applied to the test dataset. The two most used CV types are leave-one-out CV (LOOCV) and k-fold CV. In LOOCV, a single data point (e.g., a subject) is designated as the test dataset, while the remaining data points are used for training. This process is repeated for each data point in the dataset, and the model is trained and tested N times, where N is the total number of data points. In k-fold CV, the dataset is divided into k equally sized folds or subsets. The model is then trained k times, each time using $k-1$ folds for training and the remaining fold for testing. This process is repeated k times, with each of the k folds used exactly once as the test data. The k-fold CV needs to be applied in the internal loop too, where the training dataset is divided into the training and validation datasets. The k-fold CV needs to be repeated multiple times (e.g., 100 times) for the original dataset to be divided into k different folds multiple times, so the way that the original data is divided into training and test sets does not impact the results. This also has the advantage of providing an estimate of the variability in model accuracy.

Metrics used to assess the model performance

Various metrics can be used to assess the performance of the model on held-out test data. In classification problems where the labels are binary or categorical variables, the most frequently used metrics are sensitivity, specificity, accuracy, balanced accuracy, Dice coefficient, the area under the curve (AUC) of the Receiver Operating Characteristic (ROC) (AUC_{ROC}), the AUC of precision-recall (AUC_{PR}), and cross-entropy. Sensitivity and specificity are crucial measures in clinical care, as sensitivity indicates the model's ability to correctly identify true positive cases (e.g., how well true MS patients are identified when classifying MS vs controls), while specificity indicates how well true negatives are classified (e.g., how well true controls are identified when classifying MS vs controls). Accuracy provides an assessment of overall correct classifications over all classes. Balanced Accuracy is calculated as the average of sensitivity and specificity and is often used in the context of imbalanced classification problems where the number of observations in different classes is significantly unequal. The AUC_{ROC} is valuable for assessing the trade-off between sensitivity and specificity. One advantage of AUC_{ROC} is that AUC_{ROC} is threshold-independent, meaning it assesses a model's performance across all possible classification thresholds. This is advantageous when the model predicts the probability of being in one class (e.g., probability of being MS) and the optimal threshold of being in one class is not predetermined. In case of an imbalance in the number of observations in each class, the AUC_{PR} is preferred to measure the classification performance, where precision is calculated as True Positives divided by the sum of the True Positives and False Positives and recall indicates the sensitivity. AUC_{ROC} and AUC_{PR} range between 0.5 and 1, where 1 indicates perfect classification and 0.5 indicates random chance. Another metric that is frequently used when the classes are imbalanced is the Dice coefficient (also called the F1 metric). The Dice coefficient is used particularly to assess the accuracy of segmentation and is particularly useful in cases where the segmented structure is small compared to the background (e.g., lesion segmentation where lesion voxels are much fewer compared to healthy voxels). The Dice coefficient is calculated as $2 \times$ intersection/total number of observations in both sets. Intersection refers to the number of overlapping voxels between the predicted and true segmentation. The Dice coefficient ranges from 0 to 1, where 0 indicates no overlap (complete mismatch) and 1 indicates perfect overlap (perfect agreement). Another frequently used metric to evaluate the classification accuracy, particularly when training a model, is the cross-entropy loss which mainly calculates the difference between the true label (e.g., lesioned voxel) and the estimated probability (e.g., estimated probability of a voxel being a lesion). Cross-entropy loss is calculated as $L(y,p) = -(y \times \log(p) + (1-y) \times \log(1-p))$ where y is the true label and p is the estimated probability. The cross-entropy loss value is non-negative, where lower cross-entropy loss indicates better classification.

In regression problems where the labels are continuous variables (e.g., cognitive scores), the most frequently used evaluation metrics are mean squared error (MSE), R-squared (R^2), and Pearson's/Spearman's correlation coefficient. MSE is calculated as the average of the squared differences between predicted and actual values. R^2 represents the proportion of the variance in the dependent variable (i.e. output, for example, cognition) that is predictable from the independent variables (i.e. input variables, for example, age and sex). R^2 ranges from 0 to 1, with higher values indicating better model fit. Pearson's and Spearman's correlations assess linear and monotonic relationships, respectively, between two continuous variables. Correlation metrics indicate the degree to which a change in one variable corresponds to a proportional change in another.

Potential benefits of using AI and neuroimaging in MS

The integration of AI into MS neuroimaging offers a range of potential benefits to improve the diagnosis, prognosis, image acquisition, and treatment decisions of MS patients. One of the most important applications of AI in MS is the early detection of MS, given that timely interventions during the early stage of the disease can significantly enhance patient outcomes [20]. AI can help provide an earlier detection of MS by increasing the resolution of images or decreasing artifacts, which may help to better detect MS lesions. Moreover, the ability of AI to analyze and synthesize multi-modal neuroimaging features can contribute to supporting diagnostic decisions in clinical care. Another widely studied application of AI in MS neuroimaging is lesion segmentation and quantification, which provides more precise and consistent measurements for both clinical and research purposes, aiding in better monitoring of disease progression and treatment response as well as saving time for this highly time-consuming task when manually performed. In addition to lesion segmentation, AI tools are also valuable in classifying MS lesions into different categories such as lesions with a central vein sign, cortical lesions, and paramagnetic rim lesions, each of which may have different clinical and pathological implications. In this way, AI can assist neuroradiologists by automating routine tasks and allowing them to focus on more complex and nuanced aspects of MS. AI can also be used to identify treatment responses by identifying enhancing active lesions and analyzing longitudinal neuroimaging data, which can help improve treatment plans and understand the efficacy and mechanism of drugs. Additionally, AI is being used in research to identify the neuroimaging biomarkers that are the most associated with worse disease progression, which is helpful for personalized treatment planning and prognosis. Another contribution of AI in MS research is the ability to analyze large-scale neuroimaging datasets including the brain's structural and functional connectomes, which can contribute to a deeper understanding of MS's underlying disease and compensatory mechanisms. In the following section, we give details of several challenges associated with conventional and advanced neuroimaging in MS and explore how AI has been employed to enhance the application of neuroimaging techniques, thereby improving the accuracy of diagnosis, prognosis, and disease monitoring.

AI in MS neuroimaging: survey of current applications

Improving MRI acquisition protocols and image quality

Conventional MRI scans are regularly collected to monitor disease activity and to decide the most appropriate treatment plans for individuals with MS. Most MS patients are scanned yearly, so short scan duration, a lower number of clinical MRI sequences, and lower-resolution images are usually preferred to avoid patient discomfort, delays in busy clinical MRIs, motion artifacts, and high MRI scanning costs. To overcome these limitations, AI can be used to streamline MRI acquisition protocols and MRI quality. In this section, we will summarize studies that applied AI techniques to (1) decrease the amount of the gadolinium (Gd)-based contrast agents (GBCA) needed to identify the enlarging active MS lesions, (2) create DIR images which provide more accurate identification of cortical lesions, and (3) produce high-resolution images from low-resolution images using super-resolution techniques.

Reduction of gadolinium (Gd)-based contrast agents

Identifying enhancing active lesions is critical in MS as they play an important role in the assessment of ongoing inflammation and demyelination, determining the effectiveness of ongoing treatments, and making informed decisions regarding therapy adjustments. Active lesions are identified using routinely intravenously administered GBCA, an active radiotracer. However, increased signal intensity on the T1-weighted MRI scans in the dentate nucleus and globus pallidus was observed in subjects who previously scanned with the administration of GBCA and the long-term effects of the GBCA are still unknown. Therefore, the US Food and Drug Administration, the Consortium of Multiple Sclerosis Centers, and the International Society for Magnetic Resonance in Medicine suggested the judicious use of the GBCA in the clinical setting, which was an important warning for the MS field as GBCA-enhanced imaging is frequently requested by clinicians, particularly for the MS patients who experience relapses (https://www.fda.gov/drugs/drug-safety-and-availability/fda-drug-safety-communication-fda-warns-gadolinium-based-contrast-agents-gbcas-are-retained-body, http://www.mscare.org/mri, [21]).

DL algorithms have been developed that can accurately identify tissue abnormalities using images of lower levels of administered GBCA in patients with suspected or known enhancing brain abnormalities such as glioma, meningioma, and tumor [22]. Gong et al. trained an encoder-decoder CNN using 60 individuals' precontrast and 10% low-dose of gadolinium T1-weighted MRI scans to recreate the 100% full-dose T1-weighted MRI scans. Their results showed that using the encoder-decoder CNN to create synthesized full-dose images significantly improved the image quality and contrast enhancement, and reduced

3. Use of other imaging acquisition and analysis modalities in multiple sclerosis

artifact of the low-dose images. This study suggests that DL approaches could be used to detect enhancing active MS lesions with a lower dose of GBCA. Identification of enhancing active MS lesions with a low dose of gadolinium-enhanced MRI scans was not studied yet - however, Narayana et al. [15] applied a combination of pre-trained CNN (ImageNet) and fully connected networks on the T1, T2, and FLAIR images without administration of gadolinium-based contrast agents for the identification of enhancing active MS lesions and achieved an AUC of 0.82. In addition to helping to reduce the amount of GBCA, DL was also used to create fully automated segmentation algorithms for enhancing active MS lesions. Two recent studies [23,24] applied U-Net algorithms to create fully automated detection and segmentation of enhancing active MS lesions using MRI with GBCA and achieved a true positive rate of 0.90 and accuracy of 0.87, separately. All these studies collectively emphasize that AI can be successfully used to detect enhancing active MS lesions using conventional MRI with or without administration of GBCA.

Synthetic double inversion recovery imaging

Cortical lesions play an important role in understanding real-time symptoms as well as in predicting disease progression in MS. However, detecting and distinguishing cortical lesions from white matter lesions is challenging using clinically required conventional MRI sequences due to their low resolution and the small size of most cortical lesions. Therefore, DIR imaging [25] where the signals of the cerebrospinal fluid and white matter are suppressed and only signal from grey matter kept has been used as a sensitive sequence to identify cortical lesions in MS. Previous studies showed that DIR outperformed the conventional MRI sequences in detecting cortical MS lesions, particularly infratentorial and intracortical MS lesions, and in distinguishing juxtacortical and WM-GM lesions [26,27]. However, DIR images are usually unavailable due to additional scan time and technical requirements. A DL technique leveraging a common AI approach called generative adversarial networks (GAN) can create synthetic DIR images that resemble real data. Finck et al. showed that the synthetic DIR which was created by applying GAN to translate conventional MRI sequences such as T1, T2, and FLAIR images, could be used to improve the detection of MS lesions, particularly juxtacortical lesions, compared to FLAIR image [16]. Bouman et al. also showed that synthetic DIR images created via GANs applied to T1 and T2-weighted images perform similarly to conventional DIR and T1-weighted images in detecting MS lesions, with an intra-class correlation of 0.92 [28]. The same group validated these findings using data collected from seven MS centers and showed an intra-class correlation of 0.81 between lesions segmented from synthetic DIR versus conventional DIR images across centers [29]. Finck et al. also used a dataset from two centers to train and test an uncertainty-aware DL algorithm that generates synthetic DIRs in the internal dataset and validated these results using an external dataset which was collected using both 3 T and 1.5 T scanners [30]. Their multi-center study showed that the lesion number count was more accurate using synthetic DIR compared to FLAIR in both internal and external datasets [30].

Synthetic DIR also plays a critical role in detecting new cortical lesions identified via longitudinal MRI monitoring. A recent study that compared FLAIR vs synthetic DIR in longitudinal MRI scans showed that synthetic DIR improves the identification of newly appearing juxtacortical lesions, moreover, synthetic DIR could identify lesions that were missed in the FLAIR images [31]. These findings highlight how AI can help to produce synthetic DIR images to better detect cortical lesions in both cross-sectional and longitudinal studies of individuals with MS.

Creating high-resolution synthetic images: super-resolution, SynthSR, and SynthSeg

Due to the frequency of MRI use in MS patients, clinicians may prefer low-resolution and 2D anisotropic images to avoid longer scan times and motion artifacts. However, isotropic T1-weighted images are required by most quantitative neuroimaging software, including FreeSurfer, which is largely used for tissue segmentation and brain atlas construction. Moreover, high-resolution images may not always be collected since high-field MRI (3 T and 7 T) may not be available in some MS centers. Therefore, AI's ability to create high-resolution images from low-resolution images may help to increase the sensitivity of low-resolution imaging to better identify tissue abnormalities and use them for research purposes. A technique called super-resolution uses a deep CNN to increase the spatial resolution of the original images by learning an end-to-end mapping between the low and high-resolution images, essentially imputing missing voxel-level data [32,33]. Iglesias et al. [34] developed the first version of SynthSR in 2021, a regression CNN, which was trained using synthetic scans derived from clinical scans with different orientations, resolutions, and contrasts to produce 1 mm isotropic T1-weighted images. SynthSR is an open-source tool implemented in FreeSurfer that can perform tissue segmentation. While SynthSR, which was trained with a single regression CNN, is helpful for the application of healthy brain images, using an MRI scan with an abnormality such as lesions, strokes, or tumors may cause inaccuracies for the tissue segmentation. For example, MS lesions are usually painted as healthy tissue (WM or GM) to avoid biases in WM/GM segmentation or atlas registration. Therefore, in 2023, the same group proposed using a segmentation CNN that is concatenated to the output of the regression CNN, where the regression CNN (3D U-Net) is used to predict the MPRAGE intensities and the segmentation CNN (3D U-Net) is used to predict the segmentation of the image [35]. The concatenation of the regression and segmentation CNNs handles tissue abnormalities by inpainting them with normal-looking tissue which helps the segmentation CNN. This approach was used to create 1 mm isotropic MPRAGE images from the clinical scans of patients with AD, stroke, and brain tumors and performed a Pearson's correlation coefficient >0.80 between the real T1-weighted versus synthetic MPRAGE images-derived white matter, cortical, and subcortical volumes. The super-resolution technique was also used to create high-resolution T1 and T2 images using portable low-field MRI and showed highly similar morphometric measurements with real high-resolution derived measurements in stroke patients [36]. In MS, the super-resolution technique was applied to T2 FLAIR scans obtained with low-field MRI to create high-resolution T2 FLAIR images which were then used for MS lesion identification [37]. With the help of the super-resolution technique, the low-field MRI showed 94% sensitivity in detecting the lesions that were identified using 3 T (i.e., gold standard) and high correlation ($r = 0.89$, p-value < 0.05) between 3 T vs low-field MRI-derived total lesion volume in 36 MS patients, providing a potential MS diagnosis and prognosis tool where high-field MRI scans are not available.

3. Use of other imaging acquisition and analysis modalities in multiple sclerosis

The same group that developed the SynthSR approach also developed SynthSeg, the first CNN algorithm that is trained with synthetic scans created using a generative model. All parameters of the generative model (i.e. orientation, resolution, and contrast) are fully randomized to improve the generalizations with the data beyond realism [38]. The same group applied SynthSeg to create 1 mm isotropic T1 and FLAIR images from unpreprocessed T1 ($0.82 \times 0.82 \times 1.17$ mm) and FLAIR ($0.82 \times 0.82 \times 2.2$ mm) images in 15 MS patients and showed high accuracy with a Dice score of >0.68 in lesion and brain region segmentation [13]. SynthSeg can also measure atrophy in MS by using FLAIR images as opposed to the T1 weighted images which are required by FreeSurfer software. Noteboom et al. [39] compared the atrophy metrics derived from SynthSeg performed on the FLAIR images and FreeSurfer performed on T1 images and showed an intraclass correlation of 0.91. Altogether, these studies show that SynthSR and SynthSeg approaches can be used to increase the resolution of the MRI scans as well as replace the required images (e.g., T1-weighted images) with the existing images (e.g., T2-weighted or FLAIR images) for better lesion and tissue segmentation.

Tissue segmentation/volumetrics

FreeSurfer and FSL are the most widely used tools for tissue segmentation and volumetric measurements of WM, cortical, subcortical, and cerebellar volumes. However, these algorithms have some limitations as they (1) may not be always accurate, especially in areas with complex anatomical structures and in the presence of lesions, (2) may need a manual intervention to correct the segmentation which may be time-consuming for large studies, (3) may vary in different populations (e.g., different ages and ethnicities) due to differences in brain morphology, (4) are computationally intensive and time-consuming, and (5) are sensitive to the quality of the images and may provide less accurate outputs when an image with poor quality is used. Tissue segmentation tools developed using deep learning may overcome these limitations by speeding up the segmentation process and integrating different MRI modalities (e.g., T1 and T2-weighted images) to potentially increase the accuracy of tissue segmentation and decrease the time spent on manual editing.

FastSurferCNN is a technique that uses a CNN to segment the whole brain into 95 classes in 1 minute, which is much faster than FreeSurfer (up to 24 hours of processing per brain). FastSurferCNN was trained using 140 subjects that were previously processed with FreeSurfer. It was shown to produce similar outputs for brain, white matter, and gray matter volumes, and to provide better test-retest reliability results compared to FreeSurfer [4]. A recent study in MS compared multiple brain segmentation tools including FreeSurfer, FSL, and DL-trained algorithms including FastSurfer and SynthSeg in estimating the brain, white matter, and gray matter volumes using lesion-filled and nonlesion-filled T1-weighted images which were collected on three different 3 T MRI scanners (GE, Philips, and Toshiba) [2]. This study showed that FastSurfer outperforms FreeSurfer in terms of runtime, sensitivity, and test–retest reliability.

A reliable and automated tool is not only needed for the whole brain volume metrics but also for regional volumes, particularly for regions that were previously found to be the most impacted by MS and associated with motor and cognitive impairment in MS. For example, structural and functional changes in the thalamus are known to be one of the strongest predictors of cognitive impairment and disability in MS patients [8,40–43]. The tools such as FreeSurfer and FSL that are largely used to segment subcortical regions require 3D T1-weighted images; however, 40% of routine scans collect 3D T1-weighted images while 99.8% of scans are T2-FLAIR images in MS clinical setting [44]. Therefore, Dwyer et al. developed DeepGRAI (Deep Gray Rating via Artificial Intelligence), a deep learning approach based on 3D U-Net, to estimate the thalamic volume using T2 FLAIR images [3]. DeepGRAI was trained/validated/tested using 4500 2D and 3D T2 FLAIR images collected from 59 centers and the thalamic volume measured with DeepGRAI was found to be significantly correlated ($p = 0.025$) with thalamic volume derived from FSL's FIRST software [45]. In addition to the thalamus, it has been also shown that FreeSurfer-derived segmentation of choroid plexuses (CP)—which is a critical region for immunoregulation and neuroprotection, responsible for CSF production, and a part of the blood-CSF barrier—was poorly correlated (Pearson's correlation = 0.55) with the manual segmentation in MS patients [46]. To increase the segmentation accuracy of CP, a study trained a 3D U-Net on 3D T1-weighted MRI from 44 healthy controls and 97 clinically diagnosed MS patients and showed that DL-derived CP volume was highly correlated with manual segmentation ($r = 0.86$) and outperformed FreeSurfer [47].

Lesion segmentation

Lesion size, number, and location are important neuroimaging biomarkers for the diagnosis, prognosis, and monitoring of treatment response in MS. This information can be gathered by a time-consuming lesion segmentation process which is usually performed manually by neuroradiologists. In addition to being a laborious task, another limitation of lesion segmentation is the intra- and interexpert bias which may cause variability in research studies that use lesion number and volume in MS. Therefore, automated lesion segmentation is one of the areas in which MS clinicians and researchers benefit enormously. However, classical imaging analysis approaches have usually resulted in only mediocre quality lesion segmentation. A range of studies have therefore explored the use of AI for lesion segmentation in MS and presented unsupervised and supervised ML techniques which are much faster than manual segmentation, overcame the intra and interexpert variability, and were consistent in scan-rescan analysis. In this section, we introduce the studies that applied AI techniques to segment T2 FLAIR lesions, PRLs, cortical lesions, lesions with a central vein sign, and new appearing MS lesions in longitudinal scans.

T2 FLAIR lesions

One of the criteria of the MS diagnosis is the dissemination of central nervous system lesions in (1) space, i.e. appearance of a new lesion in at least 2 locations among periventricular, juxtacortical, infratentorial, and spinal cord and (2) time, i.e. the appearance of a new lesion on a scan compared to a baseline scan performed at least 30 days after the onset of the initial clinical symptoms [48]. MS diagnosis is performed by evaluating the dissemination in space and time using T2 weighted images in addition to the T1 weighted images with Gadolinium contrast agent. Therefore, one of the biggest needs in clinical practice is to create T2 FLAIR lesion masks. Several studies proposed using ML/DL techniques for automated MS lesion segmentation. Geremia at el. applied an RF algorithm on T1, T2, and FLAIR images to discriminate MS lesions from healthy tissue and achieved a true positive rate of 39% [49]. Steenwijk et al. also applied an ML approach, k-NN, to FLAIR and T1 images for the WM lesion classification in 20 MS patients using the leave-one-out CV technique and achieved a Dice coefficient of 0.75 in the original dataset [50]. However, ML techniques such as RF and k-NN are not developed for image analysis and do not consider spatial information. Therefore, the application of DL techniques such as CNN and encoder-decoder algorithms was proposed for MS lesion segmentation. Brosch et al. [51] proposed applying convolutional encoder networks to T1 and FLAIR images collected in 43 MS patients (20 for training and 23 for testing) for MS lesion segmentation and achieved a Dice coefficient of 0.68, which was not better than other studies that used RF and k-NN. Gabr et al. [52] also used CNN (U-Net) for MS lesion segmentation, but using a larger dataset this time (>1000 T1 and FLAIR scans) and achieved a dice coefficient of 0.82. The latter study shows that larger datasets are required to train the ML/DL approaches to reach higher accuracy in the validation datasets. To achieve the challenges with a low number of subjects in the training dataset, Valverde et al. [53] suggested a cascade CNN approach where two CNNs are performed: the first CNN predicts the probability of each voxel being in a lesion, and the second CNN is applied only for the voxels with >0.5 probability of being in a lesion from the first CNN. This way, the first CNN focuses on revealing possible candidate lesion voxels, while the second CNN focuses on reducing the misclassification of the first CNN. This approach was trained using T1, T2, and FLAIR images from 20 subjects and tested using 25 subjects and achieved a true positive rate of 0.68.

Paramagnetic rim lesions

Paramagnetic rim lesions (PRL), which are characterized by iron-laden active microglia and macrophages, were shown to be specific to MS [54], have more inflammation measured on 11C-PK11195-PET [55], and are associated with chronic activity and increased disability [8,56,57]. Moreover, PRL-related structural disconnectivity was found to be an earlier biomarker to identify the patients at risk for moderate or severe disability in MS [58]. Therefore, the identification of these lesions may improve our understanding of how PRL and non-PRL differently impact motor and cognitive impairment, disease progression, and response to treatment.

The manual identification of PRLs can be performed using magnetic resonance susceptibility imaging (T2*-weighted magnitude, susceptibility-weighted phase imaging, or quantitative susceptibility mapping [QSM]) [55,59–61]. However, the inter and intra-rater variability for manual PRL segmentation was found to be only moderately good [60]. Therefore, DL techniques were proposed to distinguish PRLs vs non-PRLs using multi-modal imaging sequences. Barquero et al. [62] proposed an approach called Rim-Net that applies two parallel CNNs on 3D FLAIR and EPI images to automatically classify MS lesions as PRL vs non-PRL (see Fig. 21.4). Their results showed that the combination of multi-modal images resulted in remarkably good classification accuracy with an AUC of 0.943, outperforming unimodal approaches. Another study proposed the APRL algorithm that uses an RF classifier for the detection and classification of PRLs in MS using a

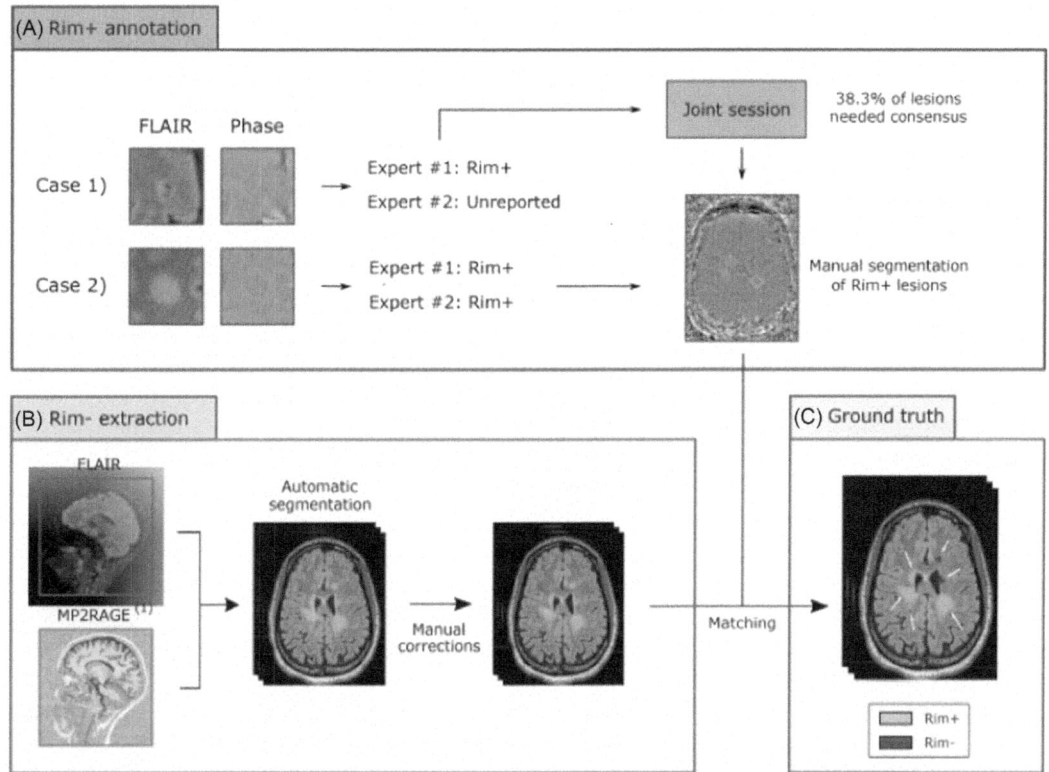

FIGURE 21.4 Description of the protocol used to label and generate the dataset in the study performed by Barquero et al. [62]. (A) For each patient, two experts visually inspected the 3D-EPI phase and 3D FLAIR images and only reported paramagnetic rim lesions (PRL, i.e., rim +). The PRL detected by one expert and undetected or considered non-PRL (i.e. rim −) by the other (unreported) went through a joint session where experts provided a final decision. (B) Lesion candidates were extracted from the automatic segmentation (corresponding to the connected components by considering a 6-connected-voxels neighborhood) and matched with the PRL annotations. In order to guarantee that one lesion candidate matched only one PRL lesion annotation, a technician manually separated the PRL lesions inside confluent ones.

combination of T1-weighted, T2-FLAIR, and T2*-phase MRI [63]. APRL is the first study that extracts lesion-level quantitative radiomic features and characterizes each lesion with intensity-based statistics on the phase contrast image. The results performed with APRL were highly correlated with the manual identification of PRLs and achieved an AUC of 0.82. A more recent study proposed QSMRim-Net, which uses QSM for the first time for the automated classification of PRL vs non-PRLs. Similar to Rim-Net which used the radiomic features from phase imaging, QSMRim-Net uses radiomic features derived from the QSM and showed 10.2% and 25.0% improvement in positive predictive value and 7.8% and 26.5% reduction in sensitivity compared to Rim-Net and APRL, respectively. QSMRim-Net also has another advantage compared to Rim-Net and APRL when there is a high imbalance between the number of PRL versus non-PRL. To overcome the class

3. Use of other imaging acquisition and analysis modalities in multiple sclerosis

imbalance problem, QSMRim-Net uses a Synthetic Majority Over-sampling Technique (SMOTE) [64] to create synthetic examples of the minority class. Overall, these studies demonstrate the potential of ML and DL techniques in identifying PRLs, which can significantly aid in the diagnosis and prognosis of MS patients.

Cortical lesions

In addition to T2FLAIR and PRLs, accurate identification of cortical lesions (CL) is also important as their count, volume, and location are associated with disability and cognitive impairment in cross-sectional and longitudinal MS studies. Their effects are independent of WM lesions, providing evidence that the information of cortical pathology needs to be included in the MS studies to better understand disease progression [65–70]. However, the visualization of cortical lesions is challenging with conventional MRI [71], and a postmortem study revealed that only a small subset of cortical lesions may be detected even with ultra-high field 7 T MRI [72]. Advanced neuroimaging sequences such as DIR [27] and phase-sensitive inversion-recovery (PSIR) [73] improved the detection of CL; however, these sequences may not always be collected in a clinical setting as they increase the scan time by 10–15 minutes. A postmortem study proposed using an automated laminar profile shape analysis technique on the histological data from one MS patient to extract feature vectors based on the shape and intensity of MRI sequences including T1, T2, and magnetization transfer ratio, followed by clustering the feature vectors using k-means. This technique resulted in 4 optimal clusters, including normal-appearing gray matter, noncortical tissue, and two cortical tissues [74]. However, this approach was not validated in vivo. In MS patients with lower disability and who are in the early stage of the disease, Fartaria et al. [75] proposed using k-NN algorithm on FLAIR and other sequences data (MPRAGE, MP2RAGE, and DIR) to detect CL, and showed that the combination of FLAIR with the advanced DIR and MP2RAGE sequences outperformed the other combinations in detecting CL (detection rate of 0.62). However, this study only included the lesions above 0.0036 mL as the partial volume effect, an imaging phenomenon where a single voxel contains a mixture of tissues, may occur in the presence of small lesions. Therefore, the same group developed an algorithm that combines k-NN and Bayesian partial volume estimation techniques to estimate the lesion concentration in each voxel and better detect small cortical lesions [76]. The sensitivity and Dice coefficient metrics were improved by 10% and 6%, respectively, when using a combination of k-NN and Bayesian partial volume estimation compared to k-NN alone. The same group also developed the MSLAST (Multiple Sclerosis Lesion Analysis at Seven Tesla) algorithm that uses a single MP2RAGE image from 7 T for WM and CL segmentation in 25 RRMS patients [75]. MSLAST follows three steps: (1) skull stripping and WM/GM segmentation, (2) computing the tissue concentration (WM, GM, and CSF) for each voxel from the MP2RAGE image to minimize partial volume, (3) creating "pseudo" lesion masks on the WM and GM by binarizing the WM, GM, and CSF concentration maps separately. The optimal thresholds of concentration maps were calculated separately for each map by maximizing the separation of the lesion mask from healthy tissue. Finally, the "pseudo" lesion masks were merged into a single, final WM and CL mask. MSLAST performed similarly to manual segmentation (Spearman's Correlation coefficient of 0.91), provided consistent results with scan-rescan lesion segmentation with an F1 score of 0.84, and detected 74% of WM and 58% of CL.

While this approach could identify only half of the CL, it is a good first step before performing manual segmentation. Two of the limitations of this approach are that the small periventricular lesions were missed and classified as CSF and the detection rate for CL was poorer due to the small size and low contrast of CL compared to WM lesions.

DL algorithms have also been proposed to identify CLs using conventional MRI. La Rosa et al. [77] applied a 3D U-Net CNN to FLAIR and MP2RAGE images collected in 3 T for WM and cortical lesion segmentation and achieved a detection rate of 76% for both types of lesions. However, this approach missed the small CL (3–10 μ L), perhaps because it did not consider partial volume effects. The same group also applied 3D U-Net to a combination of the MP2RAGE, T2*-weighted echo-planar imaging, and T2*-weighted multi-echo gradient recalled echo (GRE) sequences obtained with 7 T MRI for CL segmentation [78]. The novelty in this study was that the model could classify CL as leukocortical and intracortical/subpial and perform WM and GM segmentation. The study was performed using 60 MS patients with 2014 CL, and it demonstrated sensitivity in detecting minimum lesion sizes of 0.75 μ L which was lower than the previous studies, and achieved a detection rate of 67%. This approach also showed that 24% of the false positives detected with the DL algorithm are actually CL that were missed in the initial lesion masks created by the experts. In another study [77], the same group also compared their approach with the MSLAST algorithm in detecting CL using multi-site 7 T datasets. First, the CNN was trained using the MP2RAGE images from 60 MS patients, then domain adaptation was performed before lesion segmentation. Their detection rate of 71% outperformed the MSLAST algorithm (48%) when considering a minimum lesion volume of 6 μ L.

Central vein sign

Most MS lesions occur around central veins, thus the identification of central vein signs in MS lesions can help to improve the accuracy of MS diagnoses [79,80]. The North American Imaging in Multiple Sclerosis Cooperative recommends using the FLAIR* sequence which combines T2-FLAIR and magnetic susceptibility-weighted T2 images (i.e. T2*-weighted imaging), as FLAIR* outperforms FLAIR in identifying the MS lesions with central vein signs [81,82]. Even though detailed criteria to identify the MS lesions with a central vein sign have been proposed, multiple requirements make the identification of lesions with a central vein sign challenging, time-consuming, and likely increase inter-rater variability. These include (1) visualization in at least two perpendicular MRI planes and appearance as a thin line in at least one plane, (2) small apparent vein diameter (<2 mm), and (3) lesion not being <3 mm in diameter in any plane. To address this, Dworkin et al. [83] proposed the first fully automated approach, InterModal Segmentation Analysis (MIMoSA), using T1, T2 FLAIR, and T2*-weighted segmented echo-planar imaging (EPI) images. MIMoSA first creates a map of the veins present in the T2*-EPI using a process referred to as "vesselness filtering." Second, the vein map is co-registered to the T1 images. Third, WM lesions are segmented using T1 and T2 FLAIR images. Finally, a permutation procedure is performed to check if the probability of the vein being located in the center of the lesion is greater than the chance, which computes the probability of a lesion having a central vein sign. MIMoSA was tested on 16 MS patients and 15 subjects who did not have MS and provided a high accuracy of MS diagnosis with an AUC of 0.88. However, the false positive and negative values were not reported as this approach was

3. Use of other imaging acquisition and analysis modalities in multiple sclerosis

fully automated and the manual segmentation of MS lesions with central vein signs was lacking. Maggi et al. proposed using a DL algorithm, CVSnet, with a 3-layer CNN to detect MS lesions with a central vein sign and compare them with the manually segmented lesions [84]. It was applied to a multi-center dataset that was collected in 42 MS patients, 33 subjects with MS-mimicking disease, and 5 subjects with unknown diagnoses. CVSnet achieved a high accuracy of 81% in detecting manually segmented MS lesions with a central vein sign and provided a high-speed segmentation that was 600 times faster compared to manual segmentation (~4 seconds vs. 40 minutes per patient). CVSnet also outperformed the "centrality-corrected" vesselness filter method which was used in the first and last steps of MIMoSA with an AUC of 90 versus 77%, and provided 20 times faster segmentation procedure.

Detecting longitudinal changes in lesions

MRI is one of the key tools for identifying if MS lesions are new, diminishing, or enlarging, information which plays a critical role in disease progression and personalized treatment planning. Conventional MRI sequences such as T2 FLAIR are being collected regularly in clinical practice, however manual identification of the new, enlarging, or diminishing lesions using longitudinal MRI is challenging and time-consuming. Elliott et al. suggested a two-step process to identify new MS lesions by comparing baseline and follow-up MRI scans [85]; the first step includes a Bayesian classifier that provides a probabilistic tissue classification based on baseline and follow-up MRI, and the second step applies an RF classifier to identify new MS lesions. This approach was trained and evaluated over 250 subjects collected from multiple MS centers and provided 0.99 sensitivity for lesions greater than 0.15 cc. Another study integrated the immediate and extended neighborhood intensity of a lesion via an ML framework to identify the change in MS lesions [86]. This model was trained and evaluated using a small number of MS patients ($n = 15$) and provided a high classification performance with an AUC of 0.97, which was significantly better than the state-of-art methodology that performs a logistic regression on multi-modal data to identify the incident or enhancing lesions [87]. Salem et al. used the same approach but added deformation field information, which might be useful to take into account changes in the brain structure such as shifts and distortions in the brain tissue [88]. Their results provided better false and true positive results compared to the study that did not use deformation fields [87]. DL approaches such as 3D CNN were also performed in quantifying the new or enlarged lesions in MS. Krüger et al. [11] trained a U-Net-like encoder–decoder architecture on 1809 single and 1444 longitudinal MRI scans and validated on 185 longitudinal MRI scans from MS patients (see Fig. 21.5). The advantage of this study over the previous studies was that the MRI scans from each time point of the longitudinal data were used. The sensitivity and false positives obtained with this approach were better than the results from the Lesion Segmentation Tool (LST) (sensitivity 60 vs. 46% and false positive: 0.48 vs. 1.86). A recent study [10] performed transfer learning approach where a U-Net was trained for lesion segmentation using cross-sectional MRI scans and then used to estimate new lesions using the longitudinal MRI scans. This approach used T2 FLAIR scans from 100 MS patients with cross-sectional and longitudinal scans collected in 15 different MRI scanners and achieved a Dice coefficient of 0.543.

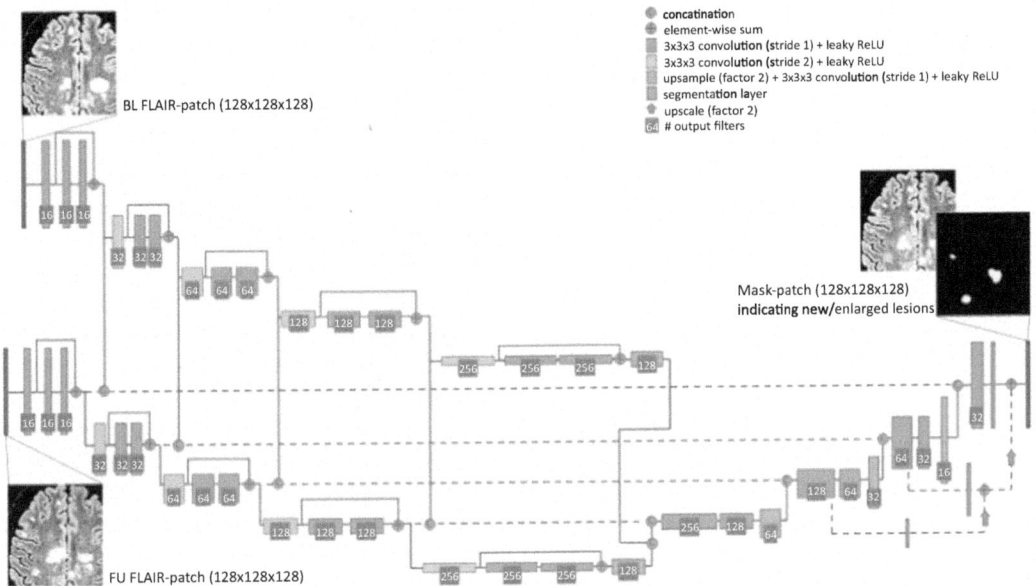

FIGURE 21.5 The proposed network by [11]: a fully convolutional encoder-decoder architecture with 3D convolutions, residual-block-connections and four reductions of the feature map size. The two input images (baseline (BL) and follow-up (FU) FLAIR-patch) are fed into the same encoder path — both visualized paths have shared weights. After each residual-block the feature maps for each input are concatenated and fed into the decoder, respectively. As an output, a segmentation mask is predicted indicating new and enlarged lesions.

Diagnosis, prognosis, and clustering

In addition to image enhancement and lesion/tissue segmentation, AI algorithms are also potentially valuable for MS diagnosis, prognosis, and neuroimaging-based disease phenotyping. In this section, we review studies that use neuroimaging and ML/DL algorithms for (1) diagnosis (i.e. classifying MS patients and healthy controls as well as differentiating MS from other similar disorders such as neuromyelitis optica spectrum disorder [NMOSD] and neurosarcoidosis), (2) the classification of MS clinical phenotypes and subtyping (i.e. PPMS, RRMS, and SPMS), and (3) predicting cross-sectional and longitudinal disability and cognitive impairment (i.e. motor, visual and verbal memory, and processing speed).

Differential diagnosis

While a detailed consensus—the McDonald criteria—is established and largely used for MS diagnosis [89], confirming a diagnosis of MS is not always straightforward and some patients may still suffer from misdiagnosis due to MS mimicking disorders, false positive/negative results, low-resolution MRI scans, or lack of MS specialists in some neurology centers. AI's ability to handle large amounts of multi-modal data including neuroimaging, clinical, serum, cerebrospinal fluid markers, and clinician reports can improve distinguishing MS patients from healthy controls and other similar disorders such as NMOSD, neurosarcoidosis, and CNS vasculitis. A review paper that analyzed 38 studies on MS revealed that AI-based predictions reach an accuracy between 81 and 100% for MS diagnosis [20]. More specifically, Eitel et al. [90] used a CNN model that was trained on MRI scans in people with Alzheimer's Disease (AD), and fine-tuned and tested in MS patients. This study applied the pre-trained model to conventional T2-weighted and FLAIR sequences collected from 76 MS patients and 71 healthy controls and achieved an AUC > 96%. These results provide an opportunity to learn across neurological disorders that have larger dataset and apply them to diseases with smaller sample sizes. Another study that trained and tested a CNN model using susceptibility-weighted MRI collected in 66 MS and 66 healthy controls achieved an accuracy of 99% in distinguishing MS patients from controls [91]. These studies show that CNNs applied to conventional MRI scans can be used to support and/or verify the MS diagnosis. AI algorithms can be also used to identify the locations in the brain that are the most informative about clinical status (MS vs healthy control) using conventional MRI. Weygandt et al. applied SVM on the normal-appearing grey matter (NAGM), normal-appearing white matter (NAWM), and lesioned areas and showed that posterior parietal WM among the lesioned locations, cerebellar regions among NAGM locations, and posterior brain among NAWM locations were the most informative areas about the clinical status with up to 96% accuracy [92]. These results show that non-lesioned locations that are not currently used for the diagnosis can also provide important information about the clinical status of MS, might help to improve the diagnostic criteria, and might be used to monitor the progression of CIS and radiologically isolated syndrome (RIS) patients who do not yet have an MS diagnosis.

While MS diagnoses are currently performed using conventional MRI, advanced neuroimaging techniques such as diffusion and functional MRI can also provide additional information about the impact of MS on brain connectivity networks [6,93]. ML algorithms such as SVM and logistic regression were previously applied on structural and functional connectivity networks in classifying MS patients and healthy controls and provided an accuracy up to 89% [5,6]. It has been also shown that structural connectivity networks better classify controls and MS patients, as expected due to lesion-related structural disruption, while dynamic functional activity, an approach that computes the functional connectivity networks using different windows of fMRI time series, better classifies MS patients based on disability level [5]. While functional connectivity metrics did not outperform structural connectivity in classifying MS vs controls, using functional MRI to distinguish MS from controls may help to uncover the reorganization/upregulation mechanism in MS patients which might be useful information for personalized treatment plans such as non-invasive approaches including transcranial magnetic stimulation.

In addition to distinguishing MS patients from healthy controls, AI algorithms were also used to differentiate MS from the disorders that mimic MS such as NMOSD, neurosarcoidosis, and CNS vasculitis. Despite significant advancements in neuroimaging, blood biomarkers, and clinical assessments, distinguishing MS from other MS-mimicking neurological disorders is still challenging and may delay an appropriate treatment. An RF approach was applied to the neuroimaging features such as cortical thickness and volume, which can be derived from conventional imaging, and provided an accuracy of 0.74 using the volume and thickness of 50 regions and an accuracy of 0.80 using only thalamic volume and lesion load [94]. The same group also applied a multi-kernel learning approach to a combination of clinical, multi-modal neuroimaging, and cognitive data to separate MS from NMOSD and reached an accuracy of 0.88 [95]. The same study showed that T1 and T2 lesion load, normal-appearing white matter integrity as measured with diffusion tensor imaging (DTI), and functional connectivity between the networks identified with independent component analysis were the most important features in classifying MS vs NMOSD. Another study applied logistic regression to the combination of spinal cord and brain images to discriminate NMOSD from multiple sclerosis. This approach was applied in 116 NMOSD and age-matched 65 MS patients and provided sensitivity and specificity metrics > 0.90 [96]. DL techniques such as 3D CNNs were also applied to FLAIR images and provided similar accuracy (0.71 for DL vs 65.9 for rater 1 and 60.7 for rater 2) but better consistency compared to human raters [97]. Rocca et al. [98] used other MS-mimicking disorders such as vasculitis and migraine in addition to NMOSD and applied a CNN to distinguish MS from these disorders using a larger sample size ($n = 268$ in total). This study provided an accuracy of 0.98 which was much higher than the previous studies [97]. This higher performance might be due to using different MS mimicking disorders in addition to the NMOSD and combining T1 and T2 images, in contrast to the previous study [97] that only used FLAIR images.

Classification, subtyping, and clustering

MS manifests in various clinical courses, each with distinct characteristics and disease trajectories or phenotypes [99]. The most common types include (1) clinically isolated syndrome (CIS) which is characterized by a first neurological symptom including vision problems, numbness or tingling of the face, body, or extremities, or difficulty in walking, (2) relapsing-remitting MS (RRMS) which is mostly followed by CIS and marked by unpredictable relapses followed by periods of partial or complete recovery, (3) secondary progressive MS (SPMS), which 85% of the RRMS patients develop and where initial relapses are followed by a gradual worsening of symptoms without distinct recovery phases, and (4) primary progressive MS (PPMS) which 15% of the MS patients develop after the first neurological symptom and is characterized by a steady progression of symptoms from onset without clear relapses. Each course presents unique challenges and treatment approaches, emphasizing the complexity of managing this neurological condition. However, the MS course is identified based on the clinical progression including the relapses and worsening in disability. Therefore, identifying the neuroimaging biomarkers that track these different clinical courses is crucial to better understanding the neurobiological mechanisms underlying disease progression to aid in clinical management.

A range of studies have explored ML approaches for classifying MS into different clinical courses. Kocevar et al. applied SVM to graph-theoretic metrics such as network density and modularity derived from structural connectivity to classify different MS clinical profiles as well as distinguish healthy controls from MS patients with an F1-score of >70% [100]. The same group also applied various ML approaches to a combination of lesion loads and metabolic features derived from spectroscopic imaging and showed that the multi-modal approach yields better results than the single-modality approach in classifying different MS clinical profiles [101]. ML algorithms such as SVM, Random Forest, and linear/logistic regression were also used to predict the conversion from CIS to MS diagnosis using neuroimaging biomarkers. Wottschel et al. [102] applied SVM to the radiological features related to the lesion characteristics derived from conventional MRI and could predict the conversion to clinically defined MS at 1 year with an accuracy of 71.4% and at 3 years with an accuracy of 68%. The same group also applied SVM to the MRI-derived metrics such as grey matter probability, white matter lesion load, cortical thickness, and volume of specific cortical and white matter regions in classifying the MS converters from CIS and achieved an accuracy of 92.9% [103].

In addition to using ML/DL algorithms to better understand differences in neuroimaging biomarkers between different clinical courses, AI algorithms were also used to cluster MS patients based on their neuroimaging features to see how well these clusters match with clinical disability. Eshaghi et al. [104] applied SuStaIn, an unsupervised ML technique that combines clustering and event-based modeling, to MRI scans collected in 19 studies (6322 MS patients for training and 3068 MS patients for validation) to define MRI-based subtypes of MS, including cortex-led, normal-appearing white matter-led, and lesion-led. This study revealed that baseline MRI subtypes showed differences in individual risk of disease progression and treatment response, and therefore might be used to better understand mechanisms underlying disease progression and to personalize treatment plans. More recently, Pontillo et al. [105] also applied SuStaIn to stratify MS patients using the MRI-derived metrics that were different between MS patients and controls. This study used a relatively lower number of subjects ($n = 425$) compared to the study performed by Eshaghi et al. and classified MS patients into "deep gray matter (DGM)-first" and "cortex-first" subtypes. These studies collectively demonstrate the potential of MRI data in identifying neuroimaging-based clusters which can be used to define the inclusion criteria when selecting patients in clinical trials, to track treatment response for each subgroup, and to better understand the neurobiological mechanisms that derive the differences in clinical disability in MS.

Prediction of motor and cognitive impairment

Lesion size, volume, and location are very heterogeneous among MS patients, as are cortical and subcortical neurodegeneration, making the prediction of disability progression challenging. Recent studies have demonstrated the potential of AI in accurately predicting cross-sectional and longitudinal motor and cognitive impairment and identifying the neuroimaging biomarkers that are the most associated with cross-sectional and longitudinal disability. Standard ML techniques are largely used in predicting disability in MS because of the shorter training time and lower computational cost compared to DL algorithms. One of the earliest studies used SVM and logistic regression on brain parenchymal fraction and lesion load information to predict worsening in disability at 5 years follow-up; they reached a sensitivity of 71% and specificity of 68% [106]. Another study classified MS patients by cognitive performance using the eXtreme Gradient Boosting (XGBoost) method applied to clinical and cortical, subcortical, and cerebellar volumetric information, reaching an AUC of 0.74 [9]. In addition to conventional MRI metrics, advanced neuroimaging techniques such as diffusion and functional MRI can provide more sensitive information about how MS lesions interrupt the brain's structural and functional architecture. Tozlu et al. applied a ridge classifier on the brain's structural and functional connectivity networks to classify the MS patients based on the expanded disability status scale (EDSS) score (EDSS <2 vs ≥2) [5,7,8]. These results showed that the application of an ML technique on brain connectivity networks provides promising results with an AUC of 0.65.

Disruptions in structural connectivity networks estimated with the Network Modification Tool [107] were also used to predict the post-rehabilitation improvement in cognition in MS. Fuchs et al. [108] showed that the structural disruption in the default mode network observed at pre-rehabilitation can successfully predict 12-week post rehabilitation cognitive score. Buyuktukoglu et al. also used ML to classify MS patients based on cognitive performance (lower vs higher performance groups) measured with the Symbol Digit Modalities Test (SDMT), using lesion volume and structural and functional connectivity metrics [109]. They achieved an AUC of 0.90, which is one of the highest accuracy metrics in the MS literature in this area.

Compared to "classical" ML techniques that can be applied to the MRI-derived features in predicting disability in MS, DL techniques allow the use of more complex and high-dimensional datasets that may also include raw MRI images. A range of studies have explored the use of conventional MRI combined with DL in predicting disability in MS. Zhang et al. applied a multi-modal DL technique that can be used with a combination of conventional neuroimaging data (pre-contrast and post-contrast T1-weighted sequences, T2-weighted sequences, proton density-weighted sequences, and FLAIR), electronic health records, and clinical notes. The multi-modal DL technique was employed to classify MS patients based on EDSS score and showed that the AUC was increased up to 19% when using multi-modal versus single modality data [110]. Another study predicting EDSS at 2 years using baseline FLAIR images and demographics showed that combining the outputs from various ML and DL techniques results in lower prediction error [111]. In addition to predicting disability, DL and ML techniques are also used to estimate the risk of developing significant disease worsening. A recent preprint [112] proposed an ensemble model approach, called subpopulation risk stratification (SunRiSe), that combines the estimated risk scores of developing significant disability worsening in 2 years computed with a DL survival model, mixture model, and gradient boosting survival model. The SunRiSe model showed that higher lesion load, lower lobar cortical grey matter, lower normal-appearing white matter T1/T2 ratio, and lower deep grey matter volumes are the most important biomarkers in estimating the risk of significant disease worsening. These studies collectively demonstrate the ability of ML and DL to increase the utility of clinically acquired conventional MRI sequences in predicting disability and significant worsening in MS patients.

In addition to predicting disability status, ML approaches can also be used to better understand the differences in the brain's functional dynamics in MS patients and controls across disability groups of MS. Tozlu et al. applied Network Control Theory to quantify the activation dynamics of the brains of people with MS and without. They applied logistic regression with ridge regularization technique to the resulting metrics to classify MS patients based on their disability level (i.e. high vs low disability groups using EDSS 3 as threshold) [113] and distinguish them from healthy controls. These results showed a possible compensatory mechanism in MS patients with lower disability (EDSS <3), i.e., increased dynamics in MS patients with lower disability compared to healthy controls. ML techniques can also be used to identify the neuroimaging biomarkers that are the most predictive of disability. Cordani et al. applied RF to predict disability level using the brain's structural connectivity networks and showed that white matter fibers connecting bilateral premotor and motor cortices were the most important neuroimaging metrics in

3. Use of other imaging acquisition and analysis modalities in multiple sclerosis

classifying MS patients by disability level (EDSS <3 vs. ≥3) and in predicting upper limb motor impairment [114]. Altogether, these studies show that ML algorithms can be used to better understand the neurobiological mechanisms that are associated with the disease and possibly compensatory mechanisms in MS.

AI pitfalls and ethical concerns

While AI has demonstrated substantial, and potentially transformative, value in many of the areas described above, it must be developed and adopted with caution. Although it has the potential to far exceed traditional approaches, it also suffers from a number of weaknesses that must be carefully considered and/or mitigated.

Generalization issues

One of the most important challenges in employing AI in neuroimaging is the difficulty of generalizing the trained algorithm to novel datasets. Factors like variability in MRI platforms, acquisition protocols, and motion in addition to shifts in age, race, and ethnicity across healthcare institutions can contribute to domain shifts that cause a drop in the accuracy of AI algorithms. Moreover, the quality of the neuroimaging data (patient positioning, distortions, artifacts, etc) might be variable between the institutions, creating discrepancies in the results. The lack of robustness when applied to a dataset from another institution can decrease the trust in AI-assisted diagnoses and make the integration of these technologies in the clinical setting difficult. To overcome these challenges, creating diverse training datasets that contain various scanners, protocols, and parameters is needed, as is the careful application of data augmentation techniques. Another important aspect to consider is an increase of collaboration among different clinics to establish common standards for imaging protocols and data sharing. Lastly, creating proper quality assessment and quality control procedures for the collected and shared data can help in harmonization.

Biases in the data

Perpetuation of biases is another challenge that refers to the tendency of AI models to replicate and potentially amplify the existing biases present in their training datasets. For example, there may be biases across various demographic groups that are present in clinical decisions including diagnostic criteria or treatment plans. These biases can be encoded into the training data, impacting the model's ability to make unbiased predictions. Similar to the recommendations for the other challenges of AI, increasing interdisciplinary collaborations between healthcare professionals and AI researchers can create more robust labeling for the training dataset, allowing more responsible and equitable deployment of AI algorithms in clinical settings. Additionally, whenever proxy outcomes rather than primary outcomes are used, potential issues should be carefully considered. Finally, there is a documented imbalance of various racial or ethnic subgroups in many datasets - this may mean that an AI algorithm trained on it may only be capturing the variance explained in one majority subgroup [115]. Efforts should be made to sample subgroups equally or to develop strategies to mitigate the effects of imbalanced data.

Quality of study description

Reproducibility is one of the key factors to ensure the accuracy of AI applications in independent datasets. However, the lack of transparency and sufficient details in many research abstracts and papers can significantly impact the reliability and validity of study findings. Examples of lack of transparency include insufficient details on (i) the training datasets including data augmentation, (ii) the description of the model, including inputs, outputs, all intermediate layers and connections, and (iii) inclusion and exclusion criteria to select the subjects used in the study. Checklists can help researchers in writing detailed documentation of experiments, methodologies, and data analyses, which then can help to increase reproducibility and ensure the proper development and evaluation of AI algorithms. An important example is the checklist created by the Radiology: Artificial Intelligence journal, called "Checklist for Artificial Intelligence in Medical Imaging (CLAIM)" (https://pubs.rsna.org/doi/10.1148/ryai.2020200029) that requires the authors to provide details in model training, parameters, approach to select the final model, training and validation dataset, and sensitivity analysis.

Ethical concerns

The integration of AI and neuroimaging presents multiple ethical concerns that demand careful attention. AI algorithms require a large number of human neuroimaging datasets to create a robust prediction model. Therefore, data sharing provides a valuable opportunity to increase the robustness and greater applicability of AI algorithms to a wider range of ages, sex/genders, races, and ethnicities. In MS, conventional neuroimaging data is regularly collected for diagnosis, disease monitoring, and treatment decision purposes; therefore a large amount of conventional neuroimaging data is available in most MS clinics. While being able to access a large amount of data is helpful for AI algorithms, one big problem is keeping sensitive neuroimaging data secure. The highly personal and intimate nature of brain-related information poses risks of unauthorized access, leading to potential privacy violations. Moreover, face removal and face scrambling algorithms may not always be a solution, as these techniques can be partially reversible, and the reconstructed images may be identifiable [116]. One solution is to use trusted data ecosystems such as federated learning, where only the models and not the data are shared between the institutions.

Explainability of AI algorithms

Another critical issue when applying AI in the clinic is the relative lack of transparency and interpretability of AI algorithms. If the AI's decision-making processes are not readily understandable by humans, this can pose challenges for both healthcare professionals and patients. In many cases, it is vital to know not just what prediction is made, but also why it is made, and most DL models cannot provide this information in a readily satisfying manner. Moreover, the inexplicability of AI algorithms may perpetuate unintentional biases present in the training dataset, leading to disparities in prediction accuracy across demographic groups. Establishing clear guidelines on how AI algorithms reach specific conclusions is crucial for fostering trust in these technologies. Partial solutions to these pitfalls include (1) developing AI models with built-in explainability features for a better understanding of the decision-making process, i.e. using techniques like LIME (Local Interpretable Model-agnostic Explanations) [117] and SHAP (SHapley Additive exPlanations) [118] and (2) validating the model using a diverse dataset to identify potential biases.

Practical challenges of AI in clinical practice

The integration of AI into healthcare itself presents several other practical challenges that need careful consideration. One of the important steps is to train healthcare professionals to effectively use and interpret AI outputs. Resistance to adopting new technologies, skepticism, and concerns about job displacement may also influence the acceptance of AI in clinical practice. Another challenge is that determining liability in cases of AI-related errors or malpractice poses a legal challenge. Clarifying responsibility and accountability for AI-driven outputs and using AI to support healthcare professionals, not to replace them, is essential for managing legal and ethical implications. Another challenge is that implementing AI technologies as well as maintaining, training, and updating these technologies may require an additional budget for the institutions, and potentially limit adoption in resource-constrained environments. Lastly, regulatory requirements and approval processes for AI tools in healthcare may vary across regions and countries, therefore, harmonizing regulatory standards globally is challenging. Furthermore, one of the greatest strengths of AI is its ability to continually learn from new data, but most regulatory environments strongly favor "locking in" technology at the time of approval. Addressing these practical challenges requires collaboration between regulatory bodies, healthcare providers, technology developers, and other stakeholders. Establishing clear guidelines, promoting transparency, addressing ethical considerations, and fostering a collaborative approach are essential for the successful integration of AI into clinical practice.

Future directions

The role of AI in the field of neuroimaging is currently in a state of rapidly expanding development, still experiencing exponential growth that began more than a decade ago and does not show immediate signs of slowing. Studies on AI and neuroimaging point to the potential benefit of AI in clinical practice. Establishing AI tools as a supportive component in regular medical routines can aid healthcare professionals including neuroradiologists and neurologists in making decisions on diagnosis, prognosis, and treatment decisions. In this section, we offer recommendations for facilitating the straightforward, safe, and ethical integration of AI into MS neuroimaging practice.

Training healthcare professionals

To accomplish a successful integration of AI in MS neuroimaging, the initial step involves training neurologists, neuroradiologists, and other healthcare professionals on commonly used AI algorithms, how to create training datasets with sufficient information to prevent overfitting, any legal or ethical issues, effectively interpreting the outcomes, and correcting misclassifications or misdiagnoses made by the AI. Fostering the collaboration of healthcare professionals with researchers specializing in ML and data science is an important step to improve the integration of AI in clinical practice. Long-term collaborations facilitate mutual learning, allowing everyone involved to understand the real clinical challenges, collaborate on developing AI algorithms, interpret their results, and assess how well models align with human biology.

Federated learning

As AI algorithms require a large amount of data to avoid overfitting and potential biases, data sharing could increase the accuracy of the prediction of AI algorithms in clinical settings. However, data sharing, particularly between different countries, is not always easy due to changing legal and regulatory implications in different institutions and countries as human data is sensitive and its identification by unauthorized people poses risks. A solution to improve AI models without sharing data is federated learning, which allows training a model across multiple decentralized devices or servers that contain the local dataset. After the model is trained in each institution using their local datasets, the model is sent to the central server where the model parameters are collected and combined (e.g., averaging the parameters). The updated model can be sent again to the institutions for further updating rounds. This loop is repeated until the model converges. However, one challenge of federated learning is that the institutions may obtain the brain imaging data using different scanners, therefore averaging the model parameters may not give an optimal prediction for new independent datasets. A specific federated learning technique such as federated disentangled representation learning (FedDis) [119] can learn disentangled representations of shared anatomical structure and local institution-specific information and only share the parameters related to the shared anatomical structure with the central server, aiming to decrease scanner effects. FedDis was shown to outperform the typical federated learning approach in segmenting brain lesions in MS using the FLAIR, T1, and T2 weighted images in two different institutions with different scanners (Siemens and Philips).

Foundational models

ML and DL models are usually trained for a single task such as classification, prediction, or segmentation problems. If the model needs to be used for a new task, it must be retrained from scratch which requires extra time and cost that may delay the individual clinical care. Moreover, if the new task may benefit a different source of information such as demographics, the model needs to be retrained. In a field like MS where multi-task models that may support diagnosis, prognosis, and image classification, foundation models [120] could provide a significant advantage as this approach trains the model in a self-supervised way once using diverse and extensive unlabeled datasets, and can then be applied to a wide range of tasks via transfer learning. Foundation models differ from transfer learning as the model is pre-trained using various datasets that are specific to a domain (e.g., only MS-related neuroimaging and serum/CSF-based biomarkers) and can perform multiple tasks, while transfer learning can be applied to a broader domain for a specific task (e.g., segmentation only). The advantages of the foundation model are that the model does not need to be trained from scratch, which can save time in the clinical setting, and can be trained with any modality of the data including electronic health records, neurologist/radiologist text reports, and genetics. Outside of MS, large language model (LLM) foundational models have contributed to tremendous recent advancements in general AI throughout society, and it is likely that such models in neuroimaging would be equally transformative.

Personalized treatment plans and digital twins

Disease progression is very heterogeneous among MS patients; moreover, the impact of brain atrophy and lesions on brain structure and function are not considered during treatment planning. However, AI can integrate large multi-modal data including neuroimaging, genomic, clinical, blood-based metrics, and patient lifestyle/expectations for use in personalized treatment plans. This digital twin approach, which creates a virtual copy of each patient using large amounts of data and makes predictions based on other similar individuals in the dataset, is another concept that can be used to support personalized treatment plans. The difference between digital twins and predictive modeling is that the digital twin is closely linked to the physical counterpart, allowing a real-time simulation, modification, and update via a mobile device or desktop. Moreover, while the goal of predictive modeling is to optimize the prediction accuracy for the test dataset, the digital twins find the best match for the specific patient and increase the accuracy for each case. Digital twins have shown promising results in neurological disorders including dementia [121] and MS [122–124]. Overall, both predictive modeling and digital twins approaches provide unique opportunities that can improve decision-making when integrated into clinical care.

Regulatory landscape

Legal frameworks are crucial to ensure safe, ethical, and useful AI applications in MS clinical care and serve to build trust between healthcare workers and patients. Several professional organizations such as the Artificial Intelligence Act of the European Union and the FDA from the United States provide approval based on different ethical and practical criteria including data profiguretection, cybersecurity, service continuity, and clear consumer consent [125]. One of the FDA-approved AI tools that are currently used in MS clinical care is Pixyl.Neuro which provides a fast, clinically relevant report of brain atrophy and lesion volume. Another FDA-approved tool Imeka uses advanced neuroimaging techniques and provides a more detailed report including microstructural changes in the 33 white matter tracts measured with diffusion MRI. Another FDA-approved AI tool that uses advanced neuroimaging techniques is BrainSpec. BrainSpec uses magnetic resonance spectroscopy scans to measure the brain metabolite concentrations in minutes, which is much shorter compared to traditional methods. These tools demonstrate that AI is currently providing fast and detailed information about the brain that can be used to monitor disease and treatment progression. While there are only a few FDA-approved tools that are currently used in MS clinical care, more are expected that can be used in MS clinics to predict motor and cognitive impairment as well as treatment response.

Conclusions

The integration of AI in MS neuroimaging provides a great opportunity to provide more personalized and optimized care for patients who are affected by this complex neurological disorder. AI has been shown to be a promising tool for creating more sensitive MRI scans that better visualize tissue abnormalities, facilitating the workflows of neuroradiologists and other healthcare workers by automatically segmenting tissue and lesions, and improving patient outcomes with more accurate diagnosis, prognosis, and treatment decisions. While AI can significantly improve clinical care in MS, there are some challenges that need careful consideration, such as generalization issues, perpetuation of biases, and privacy/ethical implications of data sharing. In addition, the training of healthcare professionals in the basics of AI and fostering their collaboration with AI experts are both crucial for a successful AI application in the clinic. Overall, AI applications in neuroimaging of individuals with MS have the potential to significantly improve future clinical care of patients with MS.

References

[1] Barkhof F. The clinico-radiological paradox in multiple sclerosis revisited. Curr Opin Neurol 2002;15:239–45.
[2] van Nederpelt DR, Amiri H, Brouwer I, Noteboom S, Mokkink LB, Barkhof F, et al. Reliability of brain atrophy measurements in multiple sclerosis using MRI: an assessment of six freely available software packages for cross-sectional analyses. Neuroradiology 2023;65:1459–72. Available from: https://doi.org/10.1007/s00234-023-03189-8.
[3] Dwyer M, Lyman C, Ferrari H, Bergsland N, Fuchs TA, Jakimovski D, et al. DeepGRAI (Deep Gray Rating via Artificial Intelligence): Fast, feasible, and clinically relevant thalamic atrophy measurement on clinical quality T2-FLAIR MRI in multiple sclerosis NeuroImage Clin 2021;30:102652. Available from: https://doi.org/10.1016/j.nicl.2021.102652.
[4] Henschel L, Conjeti S, Estrada S, Diers K, Fischl B, Reuter M. FastSurfer - A fast and accurate deep learning based neuroimaging pipeline. NeuroImage 2020;219:117012. Available from: https://doi.org/10.1016/j.neuroimage.2020.117012.
[5] Tozlu C, Jamison K, Gauthier SA, Kuceyeski A. Dynamic Functional Connectivity Better Predicts Disability Than Structural and Static Functional Connectivity in People With Multiple Sclerosis. Front Neurosci 2021;15:1683. Available from: https://doi.org/10.3389/FNINS.2021.763966/BIBTEX.
[6] Zurita M, Montalba C, Labbé T, Cruz JP, Dalboni da Rocha J, Tejos C, et al. Characterization of relapsing-remitting multiple sclerosis patients using support vector machine classifications of functional and diffusion MRI data. NeuroImage Clin 2018;20:724–30. Available from: https://doi.org/10.1016/j.nicl.2018.09.002.
[7] Tozlu C, Jamison K, Gu Z, Gauthier SA, Kuceyeski A. Estimated connectivity networks outperform observed connectivity networks when classifying people with multiple sclerosis into disability groups. NeuroImage Clin 2021;32:102827. Available from: https://doi.org/10.1016/J.NICL.2021.102827.

[8] Tozlu C, Jamison K, Nguyen T, Zinger N, Kaunzner U, Pandya S, et al. Structural disconnectivity from paramagnetic rim lesions is related to disability in multiple sclerosis. Brain Behav 2021;11:e2353. Available from: https://doi.org/10.1002/BRB3.2353.

[9] Marzi C, d'Ambrosio A, Diciotti S, Bisecco A, Altieri M, Filippi M, et al. Prediction of the information processing speed performance in multiple sclerosis using a machine learning approach in a large multicenter magnetic resonance imaging data set. Hum Brain Mapp 2023;44:186–202. Available from: https://doi.org/10.1002/hbm.26106.

[10] Kamraoui RA, Mansencal B, Manjon JV, Coupé P. Longitudinal detection of new MS lesions using deep learning. Front Neuroimaging 2022;1.

[11] Krüger J, Opfer R, Gessert N, Ostwaldt A-C, Manogaran P, Kitzler HH, et al. Fully automated longitudinal segmentation of new or enlarged multiple sclerosis lesions using 3D convolutional neural networks. NeuroImage Clin 2020;28:102445. Available from: https://doi.org/10.1016/j.nicl.2020.102445.

[12] Duarte KTN, Gobbi DG, Sidhu AS, McCreary CR, Saad F, Camicioli R, et al. Segmenting white matter hyperintensities in brain magnetic resonance images using convolution neural networks. Pattern Recognit Lett 2023;175:90–4. Available from: https://doi.org/10.1016/j.patrec.2023.07.014.

[13] Billot B, Cerri S, Van Leemput K, Dalca AV, Iglesias JE. Joint segmentation of multiple sclerosis lesions and brain anatomy in MRI scans of any contrast and resolution with CNNs. Proc IEEE Int Symp Biomed Imaging 2021;2021:1971–4. Available from: https://doi.org/10.1109/isbi48211.2021.9434127.

[14] Manso Jimeno M, Ravi KS, Jin Z, Oyekunle D, Ogbole G, Geethanath S. ArtifactID: Identifying artifacts in low-field MRI of the brain using deep learning. Magn Reson Imaging 2022;89:42–8. Available from: https://doi.org/10.1016/j.mri.2022.02.002.

[15] Narayana PA, Coronado I, Sujit SJ, Wolinsky JS, Lublin FD, Gabr RE. Deep Learning for Predicting Enhancing Lesions in Multiple Sclerosis from Noncontrast MRI. Radiology 2020;294:398–404. Available from: https://doi.org/10.1148/radiol.2019191061.

[16] Finck T, Li H, Grundl L, Eichinger P, Bussas M, Mühlau M, et al. Deep-Learning Generated Synthetic Double Inversion Recovery Images Improve Multiple Sclerosis Lesion Detection. Invest Radiol 2020;55:318–23. Available from: https://doi.org/10.1097/RLI.0000000000000640.

[17] Vaswani A, Shazeer N, Parmar N, Uszkoreit J, Jones L, Gomez AN, et al. Attention is All you Need. Advances in Neural Information Processing Systems. Curran Associates, Inc.; 2017.

[18] Bhayana R. Chatbots and Large Language Models in Radiology: A Practical Primer for Clinical and Research Applications. Radiology 2024;310:e232756. Available from: https://doi.org/10.1148/radiol.232756.

[19] Pedregosa F, Varoquaux G, Gramfort A, Michel V, Thirion B, Grisel O, et al. Scikit-learn: Machine Learning in Python. J Mach Learn Res 2011;12:2825–30.

[20] Nabizadeh F, Masrouri S, Ramezannezhad E, Ghaderi A, Sharafi AM, Soraneh S, et al. Artificial intelligence in the diagnosis of multiple sclerosis: A systematic review. Mult Scler Relat Disord 2022;59:103673. Available from: https://doi.org/10.1016/j.msard.2022.103673.

[21] Gulani V, Calamante F, Shellock FG, Kanal E, Reeder SB. Gadolinium deposition in the brain: summary of evidence and recommendations. Lancet Neurol 2017;16:564–70. Available from: https://doi.org/10.1016/S1474-4422(17)30158-8.

[22] Gong E, Pauly JM, Wintermark M, Zaharchuk G. Deep learning enables reduced gadolinium dose for contrast-enhanced brain MRI. J Magn Reson Imaging 2018;48:330–40. Available from: https://doi.org/10.1002/jmri.25970.

[23] Coronado I, Gabr RE, Narayana PA. Deep learning segmentation of gadolinium enhancing lesions in multiple sclerosis. Mult Scler Houndmills Basingstoke Engl 2021;27:519–27. Available from: https://doi.org/10.1177/1352458520921364.

[24] Gaj S, Ontaneda D, Nakamura K. Automatic segmentation of gadolinium-enhancing lesions in multiple sclerosis using deep learning from clinical MRI. PLOS ONE 2021;16:e0255939. Available from: https://doi.org/10.1371/journal.pone.0255939.

[25] Redpath TW, Smith FW. Use of a double inversion recovery pulse sequence to image selectively grey or white brain matter. Br J Radiol 1994;67:1258–63. Available from: https://doi.org/10.1259/0007-1285-67-804-1258.

[26] Wattjes MP, Lutterbey GG, Gieseke J, Träber F, Klotz L, Schmidt S, et al. Double Inversion Recovery Brain Imaging at 3T: Diagnostic Value in the Detection of Multiple Sclerosis Lesions. Am J Neuroradiol 2007;28:54–9.

[27] Geurts JJG, Pouwels PJW, Uitdehaag BMJ, Polman CH, Barkhof F, Castelijns JA. Intracortical lesions in multiple sclerosis: improved detection with 3D double inversion-recovery MR imaging. Radiology 2005;236:254–60. Available from: https://doi.org/10.1148/radiol.2361040450.
[28] Bouman PM, Steenwijk MD, Geurts JJG, Jonkman LE. Artificial double inversion recovery images can substitute conventionally acquired images: an MRI-histology study. Sci Rep 2022;12:2620. Available from: https://doi.org/10.1038/s41598-022-06546-4.
[29] Bouman PM, Noteboom S, Nobrega Santos FA, Beck ES, Bliault G, Castellaro M, et al. Multicenter Evaluation of AI-generated DIR and PSIR for Cortical and Juxtacortical Multiple Sclerosis Lesion Detection. Radiology 2023;307:e221425. Available from: https://doi.org/10.1148/radiol.221425.
[30] Finck T, Li H, Schlaeger S, Grundl L, Sollmann N, Bender B, et al. Uncertainty-Aware and Lesion-Specific Image Synthesis in Multiple Sclerosis Magnetic Resonance Imaging: A Multicentric Validation Study. Front Neurosci 2022;16.
[31] Schlaeger S, Li HB, Baum T, Zimmer C, Moosbauer J, Byas S, et al. Longitudinal Assessment of Multiple Sclerosis Lesion Load With Synthetic Magnetic Resonance Imaging—A Multicenter Validation Study. Invest Radiol 2023;58:320. Available from: https://doi.org/10.1097/RLI.0000000000000938.
[32] Dong C, Loy CC, He K, Tang X. Image Super-Resolution Using Deep Convolutional Networks. IEEE Trans Pattern Anal Mach Intell 2016;38:295–307. Available from: https://doi.org/10.1109/TPAMI.2015.2439281.
[33] Higaki T, Nakamura Y, Tatsugami F, Nakaura T, Awai K. Improvement of image quality at CT and MRI using deep learning. Jpn J Radiol 2019;37:73–80. Available from: https://doi.org/10.1007/s11604-018-0796-2.
[34] Iglesias JE, Billot B, Balbastre Y, Tabari A, Conklin J, Gilberto González R, et al. Joint super-resolution and synthesis of 1mm isotropic MP-RAGE volumes from clinical MRI exams with scans of different orientation, resolution and contrast. NeuroImage 2021;237:118206. Available from: https://doi.org/10.1016/j.neuroimage.2021.118206.
[35] Iglesias JE, Billot B, Balbastre Y, Magdamo C, Arnold SE, Das S, et al. SynthSR: A public AI tool to turn heterogeneous clinical brain scans into high-resolution T1-weighted images for 3D morphometry. Sci Adv 2023;9:eadd3607. Available from: https://doi.org/10.1126/sciadv.add3607.
[36] Iglesias JE, Schleicher R, Laguna S, Billot B, Schaefer P, McKaig B, et al. Quantitative Brain Morphometry of Portable Low-Field-Strength MRI Using Super-Resolution Machine Learning. Radiology 2023;306:e220522. Available from: https://doi.org/10.1148/radiol.220522.
[37] Arnold TC, Tu D, Okar SV, Nair G, By S, Kawatra KD, et al. Sensitivity of portable low-field magnetic resonance imaging for multiple sclerosis lesions. NeuroImage Clin 2022;35:103101. Available from: https://doi.org/10.1016/j.nicl.2022.103101.
[38] Billot B, Greve DN, Puonti O, Thielscher A, Van Leemput K, Fischl B, et al. SynthSeg: Segmentation of brain MRI scans of any contrast and resolution without retraining. Med Image Anal 2023;86:102789. Available from: https://doi.org/10.1016/j.media.2023.102789.
[39] Noteboom S, van Nederpelt DR, Bajrami A, Moraal B, Caan MWA, Barkhof F, et al. Feasibility of detecting atrophy relevant for disability and cognition in multiple sclerosis using 3D-FLAIR. J Neurol 2023;270:5201–10. Available from: https://doi.org/10.1007/s00415-023-11870-4.
[40] Zivadinov R, Havrdová E, Bergsland N, Tyblova M, Hagemeier J, Seidl Z, et al. Thalamic Atrophy Is Associated with Development of Clinically Definite Multiple Sclerosis. Radiology 2013;268:831–41. Available from: https://doi.org/10.1148/radiol.13122424.
[41] Amin M, Ontaneda D. Thalamic Injury and Cognition in Multiple Sclerosis. Front Neurol 2020;11:623914. Available from: https://doi.org/10.3389/fneur.2020.623914.
[42] Batista S, Zivadinov R, Hoogs M, Bergsland N, Heininen-Brown M, Dwyer MG, et al. Basal ganglia, thalamus and neocortical atrophy predicting slowed cognitive processing in multiple sclerosis. J Neurol 2012;259:139–46. Available from: https://doi.org/10.1007/s00415-011-6147-1.
[43] Bergsland N, Zivadinov R, Dwyer MG, Weinstock-Guttman B, Benedict RH. Localized atrophy of the thalamus and slowed cognitive processing speed in MS patients. Mult Scler J 2016;22:1327–36. Available from: https://doi.org/10.1177/1352458515616204.
[44] Zivadinov R, Bergsland N, Korn JR, Dwyer MG, Khan N, Medin J, et al. Feasibility of Brain Atrophy Measurement in Clinical Routine without Prior Standardization of the MRI Protocol: Results from MS-MRIUS, a Longitudinal Observational, Multicenter Real-World Outcome Study in Patients with Relapsing-Remitting MS. Am J Neuroradiol 2018;39:289–95. Available from: https://doi.org/10.3174/ajnr.A5442.

[45] Patenaude B, Smith SM, Kennedy DN, Jenkinson M. A Bayesian model of shape and appearance for subcortical brain segmentation. NeuroImage 2011;56:907–22. Available from: https://doi.org/10.1016/j.neuroimage.2011.02.046.

[46] Ricigliano VAG, Morena E, Colombi A, Tonietto M, Hamzaoui M, Poirion E, et al. Choroid Plexus Enlargement in Inflammatory Multiple Sclerosis: 3.0-T MRI and Translocator Protein PET Evaluation. Radiology 2021;301:166–77. Available from: https://doi.org/10.1148/radiol.2021204426.

[47] Yazdan-Panah A, Schmidt-Mengin M, Ricigliano VAG, Soulier T, Stankoff B, Colliot O. Automatic segmentation of the choroid plexuses: Method and validation in controls and patients with multiple sclerosis. NeuroImage Clin 2023;38:103368. Available from: https://doi.org/10.1016/j.nicl.2023.103368.

[48] Polman CH, Reingold SC, Banwell B, Clanet M, Cohen JA, Filippi M, Fujihara K, Havrdova E, Hutchinson M, Kappos L, Lublin FD. Diagnostic criteria for multiple sclerosis: 2010 revisions to the McDonald criteria. Ann Neurol 2011;69:292–302. Available from: https://doi.org/10.1002/ANA.22366.

[49] Geremia E, Clatz O, Menze BH, Konukoglu E, Criminisi A, Ayache N. Spatial decision forests for MS lesion segmentation in multi-channel magnetic resonance images. NeuroImage 2011;57:378–90. Available from: https://doi.org/10.1016/j.neuroimage.2011.03.080.

[50] Steenwijk MD, Pouwels PJW, Daams M, van Dalen JW, Caan MWA, Richard E, et al. Accurate white matter lesion segmentation by k nearest neighbor classification with tissue type priors (kNN-TTPs). NeuroImage Clin 2013;3:462–9. Available from: https://doi.org/10.1016/j.nicl.2013.10.003.

[51] Brosch T, Tang LYW, Yoo Y, Li DKB, Traboulsee A, Tam R. Deep 3D Convolutional Encoder Networks With Shortcuts for Multiscale Feature Integration Applied to Multiple Sclerosis Lesion Segmentation. IEEE Trans Med Imaging 2016;35:1229–39. Available from: https://doi.org/10.1109/TMI.2016.2528821.

[52] Gabr RE, Coronado I, Robinson M, Sujit SJ, Datta S, Sun X, et al. Brain and lesion segmentation in multiple sclerosis using fully convolutional neural networks: A large-scale study. Mult Scler Houndmills Basingstoke Engl 2020;26:1217–26. Available from: https://doi.org/10.1177/1352458519856843.

[53] Valverde S, Cabezas M, Roura E, González-Villà S, Pareto D, Vilanova JC, et al. Improving automated multiple sclerosis lesion segmentation with a cascaded 3D convolutional neural network approach. NeuroImage 2017;155:159–68. Available from: https://doi.org/10.1016/j.neuroimage.2017.04.034.

[54] Maggi P, Sati P, Nair G, Cortese ICM, Jacobson S, Smith BR, et al. Paramagnetic Rim Lesions are Specific to Multiple Sclerosis: An International Multicenter 3T MRI Study. Ann Neurol 2020;88:1034–42. Available from: https://doi.org/10.1002/ana.25877.

[55] Kaunzner UW, Kang Y, Zhang S, Morris E, Yao Y, Pandya S, et al. Quantitative susceptibility mapping identifies inflammation in a subset of chronic multiple sclerosis lesions. Brain 2019;142:133–45. Available from: https://doi.org/10.1093/brain/awy296.

[56] Hemond CC, Baek J, Ionete C, Reich DS. Paramagnetic rim lesions are associated with pathogenic CSF profiles and worse clinical status in multiple sclerosis: A retrospective cross-sectional study. Mult Scler J 2022. Available from: https://doi.org/10.1177/13524585221102921/ASSET/IMAGES/LARGE/10.1177_13524585221102921-FIG3.JPEG.

[57] Marcille M, Hurtado Rúa S, Tyshkov C, Jaywant A, Comunale J, Kaunzner UW, et al. Disease correlates of rim lesions on quantitative susceptibility mapping in multiple sclerosis. Sci Rep 2022;12:1–10. Available from: https://doi.org/10.1038/s41598-022-08477-6, 2022 121.

[58] Tozlu C, Olafson E, Jamison KW, Demmon E, Kaunzner U, Marcille M, et al. The sequence of regional structural disconnectivity due to multiple sclerosis lesions. Brain Commun 2023;5:fcad332. Available from: https://doi.org/10.1093/braincomms/fcad332.

[59] Yao B, Bagnato F, Matsuura E, Merkle H, van Gelderen P, Cantor FK, et al. Chronic Multiple Sclerosis Lesions: Characterization with High-Field-Strength MR Imaging. Radiology 2012;262:206–15. Available from: https://doi.org/10.1148/radiol.11110601.

[60] Absinta M, Sati P, Fechner A, Schindler MK, Nair G, Reich DS. Identification of Chronic Active Multiple Sclerosis Lesions on 3T MRI. Am J Neuroradiol 2018;39:1233–8. Available from: https://doi.org/10.3174/ajnr.A5660.

[61] Hagemeier J, Heininen-Brown M, Poloni GU, Bergsland N, Magnano CR, Durfee J, et al. Iron deposition in multiple sclerosis lesions measured by susceptibility-weighted imaging filtered phase: A case control study. J Magn Reson Imaging 2012;36:73–83. Available from: https://doi.org/10.1002/jmri.23603.

[62] Barquero G, La Rosa F, Kebiri H, Lu P-J, Rahmanzadeh R, Weigel M, et al. RimNet: A deep 3D multimodal MRI architecture for paramagnetic rim lesion assessment in multiple sclerosis. NeuroImage Clin 2020;28:102412. Available from: https://doi.org/10.1016/j.nicl.2020.102412.

[63] Lou C, Sati P, Absinta M, Clark K, Dworkin JD, Valcarcel AM, et al. Fully automated detection of paramagnetic rims in multiple sclerosis lesions on 3T susceptibility-based MR imaging. NeuroImage Clin 2021;32:102796. Available from: https://doi.org/10.1016/j.nicl.2021.102796.

[64] Chawla NV, Bowyer KW, Hall LO, Kegelmeyer WP. SMOTE: Synthetic Minority Over-sampling Technique. J Artif Intell Res 2002;16:321–57. Available from: https://doi.org/10.1613/jair.953.

[65] Treaba CA, Conti A, Klawiter EC, Barletta VT, Herranz E, Mehndiratta A, et al. Cortical and phase rim lesions on 7T MRI as markers of multiple sclerosis disease progression. Brain Commun 2021;3:fcab134. Available from: https://doi.org/10.1093/braincomms/fcab134.

[66] Treaba CA, Granberg TE, Sormani MP, Herranz E, Ouellette RA, Louapre C, et al. Longitudinal Characterization of Cortical Lesion Development and Evolution in Multiple Sclerosis with 7.0-T MRI. Radiology 2019;291:740–9. Available from: https://doi.org/10.1148/radiol.2019181719.

[67] Calabrese M, Agosta F, Rinaldi F, Mattisi I, Grossi P, Favaretto A, et al. Cortical Lesions and Atrophy Associated With Cognitive Impairment in Relapsing-Remitting Multiple Sclerosis. Arch Neurol 2009;66:1144–50. Available from: https://doi.org/10.1001/archneurol.2009.174.

[68] Harrison DM, Roy S, Oh J, Izbudak I, Pham D, Courtney S, et al. Association of Cortical Lesion Burden on 7-T Magnetic Resonance Imaging With Cognition and Disability in Multiple Sclerosis. JAMA Neurol 2015;72:1004–12. Available from: https://doi.org/10.1001/jamaneurol.2015.1241.

[69] Mainero C, Louapre C, Govindarajan S, Gianni C, Nielsen S, Cohen-Adad J, et al. A gradient in cortical pathology in multiple sclerosis by in vivo quantitative 7T imaging. Brain 2015;138:932–45. Available from: https://doi.org/10.1093/brain/awv031.

[70] Nielsen AS, Kinkel RP, Madigan N, Tinelli E, Benner T, Mainero C. Contribution of cortical lesion subtypes at 7T MRI to physical and cognitive performance in MS. Neurology 2013;81:641–9. Available from: https://doi.org/10.1212/WNL.0b013e3182a08ce8.

[71] Daams M, Geurts JJG, Barkhof F. Cortical imaging in multiple sclerosis: recent findings and 'grand challenges. Curr Opin Neurol 2013;26:345. Available from: https://doi.org/10.1097/WCO.0b013e328362a864.

[72] Kilsdonk ID, Jonkman LE, Klaver R, van Veluw SJ, Zwanenburg JJM, Kuijer JPA, et al. Increased cortical grey matter lesion detection in multiple sclerosis with 7T MRI: a post-mortem verification study. Brain 2016;139:1472–81. Available from: https://doi.org/10.1093/brain/aww037.

[73] Sethi V, Yousry TA, Muhlert N, Ron M, Golay X, Wheeler-Kingshott C, et al. Improved detection of cortical MS lesions with phase-sensitive inversion recovery MRI. J Neurol Neurosurg Psychiatry 2012;83:877–82. Available from: https://doi.org/10.1136/jnnp-2012-303023.

[74] Tardif CL, Collins DL, Eskildsen SF, Richardson JB, Pike GB. Segmentation of Cortical MS Lesions on MRI Using Automated Laminar Profile Shape Analysis. In: Jiang T, Navab N, Pluim JPW, Viergever MA, editors. Medical Image Computing and Computer-Assisted Intervention – MICCAI 2010, Lecture Notes in Computer Science. Berlin, Heidelberg: Springer; 2010. p. 181–8. Available from: https://doi.org/10.1007/978-3-642-15711-0_23.

[75] Fartaria MJ, Bonnier G, Roche A, Kober T, Meuli R, Rotzinger D, et al. Automated detection of white matter and cortical lesions in early stages of multiple sclerosis. J Magn Reson Imaging 2016;43:1445–54. Available from: https://doi.org/10.1002/jmri.25095.

[76] Fartaria MJ, Roche A, Meuli R, Granziera C, Kober T, Bach Cuadra M. Segmentation of Cortical and Subcortical Multiple Sclerosis Lesions Based on Constrained Partial Volume Modeling. In: Descoteaux M, Maier-Hein L, Franz A, Jannin P, Collins DL, Duchesne S, editors. Medical Image Computing and Computer Assisted Intervention – MICCAI 2017, Lecture Notes in Computer Science. Cham: Springer International Publishing; 2017. p. 142–9. Available from: https://doi.org/10.1007/978-3-319-66179-7_17.

[77] La Rosa F, Abdulkadir A, Fartaria MJ, Rahmanzadeh R, Lu P-J, Galbusera R, et al. Multiple sclerosis cortical and WM lesion segmentation at 3T MRI: a deep learning method based on FLAIR and MP2RAGE. NeuroImage Clin 2020;27:102335. Available from: https://doi.org/10.1016/j.nicl.2020.102335.

[78] La Rosa F, Beck ES, Abdulkadir A, Thiran J-P, Reich DS, Sati P, et al. Automated Detection of Cortical Lesions in Multiple Sclerosis Patients with 7T MRI. In: Martel AL, Abolmaesumi P, Stoyanov D, Mateus D, Zuluaga MA, Zhou SK, Racoceanu D, Joskowicz L, editors. Medical Image Computing and Computer Assisted Intervention – MICCAI 2020, Lecture Notes in Computer Science. Cham: Springer International Publishing; 2020. p. 584–93. Available from: https://doi.org/10.1007/978-3-030-59719-1_57.

[79] Cagol A, Cortese R, Barakovic M, Schaedelin S, Ruberte E, Absinta M, et al. Diagnostic Performance of Cortical Lesions and the Central Vein Sign in Multiple Sclerosis. JAMA Neurol 2023. Available from: https://doi.org/10.1001/jamaneurol.2023.4737.

[80] Ontaneda D, Sati P, Raza P, Kilbane M, Gombos E, Alvarez E, et al. Central vein sign: A diagnostic biomarker in multiple sclerosis (CAVS-MS) study protocol for a prospective multicenter trial. NeuroImage Clin 2021;32:102834. Available from: https://doi.org/10.1016/j.nicl.2021.102834.

[81] Sati P, George IC, Shea CD, Gaitán MI, Reich DS. FLAIR*: A Combined MR Contrast Technique for Visualizing White Matter Lesions and Parenchymal Veins. Radiology 2012;265:926–32. Available from: https://doi.org/10.1148/radiol.12120208.

[82] Sati P, Oh J, Constable RT, Evangelou N, Guttmann CRG, Henry RG, et al. The central vein sign and its clinical evaluation for the diagnosis of multiple sclerosis: a consensus statement from the North American Imaging in Multiple Sclerosis Cooperative. Nat Rev Neurol 2016;12:714–22. Available from: https://doi.org/10.1038/nrneurol.2016.166.

[83] Dworkin JD, Sati P, Solomon A, Pham DL, Watts R, Martin ML, et al. Automated Integration of Multimodal MRI for the Probabilistic Detection of the Central Vein Sign in White Matter Lesions. Am J Neuroradiol 2018;39:1806–13. Available from: https://doi.org/10.3174/ajnr.A5765.

[84] Maggi P, Fartaria MJ, Jorge J, La Rosa F, Absinta M, Sati P, et al. CVSnet: A machine learning approach for automated central vein sign assessment in multiple sclerosis. NMR Biomed 2020;33:e4283. Available from: https://doi.org/10.1002/nbm.4283.

[85] Elliott C, Arnold DL, Collins DL, Arbel T. Temporally Consistent Probabilistic Detection of New Multiple Sclerosis Lesions in Brain MRI. IEEE Trans Med Imaging 2013;32:1490–503. Available from: https://doi.org/10.1109/TMI.2013.2258403.

[86] Cheng M, Galimzianova A, Lesjak Ž, Špiclin Ž, Lock CB, Rubin DL. A Multi-scale Multiple Sclerosis Lesion Change Detection in a Multi-sequence MRI. In: Stoyanov D, Taylor Z, Carneiro G, Syeda-Mahmood T, Martel A, Maier-Hein L, Tavares JMRS, Bradley A, Papa JP, Belagiannis V, Nascimento JC, Lu Z, Conjeti S, Moradi M, Greenspan H, Madabhushi A, editors. Deep Learning in Medical Image Analysis and Multimodal Learning for Clinical Decision Support, Lecture Notes in Computer Science. Cham: Springer International Publishing; 2018. p. 353–60. Available from: https://doi.org/10.1007/978-3-030-00889-5_40.

[87] Sweeney EM, Shinohara RT, Shea CD, Reich DS, Crainiceanu CM. Automatic Lesion Incidence Estimation and Detection in Multiple Sclerosis Using Multisequence Longitudinal MRI. Am J Neuroradiol 2013;34:68–73. Available from: https://doi.org/10.3174/ajnr.A3172.

[88] Salem M, Cabezas M, Valverde S, Pareto D, Oliver A, Salvi J, et al. A supervised framework with intensity subtraction and deformation field features for the detection of new T2-w lesions in multiple sclerosis. NeuroImage Clin 2018;17:607–15. Available from: https://doi.org/10.1016/j.nicl.2017.11.015.

[89] Thompson AJ, Banwell BL, Barkhof F, Carroll WM, Coetzee T, Comi G, et al. Diagnosis of multiple sclerosis: 2017 revisions of the McDonald criteria. Lancet Neurol 2018;17:162–73. Available from: https://doi.org/10.1016/S1474-4422(17)30470-2.

[90] Eitel F, Soehler E, Bellmann-Strobl J, Brandt AU, Ruprecht K, Giess RM, et al. Uncovering convolutional neural network decisions for diagnosing multiple sclerosis on conventional MRI using layer-wise relevance propagation. NeuroImage Clin 2019;24:102003. Available from: https://doi.org/10.1016/j.nicl.2019.102003.

[91] Lopatina A, Ropele S, Sibgatulin R, Reichenbach JR, Güllmar D. Investigation of Deep-Learning-Driven Identification of Multiple Sclerosis Patients Based on Susceptibility-Weighted Images Using Relevance Analysis. Front Neurosci 2020;14.

[92] Weygandt M, Hackmack K, Pfüller C, Bellmann–Strobl J, Paul F, Zipp F, et al. MRI Pattern Recognition in Multiple Sclerosis Normal-Appearing Brain Areas. PLOS ONE 2011;6:e21138. Available from: https://doi.org/10.1371/journal.pone.0021138.

[93] Saccà V, Sarica A, Novellino F, Barone S, Tallarico T, Filippelli E, et al. Evaluation of machine learning algorithms performance for the prediction of early multiple sclerosis from resting-state FMRI connectivity data. Brain Imaging Behav 2018. Available from: https://doi.org/10.1007/s11682-018-9926-9.

[94] Eshaghi A, Wottschel V, Cortese R, Calabrese M, Sahraian MA, Thompson AJ, et al. Gray matter MRI differentiates neuromyelitis optica from multiple sclerosis using random forest. Neurology 2016;87:2463–70. Available from: https://doi.org/10.1212/WNL.0000000000003395.

[95] Eshaghi A, Riyahi-Alam S, Saeedi R, Roostaei T, Nazeri A, Aghsaei A, et al. Classification algorithms with multi-modal data fusion could accurately distinguish neuromyelitis optica from multiple sclerosis. NeuroImage Clin 2015;7:306–14. Available from: https://doi.org/10.1016/j.nicl.2015.01.001.

[96] Cacciaguerra L, Meani A, Mesaros S, Radaelli M, Palace J, Dujmovic-Basuroski I, et al. Brain and cord imaging features in neuromyelitis optica spectrum disorders. Ann Neurol 2019;85:371–84. Available from: https://doi.org/10.1002/ana.25411.

[97] Kim H, Lee Y, Kim Y-H, Lim Y-M, Lee JS, Woo J, et al. Deep Learning-Based Method to Differentiate Neuromyelitis Optica Spectrum Disorder From Multiple Sclerosis. Front Neurol 2020;11.

[98] Rocca MA, Anzalone N, Storelli L, Del Poggio A, Cacciaguerra L, Manfredi AA, et al. Deep Learning on Conventional Magnetic Resonance Imaging Improves the Diagnosis of Multiple Sclerosis Mimics. Invest Radiol 2021;56:252–60. Available from: https://doi.org/10.1097/RLI.0000000000000735.

[99] McDonald WI, Compston A, Edan G, Goodkin D, Hartung H-P, Lublin FD, et al. Recommended diagnostic criteria for multiple sclerosis: Guidelines from the international panel on the diagnosis of multiple sclerosis. Ann Neurol 2001;50:121–7. Available from: https://doi.org/10.1002/ana.1032.

[100] Kocevar G, Stamile C, Hannoun S, Cotton F, Vukusic S, Durand-Dubief F, et al. Graph Theory-Based Brain Connectivity for Automatic Classification of Multiple Sclerosis Clinical Courses. Front Neurosci 2016;10:478. Available from: https://doi.org/10.3389/fnins.2016.00478.

[101] Ion-Mărgineanu A, Kocevar G, Stamile C, Sima DM, Durand-Dubief F, Van Huffel S, et al. Machine Learning Approach for Classifying Multiple Sclerosis Courses by Combining Clinical Data with Lesion Loads and Magnetic Resonance Metabolic Features. Front Neurosci 2017;11. Available from: https://doi.org/10.3389/fnins.2017.00398.

[102] Wottschel V, Alexander DC, Kwok PP, Chard DT, Stromillo ML, De Stefano N, et al. Predicting outcome in clinically isolated syndrome using machine learning. NeuroImage Clin 2015;7:281–7. Available from: https://doi.org/10.1016/j.nicl.2014.11.021.

[103] Wottschel V, Chard DT, Enzinger C, Filippi M, Frederiksen JL, Gasperini C, et al. SVM recursive feature elimination analyses of structural brain MRI predicts near-term relapses in patients with clinically isolated syndromes suggestive of multiple sclerosis. NeuroImage Clin 2019;24:102011. Available from: https://doi.org/10.1016/j.nicl.2019.102011.

[104] Eshaghi A, Young AL, Wijeratne PA, Prados F, Arnold DL, Narayanan S, et al. Identifying multiple sclerosis subtypes using unsupervised machine learning and MRI data. Nat Commun 2021;12:2078. Available from: https://doi.org/10.1038/s41467-021-22265-2.

[105] Pontillo G, Penna S, Cocozza S, Quarantelli M, Gravina M, Lanzillo R, et al. Stratification of multiple sclerosis patients using unsupervised machine learning: a single-visit MRI-driven approach. Eur Radiol 2022;32:5382–91. Available from: https://doi.org/10.1007/s00330-022-08610-z.

[106] Zhao Y, Healy BC, Rotstein D, Guttmann CRG, Bakshi R, Weiner HL, et al. Exploration of machine learning techniques in predicting multiple sclerosis disease course. PLOS ONE 2017;12:e0174866. Available from: https://doi.org/10.1371/journal.pone.0174866.

[107] Kuceyeski A, Maruta J, Relkin N, Raj A. The Network Modification (NeMo) Tool: Elucidating the Effect of White Matter Integrity Changes on Cortical and Subcortical Structural Connectivity. Brain Connect 2013;3:451–63. Available from: https://doi.org/10.1089/brain.2013.0147.

[108] Fuchs TA, Ziccardi S, Benedict RHB, Bartnik A, Kuceyeski A, Charvet LE, et al. Functional Connectivity and Structural Disruption in the Default-Mode Network Predicts Cognitive Rehabilitation Outcomes in Multiple Sclerosis. J Neuroimaging 2020;30:523–30. Available from: https://doi.org/10.1111/jon.12723.

[109] Buyukturkoglu K, Zeng D, Bharadwaj S, Tozlu C, Mormina E, Igwe KC, et al. Classifying multiple sclerosis patients on the basis of SDMT performance using machine learning. Mult Scler J 2020. Available from: https://doi.org/10.1177/1352458520958362.

[110] Zhang K, Lincoln JA, Jiang X, Bernstam EV, Shams S. Predicting multiple sclerosis severity with multimodal deep neural networks. BMC Med Inf Decis Mak 2023;23:255. Available from: https://doi.org/10.1186/s12911-023-02354-6.

[111] Roca P, Attye A, Colas L, Tucholka A, Rubini P, Cackowski S, et al. Artificial intelligence to predict clinical disability in patients with multiple sclerosis using FLAIR MRI. Diagn Interv Imaging 2020;101:795–802. Available from: https://doi.org/10.1016/j.diii.2020.05.009.

3. Use of other imaging acquisition and analysis modalities in multiple sclerosis

[112] Eshaghi A, Wijeratne PA, Oxtoby NP, Arnold DL, Collins L, Narayanan S, et al. Predicting personalised risk of disability worsening in multiple sclerosis with machine learning 2022. Available from: https://doi.org/10.1101/2022.02.03.22270364.

[113] Tozlu C, Card S, Jamison K, Gauthier SA, Kuceyeski A. Larger lesion volume in people with multiple sclerosis is associated with increased transition energies between brain states and decreased entropy of brain activity. Netw Neurosci 2023;7:539–56. Available from: https://doi.org/10.1162/netn_a_00292.

[114] Cordani C, Preziosa P, Valsasina P, Meani A, Pagani E, Morozumi T, et al. MRI of Transcallosal White Matter Helps to Predict Motor Impairment in Multiple Sclerosis. Radiology 2022;302:639–49. Available from: https://doi.org/10.1148/radiol.2021210922.

[115] Ricard JA, Parker TC, Dhamala E, Kwasa J, Allsop A, Holmes AJ. Confronting racially exclusionary practices in the acquisition and analyses of neuroimaging data. Nat Neurosci 2023;26:4–11. Available from: https://doi.org/10.1038/s41593-022-01218-y.

[116] Abramian D, Eklund A. Refacing: Reconstructing Anonymized Facial Features Using GANS, in: 2019 IEEE 16th International Symposium on Biomedical Imaging (ISBI 2019). Presented at the 2019 IEEE 16th International Symposium on Biomedical Imaging (ISBI 2019), 2019. pp. 1104–1108. Available from: https://doi.org/10.1109/ISBI.2019.8759515.

[117] Ribeiro MT, Singh S, Guestrin C. "Why Should I Trust You?": Explaining the Predictions of Any Classifier 2016. Available from: https://doi.org/10.48550/arXiv.1602.04938.

[118] Lundberg S, Lee S-I. A Unified Approach to Interpreting Model Predictions 2017 Available from: https://doi.org/10.48550/arXiv.1705.07874.

[119] Bercea CI, Wiestler B, Rueckert D, Albarqouni S. Federated disentangled representation learning for unsupervised brain anomaly detection. Nat Mach Intell 2022;4:685–95. Available from: https://doi.org/10.1038/s42256-022-00515-2.

[120] Bommasani R, Hudson DA, Adeli E, Altman R, Arora S, von Arx S, et al.. On the Opportunities and Risks of Foundation Models 2022 Available from: https://doi.org/10.48550/arXiv.2108.07258.

[121] Wickramasinghe N, Ulapane N, Andargoli A, Ossai C, Shuakat N, Nguyen T, et al. Digital twins to enable better precision and personalized dementia care. JAMIA Open 2022;5:ooac072. Available from: https://doi.org/10.1093/jamiaopen/ooac072.

[122] Voigt I, Inojosa H, Dillenseger A, Haase R, Akgün K, Ziemssen T. Digital Twins for Multiple Sclerosis. Front Immunol 2021;12:669811. Available from: https://doi.org/10.3389/fimmu.2021.669811.

[123] Walsh JR, Smith AM, Pouliot Y, Li-Bland D, Loukianov A, Fisher CK, et al. Generating Digital Twins with Multiple Sclerosis Using Probabilistic Neural Networks 2020. Available from: https://doi.org/10.1101/2020.02.04.934679.

[124] Cen S, Gebregziabher M, Moazami S, Azevedo CJ, Pelletier D. Toward precision medicine using a "digital twin" approach: modeling the onset of disease-specific brain atrophy in individuals with multiple sclerosis. Sci Rep 2023;13:16279. Available from: https://doi.org/10.1038/s41598-023-43618-5.

[125] Pesapane F, Volonté C, Codari M, Sardanelli F. Artificial intelligence as a medical device in radiology: ethical and regulatory issues in Europe and the United States. Insights Imaging 2018;9:745–53. Available from: https://doi.org/10.1007/s13244-018-0645-y.

ID# Index

Note: Page numbers followed by "*f*" and "*t*" refer to figures and tables, respectively.

A

2D-CSI. *See* Two-dimensional chemical shift imaging (2D-CSI)
A1 astrocytes, 8
Abnormal veins, 208
Abnormalities, 298–299
Absolute quantification, 294
Absolute T2 lesion volume, 112
Acquisition parameters, 159–160
Active disease in MS, acute cognitive decline as marker of, 58–61
Active focal inflammatory plaques, 22
Active lesions, 104
Acute cognitive decline as marker of active disease in MS, 58–61
   ICRs, 59–61
AD. *See* Alzheimer's disease (AD)
Adaptive immune cell. *See also* Innate immune cell
  responses in multiple sclerosis, 5–18
  role of, 10–18
    B cells, 16–18
    CD4 + T cells, 10–14
    CD8 + T cells, 14–15
    unconventional T cells, 15–16
ADC. *See* Apparent diffusion coefficient (ADC)
Advanced imaging techniques, 184
Advanced network analyses, 257–258
Advanced neuroimaging techniques, 384–385, 400–401
AI. *See* Artificial intelligence (AI)
AIF. *See* Arterial input function (AIF)
Alemtuzumab, 85–88, 198–199
  alemtuzumab-treated patients, 369–370
  cladribine, 86–88
Alzheimer's disease (AD), 57–58, 384–385
Amacrine cells, 365
Angiographic images, 211
Animal models, 233, 305, 345–346
Anisotropy, 232
ANN. *See* Artificial neural network (ANN)
Annualized brain volume loss, 110–111
Anti-CD20
  monoclonal antibodies, 16
  therapies, 18
Antigen presenting cells (APCs), 5
APCs. *See* Antigen presenting cells (APCs)
Apparent diffusion coefficient (ADC), 149–150
AQP4. *See* Aquaporin-4-Ab-positive (AQP4)
AQP4-NMOSD, 179–180
Aquaporin-4-Ab-positive (AQP4), 177–179
Area under the curve (AUC), 391–392
Arrival time (AT), 268
Arterial blood water, 151
Arterial input function (AIF), 269
Arterial spin labeling (ASL), 150–151, 268, 272–274
Arterial transit time (ATT), 274
Artificial intelligence (AI), 384
  algorithms, 380
  basics of AI in medical imaging, 385–393
  decision-making processes, 409–410
  future directions, 410–413
    federated learning, 411
    foundational models, 411–412
    personalized treatment plans and digital twins, 412
    regulatory landscape, 412–413
    training healthcare professionals, 411
  in multiple sclerosis neuroimaging, 393–408
  pitfalls and ethical concerns, 408–410
    biases in data, 408–409
    ethical concerns, 409
    explainability of AI algorithms, 409–410
    generalization issues, 408
    practical challenges of AI in clinical practice, 410
    quality of study description, 409
  potential benefits of using AI in multiple sclerosis, 392–393
Artificial Intelligence Act of the European Union, 412–413
Artificial neural network (ANN), 385–386
ASL. *See* Arterial spin labeling (ASL)
Astrocytes, 345–346
AT. *See* Arrival time (AT)
Atrophied T2-lesion volume, 129–130
Atrophy
  atrophy-based segmentation, 236

Atrophy (*Continued*)
  measurements, 127
ATT. *See* Arterial transit time (ATT)
AUC. *See* Area under the curve (AUC)
Autoencoder model, 388–389
Autoimmune disease, 3–4
Autoimmune disorder, 57
Autoimmunity
  in multiple sclerosis, 4–5
  to pathology, 18–21
Axonal loss, 163

B

B cells, 13, 16–18, 83
  depletion, 82–84
    anti-CD20 drugs of taumumab, ocrelizumab, and rituximab, 83t
B lymphocyte, 18
Backpropagation process, 387
"Bagel sign" pattern, 180
"Balck holes", 198
Bandwidth (BW), 292
Baseline T1 lesional volume, 352
BBB. *See* Blood–brain barrier (BBB)
B-factor (b-factor), 149
BICAMS. *See* Brief international cognitive assessment for MS (BICAMS)
Biological water motion, 148–151
Biomarker for multiple sclerosis lesions, water content as new, 211–213
Biot-Savart Law, 140
Bipolar cells, 365
Black holes, 96, 125–126, 148, 163–164
Bloch equation, 145, 151
Blood–brain barrier (BBB), 6, 38–39, 41, 96, 126–127, 145, 164, 268, 323–324
Blood-oxygenation level dependent (BOLD), 250
Body's water motion, 148
BOLD. *See* Blood-oxygenation level dependent (BOLD)
Bone marrow-derived myeloid cells, 5–7
  dendritic cells, 6–7
  monocytes and macrophages, 5–6
BR. *See* Brain reserve (BR)
Brain
  atrophy, 109–111, 125, 279–280, 288
  imaging, 45
  MRI, 52
    findings, 181
    scans, 107
  perfusion, 267
  plasticity, 249–250
  retina as window to, 363–365
  tissue loss, 41–43
  volume, 98, 109, 352
    atrophy, 24
    loss, 109–113
Brain reserve (BR), 57–58
BrainSpec, 412–413
Brain-specific tissue atrophy, 41–43
Brainstem, 21–22, 43
BRB. *See* Brief repeatable battery (BRB)
Breastfeeding, 82
Brief international cognitive assessment for MS (BICAMS), 53–54
Brief repeatable battery (BRB), 52
Brief visuospatial memory test (BVMT), 277–278
  BVMT-R, 53
BVMT. *See* Brief visuospatial memory test (BVMT)
BW. *See* Bandwidth (BW)

C

C-11 labeled flumazenil, 356
C–11 labeled Pittsburgh compound B ([C-11]PIB), 356
CAL. *See* Chronic active lesions (CAL)
Calcium, 153, 223, 322
Calibration image, 273–274
California verbal learning test (CVLT), 277–278
  CVLT-II, 53
CALIPR. *See* Constrained, Adaptive, Low-dimensional, Intrinsically Precise Reconstruction (CALIPR)
Cardiac gating, 180–181
Carr-Purcell-Meiboom-Gill sequence (CPMG sequence), 303–304
CART. *See* Classification and regression trees (CART)
CASL. *See* Continuous arterial spin labelling (CASL)
CBF. *See* Cerebral blood flow (CBF)
CD4 + T cells, 10–14
  continuum phenotype spectrum of, 14
  Tfh cells, 13
  Th1 cells, 11–12
  Th17 cells, 12–13
  Th22 and Th9 cells, 13
  therapeutic targeting, 14
  Treg cells, 13
CD8 + T cells, 14–15
CDMS. *See* Clinically definite MS (CDMS)
CE WM lesions. *See* Contrast-enhancing WM lesions (CE WM lesions)
Cellular substrates, 228
Central atrophy, 127
Central nervous system (CNS), 3–4, 38–39, 51–52, 94, 160, 207–208, 249–250, 288, 322, 344, 363–364
Central vein, 327–328
Central vein sign (CVS), 46, 99, 160, 208, 323–324, 327–330, 328f, 401–402
  clinical significance of, 329–330

visualizing CVS using high-field imaging and 3T magnetic resonance imaging, 328–329
Cerebellum, 253–254
Cerebral arteries, 214–215
Cerebral blood flow (CBF), 150–151, 269
Cerebral hypoperfusion, 274–275, 279
Cerebral perfusion, magnetic resonance imaging techniques for estimating, 268–274
Cerebral perivascular spaces (CPVS), 5
Cerebral small vessel disease, 329
Cerebrospinal fluid (CSF), 94–96, 146–147, 163, 176–177, 194
Cerebrovascular reactivity (CVR), 274, 280–281
Cervical spinal cord volume, 161
Checklist for Artificial Intelligence in Medical Imaging (CLAIM), 409
Choline (Cho), 289–290, 290f
Choroid plexuses (CP), 396–397
Chronic active lesions (CAL), 128–129, 165–166, 234, 305–306, 323–324
  visualizing CAL on magnetic resonance imaging, 325–326
Chronic active nature, 235
Chronic autoimmune disorder, 51–52
Chronic disease process, 38
Chronic hypointense foci, 148
Chronic hypoperfusion, 280
Chronic inactive lesions, 22
Chronic inflammation, 105
Chronic inflammatory activity, 129–130
Chronic inflammatory lesions, 153
CI. See Cognitive impairment (CI)
Cingulum-hippocampus, 160
CIS. See Clinically isolated syndrome (CIS)
Cladribine, 86–88
CLAIM. See Checklist for Artificial Intelligence in Medical Imaging (CLAIM)
Classification and regression trees (CART), 386
Clinical trials
  metric in, 124
  T1 lesion load in, 125–126
Clinically definite MS (CDMS), 44
Clinically isolated syndrome (CIS), 38, 45, 71–72, 176–177, 251, 298, 327, 367, 377, 405–406
Clinical-radiological paradox, 160
Clinico-radiological paradox, 124, 249–250
Cloud-based solutions, 379–380
CLs. See Cortical lesions (CLs)
Clustering approaches, 386
CME. See Combined myelin estimation (CME)
CMSCs. See Consortium of Multiple Sclerosis Centers (CMSCs)
CNNs. See Convolutional neural networks (CNNs)

CNR. See Contrast-to-noise ratio (CNR)
CNS. See Central nervous system (CNS)
Cognition in MS, historical review of, 52–58
Cognitive assessment, 64
Cognitive change, 38
Cognitive dysfunction, 57, 279, 300
  MRI markers for predicting progression of, 62–63
  in MS, 54–58
    common cognitive profiles, 56
    consequences of cognitive impairment in MS, 56–57
    defining cognitive impairment, 55
    factors protecting against cognitive impairment in MS, 57–58
    MRI correlates of cognitive impairment in MS, 57
    neuropsychological tests, 54t
    prevalence and different phenotypes, 55–56
Cognitive impairment (CI), 51–52, 257
  defining, 55
  in MS
    consequences of, 56–57
    MRI correlates of, 57
    treatment of, 61
  prediction of, 406–408
Cognitive performance, 59, 277–279
Cognitive profiles, common, 56
Cognitive recovery, 63–64
Cognitive rehabilitation, 61–63
Cognitive relapse, 51–52, 58–59
Cognitive reserve (CR), 57–58
Cognitive systems, changes to, 251–255
Combined myelin estimation (CME), 331
Combined unique active lesions, 104
Complex autoimmune disease, 3–4
Computer tomography (CT), 162, 378
Confounding effects of myelin in deep gray matter, 231
Conscientiousness, 58
Consortium of Multiple Sclerosis Centers (CMSCs), 175–176, 393
Constrained, Adaptive, Low-dimensional, Intrinsically Precise Reconstruction (CALIPR), 308
Continuous arterial spin labelling (CASL), 273
Continuous wave irradiation, 193
Contrast-enhanced T1WI, 107–108
Contrast-enhancing lesions, 164, 165f
Contrast-enhancing WM lesions (CE WM lesions), 63
Contrast-to-noise ratio (CNR), 324
Controlled oral word association test (COWAT), 53
Conventional MRI markers in multiple sclerosis, 124–127
  atrophy measurements, 127
  gadolinium-enhancing lesion, 126–127

Conventional MRI markers in multiple sclerosis (*Continued*)
  T1 lesions, 125–126
  T2 lesions, 124–125
"Conventional" MRI metrics, 160
Convolutional layers, 387–388
Convolutional neural networks (CNNs), 387–388
Corpus callosum, 43
Correction techniques, 231
Cortical atrophy, 127
Cortical demyelination, 330
Cortical lesions (CLs), 22–23, 98–99, 129, 159–160, 166–169, 324, 400–401
  clinical relevance of, 332
  magnetic resonance imaging sequences for, 167–169
  using high-field imaging in vivo improves CLs visibility and allows detection of subpial disease, 331
Cortical pathology, 330–332
Coupled Bloch equations, 195
COWAT. *See* Controlled oral word association test (COWAT)
CP. *See* Choroid plexuses (CP)
CPMG sequence. *See* Carr-Purcell-Meiboom-Gill sequence (CPMG sequence)
CPVS. *See* Cerebral perivascular spaces (CPVS)
CR. *See* Cognitive reserve (CR)
Creatine (Cr), 289–291
Cross-sectional measures, 112
Cross-sectional optical coherence tomography, 368
Cross-sectional studies, 280
Cross-validation (CV), 390–391, 399f
CSF. *See* Cerebrospinal fluid (CSF)
CT. *See* Computer tomography (CT)
CV. *See* Cross-validation (CV)
CVLT. *See* California verbal learning test (CVLT)
CVR. *See* Cerebrovascular reactivity (CVR)
CVS. *See* Central vein sign (CVS)
CVSnet, 401–402
CXCR2, 9
Cytokines, 38–39

D

DAN. *See* Dorsal attention network (DAN)
DARE. *See* Discriminative analysis of regional evolution (DARE)
Data, 84–85
  data-sharing, 384–385
  sharing, 409
DAWM. *See* Diffusely abnormal white matter (DAWM)
DCE. *See* Dynamic contrast enhancement (DCE)
DCs. *See* Dendritic cells (DCs)
Decision-making process, 409–410

Deep gray matter (DGM), 222
  confounding effects of myelin in, 231
  gadolinium retention in, 230
  nonheme iron concentrations in, 227–229
  regional content of nonheme iron in, 229
  segmentation, 236
  subvoxel distribution of iron in, 230
Deep gray rating via artificial intelligence (DeepGRAI), 396–397
Deep learning (DL), 385–386
  architectures for image analysis, 387–389, 389f
Deep neural network architectures, 387
DeepGRAI. *See* Deep gray rating via artificial intelligence (DeepGRAI)
Default-mode network (DMN), 252
Delis-Kaplan executive functioning system sorting test (DKEFS Sorting), 53
DELIVER-MS. *See* Determining Effectiveness of earLy Intensive *Vs.* Escalation Approaches for Treatment of Relapsing-Remitting Multiple Sclerosis (DELIVER-MS)
Dementia, 384–385
Demyelinating plaques, 207–208
Demyelination, 163, 302
Dendritic cells (DCs), 5–7
Depression, 350
Determining Effectiveness of earLy Intensive *Vs.* Escalation Approaches for Treatment of Relapsing-Remitting Multiple Sclerosis (DELIVER-MS), 74
DGM. *See* Deep gray matter (DGM)
Dice coefficient, 391–392
Differential diagnosis of multiple sclerosis, magnetic resonance imaging in, 98–100
Diffuse spinal cord lesions, 97
Diffusely abnormal white matter (DAWM), 306–307
Diffusion, 148–151
  effect, 149
Diffusion MRI (dMRI), 232
Diffusion tensor imaging (DTI), 149, 183–184
Diffusion weighted imaging (DWI), 96, 149
Digital twins, 412
  approach, 412
Dimethyl fumarate (DMF), 78–79, 235
DIR. *See* Double inversion recovery (DIR)
Diroximel fumarate (DOF), 78–79
DIS. *See* Disease activity in space (DIS)
Disability, 253–254
  worsening, 276–277
Discriminative analysis of regional evolution (DARE), 231
Disease activity, magnetic resonance imaging in prediction of, 100

Disease activity in space (DIS), 94
Disease activity in time (DIT), 94
Disease modifying therapy (DMT), 45, 124
   magnetic resonance spectroscopy in DMT evaluation, 300–301
Disease modifying treatments, 307–308
Disease monitoring, spinal cord MRI assessment for, 182–183
Disease progression, 166, 412
Disease-modifying therapies (DMTs), 61, 175–176, 300–301, 353–355, 369
Disease-modifying treatments (DMTs), 71
Distribution volume ratio (DVR), 355
DIT. *See* Disease activity in time (DIT)
DKEFS Sorting. *See* Delis-Kaplan executive functioning system sorting test (DKEFS Sorting)
DL. *See* Deep learning (DL)
DMF. *See* Dimethyl fumarate (DMF)
DMN. *See* Default-mode network (DMN)
dMRI. *See* Diffusion MRI (dMRI)
DMT. *See* Disease modifying therapy (DMT)
DMTs. *See* Disease-modifying therapies (DMTs); Disease-modifying treatments (DMTs)
DOF. *See* Diroximel fumarate (DOF)
Dorsal attention network (DAN), 255
Double inversion recovery (DIR), 94–96, 167, 384
"Downstream" tasks, 384
Drugs, 72
   MRI in drug safety surveillance, 106–108, 107t
DSC. *See* Dynamic susceptibility contrast (DSC)
DTI. *See* Diffusion tensor imaging (DTI)
DVR. *See* Distribution volume ratio (DVR)
DWI. *See* Diffusion weighted imaging (DWI)
Dynamic contrast enhancement (DCE), 150–151, 268
   magnetic resonance imaging, 270–272, 271f
Dynamic functional connectivity, 257
Dynamic susceptibility contrast (DSC), 150–151, 268
   magnetic resonance imaging, 268–270, 269f
Dysmenorrheal, 75–77

E

EAE. *See* Experimental autoimmune encephalomyelitis (EAE)
EBNA1. *See* Epstein–Barr virus nuclear antigen 1 (EBNA1)
EBV. *See* Epstein–Barr virus (EBV)
Echo, 142
Echo time (TE), 142, 163, 209–210, 223–224, 291–292
Echo-planar imaging (EPI), 401–402
Ectopic B-cell follicles, 332
EDSS. *See* Expanded disability status scale (EDSS)
Electro-Magnetic Tissue Properties (EMTP), 225
Electromotive force, 140–141

EMA. *See* European Medicines Agency (EMA)
EMTP. *See* Electro-Magnetic Tissue Properties (EMTP)
Encoded envelope (Env), 20
Encoder–decoder model, 388–389
Endothelin-1 vasoconstricting peptide, 275–276
Env. *See* Encoded envelope (Env)
EPI. *See* Echo-planar imaging (EPI)
Epstein–Barr virus (EBV), 4, 39
Epstein–Barr virus nuclear antigen 1 (EBNA1), 16
Escalation approach, 72–74
   graphical presentation of, 73f
Essential nutrients, 150
European Medicines Agency (EMA), 72, 384–385
Ex vivo studies, evidence from, 331
Excitation process, 144
Expanded disability status scale (EDSS), 71–72, 160, 182–183, 276–277, 298, 332, 347
Experimental autoimmune encephalomyelitis (EAE), 6
External capsule, 160
eXtreme Gradient Boosting method, 406

F

F-18 florbetaben, 356
FA. *See* Fractional anisotropy (FA)
Faraday's law, 140–141
Fast low angle shot (FLASH), 193
Fast spin-echo (FSE), 167, 177–179
FastSurferCNN, 396
Fatigue, 350
FC. *See* Functional connectivity (FC)
fcMRI. *See* Functional connectivity magnetic resonance imaging (fcMRI)
FDA. *See* United States Food and Drug Administration (FDA)
FDG. *See* Fluorodeoxyglucose (FDG)
Federated disentangled representation learning (FedDis), 411
Federated learning, 411
Ferumoxytol, 214–215
Fibrin deposition, 23
Fick's law, 149
Fick's second law, 149
FID. *See* Free induction decay (FID)
Field strength, 159–160
Fingolimod, 80–82, 126–127, 355
FIRST. *See* FMRIB Integrated Registration and Segmentation Tool (FIRST)
FLAIR. *See* Fluid-attenuated inversion recovery (FLAIR)
FLASH. *See* Fast low angle shot (FLASH)
Fluid shifts, 111
Fluid-attenuated inversion recovery (FLAIR), 41, 96, 146–147, 147f, 163, 209–210, 324

Fluid-attenuated inversion recovery (FLAIR) (*Continued*)
   FLAIR*, 209–210
Fluorodeoxyglucose (FDG), 350
Flutemetamol, 356
fMRI. *See* Functional magnetic resonance imaging (fMRI)
FMRIB Integrated Registration and Segmentation Tool (FIRST), 111
FMRIB Software Library for Mrs (FSL for Mrs), 294
Focal white matter damage, 232–235
Former technique, 274–275
Fornix-stria terminalis, 160
Forward exchange rate, 200
Foundational models, 411–412
Fourier phase factor, 141
Fourier transform, 141, 152–153
Fourier transformation, 362
FPN. *See* Frontoparietal network (FPN)
Fractional anisotropy (FA), 149–150
Free induction decay (FID), 142–144
FreeSurfer, 395–396
Frontoparietal network (FPN), 253–254
FSE. *See* Fast spin-echo (FSE)
Fully-connected layers, 387–388
Functional brain changes over time, 255–256
Functional connectivity (FC), 184, 250, 253–255
   change
      to cognitive systems, 254–255
      to sensorimotor systems, 253–254
Functional connectivity magnetic resonance imaging (fcMRI), 350
Functional disconnection hypothesis, 350
Functional magnetic resonance imaging (fMRI), 153, 183–184, 249–250
   advanced network analyses, 257–258, 258f
   dynamic functional connectivity, 257
   functional brain changes over time, 255–256
      longitudinal resting-state fMRI, 256
      longitudinal task-based fMRI, 255–256
   functional connectivity, 253–255
   functional reorganization, 250–253
      changes to cognitive systems, 251–253
      changes to sensorimotor systems, 250–251
      default-mode network, 252f
   network collapse, 257
Functional networks, 253–254

G

GA. *See* Glatiramer acetate (GA)
GABA. *See* Gamma-aminobutyric acid (GABA)
Gadolinium (Gd), 182–183, 230, 393
   Gd-based chelates, 268
   retention in deep gray matter, 230
Gadolinium diethylenetriamine penta-acetic acid (Gd-DTPA), 322
Gadolinium-based contrast agents (GBCAs), 44, 106, 181, 230, 393
   reduction of, 393–394, 403f
Gadolinium-enhancing lesions, 126–127, 164
   identification of, 233
Gamma-aminobutyric acid (GABA), 293
Ganglion cell layer (GCL), 364–365
Ganglion cell-inner plexiform layer (GCIPL), 364–365
Ganglion cells, 364–365
GANs. *See* Generative adversarial networks (GANs)
Gastrointestinal tolerability, 78
GBCAs. *See* Gadolinium-based contrast agents (GBCAs)
GCIPL. *See* Ganglion cell-inner plexiform layer (GCIPL)
GCL. *See* Ganglion cell layer (GCL)
Generative adversarial networks (GANs), 394
Genetic polymorphism, 346–347
Glatiramer acetate (GA), 77, 235
Glial cell adhesion molecule (GlialCAM), 16
Glial imaging targets, 356
GlialCAM. *See* Glial cell adhesion molecule (GlialCAM)
Glioma, 393–394
Globus pallidus, 228
Glucose hypometabolism, 356
Glutamate (Glu), 289, 291
Glutamine (Gln), 289, 291
GM. *See* Gray matter (GM)
GM-CSF. *See* Granulocyte–macrophage colony stimulating factor (GM-CSF)
Go-NoGo task, 252–253
Gradient echo (GRE), 41, 142–144
   sequences, 325
Gradient echoes to each spin echo (GRASE), 304
Gradient field (G), 140–141
Gradient system, 141
Gradient-echo sequences, 270
Gradient–recalled echo (GRE), 222
Granulocyte–macrophage colony stimulating factor (GM-CSF), 6
Graph theoretical analysis, 257–258
Graph theory, 258
GRASE. *See* Gradient echoes to each spin echo (GRASE)
Gray matter (GM), 57, 94–96, 146–147, 161, 199–200, 222, 268, 291, 323
   pathology in multiple sclerosis, 347
GRE. *See* Gradient echo (GRE); Gradient–recalled echo (GRE)

Guinea pigs, 305
Gyromagnetic ratio, 140

H

Hazard ratio (HR), 369
HC. *See* Healthy controls (HC)
Health system, 73
Healthcare professionals, training, 411
Healthy controls (HC), 307
Healthy volunteers, 250–251
Hepatitis B virus, 84
HERVs. *See* Human endogenous retroviruses (HERVs)
HFI. *See* High-field imaging (HFI)
High-efficacy therapies, 74
High-field imaging (HFI), 322
    clinical application of, 333–334
    multiple sclerosis–induced disease under microscope of, 323–333
    technical challenges associated with use of, 334
    visualizing central vein sign using, 328–329
    in vivo improves cortical lesions visibility and allows detection of subpial disease, 331
High-resolution synthetic images, creating, 395–396
Horizontal cells, 365
HR. *See* Hazard ratio (HR)
Hub-analyses, 258
Human endogenous retroviruses (HERVs), 20
Hydrogen, 288–289
Hyperconnectivity, 254–255
Hyperintense T2 lesions, 96
Hypoconnectivity, 254–255
Hypoperfusion, 275, 279–280

I

ICRs. *See* Isolated cognitive relapses (ICRs)
IFN-γ. *See* Interferon-γ (IFN-γ)
ihMT. *See* Inhomogeneous magnetization transfer (ihMT)
ihMTR. *See* Inhomogeneous magnetization transfer ratio (ihMTR)
IL-121. *See* Interleukin-121 (IL-121)
ILCs. *See* Innate lymphoid cells (ILCs)
Image analysis, deep learning architectures for, 387–389
Image formation, 141
Imaging, 192
    biomarkers in multiple sclerosis, 208
    technologies, 378
Immune cells, 14–15, 23
Immune reconstitution inflammatory syndrome (IRIS), 107–108
Immune therapies, 82
Immunoglobulin-dependent effects, 16
Immunology of multiple sclerosis, 4–18
Immunomodulatory treatment, 106–107, 109
In vivo imaging, 41–43
In vivo visualization of meningeal inflammation, 332
Induction approach, 72–74
    graphical presentation of, 73f
    personalized decision, 73–74
Inflammatory cells, 349
Inflammatory lesion formation and evolution, 21–22
Inflammatory processes in MS, 38–39
Infusion-associated reactions, 86
Infusion-related reactions, 84
Inhomogeneous magnetization transfer (ihMT), 197
    in multiple sclerosis, 200–201
Inhomogeneous magnetization transfer ratio (ihMTR), 197
INL. *See* Inner nuclear layer (INL)
Innate immune cell. *See also* Adaptive immune cell
    responses in multiple sclerosis, 5–18, 11f
    role of, 5–10
        bone marrow-derived myeloid cells, 5–7
        microglia, 7–8
        neutrophils, 9–10
        NK cells and ILCs, 8–9
Innate immune system, 5
Innate lymphoid cells (ILCs), 8–9
Inner nuclear layer (INL), 364–365
Inner plexiform layer (IPL), 364–365
Interferon beta 1-a, 74–77
    disease-modifying therapies important, 75f
    MS therapies modified from multiple sclerosis therapy consensus group, 76t
Interferon beta 1-b, 74–77
    disease-modifying therapies important, 75f
    MS therapies modified from multiple sclerosis therapy consensus group, 76t
Interferons, 75
Interferon-β, 123–124
Interferon-γ (IFN-γ), 6
Interleukin-121 (IL-121), 5
InterModal Segmentation Analysis (MIMoSA), 401–402
International Society for Magnetic Resonance in Medicine, 393
Interplexiform cells, 365
Intracortical lesions, 22–23
Inversion recovery sequences, 271
Inversion time (TI), 163
IPL. *See* Inner plexiform layer (IPL)
IRIS. *See* Immune reconstitution inflammatory syndrome (IRIS)
Iron, 153, 308, 322
    concentrations, 226, 229

Iron (*Continued*)
  localization, 325
  rim, 325–326
  subvoxel distribution of iron in deep gray matter, 230
Ischemic lesions, 208
Isolated cognitive relapses (ICRs), 59–61

J

JCV. *See* John Cunningham virus (JCV)
JLO. *See* Judgment of line orientation test (JLO)
John Cunningham virus (JCV), 107
Judgment of line orientation test (JLO), 53
Juxtacortical lesions, 22–23

K

Kernels, 387–388
Kety's equation, 151
K-nearest neighbours (k-NNs), 386–387
k-NNs. *See* K-nearest neighbours (k-NNs)
K-space vector, 141

L

Laboratory screening, 79
Lactate (Lac), 289, 291, 298
Langevin diamagnetism, 223
Large language models (LLMs), 388–389
Larmor frequency, 140–142, 145
  modulation, 141
Larmor precession, 140–141
Late onset MS (LOMS), 39
Lattice, 144
LCModel. *See* Linear combination of model spectra (LCModel)
Leave-one-out cross-validation (LOOCV), 390–391
Leflunomide, 79
Leptomeningeal enhancement (LME), 46, 324, 332–333
  histopathologic and clinical significance of, 332–333
  in vivo visualization of meningeal inflammation, 332
Lesions, 124, 148, 161, 198, 208, 232, 305–307
  evolution, 279
  segmentation, 397–402
    central vein sign, 401–402
    cortical lesions, 400–401
    detecting longitudinal changes in lesions, 402
    paramagnetic rim lesions, 398–400
    T2 FLAIR lesions, 397–398
  volume, 288
LFB. *See* Luxol Fast Blue (LFB)
LIME. *See* Local Interpretable Model-agnostic Explanations (LIME)
Linear combination of model spectra (LCModel), 293–294

Liver
  function tests, 79
  parameters, 77
LLMs. *See* Large language models (LLMs)
LME. *See* Leptomeningeal enhancement (LME)
LMICs. *See* Low–and middle-income countries (LMICs)
Local Interpretable Model-agnostic Explanations (LIME), 409–410
Logistic regression, 404
LOMS. *See* Late onset MS (LOMS)
Longitudinal optical coherence tomography, 368
Longitudinal relaxation (T1 relaxation), 144–145
Longitudinal resting-state functional magnetic resonance imaging, 256
Longitudinal task-based functional magnetic resonance imaging, 255–256
LOOCV. *See* Leave-one-out cross-validation (LOOCV)
Low–and middle-income countries (LMICs), 377
Luxol Fast Blue (LFB), 305
Lymphocytes, 87
Lymphopenia, 84, 88

M

MACFIMS. *See* Minimal Assessment of Cognitive Function in MS (MACFIMS)
Machine learning (ML), 308, 385–386
  algorithms, 380
  techniques, 397–398
Macromolecular, 198
  pool fraction, 198
  protons, 192
  fraction, 200
Macromolecules, 192, 194
Macrophages, 5–6, 325
Magnesium, 322
Magnetic fields (B_0), 140–141
Magnetic resonance (MR), 97
Magnetic resonance imaging (MRI), 2, 4, 41, 52, 71–72, 94, 123–124, 139–140, 159–160, 175–176, 192, 222, 249–250, 267, 288, 322, 344, 377–378
  black holes, 163–164
  chronic active lesions, 165–166
  contrast-enhancing lesions, 164
  cortical lesions, 166–169
  conventional MRI markers in multiple sclerosis, 124–127
  correlates of cognitive impairment in MS, 57
  in differential diagnosis of multiple sclerosis, 98–100
  in drug safety surveillance, 106–108, 107t
  emerging imaging biomarkers, 128–130
    atrophied T2-lesion volume, 129–130
    chronic active lesions, 128–129

cortical lesions, 129
image formation, 140–144
  basics of MRI pulse sequences, 142–144
  gradient echo pulse sequence, 143f
  gradient field, 141
  main magnetic field, 140–141
  radiofrequency field, 141–142
  spin echo pulse sequence, 143f
improving MRI acquisition protocols and image quality, 393–396
  creating high-resolution synthetic images, 395–396
  reduction of gadolinium-based contrast agents, 393–394
  synthetic double inversion recovery imaging, 394–395
markers for predicting progression of cognitive dysfunction and patient outcomes, 62–63
methods, 207–208
for monitoring disease activity, 100–106, 101t
  conventional MRI predictors of future disease activity, 103t
  recommendations and challenges of, 105t
perfusion studies in multiple sclerosis, 274–281
  cerebrovascular reactivity, 280–281
  cognitive performance, 277–279
  hypoperfusion and brain atrophy, 279–280
  lesion evolution, 279
  perfusion differences across multiple sclerosis phenotypes, 276
  physical disability, 276–277
in prediction of disease activity, 100
relationship between MRI signal and tissue properties, 144–154
  biological water motion, 148–151
  relaxation, 144–145
  T1-, T2-, and proton density weighting in MR imaging, 146–148
  tissue magnetism, 152–154
sequences for cortical lesion detection, 167–169, 168f
signal detection, 140–141
techniques, 384
  arterial spin labelling, 272–274
  dynamic contrast enhancement MRI, 270–272
  dynamic susceptibility contrast MRI, 268–270
  for estimating cerebral perfusion, 268–274
  recommendations on use of MRI techniques for evaluation of MS, 46
visualizing chronic active lesions on, 325–326
white matter lesions, 160–163
Magnetic resonance imaging in multiple sclerosis (MAGNIMS), 175–176, 327
diagnosis, 94–98, 95t

Magnetic resonance sequences for T2 white matter lesions imaging, 162–163, 162f
Magnetic resonance spectroscopy (MRS), 183–184, 288–289, 299–300, 300f, 378
  data acquisition, 292–293
  data analysis, 293–294
  factors affecting reproducibility of Mrs and consensus protocols, 294–295
  findings in MS
    clinical trials and disease modifying therapy evaluation, 300–301
    lesions, 295–298, 296f
    normal-appearing white matter and gray matter, 298–299
  future directions in Mrs for MS research and clinical care, 301–302
  limitations of Mrs studies in MS, 301
Magnetic susceptibility, 227, 233
  of tissue, 223
Magnetization prepared two rapid acquisition gradient echoes (MP2RAGE), 164, 168
Magnetization transfer (MT), 192
  clinical use of, 193
  experiment, 192–193
    qualitative representation of spectra of free water and bound protons, 193f
  history, 192
  imaging, 231
  inhomogeneous MT in multiple sclerosis, 200–201
  MTR in multiple sclerosis, 198–200
  quantifying MT effect, 193–197
    ihMT, 197
    magnetization transfer saturation, 196–197
    MTR, 194
    qMT models, 194–196
  quantitative MT in multiple sclerosis, 200
  saturation in multiple sclerosis, 200
  in spinal cord and optic nerve, 201
  validation of MT–derived parameters as myelin markers, 198
Magnetization transfer ratio (MTR), 183–184, 194
  in multiple sclerosis, 198–200
Magnetization-prepared fluid-attenuated inversion recovery (MPFLAIR), 333
Magnetization-prepared rapid gradient echo (MPRAGE), 163–164, 324–325
Magnetostatic theory, 225
MAGNIMS. See Magnetic resonance imaging in multiple sclerosis (MAGNIMS)
MAIT cells. See Mr1-restricted mucosal associated invariant T cells (MAIT cells)
Major histocompatibility complex (MHC), 4
Mass effect, 107–108

Maxwell equations, 153
McDonald criteria, 43–44
MD. See Mean diffusivity (MD)
Mean diffusivity (MD), 149–150
Mean squared error (MSE), 390
Mean transit time (MTT), 269
Medical imaging, artificial intelligence in, 385–393
    assessment of model performance, 390–392
        cross-validation, 390–391
            metrics used to assess model performance, 391–392
    deep learning architectures for image analysis, 387–389, 390f
    frequently used machine learning algorithms, 388f
    potential benefits of using AI and neuroimaging in MS, 392–393
    supervised vs. unsupervised machine learning, 386–387
Membrane transport, 290–291
Meningeal inflammation, in vivo visualization of, 332
Meningeal lymphatics, 332
Meningioma, 393–394
Menstrual disorders, 75–77
Metaanalysis, 365
Metabolic changes, 184
Metabolite monomethyl fumarate (MMF), 78
Metabolites, 293, 299
    abnormalities, 298
    concentrations, 294
    of interest for multiple sclerosis studies, 289–291
        choline, 289–290
        creatine, 290–291
        glutamate and glutamine, 291
        lactate, 291
        myo-inositol, 291
        N-acetylaspartate, 289
Metanalytic data, 332
Metrorrhagia, 75–77
MHC. See Major histocompatibility complex (MHC)
MICRO. See Microscopic in vivo contrast revealed origins (MICRO)
Microbleeds, 208, 235
Microcystic macular edema (MME), 366–367
Microdistribution, 230
Microglia, 7–8, 325, 345–346
Microglial activation and translocator protein–positron emission tomography, 345–346
Microscope of high-field imaging, multiple sclerosis–induced disease under, 323–333
Microscopic in vivo contrast revealed origins (MICRO), 214
Microvascular in vivo contrast revealed origins, 214–215

Midbrain, 94–96
Middle cerebellar peduncle, 43
Migraines, 329
Minimal Assessment of Cognitive Function in MS (MACFIMS), 53, 277–278
ML. See Machine learning (ML)
MME. See Microcystic macular edema (MME)
MMF. See Metabolite monomethyl fumarate (MMF)
MOGAD. See Myelin oligodendrocyte antibody disease (MOGAD); Myelin oligodendrocyte glycoprotein antibody associated disease (MOGAD)
Monitoring disease activity
    conventional MRI predictors of future disease activity, 103t
    magnetic resonance imaging for, 100–106, 101t
    recommendations and challenges of MRI monitoring of MS disease activity, 105t
Monoamine oxidase B, 356
Monoclonal antibody medication natalizumab, 353–355
Monocytes, 5–6
    derived DCs, 6
Monomethyl fumarate, 78–79
Motor impairment, prediction of, 406–408
"Motor neuron" pattern, 180
MP2RAGE. See Magnetization prepared two rapid acquisition gradient echoes (MP2RAGE)
MPFLAIR. See Magnetization-prepared fluid-attenuated inversion recovery (MPFLAIR)
MPM. See Multiparameter mapping (MPM)
MPRAGE. See Magnetization-prepared rapid gradient echo (MPRAGE)
MR. See Magnetic resonance (MR)
Mr1-restricted mucosal associated invariant T cells (MAIT cells), 15–16
MRI. See Magnetic resonance imaging (MRI)
MRS. See Magnetic resonance spectroscopy (MRS)
MS. See Multiple sclerosis (MS)
MSE. See Mean squared error (MSE)
MSFC. See Multiple sclerosis functional composite (MSFC)
MSLAST algorithm. See Multiple Sclerosis Lesion Analysis at Seven Tesla algorithm (MSLAST algorithm)
MSNQ. See Multiple sclerosis neuropsychological screening questionnaire (MSNQ)
MSSS. See Multiple sclerosis severity score (MSSS)
MT. See Magnetization transfer (MT)
MTR. See Magnetization transfer ratio (MTR)
MTT. See Mean transit time (MTT)
Müller cells, 365–366
Multiband sequences, 270

Multichannel detectors, 322
Multicontrast deformation approach, 236
Multidelay sequences, 272
Multiecho data, 225–226
Multiecho sequences, 223–224, 270
Multimodal DL technique, 407
Multimodal perfusion-diffusion MRI study, 280
Multimodal studies, 275–276
Multiparameter mapping (MPM), 196–197
Multiple imaging studies, 232–233
Multiple sclerosis (MS), 1, 3–4, 38, 51–52, 71, 94, 123–124, 139–140, 159–160, 175–176, 193–194, 207–208, 249–250, 267, 288, 322, 344, 362, 377, 384
   acute cognitive decline as marker of active disease in, 58–61
   applications of quantitative susceptibility mapping in, 235–236
     deep gray matter segmentation, 236
     microbleeds, 235
     oxygen extraction fraction measurements, 235
   artificial intelligence in MS neuroimaging, 393–408
   central vein sign, 327–330
   characteristics of spinal cord lesions in, 176–180, 178t
   clinical course of, 40–41
   clinical symptoms of, 38
   cognitive impairment, 299–300, 300f
   consequences of cognitive impairment in, 56–57
   conventional MRI markers in, 124–127
   cortical pathology, 330–332
     clinical relevance of cortical lesions, 332
     cortical disease in multiple sclerosis, 330f
     evidence from ex vivo studies, 331
     using HFI in vivo improves cortical lesions visibility and allows detection of subpial disease, 331
   degree of cognitive change during and after, 63–64
   development and mechanisms of, 38–39
   diagnosis, prognosis, clustering, 403–408
     classification, subtyping, and clustering, 405–406
     differential diagnosis, 403–405
     prediction of motor and cognitive impairment, 406–408
   diagnostic criteria, 43–44
   ethnic and racial disparities in, 39–40
   factors protecting against cognitive impairment in, 57–58
   findings in, 227–228
   future directions in Mrs for MS research and clinical care, 301–302
   historical review of cognition in, 52–58
     cognitive dysfunction in, 54–58

     neuropsychological assessment, 52–54
   imaging biomarkers in, 208, 209f
   immunology of, 4–18
     adaptive and innate immune cell responses in, 5–18
     autoimmunity in, 4–5
     meet players, 5–18
   improved visibility of white matter lesions, 324–325
   improving MRI acquisition protocols and image quality, 393–396
   inhomogeneous magnetization transfer in, 200–201
   leptomeningeal enhancement, 332–333
   lesions
     evolution, 323f
     Mrs findings in, 295–298, 297f
     segmentation, 397–402
     water content as new biomarker for, 211–213
   limitations of magnetic resonance spectroscopy studies in, 301
   linking MS pathophysiology and in vivo imaging, 41–43
   magnetic resonance imaging in, 94–98, 95t
   magnetic resonance spectroscopy in MS clinical trials, 300–301
   magnetization transfer saturation in, 200
   metabolites of interest for MS studies, 289–291
   Mrs findings in MS normal-appearing white matter and gray matter, 298–299
   MS–induced disease under microscope of high-field imaging, 323–333
   MTR in, 198–200
   MRI
     correlates of cognitive impairment in, 57
     gray matter, 199–200
     in differential diagnosis of, 98–100
     lesions, 198
     markers for predicting progression of cognitive dysfunction and patient outcomes, 62–63
     NAWM, 198–199
     perfusion studies in, 274–281
   myelin water fraction
     in different MS tissues, 305–307
     in MS subtypes, 307
   OCT
     cross-sectional OCT, 368
     longitudinal OCT, 368
     marker of MS–associated neuroaxonal damage, 367–369
     in progressive multiple sclerosis, 368–369
   open questions and future directions, 61
   optimally define and recognize cognitive change in people with, 61–62
   paramagnetic rim lesions, 325–327

Multiple sclerosis (MS) (*Continued*)
  pathophysiology of, 18–24
    from autoimmunity to pathology, 18–21
      cortical lesions in, 23f
      inflammatory lesion formation and evolution, 21–22
      pathogenesis in context of progressive MS, 22–24
      remyelination, 24
  perfusion differences across MS phenotypes, 276
  positron emission tomography imaging of nonimmune mechanisms in, 356
  potential benefits of using artificial intelligence and neuroimaging in, 392–393
  practical issues and limitations affecting clinical application of OCT in, 370–371
  precursors to, 45
  quantitative magnetization transfer in, 200
  recommendations
    on timing of imaging for monitoring of MS disease activity and progression, 45
    on use of newer MRI techniques for evaluation, 46
  tissue segmentation/volumetrics, 396–397
  translate lessons from literature to clinical practice, 64
  translocator protein–positron emission tomography
    and gray matter pathology in, 347
    and prognostication in, 351–352
  treatment
    alemtuzumab, 85–88
    B-cell depletion, 82–84
    conflict of interest, 73–74
    dimethyl fumarate and diroximel fumarate, 78–79
    escalation *vs.* induction approach, 72–74
    glatiramer acetate, 77
    interferon beta 1-a and interferon beta 1-b, 74–77
    natalizumab, 84–85
    of cognitive impairment in, 61
    sphingosine 1-phosphate–receptor modulators, 80–82
    teriflunomide, 79
Multiple sclerosis functional composite (MSFC), 276–277, 298–299
Multiple Sclerosis Lesion Analysis at Seven Tesla algorithm (MSLAST algorithm), 400–401
Multiple sclerosis neuropsychological screening questionnaire (MSNQ), 58
Multiple sclerosis severity score (MSSS), 276–277
Multislice imaging, 334
Multivoxel spectroscopy, 292
Musculoskeletal imaging, 193
MWF. *See* Myelin water fraction (MWF)
MWI. *See* Myelin water imaging (MWI)
Myelin, 223, 308
  confounding effects of myelin in deep gray matter, 231
  content, 184
  debris, 8
  myelin-sensitive imaging techniques, 228
  sheaths, 150
  validation of magnetization transfer–derived parameters as, 198
  visualization, 183–184
  water, 302, 303f
Myelin oligodendrocyte antibody disease (MOGAD), 177–179
Myelin oligodendrocyte glycoprotein antibody associated disease (MOGAD), 97, 366–367
Myelin water fraction (MWF), 302, 307–308
  correlations with clinical measures, 307
  in different multiple sclerosis tissues, 305–307
  measurement of, 303–305, 304f
  in multiple sclerosis subtypes, 307
  validation, 305
Myelin water imaging (MWI), 184, 302
Myeloid cells, 12, 20
Myo-inositol (mI), 289, 291

N

NA. *See* Number of averages (NA)
NAA. *See* N-acetylaspartate (NAA)
N-acetyl cysteine, 279
N-acetylaspartate (NAA), 289
NADPH. *See* Nicotinamide Adenine Dinucleotide Phosphate (NADPH)
NAGM. *See* Normal-appearing gray matter (NAGM)
NAIMS. *See* North American Imaging in Multiple Sclerosis (NAIMS)
Natalizumab, 84–85, 126–127
Natural killer (NK) cells, 8–9
NAWM. *See* Normal-appearing white matter (NAWM)
N-back task, 252
NEDA. *See* No evidence of disease activity (NEDA)
Neoplastic processes, 106–107
Network
  collapse, 257
  efficiency, 257–258
Neural tissue, 125, 127
Neuroaxonal damage, retinal thinning indicates ongoing, 368
Neuroaxonal degeneration, 362
Neuroaxonal reserve, retinal thickness indicates, 368
Neuro-Behçet's disease, 180
Neurochemical abnormalities, 350, 356
Neurocognitive disorder, 61
Neurodegeneration, 41–43, 109, 324, 368
Neurodegenerative processes, 161

Neuroendocrine, 350
Neurofilament light chains (NfL chains), 45
Neuroimaging, 43
  biomarkers, 405
  potential benefits of using neuroimaging in multiple sclerosis, 392–393
Neuroinflammation, 111
Neurologic disorders, 140
Neurologic reserve, 63–64
Neurologic Software Tool for REliable Atrophy Measurement (NeuroSTREAM), 109–110
Neurological disability, 1
Neurological symptoms, 21
Neuromyelitis optica (NMO), 297–298, 329–330
Neuromyelitis optica spectrum disorders (NMOSD), 97, 177–179, 208, 329–330, 366–367, 403
Neuropathological studies, 349
Neuropsychological assessment, 52–54
Neuropsychological tests, 54
Neuropsychology, 55
Neurosarcoidosis, 403
NeuroSTREAM. See Neurologic Software Tool for REliable Atrophy Measurement (NeuroSTREAM)
Neurotransmission, 290–291
Neurovascular coupling integrity, 280
Neutrophils, 9–10
NfL chains. See Neurofilament light chains (NfL chains)
Nicotinamide Adenine Dinucleotide Phosphate (NADPH), 8
n-IRLs. See Noniron rim containing lesions (n-IRLs)
NMO. See Neuromyelitis optica (NMO)
NMOSD. See Neuromyelitis optica spectrum disorders (NMOSD)
NMR. See Nuclear magnetic resonance (NMR)
NNLS method. See Nonnegative least-squares method (NNLS method)
No evidence of disease activity (NEDA), 369
Nonconventional sequences, 129
Nonheme iron, 223
  concentrations in deep gray matter, 227–229
    comparison with methods, 228–229
    findings in multiple sclerosis, 227–228
    regional content of nonheme iron in deep gray matter, 229
Nonimmune mechanisms in multiple sclerosis, positron emission tomography imaging of, 356
Noninvasive assessment, 345
Noniron rim containing lesions (n-IRLs), 128–129
Nonnegative least-squares method (NNLS method), 303–304

Normal-appearing gray matter (NAGM), 166–167, 288–289, 324, 403–404
Normal-appearing white matter (NAWM), 23, 150, 198–199, 231–232, 269, 288–289, 306, 324, 345–346, 403–404
Normalized brain volume, 112
North American Imaging in Multiple Sclerosis (NAIMS), 235
Nrf2. See Nuclear factor (erythroid-derived 2)-like 2 (Nrf2)
Nuclear factor (erythroid-derived 2)-like 2 (Nrf2), 78
Nuclear magnetic resonance (NMR), 192
Number of averages (NA), 292

O

Ocrelizumab, 82–83, 126–127
OCT. See Optical coherence tomography (OCT)
Odds ratio (OR), 365–366
OEF. See Oxygen extraction fraction (OEF)
Ofatumumab, 82
  efficacy, 83
Off-resonance
  pulses, 192
  saturation, 193
Oligodendrocytes, 325
ON. See Optic neuritis (ON)
ONL. See Outer nuclear layer (ONL)
ONs. See Opticnerves (ONs)
^{15}O-positron emission tomography (PET), 235
Optic nerve, 201
  lesions, 99
  magnetization transfer in, 201
Optic neuritis (ON), 59, 362
  optical coherence tomography in, 364f, 365–367
Optical coherence tomography (OCT), 361–362, 378
  marker of MS-associated neuroaxonal damage, 366f, 367–369
  in optic neuritis, 365–367
  practical issues and limitations affecting clinical application of OCT in MS, 370–371
  in progressive multiple sclerosis, 368–369
  retina as window to brain, 363–365
  technical principles of, 362–363, 363f
  in treatment monitoring, 369–370
Opticnerves (ONs), 43
Optional sequences, 177–179
OR. See Odds ratio (OR)
Outer nuclear layer (ONL), 364–365
Oxidative stress, 228
Oxygen, 150
Oxygen extraction fraction (OEF), 235
  measurements, 235
Ozanimod, 80–82

P

Paced auditory serial addition test (PASAT), 53, 251
Parallel imaging, 334
Paramagnetic metals, 322
Paramagnetic rim, 46
Paramagnetic rim lesions (PRLs), 62, 128, 234–235, 234f, 325–327, 326f, 392–393, 398–400
    clinical significance of, 326–327
    visualizing chronic active lesions on magnetic resonance imaging, 325–326
PASAT. See Paced auditory serial addition test (PASAT)
pASL. See Pulsed arterial spin labelling (pASL)
Pathophysiological MS processes, 2
Patients, 126–127
    data, 310
    MRI markers for predicting progression of patient outcomes, 62–63
    spinal cord MRI assessment for patient's prognosis, 182–183
pCASL. See Pseudo-continuous arterial spin labelling (pCASL)
PDw. See Proton density-weighted (PDw)
Pearson's correlations, 392
People with multiple sclerosis (pwMS), 51–52, 322
People with primary progressive multiple sclerosis (pwPPMS), 333
People with secondary progressive form of multiple sclerosis (pwSPMS), 331
Perfusion, 148–151, 160
    differences across multiple sclerosis phenotypes, 276
    indices, 277–278
    MRI
        perfusion studies in multiple sclerosis, 274–281
        techniques for estimating cerebral perfusion, 268–274
    perfusion-weighted images, 151, 273–274
Perivascular fibrin deposition, 23
Periventricular, 43
    lesions, 94–96
Personalized decision, 73–74
Personalized treatment plans, 412
PET. See Positron emission tomography (PET)
Phase unwrapping process, 225–226
Phase wraps, 225–226
Phase-sensitive inversion recovery (PSIR), 94–96, 167–168
Photoreceptors, 364
Physical disability, 276–277
PIRA. See Progression independent of relapse activity (PIRA)
Plasmacytoid DCs, 7
Plexiform layers, 366
PML. See Progressive multifocal leukoencephalopathy (PML)
PMS. See Progressive multiple sclerosis (PMS)
Point resolved spectroscopy (PRESS), 292
Ponesimod, 80–82
Pooling layers, 387–388
Population-level findings, 228
Positron emission tomography (PET), 151, 344–345
    future directions, 356–357
    imaging, 344–345
        of nonimmune mechanisms in multiple sclerosis, 356
    microglial activation and translocator protein, 345–346
    translocator protein–PET, 356
        and gray matter pathology in multiple sclerosis, 347
        ligands, 346–347
        and prognostication in multiple sclerosis, 351–352
        and symptom pathogenesis, 350
        and treatment effects, 353–355
        and white matter, 347–350
Positron emitting radioisotopes, 344–345
Post labeling delay (T_{PLD}), 151, 272
PP. See Primary progressive (PP)
PPMS. See Primary progressive multiple sclerosis (PPMS)
Predictive modeling approaches, 412
Pregnancy, 82
PRESS. See Point resolved spectroscopy (PRESS)
Prevalence phenotypes, 55–56
Primary progressive (PP), 328
Primary progressive multiple sclerosis (PPMS), 3–4, 72, 176–177, 198–199, 276, 295, 344, 386, 405
Priming, 9
PRLs. See Paramagnetic rim lesions (PRLs)
pRNFL, 364–365
Progression independent of relapse activity (PIRA), 128
Progressive MS patients, 41
Progressive multifocal leukoencephalopathy (PML), 58, 84, 108f
Progressive multiple sclerosis (PMS), 161, 307
Proof-of-concept, 14
Proton, 288–289
    Mrs, 183–184
    proton-density acquisitions, 194
Proton density-weighted (PDw), 146, 163
    in MR imaging, 146–148, 146f
Proton spin density (PSD), 211
PSD. See Proton spin density (PSD)
"Pseudo" lesion, 400–401
Pseudoatrophy, 109

Pseudo-continuous arterial spin labelling (pCASL), 273
PSIR. *See* Phase-sensitive inversion recovery (PSIR)
Public health programs, 377
Pulse sequence, 140, 142–144, 159–160
Pulsed arterial spin labelling (pASL), 273
Pulsed MT, 193
pwMS. *See* People with multiple sclerosis (pwMS)
pwPPMS. *See* People with primary progressive multiple sclerosis (pwPPMS)
pwSPMS. *See* People with secondary progressive form of multiple sclerosis (pwSPMS)

Q
qBOLD. *See* Quantitative BOLD (qBOLD)
qMT. *See* Quantitative magnetization transfer (qMT)
QoL. *See* Quality of life (QoL)
QSM. *See* Quantitative susceptibility mapping (QSM)
QSMRim-Net, 398–400
Quality of life (QoL), 51–52
Quantitative approaches, 201
Quantitative BOLD (qBOLD), 235
Quantitative magnetic resonance imaging reports in clinical care, 109–113
  modifiers of brain volume measures, 113t
Quantitative magnetization transfer (qMT), 194–196, 196f
  in multiple sclerosis, 200
Quantitative MRI techniques, 46
Quantitative susceptibility mapping (QSM), 62, 99, 144, 152–154, 154f, 210–211, 222, 398–400
  applications of quantitative susceptibility mapping in multiple sclerosis, 235–236
  deep gray matter
    confounding effects of myelin in, 231
    gadolinium retention in, 230
    nonheme iron concentrations in, 227–229
    regional content of nonheme iron in, 229
    subvoxel distribution of iron in, 230
  focal white matter damage, 232–235
    identification of gadolinium-enhancing lesions, 233
    paramagnetic rim lesions, 234–235
  fundamentals of, 223–226
    imaging, 223–225, 224f
    interpretation of changes in tissue susceptibility, 226
    magnetic susceptibility of tissue, 223
    processing, 225–226
  normal appearing white matter, 231–232

R
Radio frequency (RF), 140, 192
  field, 141–142
  pulses, 272
  RF-echo blocks, 144
Radioactive decay, 344–345
Radiological disease activity, 104
Radiologically isolated syndrome (RIS), 38, 45, 97
Radionucleotides, 344–345
Radiopharmaceuticals, 344–345
Random forest, 386–387
Random thermal motion, 149
Randomized controlled trials (RCTs), 124
rCBV. *See* Relative cerebral blood volume (rCBV)
RCTs. *See* Randomized controlled trials (RCTs)
Reactive oxygen species (ROS), 5
Receiver operating characteristic (ROC), 391–392
Region of interest (ROI), 201, 226
Regional content of nonheme iron in deep gray matter, 229
Regression problems, 392
Regulatory landscape, 412–413
Regulatory T cells (Treg cells), 6, 13
Relapsing–remitting multiple sclerosis (RRMS), 3–4, 40–41, 51–52, 72, 128, 161, 179–180, 198–199, 251, 275–276, 295, 328, 347–349, 368, 386, 405
Relative cerebral blood volume (rCBV), 269
Relative mean transit time (rMTT), 269
Relaxation, 142, 144–145
Remission period, 59
Remyelination, 24
Renal parameters, 77
Repetition time (TR), 144, 163, 292
Resource–poor settings
  optimization strategies, 379–380
  cost-reduction strategies, 379–380
Resting-state (RS), 250
  functional magnetic resonance imaging, 253–255
Restorative training targets, 61
Retina as window to brain, 363–365
Retinal layers, 364–365
Retinal nerve fiber layer (RNFL), 364–365
Retinal thickness indicates neuroaxonal reserve, 368
Retinal thinning, 367
  indicates ongoing neuroaxonal damage, 368
Retrospective metanalyses, 325–326
RF. *See* Radio frequency (RF)
Rim-Net, 398–400
RIS. *See* Radiologically isolated syndrome (RIS)
Rituximab, 82–83
  rituximab-treated patients, 369–370
rMTT. *See* Relative mean transit time (rMTT)
RNFL. *See* Retinal nerve fiber layer (RNFL)
ROC. *See* Receiver operating characteristic (ROC)
ROI. *See* Region of interest (ROI)
ROS. *See* Reactive oxygen species (ROS)

RRMS. *See* Relapsing–remitting multiple sclerosis (RRMS)
RS. *See* Resting-state (RS)
R-squared (R^2), 392

S

S1P. *See* Sphingosine 1-phosphate (S1P)
Safety of Tysabri re-dosing and treatment (STRATA), 58
SAR. *See* Specific absorption rates (SAR)
SC. *See* Spinal cord (SC)
Scanner-based phase reconstruction, 225
SDMT. *See* Symbol Digit Modalities Test (SDMT)
SD-OCT. *See* Spectral domain optical coherence tomography (SD-OCT)
SE. *See* Spin echo (SE)
Secondary progressive (SP), 328
Secondary progressive multiple sclerosis (SPMS), 3–4, 51–52, 71–72, 166, 198–199, 276, 295–297, 344, 386, 405
Segmentation-based software, 104
Segmented EPI (SEPI), 210
Selective reminding test (SRT), 53
SELs. *See* Slowly-expanding lesions (SELs)
semi-LASER. *See* Semi–localization by adiabatic selective refocusing (semi-LASER)
Semi–localization by adiabatic selective refocusing (semi-LASER), 293
Sensorimotor systems, changes to, 250–251, 253–254
SEPI. *See* Segmented EPI (SEPI)
Shadow plaques, 24
SHAP. *See* SHapley Additive exPlanations (SHAP)
SHapley Additive exPlanations (SHAP), 409–410
Shimming method, 292
Short tau inversion recovery (STIR), 176
SIENAX. *See* Structural Image Evaluation using Normalization of Atrophy Cross-sectional (SIENAX)
Signal abnormalities, 176
Signal-time curve, 270–271
Signal-to-noise ratio (SNR), 201, 292, 322
"Similarity index", 109–110
Single-cell RNA-sequencing, 6
Single-photon emission-computed tomography study, 279
Single-voxel spectroscopy, 292
Siponimod, 80–82
Slowly-expanding lesions (SELs), 128
Smoldering inflammation, 128
SMOTE. *See* Synthetic Majority Over-sampling Technique (SMOTE)
SNR. *See* Signal-to-noise ratio (SNR)
Software toolboxes, 225
Solitary sclerosis, 97
SP. *See* Secondary progressive (SP)

SPART. *See* Spatial recall test (SPART)
Spatial localization methods, 292
Spatial recall test (SPART), 53
Spearman's correlations, 392
Specific absorption rates (SAR), 324
Spectral domain optical coherence tomography (SD-OCT), 362
Spectral editing, 293
SPGR. *See* Spoiled gradient-recalled (SPGR)
Sphingosine 1-phosphate (S1P), 81
 S1P–receptor modulators, 80–82, 81t
Spin, 140
Spin dynamics and excitation, 141–142
Spin echo (SE), 142–144
 sequences, 324
Spinal cord (SC), 21–22, 43, 113, 175–176, 201
 area, 161
 atrophy, 113, 183–184
 characteristics of SC lesions in multiple sclerosis, 176–180, 178t
 lesions, 97, 161
 magnetization transfer in, 201
 MRIs, 97
 MWF, 307
Spinal cord imaging, 45, 334. *See also* Lesions imaging
 characteristics of spinal cord lesions in multiple sclerosis, 176–180
 future directions, 183–185
 practical considerations in, 180–181
 spinal cord MRI assessment for patient's prognosis and disease monitoring, 182–183
Spinal imaging, 334
Spin-echo sequences, 270
Spin–spin interactions, 153
SPMS. *See* Secondary progressive multiple sclerosis (SPMS)
Spoiled gradient-recalled (SPGR), 193
SRT. *See* Selective reminding test (SRT)
STAGE. *See* STrategically Acquired Gradient Echo (STAGE) imaging
STEAM. *See* Stimulated echo acquisition mode (STEAM)
Steroids, 123–124
Stimulated echo acquisition mode (STEAM), 293
STIR. *See* Short tau inversion recovery (STIR)
STRATA. *See* Safety of Tysabri re-dosing and treatment (STRATA)
STrategically Acquired Gradient Echo (STAGE) imaging, 211, 213f
Structural Image Evaluation using Normalization of Atrophy Cross-sectional (SIENAX), 111
Study neuroaxonal integrity (DTI), 184
STurbo-field-echo sequences (TFE sequences), 331
Subpial disease, using high-field imaging in vivo improves cortical lesions visibility and allows detection of, 331

Index

Subpopulation risk stratification (SunRiSe), 407
Subvoxel distribution of iron in deep gray matter, 230
Such pulses, 197
SunRiSe. *See* Subpopulation risk stratification (SunRiSe)
Super-resolution, 395–396
Superresolution technique, 395
Supervised learning, 386
Supervised machine learning, 386–387
Support vector machines (SVMs), 386, 404
Supratentorial brain, 161
Susceptibility field, 152–153
Susceptibility weighted imaging (SWI), 41, 128–129, 208, 222
   imaging biomarkers in multiple sclerosis, 208
   microvascular in vivo contrast revealed origins, 214–215
   quantitative susceptibility mapping, 210–211
   STrategically Acquired Gradient Echo, 211
   SWI-FLAIR, 209–210
   SWI-FLAIR or FLAIR*, 209–210
   water content as new biomarker for multiple sclerosis lesions, 211–213
SuStaIn, 406
SVMs. *See* Support vector machines (SVMs)
SWI. *See* Susceptibility weighted imaging (SWI)
Symbol Digit Modalities Test (SDMT), 52
Synthetic double inversion recovery imaging, 394–395
Synthetic Majority Over-sampling Technique (SMOTE), 398–400
SynthSeg, 395–396
SynthSR, 395–396

T

T cells, 5, 13
T follicular helper cells (Tfh cells), 13
T helper 1 (Th1) cells, 5, 11–12
T lymphocytes, 18
T1 hypointense lesions, 107–108, 125–126, 349–350
T1 lesions, 125–126
T1-isointense lesions, 279
T1-lesion loads, 124
T1w. *See* T1-weighting (T1w)
T1-weighting (T1w), 146
   acquisitions, 163
   images, 236
   imaging T1WI, 96
   MPRAGE, 168
   in MR imaging, 146–148
   sequences, 322
T2 FLAIR lesions, 397–398
T2 lesions, 124–125, 162
   activity, 124–125
T2* relaxation, 152–154, 154f
T2*w. *See* T2*-weighted (T2*w)
T2*w sequence, 168
T2*-weighted (T2*w), 167
T2*-weighted imaging (T2*-WI), 99
T2*-WI. *See* T2*-weighted imaging (T2*-WI)
T2-lesion loads, 124
T2w. *See* T2-weighting (T2w)
T2-weighting (T2w), 146
   axial echo, 177–179
   images, 94–97, 163
   magnetic resonance sequences for T2 white matter lesions imaging, 162–163, 162f
   in MR imaging, 146–148
   MRI scans, 124
   sequences, 322
Task shifting, 380
Task-based functional magnetic resonance imaging, 250–253
"Task-negative" network, 252
"Task-positive" network, 255
T-cell receptor (TCR), 10–11
T-cells, 83
TCR. *See* T-cell receptor (TCR)
TD-OCT. *See* Time domain optical coherence tomography (TD-OCT)
TE. *See* Echo time (TE)
Technologies, 2
Telemedicine, 379–380
Telemonitoring, 379–380
Teriflunomide, 79
TFE sequences. *See* STurbo-field-echo sequences (TFE sequences)
Tfh cells. *See* T follicular helper cells (Tfh cells)
Th1-like Th17 cells, 12–13
Th9 cells, 13
Th17 cells, 12–13
Th22 cells, 13
Thalamic radiotracer uptake, 352
Thalamus, 228, 253–254
Theiler's Murine encephalomyelitis virus (TMEV), 19–20
Therapeutic options, 71
Therapeutic targeting, 14
Therapy, 77
Three dimension (3D)
   3D-FLAIR sequence, 98
   acquisition, 98–99
   image, 142, 331, 362
   isotropic imaging, 106
   MPRAGE, 163–164
3T magnetic resonance imaging, visualizing central vein sign using, 328–329
Three-pool model, 195
Thymic stromal lymphopoietin receptor, 6
Thyroid disorders, 86
TI. *See* Inversion time (TI)

Time domain optical coherence tomography (TD-OCT), 362
Time to peak (TTP), 268
Tip-DCs, 14–15
Tissue
  interpretation of changes in tissue susceptibility, 226
  iron concentration, 236
  magnetic susceptibility, 223, 225
  magnetism, 150–154, 154f
  relationship between tissue properties and MRI signal, 144–154
    biological water motion, 148–151
    relaxation, 144–145
    T1-, T2-, and proton density weighting in MR imaging, 146–148
    tissue magnetism, 152–154
  segmentation
    tools, 396
    volumetrics, 396–397
  tissue–air interfaces, 226
  tissue–bone interfaces, 226
  tissue-resident memory cells, 14–15
TMEV. See Theiler's murine encephalomyelitis virus (TMEV)
tNAA. See Total N-acetylaspartate (tNAA)
Torrey–Bloch equation, 149
Total N-acetylaspartate (tNAA), 289
Totally automatic robust quantitation in NMR (Tarquin), 294
T_{PLD}. See Post labeling delay (T_{PLD})
TR. See Repetition time (TR)
Tracer kinetics model, 271
Traditional clinical predictors, 100–104
Traditional vs. Early Aggressive Therapy for MS (TREAT-MS), 74
Transit time, 276
Translocator protein (TSPO), 345–346
  translocator protein–PET
    beyond, 356
    ligands, 346–347
    microglial activation and, 345–346
    in multiple sclerosis, 347
    and prognostication in multiple sclerosis, 351–352
    and symptom pathogenesis, 350, 351f
    and treatment effects, 353–355, 354t
    and white matter, 347–350
Transverse relaxation (T2 relaxation), 144–145
Treatment of MS, 71
TREAT-MS. See TRaditional vs. Early Aggressive Therapy for MS (TREAT-MS)
Tree-based algorithms, 386–387
Treg cells. See Regulatory T cells (Treg cells)
TSE. See Turbo spin echo (TSE)

TTP. See Time to peak (TTP)
Tumor, 393–394
Turbo spin echo (TSE), 167, 177–179
Two-dimensional chemical shift imaging (2D-CSI), 293
Two–pool model, 195

U
Unconventional T cells, 15–16
U-Net, 388–389
  algorithms, 393–394
United States Food and Drug Administration (FDA), 72, 344–345, 384–385, 393, 412–413
Unsupervised learning, 386
Unsupervised machine learning, 386–387
"Upstream" tasks, 384–385

V
VAN. See Ventral attention network (VAN)
Vascular cell adhesion molecule-1 (VCAM-1), 84
Vascular links, 1
VCAM-1. See Vascular cell adhesion molecule-1 (VCAM-1)
Veins, 208
Venous blood volume, 214–215
Ventral attention network (VAN), 256
"Vesselness filtering", 401–402
Volumetric analyses, 111
Volumetric assessments, 344
Volumetric data, 112
Volumetric MRI techniques, 46
Volumetric scans, 363
Voxel-wise analyses, 232, 277–278

W
Water
  content as new biomarker for multiple sclerosis lesions, 211–213, 212f
  motion, 148
  protons, 192, 198
Wavelength spectrum, 362
White matter (WM), 57, 97, 159–160, 208, 222, 268, 303–304, 322
  lesions, 160–163
    improved visibility of, 324–325
    magnetic resonance sequences for T2 white matter lesions imaging, 162–163
  translocator protein–PET and, 347–350, 348t
White matter hyperintensities (WMHs), 209–210
Whole-brain analysis methods, 232
Whole-brain atrophy, 127
WM. See White matter (WM)
WMHs. See White matter hyperintensities (WMHs)